REA

S0-BVN-429

mathematics for elementary teachers

Ruth E. Heintz *State University College at Buffalo*

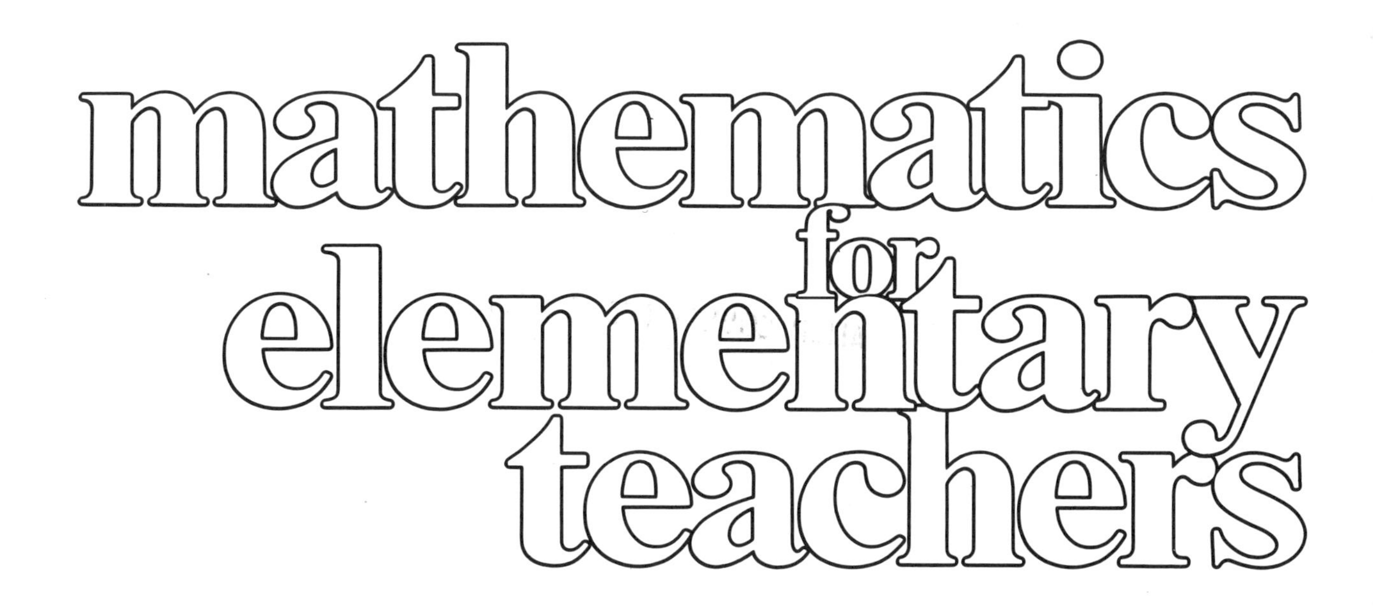

A Content Approach

Addison-Wesley
Publishing Company

Reading, Massachusetts
Menlo Park, California • London • Amsterdam • Don Mills, Ontario • Sydney

Photographs of classroom activities on pages 3, 13, 65, 81, 97, 113, 118, 156, 171, 187, 246, 263, 311, 356, 358, 376, 386, 426 were supplied by the courtesy of photographer Robert B. Ludwig of the State University College, Buffalo, New York.

Library of Congress Cataloging in Publication Data

Heintz, Ruth E. 1921–
Mathematics for elementary teachers.

Includes index.
1. Mathematics—1961– I. Title.
QA39.2.H395 372.7 79-18727
ISBN 0-201-03227-9

Second printing, November 1980

ISBN 0-201-03227-9
ABCDEFGHIJ-MA-89876543210

preface

This text is intended for a one- or two-semester basic content course for students who are planning to teach in grades K to 9. The book provides a prospective elementary-school teacher with the mathematics needed for that purpose and hence is more than just a review of the computational skills learned in the elementary school. It includes a rather complete development of the rational-number system, some development of the real numbers, and material on geometry, measurement, and probability and statistics.

It is assumed that the reader has had two years of high-school mathematics. However, the book is organized so that it can be used successfully by students of varying backgrounds, with either less or more experience.

The margin exercises are intended to be done in conjunction with the reading of the text. Since all of the answers for these exercises are included at the end of the book, they give both practice and immediate feedback on the material being developed. Exercise Sets offering routine practice exercises (as well as others) are provided for each section. Exercises marked ■ are challenge exercises either because they are actually more difficult or simply go a little beyond the list of aims stated at the beginning of the section. Many problem-solving techniques are carefully examined and demonstrated, and opportunities for practicing these skills are provided. The emphasis is not on memorizing certain fixed rules for solving problems but rather on using common sense and thinking through a variety of possibilities.

For those exercises marked 🖩 it is recommended that a hand calculator be used; there are others not so marked in which a calculator can also be found helpful. These exercises are intended to explore the uses to which a calculator can be put to expand understanding of mathematical concepts rather than to merely facilitate computation.

Because we have paid so much attention to problem solving and effective use of the calculator, the text can be employed to advantage for in-service courses for teachers or for a freshman course in mathematics for liberal-arts students.

Some proofs are included in the text, but they are well balanced with numerous worked-out examples and illustrations in the development of each concept. We hope that the included proofs will be regarded as the logical explanations which they actually are and that they will contribute toward the development of a new perspective of elementary-school mathematics. The approach in the book, while mathematically sound, has been carefully thought out to keep the needs of the prospective teacher in mind. We hope some of the excitement of working out a mathematical problem can be carried into the elementary classroom.

It goes without saying that I am indebted to the many fine teachers and colleagues I have been privileged to have contact with in the course of my career. I should also like to thank the many students of "Math 121" whom I have had the good fortune of both teaching and learning from. Many of them are now faculty members of elementary schools themselves.

I am also indebted to my friend Dr. Mervin L. Keedy, professor of mathematics at Purdue University, who was initially responsible for my embarking on the project of writing this textbook and was part of its initial planning and early development. The numerous illustrations and excerpts from the elementary series *Mathematics in Our World* by R. E. Eicholz, P. G. O'Daffer, and C. R. Fleenor, 1978, appear throughout the text due to the gracious courtesy of the Addison-Wesley School Book Division. Last, I am grateful to my patient, cooperative, and supportive editor Steve Quigley who was involved at all stages of producing this text.

I should like to thank Mrs. Susanne M. Reeder, Supervisor of mathematics for the Buffalo Board of Education and Dr. Russell Macaluso, Director of the College Learning Lab on the State College campus at Buffalo, for their cooperation and assistance in making it possible for photographs to be taken in actual schoolrooms.

My appreciation is extended to the following persons who reviewed the manuscript: Sue Brown, University of Houston at Clear Lake City; Lowell Carmony, Southern Illinois University; Phillip M. Eastman, Boise State University; David Gay, University of Arizona; Norman G. Gunderson, The University of Rochester; Alan R. Hoffer, University of Oregon; Leo F. Kuczynski, Southern Connecticut State College; Marjorie McLean, University of Tennessee; Katherine Pedersen, Southern Illinois University at Carbondale, and Holly B. Puterbaugh, University of Vermont.

Buffalo, New York
February 1980

R. E. H.

contents

chapter 1 what is mathematics?

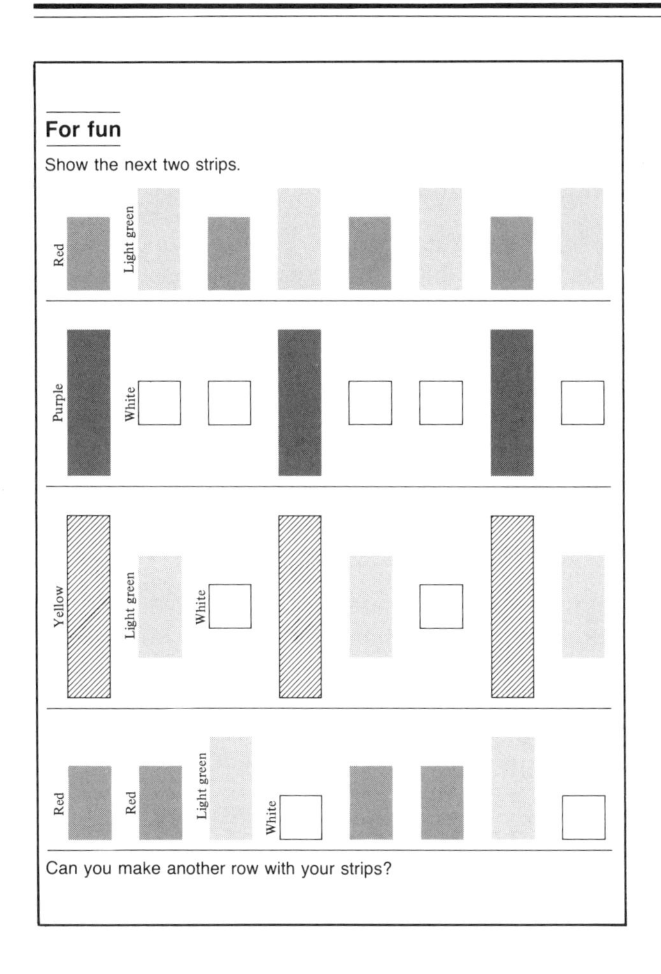

Youngsters in kindergarten can investigate patterns of shape, size, and color before they consider those that involve numbers. These prenumber experiences are extremely important.

1.1 BEING CREATIVE—PATTERN FINDING

Human beings have been fascinated with mathematics for over 4000 years, and we see each succeeding generation building upon what the previous ones have created, to add another story to a building that already exists.

All children, then, should have a chance to come into contact with this important part of their cultural heritage because the number history of mankind is just as important as that of the literature or the arts, and a child would be deprived, not to say handicapped, without any knowledge of mathematics whatsoever.

Mathematics can be regarded from a variety of viewpoints; unfortunately, many people think only in terms of the basic arithmetic skills that are needed to get along on a day-to-day basis in taking care of personal business, money, budgeting, and understanding current affairs.

Number concepts are useful in all of these activities, of course, but there is much more to mathematics. It can be challenging and command interest in the same way that puzzles do and thus can be enjoyed for its own sake as recreation.

Mathematics as a mode of thought, a way of looking at the world, and a means of solving problems, was early adopted by sciences such as physics, chemistry, and engineering and more recently by psychology, economics, and sociology.

The apparent emphasis on reasoning and number makes many people unaware that there are also important creative and aesthetic aspects of mathematics. Mathematicians speak of a beautiful proof or elegant solution to a problem in much the same way that an artist might speak of a particularly successful painting.

For the professional mathematician, creativity and problem solving are closely related, but it is possible to be creative, to use intuition, and to develop problem-solving skills at any level of learning in the subject. Mathematics provides a precise language and other tools which can be applied in diverse and ingenious ways in the solution of problems, not all of which are necessarily practical ones.

Problem solvers in any situation, not just a mathematical one, frequently get hunches and make guesses about a possible solution to a problem on the basis of a study of special cases. They are using *inductive reasoning*, which is the drawing of conclusions on the basis of examples or experience. Young children are inveterate problem solvers and solve inductively such problems as distinguishing between animals called cats and others called dogs.

Similarly, mathematical discoveries are often found inductively on the basis of observations made on a series of examples, all of which suggest a particular pattern to the problem solver.

1.1 AIMS

You should be able to:

a) Distinguish between inductive and deductive reasoning.
b) Describe the relationship between inductive reasoning and pattern finding.
c) Describe a counterexample and explain its role in the proving process.

”

At the heart of mathematics is a constant search for simpler and simpler ways to prove theorems and solve problems . . .—hunches that lead to short, elegant solutions of problems—are now called by psychologists ‘aha! reactions.’

Martin Gardner in “*aha! Insight*”, Scientific American, Inc., W. H. Freeman and Company. New York City, San Francisco, 1978.

Working with pattern blocks

Pattern Finding

PATTERNS IN NUMBER SEQUENCES. Since pattern finding has been so effective in mathematics, we deliberately look for patterns to help us formulate conclusions or generate a solution. Number sequences provide some opportunity for practice in being creative in pattern recognition.

Example 1 Find the next number in the sequence

$$2, 4, 6, 8, \ldots$$

Solution We look for a pattern and see that each number in the sequence is 2 more than the preceding number. If this pattern continues, the next number should then be 10.

Try Exercise 1

Someone else might see an entirely different pattern in Example 1 and think of a simple repetition

$$2, 4, 6, 8, 2, 4, 6, 8, \ldots$$

1. Find the next number in the sequence

$$1, 3, 5, 7, 9, \ldots$$

2.

a) Find and describe a pattern in the sequence

$$1, 3, 5, 3, \ldots$$

Then find the next number in the sequence.

b) Find a different pattern in the above sequence. Then find the next number in the sequence on the basis of your new pattern.

The results obtained by inductive reasoning from specific cases are not always certain because different people may study a small number of cases and conclude that quite different patterns are present. The next example shows this.

Example 2 Find the next number in the sequence

$$2, 4, 8, \ldots$$

a) Person A thinks in terms of multiplication and observes that each number is twice the preceding number. On the basis of this pattern, A says the next number is 16.

b) Person B thinks in terms of addition and sees that 2 must be added to 2 to get 4. Then 4 must be added to 4 to get the next term 8. Then B concludes that 6 should be added to get the next term, which will then be 14.

You might use your imagination and find still another pattern different from either of these.

Try Exercise 2

PATTERNS IN GEOMETRY. Pattern search is not restricted to number sequences, since it is possible to study a series of particular cases or examples in geometry in a similar way. In fact, to look for a pattern is a method that can be applied in many different problems in mathematics. It is useful in the following geometry problem.

Example 3 A line segment joining the two vertices of a polygon may be a side of a polygon. If not, it is called a *diagonal.* Find a formula for determining the number of diagonals of a polygon, when the number of sides is known.

Idea: For a start we make some sketches of particular polygons, draw the diagonals for each, and count them.

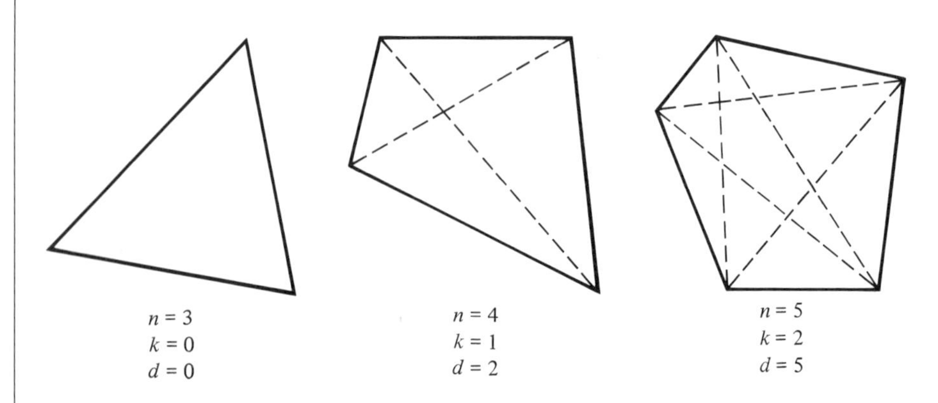

If we call the number of sides n, the number of diagonals through one vertex k, and the total number of diagonals d, then the results can be summarized in a table:

Number of sides, n	3	4	5	6	7
Number of diagonals through one vertex, k	0	1	2	?	?
Total number of diagonals, d	0	2	5	?	?

Try Exercise 3

3. Draw some more polygons and their diagonals. Then extend the table in the text.

How can the table be used to find a formula that will work for each polygon, no matter how many sides it has? What relationships can be seen?

Did you notice that each value of k is 3 less than the corresponding n? This means $k = n - 3$.

Did you notice that $\frac{(n)(k)}{2} = d$?

These combined observations lead to the tentative formula for finding the total number of diagonals:

$$d = \frac{(n)(n-3)}{2}.$$

Try Exercise 4

4. Check the tentative formula for all the entries that you found in extending the table.

Although the formula is verified for every case in the table, it has not yet been shown that the result holds for a polygon of any arbitrary number of sides. In this problem about diagonals, simple cases were studied in an orderly way and the information was organized so that some sense could be made out of the results. These same methods are helpful in solving many different kinds of problems, and one can use them to become increasingly creative in mathematics.

The demonstration of a few examples does not mean that a mathematical principle has been verified; even though a pattern may suggest very strongly that a conclusion is correct, it is still necessary to provide a proof that the result is true.

The Mathematical Method

INDUCTION AND PROOF. Since it is often possible to see more than one pattern in a given situation, there is some uncertainty in the conclusions that are drawn from the inductive evidence of specific cases. Unless some proof can

5. How are mathematical "truths" often discovered?

6. What do we call a single example that shows some statement is not true?

7. Someone says to you: "I don't think it is possible to find any value for x that will make $x^2 - x + 13$ a perfect square." If you could find even one such value for x you would prove this friend wrong. Try to find such a value.

be developed, it is not possible to know for sure if the pattern about the diagonals in Example 3 will continue to hold. Although many mathematical discoveries are made by reasoning from observations made from specific examples, it is still necessary to prove or verify that such discoveries are correct in all cases. Here is an example.

Example 4 Imagine that you have "discovered," by trying 17 cases, that the sum of any two even numbers is even. This is not enough, since if someone should produce a single example of a case to the contrary, your conclusion will be proved wrong. For a proof you must reason in such a way that what you say applies to any pair of even numbers. Your argument might be something like this: An even number is one that has a factor of 2, so an even number is $2 \cdot x$ for some number x. Consider any two even numbers $2 \cdot m$ and $2 \cdot n$. Their sum $2 \cdot m + 2 \cdot n$ can be written as $2 \cdot (m + n)$, therefore it also has a factor of 2, and hence the sum is even.

The above argument is a good example of a simple proof. It is convincing to anybody with a little algebraic knowledge and thus conclusively verifies the "discovery" that was made on the basis of your inspection of 17 cases. In this proof, the conclusion is drawn in a logical way from a set of assumptions about even numbers. This proof would not be convincing to a child with no knowledge of algebra. Does this mean that children must wait until they know algebra before thinking about a proof? The answer is NO.

An elementary student, who could not appreciate the algebraic representation $2 \cdot n$, might understand very well that any even number of counters could be lined up by "twos." Thus the sum of two such even number of counters could be represented by a pile of counters that could also be lined up by "twos." Hence, the represented sum would also be an even number.

With the mathematical method one is free to look for patterns, get hunches, and so on, in order to discover possible "truths" about a situation, but these truths must be considered tentative until they are proved. The second step in the mathematical method is a search for a proof, which generally consists of a deductive argument showing that the conclusion follows logically from the assumed facts or circumstances.

COUNTEREXAMPLES. Not all "truths" stemming from the examination of specific cases are correct; sometimes a single example will turn up later and show that the conclusion is not true, and this is enough for us to know that no proof will ever be found. Such an example is called a *counterexample.* A counterexample may be difficult to find, but it does provide a disproof and can settle the question of whether something is true or not. Thus a disproof is one type of proof.

Try Exercises 5, 6, and 7

EXERCISE SET 1.1

Find a pattern for each of the following sequences. On the basis of the rule for the pattern, find the next few members.

1. 3, 6, 9, 12, . . .

2. 4, 8, 16, . . .

3. 2, 3, 5, 7, 11, 13, . . .

4. 1, 3, 6, 10, 15, . . .

5. 2, 4, 11, 23, . . .

6. 9, 8, 10, 11, . . .

7–12. For each sequence of Exercises 1 through 6 find a pattern different from the one you found before. On the basis of the new pattern rule, find the next few members.

13. Study the exercises found on the page from a first-grade textbook shown at the beginning of the chapter. How might the pattern there be described? What are the next two strips?

14. The following triangular configuration arises in algebra.

$$\begin{array}{ccccccccc} &&&&1&&&& \\ &&&1&&1&&& \\ &&1&&2&&1&& \\ &1&&3&&3&&1& \\ 1&&4&&6&&4&&1 \end{array}$$

Find a pattern and then write the next three lines of the "triangle."

Study the following and add to the list if you wish. What is the pattern being developed on the left? On the right? Can you guess what the next line might be?

15.

$$\begin{aligned} 1 &= 0 + 1 \\ 2 + 3 + 4 &= 1 + 8 \\ 5 + 6 + 7 + 8 + 9 &= 8 + 27 \\ 10 + 11 + 12 + 13 + 14 + 15 + 16 &= 27 + 64 \end{aligned}$$

16.

$$\begin{aligned} 1 + 2 + 1 &= 4 \\ 1 + 4 + 4 &= 9 \\ 1 + 6 + 9 &= 16 \\ 1 + 8 + 16 &= 25 \\ 1 + 10 + 25 &= 36 \end{aligned}$$

17. Investigate for patterns the sum of the first n consecutive odd numbers. Here is a start:

$$\begin{aligned} n = 1&: \quad 1 = 1 \\ n = 2&: \quad 1 + 3 = 4 \\ n = 3&: \quad 1 + 3 + 5 = 9 \end{aligned}$$

■ **18.** A circle separates a plane into two regions: the inside and the outside. Two circles separate a plane into three regions if they do not intersect, four if they do (a maximum of four regions).

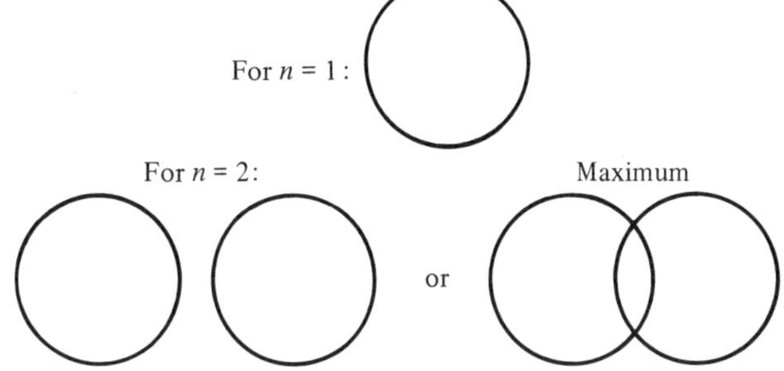

Three circles separate the plane into a maximum of eight regions. Thus, if n is the number of circles and R is the maximum number of regions, the relation is as follows:

n	1	2	3
R	2	4	8

a) From this table, make a conjecture about the case of four circles. Test your conjecture and revise it if necessary.
b) State a tentative general rule or formula for any number of circles.

19. Examine particular cases and make an "inductive" discovery about the sum of two odd numbers, using the sequence 1, 3, 5, 7, 9, ... Find an informal nonalgebraic proof or demonstration with a physical model such as chips. Write an algebraic proof of your result.

20. Repeat Exercise 19 for the multiplication of two odd numbers.

21. Your friend has noticed that

$$4^2 + 1 = 17,$$
$$6^2 + 1 = 37,$$
$$8^2 + 1 = 65,$$

and speculates that any number that can be written $n^2 + 1$ is an odd number.
a) Prove that this is not the case. Hint: look for a counterexample.
b) Is any further proof necessary?

22. Mrs. Hilbert has three daughters: Aline, Bea, and Cami, all of whom are away at school. The one who is an art major is in California, a second is in the Midwest, and the third attends an Eastern junior college. They have quite different career plans. One is planning to be an elementary teacher, one is an artist, and one is very involved with computers. Aline is not in California, Cami is not in the Midwest, and the one who is involved with computers is not in the East. If the art major is not Cami, who is working with computers and where?

23. Suppose that in your math class there are 25 students. On your campus there are 15 different physical-education sections in which both men and women are enrolled. Each of the students in your class is enrolled in one of these sections. Is it true that at least one of the sections contains two or more people from your math class?

■ **24.** Write an informal proof of the tentative formula established in Example 3 of this section.

1.2 AIMS

You should be able to:
a) Describe and use the five-step strategy* to solve appropriate problems.
b) Begin to use the following techniques:
   i. Make sketches or diagrams.
   ii. Collect and organize data.

* This strategy is an adaptation for use in mathematics of a general problem-solving model proposed by Osborn and Parnes. See the following:
Applied Imagination by Alex F. Osborn. Charles Scribner's Sons, New York, Revised Edition (1963);
Creative Actionbook, Revised Edition of *Creative Behavior Workbook* by Ruth B. Noller, Sidney J. Parnes, Angelo M. Biondi. Charles Scribner's Sons, New York (1976).

1.2 PROBLEM SOLVING—A STRATEGY

Although you may now be somewhat more aware of the methods used in mathematics, how does one go about solving a mathematical problem? There are no cut-and-dried rules for such things, but there are general strategies that are helpful. Here is one that has proven useful in solving a large variety of problems.

1. *Identify the relevant facts.* This means that all the known facts are to be accumulated and listed. It may be necessary to derive new facts to add to the list later, since all of the relevant facts are not always "given" to you; on the other hand, there may be irrelevant information provided.

2. *Decide what the problem actually is.* This one may seem obvious, but it is often clarifying to restate the problem in several different ways. One of the

restatements may trigger some ideas for a solution or it may reveal a related problem that should be solved first.

3. *Collect ideas.* List any ideas that come to mind that might possibly help in solving the problem. These ideas should not be pursued or discarded immediately; the first one that comes to you is often not the best. After you have accumulated your ideas, choose one of them and proceed.

4. *Find a possible solution.* Work with the idea you have selected and see if it does lead to the solution. You may find that you need new facts or the idea may lead nowhere, in which case you go back to the idea list and try another.

5. *Check the result.* Once you have found what seems to be a solution, you should check to see that it actually is and that it really does satisfy the conditions of the problem. Often in mathematics it is necessary to write out a convincing argument or proof that you indeed have a solution to the problem.

This five-step process is a strategy that can be applied to all types of mathematical problems.

Example 1 You have four pieces of chain, each three links long. It costs 10 cents to open a link and 15 cents to close a link. All links are closed at the beginning. How can you make a single closed chain using all of the links at a cost of no more than $0.75?

The above steps are applied:

1. The *facts*:

Four pieces of chain
Three links in each
10 cents to open a link
15 cents to close a link

2. The *problem*: Obtain a single *closed* chain. Do not spend more than $0.75.

What does *closed* mean here? It could mean
a) a linear piece of chain when finished,
b) a circle of chain when finished.

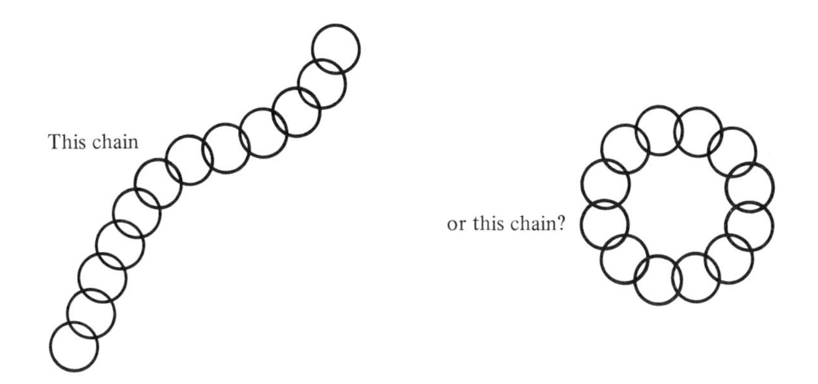

8. Construct a table of cuts and costs, as suggested in Idea (e).

Both interpretations are possible. We select interpretation (a) for the present.

3. *Ideas*:

a) Cut all the links and resolder.
b) Cut half the links and resolder.
c) Cut one link in each piece and resolder.
d) Figure out how many links can be cut and resoldered for $0.75.
e) Make a table of cuts and costs using the ideas of (a) through (d).

Try Exercises 8 and 9

Note in the above example that the ideas (generated in an actual classroom) seemed to improve going down the list. Sketches were used and a table was made, so that it became easier to see a pattern or relationships. These are helpful aids in problem solving, and you should practice using them.

Each step of the five-step process requires practice, particularly if your own problem-solving methods have tended to be haphazard. The first two steps are illustrated in the next example, which is a more traditional problem of a type that could be solved by elementary pupils.

9. Solve the problem of the example. Check your answer to see that you have satisfied the conditions of the problem.

Example 2 Two race cars went two laps around the racetrack. The first car on the first lap averaged 215 km/hr and 235 km/hr on the second lap. The other car averaged 225 km/hr for both laps. What was the result of the race?

The *facts* are:

First car	1st lap	215 km/hr
	2nd lap	235 km/hr
Second car	1st lap	225 km/hr
	2nd lap	225 km/hr

Exactly what is the *problem*? As stated above it is:

What was the result of the race?

Restating the problem in a few different ways may generate some ideas for solving it. Some possible restatements are:

1. Which car won the race?
2. Is there a winner in the race?
3. Did it take longer for the first car to complete the two laps?

10. Think of two more ways to restate the problem of Example 2.

Try Exercise 10

Rewording the problem in this way may stimulate some new ideas in your mind. Until you read restatement (2) above, had it occurred to you that the race might be a tie? Had it occurred to you that the race might be analyzed one lap at a time? Get into the habit of writing down all of the ideas you get as a result of restating the problem.

Try Exercise 11

11. Write down a list of at least two different things that you think could be used to solve the problem. Do not solve it yet.

A glance toward the classroom

It is just as important for children to know about the process of problem solving as it is for them to learn the basic facts of addition. Addition facts are not useful if a child cannot decide when it is appropriate to perform the addition.

EXERCISE SET 1.2

1. In Example 1 of the text, a possible interpretation of *closed* was that the final result was a circle of chain. Solve the problem using this interpretation.

2. Refer to Example 2 of the text and choose a problem statement to work with. Select one of your ideas, find a solution, and check it.

Problems 3–5 are similar to those given to elementary pupils. In addition to solving the problems think about how they might be presented to such students.

3. You and a friend have agreed to the following payoffs to each other when playing tennis. If you lose you pay your friend $2, but if you win you get $3 from your friend. (You are not equally skilled, hence the different payoffs.) At the end of 15 games you discover that you have broken exactly even. How many games did you win?

4 A family went strawberry picking. Their three daughters picked 18 quarts. Sue, who was older than Pam, picked twice as many berries as Pam. Sandy, the oldest, picked three times as many berries as Sue. How many quarts of berries did Pam pick?

5. A discount house charges $19.97 for a set of small wrenches; there are 14 in the set. The same wrenches cost $1.65 separately. If you really need only 12 of the wrenches, would you save any money by buying just the ones you need?

For the following problems use the strategy of the text and label each step so you do not omit one. At the idea stage do not stop at one idea unless you happen to see the solution immediately. Even then it is worthwhile to try to think of more ways to attack the problem.

6. You are making paper chains to use as decorations for a party. They are formed from strips of paper that are pasted to make links. You have five sections of 30 links each and you need one long chain of 150 links. What is the smallest number of links you will need to cut and repaste to make the long chain?

7. You friends tell you that their combined change is all in dimes and quarters and that they have 20 coins altogether. The total value of these coins is $4.10. Can they provide you with $1.00 in dimes? Is there enough information given in this problem to solve it?

8. This problem is similar to one that appeared in an 8th grade arithmetic book of the early 1920's:

> A curious boy at a circus sideshow says: "How many animals and birds are in this show?"
>
> The circus attendant replies: "I just noticed that there are 30 heads and 100 feet. You can figure out for yourself how many of each there are."

Help the curious boy, who, it turns out, was not too good in arithmetic.

9. Jean has three coins that look identical. However, one of them is heavier than each of the other two, which weigh exactly the same. She has a balance scale that will indicate which of two objects is heavier or lighter. (It will balance if they are the same.) How can Jean find which is the heavy coin if she is allowed to use the scale just once?

■ **10.** You have eight coins that look exactly alike. It is possible that one of these coins is counterfeit and thus weighs less than the others. You have only a balance scale to use. How can this scale be used only twice to:

a) Find out if there is an underweight coin?
b) Find exactly which coin it is, if there is an underweight one?

11.

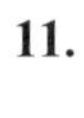

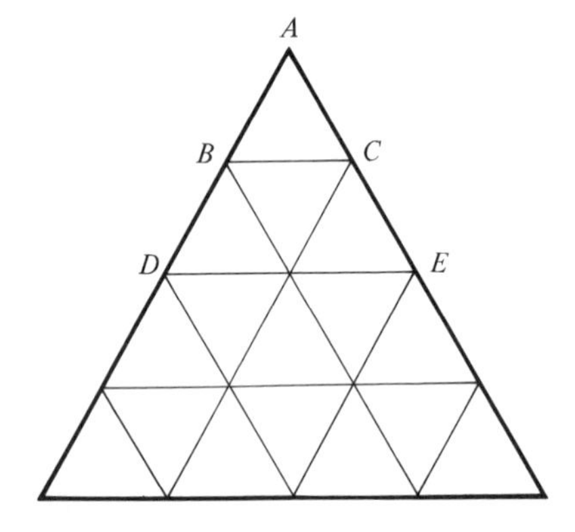

a) How many triangles that "point up" (△) can you find in the given figure? There are triangles of different sizes: a "one-row" triangle such as triangle *ABC*, a "two-row" triangle such as *ADE*, and so on. Organize the information you collect and look for a pattern.

■ b) Generalize and find a formula or rule that will work for any sized figure of this type. Prove your formula is correct.

12. If the letters of the alphabet are assigned values in order, so that *a* is 1, *b* is 2, *c* is 3, and so on, with *z* representing 26, then words are assigned the total value of the letters used in them. Thus the word *the* has the value $20 + 8 + 5$, or 33. How many words can you find that have the value 24?

1.3 AIMS

You should be able to:

a) Use the addition operation to compare the three basic notation schemes used in various hand calculators.
b) Write a flowchart for any basic computational procedure for at least one type of hand calculator.

1.3 COMPUTATION AND ALGORITHMS—THE CALCULATOR

The widespread use of small calculators and computers makes many people feel that all the mathematical problems of the future will somehow be solved electronically. Actually you need to know just as much mathematics to solve problems with a calculator as you would without it—in fact, you may need to know more. A calculator will not really *solve* a problem for you; it will at most provide a quick answer after *you* have figured out the method of solution, which is quite a different thing.

Various calculators

There is no question as to whether hand calculators will be found in the schoolroom; they are already there! From time to time in this book it is recommended that certain exercises are to be done with a calculator. In many cases these add to your mathematical insight rather than just give you a chance to work on some lengthy or tedious problem. There are many varieties of small calculators on the market, and while we do not know the exact scheme used in yours, it is almost certain that it is one of the schemes briefly outlined below. A simple addition problem will show the basic differences for three types of calculators; there are certain advantages and disadvantages to each system.

Calculator Schemes for Computation

Example 1 Use a calculator to find the sum of 125 and 371.

Solution

A CALCULATOR THAT USES ALGEBRAIC NOTATION. With the algebraic system, numbers and operations are entered in the same order as they occur in the mathematical statement: $125 + 371 = \underline{?}$

The [CLEAR] key is pressed to make sure that there is nothing left from a previous operation. 125 is entered on the keyboard, the [+] key is pressed, 371 is entered on the keyboard, the [=] key is pressed, the answer 496 shows on the display. (The pressing of the [=] key automatically prepares the calculator for the next operation.)

In this system, the [+] key is pressed *before* the second number, 371, is entered. This is the basic difference between this system and the next type discussed. (If an error in an entry is made, it can be corrected by pressing the [CE] key.)

12. Which of these schemes does your own calculator use?

A CALCULATOR THAT USES REVERSE POLISH NOTATION. In this type of calculator, both numbers must be entered before the calculator is instructed to perform the addition. When more than one number is keyed into a calculator of this type the [ENTER↑] key is used.

The same addition goes as follows. The number 125 is entered on the keyboard, the [ENTER↑] key is pressed. (This separates the first number from the next one to be entered.) The number 371 is entered on the keyboard. The [+] key is pressed to perform the operation. The same answer 496 is then on the display. This system has the advantage that parentheses do not need to be used no matter how complex the computation.

A CALCULATOR THAT USES THE ARITHMETIC SYSTEM. This third scheme has certain features of each of the two given above. This type of calculator has two different keys: a [+/=] key and a [−/=] key.

To do the same addition, it is necessary to press the [CLEAR] key and then enter the first number, 125, on the keyboard. The [+/=] key is then pressed. The second number, 371, is entered and the [+/=] key is pressed again. The display shows 496.

Try Exercise 12

Obviously you must get acquainted with your own calculator and investigate the potential it has. Different calculators may also require different procedures when a constant is used in connection with certain operations. You will have to investigate this on your own with the help of your instruction book.

Flowcharts

Instructions in the use of a calculator can often be made more understandable in terms of a flowchart or diagram instead of a paragraph of description. Simple flowcharts are being used increasingly in elementary school textbooks to clarify operations and procedures. Flowcharts are, of course, a tremendous help when one is writing a program that instructs a calculator or computer what to do to solve a particular problem.

A flowchart can be used to describe the addition operation just discussed. There are certain symbols and conventions that must be learned.

A circle is used to mark the *beginning* and the *end* of a problem.

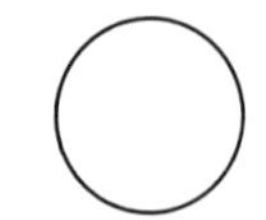

A diamond is used to indicate at what point a *decision* must be made (a question is asked).

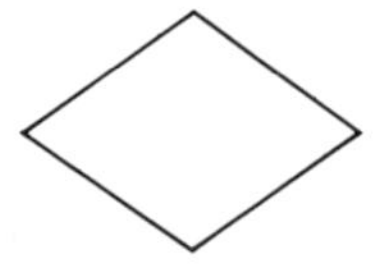

“Following closely on the heels of the ‘new math’ and ‘metrics,’ the ‘minicalculator’ is creating greater controversy than either of its predecessors. . . . teachers, parents, and school administrators are interested in and concerned about its impact on the curriculum, the classroom, and the student.

Jesse A. Rudnick and Stephen Krulik, “The minicalculator: friend or foe?” *The Arithmetic Teacher*, 23:8, December 1976, pp. 654–656.

A rectangle indicates an *operation* to be performed.

An arrow indicates the direction of the *flow*; it shows how the diagram is to be read.

A flowchart for a calculator that uses algebraic notation follows.

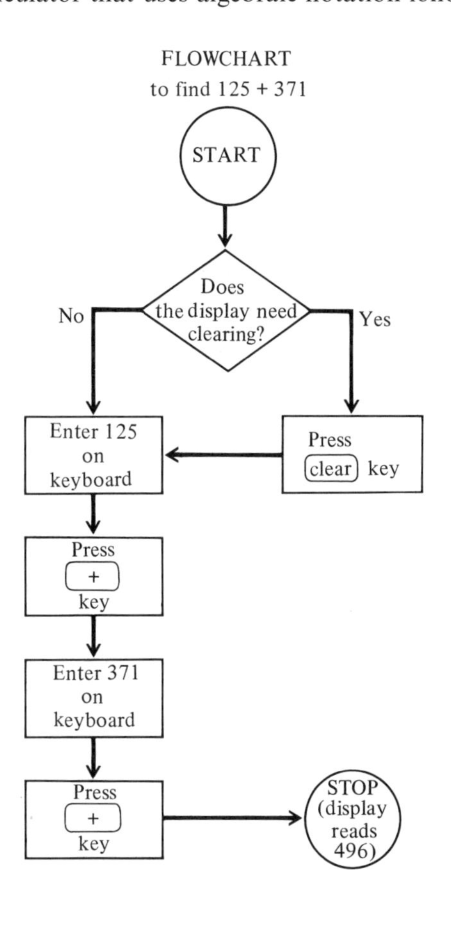

Try Exercise 13

13. Write the flowchart for 125 + 371 for a calculator that uses Reverse Polish Notation.

EXERCISE SET 1.3

1. Get acquainted with your own calculator. Write a flowchart for each of the following:

a) Subtraction: 756 − 139
b) Multiplication: 15 × 681
c) Division: 306/17

2. If it is possible to multiply a series of numbers by a constant with your calculator, learn how this is done. Write a flowchart that describes the procedure.

3. Because of the computation, you might not ordinarily want to investigate certain number patterns. Here is one:

$$7 + (9 \times 9) = 88$$
$$6 + (98 \times 9) = 888$$
$$\text{Your prediction for: } 5 + (987 \times 9) = \underline{?}$$

How far does this pattern carry on? ■ Can you think of an explanation?

4. How far does this pattern carry on? ■ Find an explanation.

$$2 + (1 \times 9) = 11$$
$$3 + (12 \times 9) = 111$$
$$4 + (123 \times 9) = ?$$

5. If the number 81 is divided by the sum of its digits $(8 + 1)$, the quotient is 9 $(81/9 = 9)$. Suppose that any two-digit number (from 10 to 99) is treated in this way. What is the smallest answer you can get? What is the largest answer you can get?

6. Investigate the pattern formed by the sum of the first n cubes. Remember that n^3 means $n \times n \times n$; thus $2^3 = 2 \times 2 \times 2 = 8$. Here is a start:

$$n = 1: \quad 1^3 = 1$$
$$n = 2: \quad 1^3 + 2^3 = 1 + 8 = 9$$

Make a tentative general statement about the pattern after you examine more cases.

1.4 PRECISE LANGUAGE—STATEMENTS

1.4 AIMS

You should be able to:

a) Recognize and provide examples of statements and translate from English to symbolic form and vice versa.
b) Use the existential and the universal quantifiers to convert an open sentence to a statement.
c) Find the truth values for compound statements.

One aspect of mathematics is the precise language it offers in the solution of a wide variety of practical problems. In particular, logic can be thought of as the study of this careful kind of communication. We will merely scratch the surface of the study of mathematical language here, but as a minimum you want to at least become aware of the structure of the mathematical statements you read. Some basic ideas of logic will help to keep you from making ambiguous statements yourself. In addition, you want to be able to follow the steps of a proof and see that it is correct.

Statements

Ideas are expressed by sentences, and it is instructive to see how the sentences of mathematics and logic, while related to those of ordinary speech, are read and interpreted much more carefully; we start with simple sentences that are called *statements*.

A *statement* is a declarative sentence that is either true or false, but not both.

Here are some examples of statements:

a) $1 + 6 = 7$ (a true statement)
b) $8 - 5 = 2$ (a false statement)
c) New York City is east of Chicago.
d) George Washington was President of the United States during the 1960's.

Try Exercise 14

14. Which of the following are statements?

a) $1 + 3 = 4$
b) $1 + 4 = 6$
c) Why did you come?
d) Please close the window.

There are other sentences for which we cannot determine, as they stand, if they are statements or not. Here are some examples of such sentences:

a) $12x = 60$ b) $x - 3 = 5$

If we replace the variable x in the first sentence by the whole number 5, then the sentence is true, but for any other whole number the sentence is false. In the second sentence, if the variable x is replaced by 8, the statement is true, and any other whole number will make the statement false. If we restrict x to some set of values, each of these sentences can be made to be either true or false. If the truth value of a sentence depends on what replacement is made for a variable, it is an open sentence.

An *open sentence* is one that will become a statement when replacements for the variable are made from the set of objects under discussion.

It is also possible to use the quantifiers to accomplish the same result for an open sentence. We use one of the following quantifiers:

1. The *existential quantifier* $\exists$:

$$\exists x \text{ means "there exists at least one } x\text{"}$$

2. The *universal quantifier* $\forall$:

$$\forall x \text{ means "for all } x\text{"}$$

Thus we can now write

$$\text{There exists at least one } x \text{ such that } 12x = 60$$

in a symbolic form as

$$\exists x\colon 12x = 60.$$

(If x is restricted to the set of whole numbers, this is a true statement, since $x = 5$ is such a number.)

If we use the same open sentence with the universal quantifier, we have:

$$\text{For all } x,\ 12x = 60;$$

symbolized, this becomes:

$$\forall x\colon 12x = 60.$$

(If x can be any whole number, this is a false statement.)

Try Exercise 15

In mathematics it is assumed that the universal quantifier is intended when neither one is mentioned. The sentence:

$$\text{If } x \text{ is greater than 2, then } x^2 \text{ is greater than 4}$$

means that for all x, if x is greater than 2, then x^2 is greater than 4.

Restrict x to the set of whole numbers:

15.

a) Use the existential quantifier with the open sentence $x - 3 = 5$. Is the resulting statement true or false?

b) Use the universal quantifier with the open sentence $x - 3 = 5$. Is the resulting statement true or false?

16. Write the conjunction for the following pairs of sentences and state the truth value of each conjunction:

a) $1 + 6 = 2$.
   $5 > 3$.

b) July has 15 days.
   Thanksgiving Day is in November.

Compound Sentences

CONJUNCTION ($\wedge$). Because different combinations of statements must be analyzed in the process of drawing conclusions, we must consider a variety of compound sentences. The *conjunction* of several statements is usually formed by joining them with the connective word *and*; thus if p and q are sentences, then their conjunction is p *and* q, or $p \wedge q$.

Example 1 Let p and q be as follows. Write out their conjunction $p \wedge q$.

p: Five is a prime number,

q: $2 + 3 = 6$.

Solution $p \wedge q$: Five is a prime number *and* $2 + 3 = 6$.

A conjunction is true when both parts of it are true, and it is false if any part of it is false. This agrees with the usual usage of everyday language since the sentence, "I have $12.00 *and* I am going to buy a wallet," is regarded as true only if both parts are true. The truth-table definition for $\wedge$ is given by:

p	q	$p \wedge q$
T	T	T
T	F	F
F	T	F
F	F	F

To illustrate the reading of this table, look at the second line, which is the case when p is true (T) and q is false (F). In this case $p \wedge q$ is false (F) because the parts are not both true.

Try Exercise 16

DISJUNCTION ($\vee$). When two statements are joined by the connective word *or*, the compound sentence that results is called a *disjunction.* If p and q are sentences, then their disjunction is p *or* q, usually written as $p \vee q$.

Example 2 Let p and q be as follows. Write out their disjunction $p \vee q$.

p: Nine is an odd number,

q: $2 \times 4 = 9$.

Solution $p \vee q$: Nine is an odd number *or* $2 \times 4 = 9$.

The word *or* in mathematics and logic is used more precisely than in ordinary conversation, as we see in these examples:

A. Do you expect to get an A in English or mathematics? If a person answered YES to this, it could mean an A was expected in English, or math, or both. This is the *inclusive* use of the word *or*, and is the way *or* is used in mathematics and logic.

B. Are you going to the movie or to the basketball game tonight? For this question a YES answer is not appropriate, since you ordinarily expect that the person will go to one place or the other but not to both. This is the *exclusive* use of *or*.

Because the inclusive *or* is used in mathematics and logic, a disjunction is true if at least one of the sentences is true, and false only in the case that both sentences are false. This is shown in the following truth-table definition of $\vee$.

p	q	$p \vee q$
T	T	T
T	F	T
F	T	T
F	F	F

Note that in the third line of this table p is false (F), q is true (T), and $p \vee q$ is true because at least one of the simple sentences is true.

Try Exercise 17

17. Write the disjunction of the following pairs of sentences; then use the definition of $\vee$ to find the truth value of each disjunction.

a) $2 + 5 = 4 + 3$.
   $9 - 5 = 2 \times 3$.

b) Four is two more than three.
   Eight is three times as large as three.

NEGATION ($^{-}$). Negation may involve a single simple sentence and is usually formed by using the word *not*; symbolically, the negation of a sentence p is $\overline{p}$. The truth-table definition of negation is obvious because p and $\overline{p}$ always have the opposite truth values:

p	$\overline{p}$
T	F
F	T

This is consistent with ordinary English usage, since if a statement is true then we intend that the negation of it be interpreted as false.

Example 3 Let p be as follows. Write out its negation $\overline{p}$.

p: She won the tennis match today.

$\overline{p}$: She did *not* win the tennis match today.

Try Exercise 18

18. Let q be as follows. Write out its negation $\overline{q}$.

a) q: Two added to five is seven.

b) q: $3 + 6 = 9$.

An open sentence that has been made into a statement by means of a quantifier may cause difficulty when it is necessary to think of its negation. How should the following sentence be negated?

All roses are red.

19. Write the negation of each of the following:
a) All college students study a lot.
b) For all x, x^2 is greater than 0.

20. Write the negation of each of the following:
a) Some TV programs are very informative.
b) At least one of the students in my class is at home with the flu.

We could say:

It is not the case that all roses are red

but it is probably more natural to use:

There are some roses that are not red.

The original statement meant that every rose under discussion was red and the negation indicates with *some* that at least one rose exists that is not red. A common error is to believe the negation is: "No roses are red," but this is incorrect, since we need only the existence of one white rose to make false (negate) the original statement.

Example 4 Write the negation of each of the following:
a) All triangles have three vertices.
b) Every even number greater than 2 is the sum of two prime numbers.

Solution The negations are:
a) It is not the case that all triangles have three vertices, or:
There are some triangles that do not have three vertices.
b) It is not the case that every even number greater than 2 is the sum of two prime numbers, or:
There are some (at least one) even numbers greater than 2 that are not the sum of two prime numbers.

Try Exercise 19

In the statements that involve the words *some* or *there exists* there are also various ways to state the negation. For example:

Some polygons are convex

means that there exists at least one polygon that is convex, and this can be negated in the following ways:

It is not the case that some polygons are convex.
There are no polygons that are convex.
Every polygon has the property that it is not convex.

Try Exercise 20

CONDITIONAL SENTENCES (→). Much ordinary conversation consists of sentences that follow the form of these examples:

1. If his shoulder continues to pain him, then he will go to see the doctor.
2. If the tickets are not sold out, then we will go to the concert.
3. If my check arrives in the mail, then I will go to the bank later today.

Such a sentence: "If p, then q" is called a *conditional sentence* or just a *conditional* and is symbolized by $p \rightarrow q$; the arrow symbol represents another connective.

In the conditional $p \to q$, p is called the *antecedent* and q is called the *consequent*. Frequently in mathematics the antecedent is called the *hypothesis* and the consequent is called the *conclusion*. In the theorems of mathematics there is an apparent connection between the parts p and q, but this need not be the case.

Example 5 If $1 + 1 = 2$, then $3 + 3 = 6$.

Example 6 If England has a cold winter, then the price of eggs in Buffalo will be higher the following year.

In Example 5 there seems to be a direct connection between the antecedent $(1 + 1 = 2)$ and the consequent $(3 + 3 = 6)$; however, in Example 6 it may be difficult to decide whether a connection exists at all. Rather than make it necessary to decide on the existence of a connection, the truth value of $p \to q$ is determined only by the truth or falsity of the statements, whether they seem related or not. This agrees with ordinary usage more often than you might think.

In English and in mathematics, conditionals may take many forms, so that it takes practice to be able to recognize them; the theorems of mathematics are often expressed with this type of statement.

Example 7 In the following sentence determine the antecedent and consequent and rewrite the sentence in the *if*, *then* form:

John takes pictures only when the sun is shining.

To solve this, represent

John takes pictures with p,
The sun is shining with q.

Thus we have $p \to q$, since the sentence means: if you find John taking pictures, then the sun is shining, and the sentence has been symbolized in this way. Observe that the original sentence does not say that John takes pictures every time the sun shines.

The word *only* very often precedes and hence identifies the consequent as it does here in this example.

Try Exercise 21

Under what combination of truth values does it make sense that a conditional be called false? Think about this statement that has been made for you:

If it is warm, then I shall swim before lunch.

Has the speaker told the truth:

1. If it is warm, and the speaker swims before lunch?
2. If it is warm, and the speaker does not swim before lunch?
3. If it is not warm, and the speaker swims anyway?
4. If it is not warm, and the speaker does not swim?

21. Determine the antecedent and consequent and rewrite in the *if*, *then* form.

a) This hat will be ruined in case of rain.
b) A prime number is odd only if it is greater than 2.

22.
a) In which case(s) would you say the speaker told the truth?
b) In which case(s) would you say the speaker did not tell the truth?

Try Exercise 22

In this example, the only time it would be correct to say the speaker did not tell the truth is in the case in which the day was warm and the speaker failed to swim before lunch. These ideas can be summarized in the truth-table definition for $\rightarrow$:

p	q	$p \rightarrow q$
T	T	T
T	F	F
F	T	T
F	F	T

The last two lines of the truth table for the conditional indicate that from a false antecedent either a false or true consequent can follow, which is consistent with ordinary speech.

Consult the table for the conditional as you think about the following statement:

If you give me $5.00 now, then I will get you
a ticket for the rock concert on Saturday.

Have you been lied to:

1. If you give the speaker $5.00, and get a ticket?
2. If you give the speaker $5.00, and do not get a ticket?
3. If you do not give the speaker $5.00, and get a ticket?
4. If you do not give the speaker $5.00, and do not get a ticket?

The only case you should feel unhappy about is (2) since the speaker failed to deliver the ticket even though you supplied $5.00.

With the truth-table definition these are all true statements:

If $1 = 0$, then $2 \times 2 = 4$.
If $5 \times 6 = 27$, then the cow jumped over the moon.
If $2 \times 3 = 13$, then $6 \times 7 = 42$.

Example 8 Give the truth value for each of the following:

a) If $6 + 2 = 8$, then $12 + 4 = 16$.
b) If $8 + 7 = 15$, then $8 + 8 = 21$.
c) If $8 \times 8 = 63$, then $7 \times 8 = 56$.
d) If $8 \times 8 = 63$, then $8 \times 9 = 92$.

Solution We use in each case the truth-table definition for $\rightarrow$.

a) $\left.\begin{array}{l} p\colon\ 6 + 2 = \ \ 8 \text{ (true)} \\ q\colon 12 + 4 = 16 \text{ (true)} \end{array}\right\}$ we conclude $p \rightarrow q$ is true.

b) $\left.\begin{array}{l} p\colon\ 8 + 7 = 15 \text{ (true)} \\ q\colon\ 8 + 8 = 21 \text{ (false)} \end{array}\right\}$ we conclude $p \rightarrow q$ is false.

c) $\left.\begin{array}{l} p\colon\ 8 \times 8 = 63 \text{ (false)} \\ q\colon\ 7 \times 8 = 56 \text{ (true)} \end{array}\right\}$ we conclude $p \rightarrow q$ is true.

d) $\left.\begin{array}{l} p\colon\ 8 \times 8 = 63 \text{ (false)} \\ q\colon\ 8 \times 9 = 92 \text{ (false)} \end{array}\right\}$ we conclude $p \rightarrow q$ is true.

Try Exercise 23

23. Determine the truth value of each conditional:

a) If 6 is less than 7, it is not divisible by 9.
b) If 2×2 is 3, then 3×3 is 7.
c) If $6 + 2$ is 8, then $6 + 3$ is 10.
d) If $13 = 7$, then $0 \times 13 = 0 \times 7$.

MORE THAN ONE CONNECTIVE. The statements which use just one connective can be combined in various ways to form more complex statements, and the truth tables already established can be used to determine the truth values for them.

Example 9 What is the truth value of the compound sentence $\overline{p \wedge q}$?

Solution To find this we list all the possible combinations of truth values for p and for q, then compute systematically the truth values of the various parts. In this case we first find the truth values for $p \wedge q$ (see column 1) and then find the negation of $p \wedge q$ (see column 2).

p	q	$p \wedge q$	$\overline{p \wedge q}$
T	T	T	F
T	F	F	T
F	T	F	T
F	F	F	T
		(1)	(2)

This truth table shows for all possible combinations of truth values for p and q when the entire sentence will be true and when it will be false.

Example 10 Find the truth value of the sentence $\overline{p \vee \bar{q}}$. To do this we need columns for $\bar{q}$, $p \vee \bar{q}$, and finally $\overline{p \vee \bar{q}}$.

p	q	$\bar{q}$	$p \vee \bar{q}$	$\overline{p \vee \bar{q}}$
T	T	F	T	F
T	F	T	T	F
F	T	F	F	T
F	F	T	T	F
		(1)	(2)	(3)

24. Find the truth value of the statement $\bar{q} \to \bar{p}$.

25. Write the two conditional statements that can be derived from the following biconditional:

Triangle *ABC* is equiangular if and only if it is equilateral.

Try Exercise 24

THE BICONDITIONAL (↔). Theorems and definitions in mathematics often involve two ideas, each of which can be expressed by two related conditional statements. For example, if the two ideas to be conveyed are

$p \to q$: If a triangle is isosceles, then two of its sides are the same length

and

$q \to p$: If two sides of a triangle are the same length, then it is isosceles,

then it is possible to write instead of $(p \to q) \wedge (q \to p)$ the following biconditional:

A triangle is called isosceles if and only if two sides of the triangle are the same length.

The usual form of such a biconditional statement, symbolized $p \leftrightarrow q$, contains the words *if and only if*, and we see that such statements are not limited to mathematics.

Example 11 Analyze and write in symbols:

I shall ski this weekend if and only if it snows on Friday.

Solution Let p be: I shall ski this weekend and let q be: It snows on Friday. Thus: $p \leftrightarrow q$. There are two ideas conveyed by this sentence:

$p \to q$: If I ski this weekend, then it will have snowed on Friday

and

$q \to p$: If it snows on Friday, then I shall ski this weekend.

Try Exercise 25

It makes sense to have the biconditional true when and only when the two conditional statements, $p \to q$ and $q \to p$, are both true. We find these instances by looking at the truth table for the compound statement

$$(p \to q) \wedge (q \to p):$$

p	q	$p \to q$	$\wedge$	$q \to p$
T	T	T	T	T
T	F	F	F	T
F	T	T	F	F
F	F	T	T	T
		(1)	(3)	(2)

The column labeled (1) is done first by using the truth table for $p \to q$; the one labeled (2) is done next by applying the definition for $\to$ with $q \to p$, and finally the column marked (3) is completed by combining the results of (1), (2), and the truth-table definition for $\wedge$.

We see that both of the conditionals are true in those lines of the table in which p and q have the same truth values, and that is the only time when both conditionals are true.

The truth-table definition for the biconditional reflects this result:

p	q	$p \leftrightarrow q$
T	T	T
T	F	F
F	T	F
F	F	T

Try Exercise 26

Mathematics is a precise language not only because of the symbolism that is frequently employed but also because careful attention is paid to the correct usage of the connectives discussed in this section. A change from the connective *and* to the connective *or*, for example, can result in a drastic change in the meaning of a definition. Furthermore, statements must be correctly organized, that is, show correct reasoning in a deductive argument or proof, as we shall see in the next section.

26. Find the truth value of the following statements:

a) $(p \leftrightarrow q) \to q$

b) $(p \to q) \leftrightarrow q$

EXERCISE SET 1.4

1. Which of the following are statements and which are open sentences? Use a quantifier to convert each open sentence to a statement.

a) $3 + 5 = 35$

b) $x^2 = 17$

c) The points (1, 1) and (2, 2) determine a straight line.

d) Tell me that you really mean that.

e) π is an irrational number.

f) Do you think it will rain?

g) Some college students on our campus work at full-time jobs.

h) There is an x such that $x^2 - 3x + 2 = 0$.

i) $17 - y = 9$

j) If $5 = 37$, then $0 \times 5 = 0 \times 37$.

2. Let p stand for: You exercise regularly.
q stand for: You will keep healthy.
r stand for: You don't eat junk food.

Translate the following symbolic forms into English:

a) $p \to r$

b) $p \vee \overline{q}$

c) $\overline{q} \to \overline{p}$

d) $\overline{p \wedge q}$

e) $\overline{p} \vee \overline{q}$

f) $(p \wedge r) \to q$

g) $\overline{r} \wedge \overline{p}$

h) $(p \wedge r) \wedge q$

i) $\overline{r} \to \overline{q}$

j) $\overline{q} \to (\overline{p} \vee \overline{r})$

Each of the statements 3–10 can be regarded as a compound sentence. Write the statements symbolically.

3. Rick said he would mail a letter today or call you next week.

4. Martha went to the state fair yesterday and David went fishing at the lake.

5. If Sue gets the report finished, then she will bring it to you tomorrow.

6. I will get an A in math if and only if I get an A on all of the tests.

7. All professors expect you to do some work outside of class.

8. It is not true that I don't spend time in the library.

9. There are some problems in mathematics that are very interesting.

10. Perhaps I shall go and perhaps not.

11. Use at least two different forms to state in words the negation of each statement:
a) There is one prime number that is divisible by 2.
b) Some violets are not purple.
c) All numbers are integers.
d) All potatoes are a source of vitamin C.

12. For any statements p and q:
a) If $(p \wedge q)$ is true and p is true, can we conclude that q is also true?
b) If $(\overline{p \vee q})$ is true, then must $\overline{p}$ always be true?
c) Suppose that p is false and q is true, then can we conclude that $p \vee q$ is true?
d) Suppose that $(p \vee q)$ is true, does it mean that $(p \wedge q)$ is also true?

Determine the truth value of each of the statements 13–16.

13. a) $(3 + 12 = 15) \vee (3 \cdot 4 = 13)$
b) $(4 + 7 = 10) \vee (6 + 3 = 8)$

14. a) $(2 \cdot 4 = 8) \wedge (4 \cdot 4 = 16)$
b) $(2 + 3 = 5) \wedge (3 + 4 = 6)$

15. a) $(3 + 2 = 10) \rightarrow (2 \cdot 3 + 2 \cdot 2 = 20)$
b) $(2 + 3 = 5) \rightarrow (3 + 4 = 7)$

16. a) $4 \cdot 2 = 8 \leftrightarrow 2^3 = 8$
b) $(14 \div 3 = 5) \leftrightarrow (28 \div 6 = 10)$

17. You promise your parents: "If it is a sunny day on Saturday, I will mow the lawn." Under which of the following circumstances will your parents believe that you have told the truth?
a) It is sunny on Saturday; you mow the lawn.
b) It is sunny on Saturday; you don't mow the lawn.
c) It is cloudy on Saturday; you mow the lawn.
d) It is cloudy on Saturday; you don't mow the lawn.

18. If p is true and q is false, which of the following are true?
a) $\overline{p} \wedge q$ b) $p \rightarrow q$ c) $q \wedge (p \vee q)$
d) $p \rightarrow (\overline{p} \vee \overline{q})$ e) $\overline{p} \wedge \overline{q}$

19. Suppose that the truth value of $p \rightarrow q$ is T. What can be said about the truth value of $(\overline{p} \wedge q) \leftrightarrow (p \vee q)$?

20. Suppose the truth value of $p \leftrightarrow q$ is T. What can be said about the truth value of $p \leftrightarrow \overline{q}$? about $\overline{p} \leftrightarrow q$?

21. Determine the truth value:
a) $p \wedge \overline{p}$ b) $p \vee \overline{p}$

■ **22.** Determine the truth value of

$$\{[(p \rightarrow q) \wedge (r \rightarrow \overline{q})] \wedge r\} \rightarrow \overline{p}.$$

■ **23.** For each of the following determine whether the given information is sufficient to decide the truth value of the statement. Either state the correct truth value or show that both values are possible.

a) $(p \rightarrow q) \rightarrow \underset{\text{T}}{r}$

b) $(\overline{p \underset{\text{T}}{\vee} q}) \leftrightarrow (p \wedge q)$

c) $p \wedge (q \underset{\text{T}}{\wedge} r)$

d) $(p \underset{\text{T}}{\rightarrow} q) \rightarrow (\overline{q} \rightarrow \overline{p})$

1.5 MAKING PROOFS—DEDUCTION AND LOGIC

It is not necessary to ever have studied logic formally in order to be able to reason correctly, and we expect children to gradually acquire the accepted logical patterns of thought and reasoning as a result of their daily activities and interaction with adults.

Some children, however, may have trouble in mathematics because even in everyday language they do not understand the logical connectives discussed above or do not use them properly. It may be necessary to provide such children with a chance to learn the correct use of *or*, *and*, *if*, *then*, and so on, and to help them think about deductions that can be made from given facts.

In a practice session children might be using some colored discs: two large yellow ones, one small yellow one, and two small blue ones.

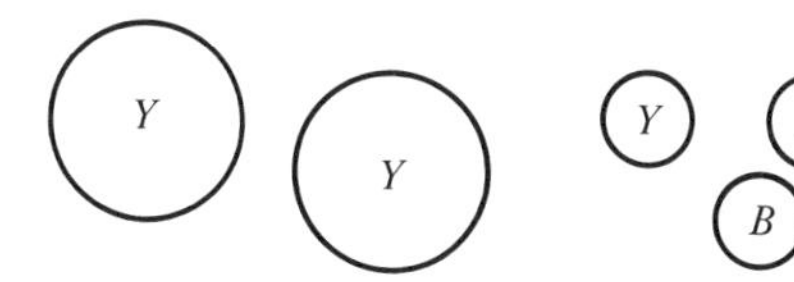

They could be asked:

> Choose a shape that is yellow and small.
> Choose one that is large and blue.
> Choose one that is small and not yellow.
> If the shape is large, then is it yellow?
> If the shape is yellow, then is it large?

The child who answers YES to the last question has either observed the objects incorrectly or has simply assumed that because the statement

> If a shape is large, then it is yellow

is true, then the reverse of that statement is true, a supposition which is clearly not the case here.

STATEMENTS DERIVED FROM A CONDITIONAL. Reversing the statements in a conditional $p \rightarrow q$ to form a new sentence $q \rightarrow p$ gives a new statement

1.5 AIMS

You should be able to:

a) Given a conditional sentence, write its converse, inverse, and contrapositive.
b) Use a truth table to determine if two sentences are equivalent.
c) Use a truth table to test the validity of an argument.
d) Explain the difference between the validity of an argument and the truth of the conclusion.

27. For the following shapes, G means green and W means white.

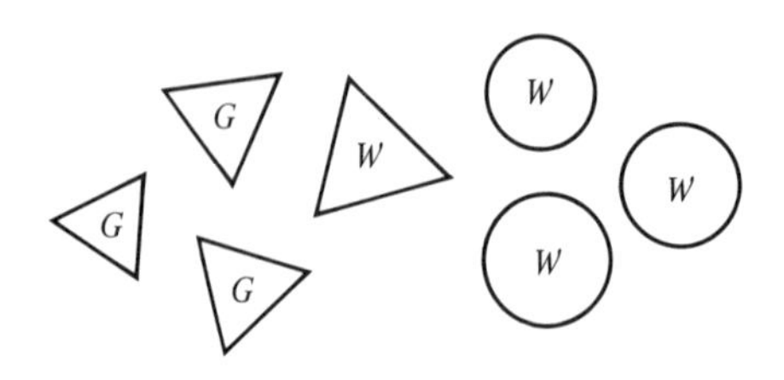

a) Write the converse of the statement: If the shape is green, it is triangular.
b) Is the original statement true? Is the converse true?

28. Give the converse, inverse, and contrapositive of the following statement:

If a number x is larger than 15, then it is greater than 10.

Here is one to think about:
I've seen Mary standing at the edge of the skating rink with skates on her feet. I've often seen her walking and running to school and standing about. I've seen her watching the skating at the rink, and she is always talking about skating.

1. Can Mary skate?
2. Why do you say that?

Orbit, Vol. 6, No. 3, June 1975.

called the *converse*, and $p \to q$ and $q \to p$ are then said to be converses of each other.

Example 1 Write the converse of the following sentence:

If the disc you took is blue, then it is small.

Solution Let p be: The disc you took is blue, and let q be: The disc is small. The original statement is $p \to q$, and its converse is $q \to p$: If the disc you took is small, then it is blue.

Try Exercise 27

The error that the child made in the case of the discs is exactly the one a manufacturer hopes the customer will make when an advertisement states:

If you want to be beautiful, use Glamma cream.

The hope of the manufacturer is that the customer will think instead:

If I use Glamma cream, then I will be beautiful.

That these two statements do not mean the same thing, we shall prove shortly.

There are four different statements that can be associated with any conditional $p \to q$, the converse being but one of these:

original statement	$p \to q$
converse	$q \to p$
inverse	$\overline{p} \to \overline{q}$
contrapositive	$\overline{q} \to \overline{p}$

Example 2 Give the converse, inverse, and contrapositive of the following statement:

If Bill drove the white car, then his brother took a bus to work.

Solution The three other conditionals are:

Converse: If Bill's brother took a bus to work, then Bill drove the white car.

Inverse: If Bill did not drive the white car, then his brother did not take the bus to work.

Contrapositive: If Bill's brother did not take the bus to work, then Bill did not drive the white car.

Try Exercise 28

LOGICAL EQUIVALENCE. From appropriate truth tables we can see exactly how these statements are related; we look at a conditional and its contrapositive first:

p	q	$p \to q$	$\bar{q}$	$\bar{p}$	$\bar{q} \to \bar{p}$
T	T	T	F	F	T
T	F	F	T	F	F
F	T	T	F	T	T
F	F	T	T	T	T

↑——Identical——↑

The truth-table definition for a conditional statement and its contrapositive are identical, and it is correct to replace one by the other, since it is not possible for one to be true at the same time that the other is false.

DEFINITION. Two statements p and q are logically equivalent if and only if they have the same truth values for all possible combinations of the truth values of the simple statements in them.

Thus we have proved in the table just above that a conditional and its contrapositive are logically equivalent and write symbolically:

$$(p \to q) \equiv (\bar{q} \to \bar{p}).$$

Try Exercise 29

On the other hand, we can now easily prove that a conditional and its converse are not logically equivalent.

Example 3 Prove that $p \to q$ and $q \to p$ are not logically equivalent.

Solution We must show that the truth value for $p \to q$ is not identical to that of $q \to p$:

p	q	$p \to q$	$q \to p$
T	T	T	T
T	F	F	T
F	T	T	F
F	F	T	T
		(1)	(2)

Column (1) is not the same as (2), so $p \to q \not\equiv q \to p$; we cannot replace one by the other, and the statements do not mean the same thing.

Try Exercise 30

29. Is the statement $p \to q$ logically equivalent to the statement $\bar{p} \vee q$?

p	q	$p \to q$	$\bar{p} \vee q$
T	T		
T	F		
F	T		
F	F		

30. Is the statement $\overline{p \vee q}$ logically equivalent to the statement $\bar{p} \vee \bar{q}$?

p	q	$p \vee q$	$\overline{p \vee q}$	$\bar{p} \vee \bar{q}$
T	T			
T	F			
F	T			
F	F			

VALID ARGUMENTS. In mathematics beyond the elementary-school level, there is an increasing emphasis on the formal proof of theorems, but even informal proofs and explanations must be done just as clearly and be based on correct reasoning.

Example 4 A child might be using the following blocks in various colors and sizes; the blue are marked B and the yellow are marked Y in the sketch.

The child is told: I have secretly chosen one of the small blocks; can you tell me more about it?

The child can observe: If the block is small, then it is yellow. I know the chosen block is small.

And conclude aloud: Your block must be yellow.

This child has assumed several statements and then offered a conclusion that followed from them. An argument usually consists of such a series of assumed statements called *premises* and a final statement called the *conclusion.* Symbolically, an argument with three premises and a conclusion q is:

$$[p_1 \wedge p_2 \wedge p_3] \to q.$$

An *argument* is *valid* if and only if the conclusion is true whenever all the premises are true. This is the same as saying for the case of three premises that the conditional $[p_1 \wedge p_2 \wedge p_3] \to q$ must be true in every case.

A statement that is true in every case is said to be *logically true.* A truth table is used to check the validity of an argument.

Example 5 Check the validity of the argument used by the child in Example 4.

Solution Given $p \to q$: If the block is small, it is yellow.
p: A small shape has been chosen.
Conclusion q: The shape is yellow.

The argument is valid if and only if q is true in every case for which both premises are true. This is the same as saying that the conditional $[(p \to q) \wedge p] \to q$ must be logically true, that is, must be true in every case. This can be checked in a truth table:

p	q	$[(p \to q)$	$\wedge$	$p]$	$\to$	q
T	T	T	T	T	T	T
T	F	F	F	T	T	F
F	T	T	F	F	T	T
F	F	T	F	F	T	F
		(1)	(2)		(3)	

← All premises are true; q is true.

Since column (3) shows the conditional statement is logically true, the argument of the example is valid.

This particular rule of inference or logical reasoning $[(p \to q) \wedge p] \to q$ is an important one called

THE LAW OF DETACHMENT.′ If the conditional $p \to q$ is assumed to be true and p is true, then q must also be true.

Some other rules of inference will be investigated in the exercises.

Example 6 Determine if the following argument is valid.

Given: If you got all A's on the tests, you must have studied long hours.
You did not get all A's on the tests.
Conclusion: You did not study long hours.

Solution The form of the argument is $[(p \to q) \wedge \overline{p}] \to \overline{q}$. This is checked with a truth table:

p	q	$[(p \to q)$	$\wedge$	$\overline{p}]$	$\to$	$\overline{q}$
T	T	T	F	F	T	F
T	F	F	T	F	T	T
F	T	T	T	T	F	F
F	F	T	T	T	T	T
		(1)	(3)	(2)	(5)	(4)

We see that this conditional is not logically true, since not all truth values are T in column 5; the argument is *not valid* and is called a *fallacy*. Note that there is a line in which both premises are true and yet the conclusion $\overline{q}$ is not true.

Try Exercise 31

31. Determine if the following argument is valid; that is, write the argument symbolically and check by means of a truth table.

If you like to swim then you would enjoy a trip to the Gulf coast. You don't like to swim. Therefore you would not enjoy a trip to the Gulf coast.

The sequencing of statements in a logically correct manner in a proof or deductive argument is an important part of the language of mathematics. The discussion of this section has been intended to clarify the deductive arguments, presented both formally and informally in the remainder of the book. Whether or not you ever write formal arguments or deductive proofs yourself, it is still necessary in teaching elementary-school pupils to be able to provide logically sound explanations for them.

LAWS FOR STATEMENTS—A SUMMARY. We have used a truth table to decide if an argument is valid, if a statement is logically true, and if two statements are logically equivalent; in each case, a truth table provides the proof or disproof of some property about logical statements and the connectives used with them.

Some of the theorems proved in this way are so important and used so often that special names have been given to them. Those laws that will be most useful in the next chapters are given below; each of them is proved either in the text or in an Exercise Set of this chapter.

Commutative laws: $p \wedge q \equiv q \wedge p$
$p \vee q \equiv q \vee p$

Associative laws: $p \wedge (q \wedge r) \equiv (p \wedge q) \wedge r$
$p \vee (q \vee r) \equiv (p \vee q) \vee r$

Distributive laws: $p \wedge (q \vee r) \equiv (p \wedge q) \vee (p \wedge r)$
$p \vee (q \wedge r) \equiv (p \vee q) \wedge (p \vee r)$

De Morgan's laws: $\overline{p \wedge q} \equiv \overline{p} \vee \overline{q}$
$\overline{p \vee q} \equiv \overline{p} \wedge \overline{q}$

Negation: $p \equiv \overline{\overline{p}}$

Rules for inference: $[(p \rightarrow q) \wedge p] \rightarrow q$
$[(p \rightarrow q) \wedge \overline{q}] \rightarrow \overline{p}$
$[(p \rightarrow q) \wedge (q \rightarrow r) \wedge p] \rightarrow r$

EXERCISE SET 1.5

Write the converse, inverse, and contrapositive of the given sentences 1–4.

1. If a number is a multiple of 14, then it is a multiple of 7.

2. If 16 is a divisor of a number, then 4 is a divisor of the number.

3. If I stay home from the party, I will be able to get my term paper finished.

4. If you are a ballet dancer, then you must practice long hours.

5. If two triangles have their sides respectively parallel, then they are similar. From logical considerations alone, which of the following statements is equivalent to the one just given?

a) If two triangles are similar, then they have their sides respectively parallel.
b) If two triangles are not similar, then they do not have their sides respectively parallel.
c) If two triangles do not have sides respectively parallel, then they are not similar.

6. The following are known as de Morgan's laws. Use a truth table to prove that the equivalence holds in each case:

a) $\overline{p \vee q} \equiv \overline{p} \wedge \overline{q}$ b) $\overline{p \wedge q} \equiv \overline{p} \vee \overline{q}$

7. Determine whether the following pairs of statements are equivalent:

a) $p \wedge q, q \wedge p$ b) $p \vee q, q \vee p$
c) $p \wedge (q \wedge r), (p \wedge q) \wedge r$ d) $p \vee (q \vee r), (p \vee q) \vee r$
e) $p, \overline{\overline{p}}$ f) $\overline{p} \rightarrow q, p \vee q$

8. The following are known as the distributive laws. Use a truth table to prove that the equivalence holds in each case:

a) $p \wedge (q \vee r) \equiv (p \wedge q) \vee (p \wedge r)$
b) $p \vee (q \wedge r) \equiv (p \vee q) \wedge (p \vee r)$

9. Use truth tables to decide which of the following arguments are valid:

a) If it snows, then I shall go skiing. It snows. Therefore I shall go skiing.
b) If it snows, then I shall go skiing. It doesn't snow. Therefore I shall not go skiing.
c) If I fall from a ladder, I shall be injured. I am injured. Therefore I fell from a ladder.

10. Show whether or not the following arguments justify the conclusion.

a) If I go to Toronto, then I shall travel by plane. I am traveling by train. Therefore I am not going to Toronto.
b) If I do not get a haircut, then I shall be at home this afternoon. I went shopping this afternoon. Therefore I got a haircut.
c) If I have a bad cold, then I do not go to work. I do not have a bad cold. Therefore I am at work.
d) If I am not hungry, then I will not eat dessert. I will eat dessert. Therefore I am hungry.

In 11 and 12 the premises (P) are assumed to be true. Determine if the conclusion (C) is correct.

11.
P: If this figure is a triangle, then it is a polygon.
P: If this figure is a polygon, then it is a closed figure.
P: This figure is a triangle.
C: Therefore, this figure is a closed figure.

12.
P: If the day is sunny on the 4th of July, we will have a picnic.
P: On the 4th of July we had a picnic.
C: Therefore the day was sunny.

The following questions 13–16 are similar to those asked of fourth graders to test logical ability. Answer YES or NO to each question:

13.
P: If the bicycle in the yard is yellow, then it is Ann's.
P: If the bicycle has a transistor radio on it, then it is not Ann's.
P: The bicycle does have a radio on it.
C: Is the bicycle yellow?

14.
P: If the peg is not a small one, it will not fit into this hole.
P: The peg does fit into the hole.
C: Is the peg a small one?

15.
P: If that gift is in a large package, then it has a toy in it.
P: If it is a toy, then it is not meant for your mother.
P: The gift is in a large package.
C: Is the gift for your mother?

16.
P: If the family does not get the station wagon back from being repaired, then they will not drive on their vacation.
P: If the family does not drive on their vacation, then they will not take their two cats along.
P: The family does take the two cats along.
C: Did the family get the station wagon back from being repaired?

17. Is it possible to arrive at a false conclusion although the pattern of reasoning is a valid one? Why or why not?

18. What is the difference between an invalid argument and a false conclusion?

REVIEW TEST

1. Find and describe a pattern in the following sequences. Use your pattern to find the next three numbers in the sequence.
a) 1, 4, 9, . . .   b) 1, 2, 4, 7, 11, . . .

2.
a) Find and describe a pattern in the following sequence of poker chips (B represents a blue chip, W a white one);

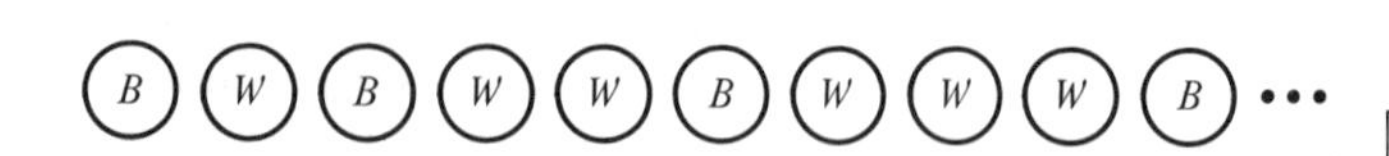

b) With your pattern what should the next three chips be?
c) Find another pattern in the sequence in (a).

3. Make a conjecture about each of the following patterns:
a) $2^2 - 2 = 2$, $3^2 - 3 = 6$, $4^2 - 4 = 12$, $5^2 - 5 = ?$
b) $2^3 - 2 = 6$, $3^3 - 3 = 24$, $4^3 - 4 = 60$, $5^3 - 5 = ?$
■ c) Prove your conjecture for part (a).
■ d) Prove your conjecture for part (b).

4. A silversmith makes a certain type of bracelet from a pre-cut silver blank. The scrap silver from five such bracelets is enough to remelt into another blank. If the smith starts with 25 precut silver blanks, how many bracelets can be made?

5. A jogger who lives in a six-story apartment building usually runs in the park across the street but on blizzardy days runs up and down the stairs of the apartment building instead. To go from the first floor to the third takes the jogger 20 seconds. How much time is needed for a trip from the first floor to the sixth and down again? (Assume a constant rate of speed for the jogger.)

6. Prove that:
a) $\exists x: x^2 = 2x$ is a true statement
b) $\forall x: x^2 = 2x$ is a false statement
Assume that the x values under discussion are whole numbers.

7. Show that no pair of whole numbers (0, 1, 2, 3, 4, . . .) satisfies the equation $6x + 9y = 35$. There are several ways to solve this problem; by means of a calculator it can be done without any sophisticated knowledge of mathematics. The orderly organization of your results in a table may prove helpful. Don't forget to use the five-step process.

8. Children are often given practice in reasoning by means of small shapes called attribute blocks. Suppose you have three each of large square, triangular, and circular shapes and the same number of smaller ones to correspond. Each of the different shapes and sizes comes in red, blue, and yellow. Children are shown one of the pieces covered by a clue card. They are to determine what is under the clue card without looking.
a) Clue card: It is not large. It is neither red nor blue. It is not a triangular or circular shape. What is it?
b) Clue card: If it is blue, then it is small. If it is small then it is triangular. It is not red or yellow. What is it?

c) Clue card: It is either a large triangular shape or a circular one. If it is blue, then it is circular. It is not red and it is not yellow. What is it?
d) Symbolize the arguments used in (a), (b), and (c).

9. State in words the negation of the following in two different ways.
a) All numbers are prime.
b) There are integers that are not whole numbers.

10. For the given statement $s \rightarrow \bar{t}$, state which of the following are true (T) and which are false (F).
a) $t \rightarrow s$ is the converse.
b) $s \rightarrow t$ is the inverse.
c) $\bar{s} \rightarrow t$ is the contrapositive.
d) $t \rightarrow \bar{s}$ is the contrapositive.

11.
a) Construct the truth table for $p \wedge \bar{p}$, for $p \vee \bar{p}$.
b) Describe the two principles that are established by means of these truth tables.

12. Prove by means of a truth table that the following two sentences mean the same thing:
a) If she earned some extra money this week, then she would purchase the camera.
b) Either she did not earn the extra money this week or she purchased the camera.

13. Which of the following are logically true?
a) $p \rightarrow (p \wedge q)$ b) $p \rightarrow (p \vee q)$ c) $[(p \rightarrow q) \wedge \bar{q}] \rightarrow \bar{p}$

14. Which of the following pairs of statements are logically equivalent?
a) $\bar{p} \leftrightarrow q$, $\overline{p \leftrightarrow q}$ b) $p \leftrightarrow q$, $(p \wedge q) \vee (\bar{p} \wedge \bar{q})$
c) $p \rightarrow \bar{q}$, $\bar{p} \vee \bar{q}$ d) p, $p \vee p$

15. Write the following statements in symbolic form. The statements p and q are defined as follows:

p: I am enrolled in Math 121.
q: I drive my car to school.

a) I am enrolled in Math 121 and I don't drive my car to school.
b) I am neither enrolled in Math 121 nor do I drive my car to school.
c) I drive my car to school only if I am enrolled in Math 121.
d) Being enrolled in Math 121 is sufficient for my driving my car to school.

16. If the information given is sufficient to determine the truth value of the statements below, do so; otherwise show that both truth values are possible:
a) $p \vee (q \rightarrow r)$, with p = T
b) $(p \wedge q) \rightarrow (p \vee s)$, with $(p \wedge q)$ = T, $(p \vee s)$ = F

17. A connective called *Sheffer's stroke* (/) has been defined for two statements. The truth table for this connective is given:

p	q	p/q
T	T	F
T	F	T
F	T	T
F	F	T

a) Show that $p/q \equiv \overline{p \wedge q}$.
■ b) How would the statement $p \vee q$ be represented in terms of the Sheffer stroke?

FOR FURTHER READING

1. Two mathematicians speak from personal experience about creativity in mathematics in *Mathematics in the Modern World: Readings from Scientific American*, W. H. Freeman and Company, San Francisco and London (1949–1958) by Scientific American Inc.: Paul R. Halmos writes on "Innovation in mathematics" in Chapter 1 and Henri Poincaré on "Mathematical creation" in Chapter 2.

2. Edith Robinson discusses ". . . some kinds of problems we should seek out if our aim is to be realized — that the student be able to solve problems whenever the need arises." Her ideas can be found in the article "On the uniqueness of problems in mathematics," *Arithmetic Teacher*, 25:2, November 1977.

TEXTS

Fehr, Howard F., and Jo McKeeby Phillips. *Teaching Modern Mathematics in the Elementary School.* Addison–Wesley Publishing Company, Reading, Massachusetts (1967, 1972). This book has a chapter on problem solving.

Johnson, Paul B. *From Sticks and Stones: Personal Adventures in Mathematics.* Science Research Associates, Inc., Chicago, Illinois (1975). Contains information on both problem solving and number patterns.

chapter 2 sets, relations, and functions

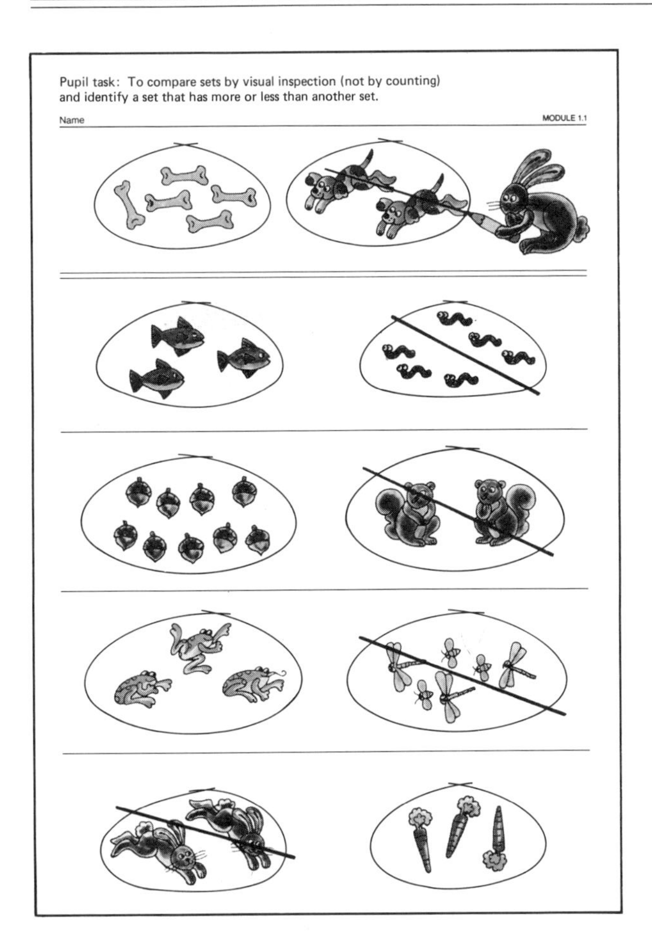
Pupil task: To compare sets by visual inspection (not by counting) and identify a set that has more or less than another set.

Name MODULE 1.1

Not all children have a firm grasp of the concepts of "less than" and "more than" when they reach first grade; they need some practice, such as the examples provided in this lesson.

2.1 AIMS

You should be able to:

a) Describe a set by using the roster method or the standard description.
b) Apply the terms universe, empty set, subset, set equality, and proper subset correctly.
c) Distinguish between:
   an element and a subset;
   a subset and a proper subset.

2.1 SETS AND SET NOTATION

The mathematician Georg Cantor* can either be blamed or credited, depending upon your viewpoint, for the fact that sets appear at all in a textbook for elementary teachers. Almost all of Cantor's published work dealt with his set theory, which had a tremendous impact on 20th-century mathematicians and thus was an important force in shaping a significant part of our culture.

Some Language and Notation

One or more of the concepts of set, relation, and function will form the basis of the subsequent material that will be developed in this book. These ideas serve to unify the subject of mathematics, so that once the properties of a particular operation or relation are understood, it is possible to investigate similar properties in an entirely different context. Whereas the formal ideas about sets are not necessarily taught to children, their work with concrete models for sets such as blocks or counters does help them to develop the abstract idea of number. Although the terminology in the elementary school may be less sophisticated, the same concepts that provide unity here can also be thought of as the main themes of elementary-school mathematics.

In higher mathematics the word *set* is taken as an undefined term, as we do here, with the presumption that it is commonly understood. In everyday language we can be somewhat casual about the use of the word *set*, but in mathematics it must be possible to definitely decide if something is an element of a set or not. Take, for example, the set of all the two-digit whole numbers used in a problem found earlier in this book. The members of this set are the numbers 10, 11, 12, 13, 14, 15, and so on up to 99. Because we can tell exactly which whole numbers get into the set and which do not, this is a *well-defined set*.

The following are not well-defined sets:

The set of good trumpet players.
The set of handsome hockey players.
The set of very creative persons.

It is important that the sets used in mathematics be well-defined.

ROSTER NOTATION. One way to describe a set is to simply list the members and enclose the entire list in a pair of braces. This is called the *roster notation* for a set.

Example 1 Give the roster notation for the set of whole numbers less than 8.

Solution $\{0, 1, 2, 3, 4, 5, 6, 7\}$

* Georg Cantor (1845–1918) was born in Petrograd, Russia, and spent the years 1856–1863 in southern Germany. He studied mathematics in Berlin from 1863–1869 and then went on to teach and do research in mathematics. His original work in set theory did a great deal to clarify many concepts about numbers.

Example 2 Give the roster notation for the set of two-digit whole numbers.

Solution $\{10, 11, 12, \ldots, 98, 99\}$

The three dots indicate that some of the list has been omitted, but it is clear what is intended. The order in which the members of the set are listed ordinarily makes no difference although in Example 2 it would present some difficulty in interpretation if the order were changed from what appears there.

SET-BUILDER NOTATION. Instead of a list, the braces may enclose a statement or open sentence which describes the properties that an element must satisfy before it can qualify as a member of the set. This notation is called *set-builder* notation or the *standard description* of the set.

Example 3 What is the set-builder notation for the set of whole numbers less than 8?

Solution $\{x \mid x$ is a whole number and x is less than $8\}$. This is read: The set of all x such that x is a whole number and x is less than 8. The only elements that are members of the set are those that make a true statement of the open sentence.

Example 4 Write roster and set-builder notation for the set of natural numbers between 3 and 10.

Solution

$$\text{Roster: } \{4, 5, 6, 7, 8, 9\} \text{ or } \{4, 6, 7, 9, 5, 8\}$$

$$\text{Set-builder: } \{x \mid 3 < x < 10 \text{ and } x \text{ is a natural number}\}$$

When the number system in question is clearly understood, reference to it is often omitted, so in this case it might not be specified that the numbers were natural numbers. Instead it would be written $\{x \mid 3 < x < 10\}$.

Often a single letter, usually upper case, is used to denote a particular set. For example, set A consists of the following elements or members: 4, 6, 9, and 72; thus

$$A = \{4, 6, 9, 72\}$$

Try Exercises 1 and 2

MEMBERSHIP. The symbol $\in$ means *is a member of*, or *is an element of*, or *belongs to*. Thus the sentence $a \in A$ means a is an element of set A, or belongs to set A; to indicate that an element *does not* belong to a set A we can write $a \notin A$.

1. Write roster notation for the set of whole numbers that are greater than 10 and less than 17.

2. Write the standard description for the set of Exercise 1.

3. Explain what the following sentences say:
a) $p \notin \varnothing$,
b) $q \in \mathscr{U}$.

Example 5 If set B is $\{7, 3, 11, 15, 1, 8\}$, then

$$7 \in B \text{ is a true statement,}$$
$$67 \notin B \text{ is a true statement,}$$
$$\{7, 3, 11\} \in B \text{ is not a true statement.}$$

In the last case there is no set that is an element of set B.

UNIVERSAL SETS. In mathematics it is either stated specifically or clear from the context that only certain elements will enter into a discussion. The universal set, often named $\mathscr{U}$, is the set of all elements involved in a particular situation. In a problem in Chapter 1, it was specified that the values for x and for y were to be selected from the set of whole numbers, so for that problem the universal set for possible values of x and y was the set of whole numbers.

THE EMPTY SET. The set without any members at all is called the *empty*, or *null* set; it is indicated by $\varnothing$ or by a pair of braces $\{\ \}$, the latter showing clearly that there are no members in the set. There are many ways to describe the empty set, for example:

The set of whole numbers between 3 and 4.
The set of people in your class who have visited the planet Mars.
The set of purple cows that have jumped over the moon.

For each of these statements it is impossible to find a single element of the universe under discussion that satisfies the stated property.

Try Exercise 3

Set Relations

SUBSET. Sets can be related to each other in various ways. Because in some cases we find that each and every element of a set A is also an element of B, in this situation A is related to B in that it is a *subset* of B, or $A \subseteq B$.

DEFINITION. ***A* is a *subset* of *B* ($A \subseteq B$) if and only if every element *a* of set *A* is also a member of set *B*.**

More briefly stated:

$$A \subseteq B \leftrightarrow (\text{If } \forall a,\ a \in A, \text{ then } a \in B.)$$

Each element found in A must also be found in B; this is equivalent to stating that if an element is not in B, then it is not in A.

It follows logically that every set A is a subset of itself, that is, for any set A, $A \subseteq A$. The definition provides the means for a simple proof of this conclusion.

Every element in set A is also then in set A, since every element of A is a member of A, and we conclude that $A \subseteq A$ for any set A.

The empty set is unusual in that it is a subset of every set because all of the elements the empty set has (it has none) are members of every set. We could also prove this by reasoning that if an element is not in some set A, then it is also not in the empty set, a statement that is logically equivalent to the definition of subset, as we have seen.

As a result of either argument we conclude that for any set A

$$\varnothing \subseteq A.$$

For any set A it holds that

$$A \subseteq A, \qquad \varnothing \subseteq A$$

Example 6 If $\mathscr{U} = \{x \mid x \text{ is a whole number less than } 10\}$ and $A = \{1\}$, $B = \{1, 2, 3\}$, then

$$A \subseteq B \text{ (} A \text{ is a subset of } B\text{)},$$
$$B \nsubseteq A \text{ (} B \text{ is not a subset of } A\text{)}.$$

Notice that a subset of a particular set is different from an element of that set. In Example 6 above:

$$1 \in A, \qquad 6 \notin B$$

These are statements about an element and whether or not it is an element of a set. On the other hand,

$$A = \{1\}$$

means that A is a set that has 1 as its only element. The phrase "subset of" describes the relationship between two sets and the symbol $\subseteq$ must appear between two sets as in

$$A \subseteq \{1, 2, 3\}.$$

It is correct to say that

$$\{1, 2\} \subseteq \{1, 2, 3\} \qquad \text{or} \qquad 2 \in \{1, 2, 3\},$$

but incorrect to say that

$$\{1, 2\} \in \{1, 2, 3\} \qquad \text{or} \qquad 2 \subseteq \{1, 2, 3\}.$$

EQUALITY OF SETS. If we say that set A equals set B, we mean that the sets are really a single set that has been given two different names. If A is $\{a, b, c\}$ and B is $\{c, a, b\}$, then A is the same set as B.

DEFINITION. Two sets A and B are *equal* if and only if A is a subset of B and B is a subset of A. That is,

$$A = B \leftrightarrow (A \subseteq B \quad \text{and} \quad B \subseteq A).$$

If A and B are sets, the sentence $A = B$ states that A and B are the same set, that is, their members are identical.

4. Write a proper symbol $\subseteq$, $\not\subseteq$, $\subset$, $\in$, $\notin$, or $=$ between the pairs of symbols to make a true statement:

a) $\{1, 2, 3\}$ $\{4, 5, 6\}$
b) 4 $\{4, 5, 6\}$
c) $\{a, b\}$ $\{a, b, c\}$
d) 7 $\{1, 2, 3\}$
e) $\{a, b, c\}$ $\{a, b, c\}$
f) A $\mathcal{U}$
g) $\varnothing$ $\mathcal{U}$

PROPER SUBSET. If A is a proper subset of B, then A must first of all be a subset of B, but there must be at least one element in B that is not an element of A.

DEFINITION. ***A* is a *proper subset* of *B* ($A \subset B$) if and only if *A* is a subset of *B* and *A* is not equal to *B*. That is,**

$$A \subset B \leftrightarrow A \subseteq B \quad \text{and} \quad A \neq B.$$

In order to be a *proper subset*, set A must satisfy two conditions.

Example 7 How are sets $A = \{4, 5, 6\}$ and $B = \{4, 5, 6, 7, 9\}$ related?

Solution $A \subseteq B$ is a true statement.
$A \neq B$ is a true statement.
Thus $A \subset B$ is also a true statement.

Try Exercise 4

EXERCISE SET 2.1

Write roster notation for the following sets:

1. The set of whole numbers that are less than 12.

2. The set of people you know who are over 35 feet tall.

3. The set of United States coins that are worth $1.00 or less.

4. The set of all your brothers and sisters.

Write the standard description for the following sets.

5. The set of all even integers.

6. The set of all odd integers.

7. The set of all rational numbers greater than 2.

8. The set of all real numbers less than -5.

Write a symbol, $\in$, $\notin$, $\subseteq$, $\not\subseteq$, $\subset$, $\not\subset$, or $=$ between the pairs of symbols to make a true statement:

9. $\{1, 2, 3\}$ $\{3, 1, 2\}$

10. $\{5, 3\}$ $\{3, 5\}$

11. 2 $\{4, 5\}$

12. 3 $\{5, 3, 2\}$

13. $\{2\}$ $\{4, 5\}$

14. $\{3\}$ $\{5, 3, 2\}$

15. State which of the following statements are true (T) and which are false (F) for any sets A and B:

a) $A \subset A$
b) If $A = B$, then $A \subset B$.
c) $\varnothing = \{\ \}$
d) $\{a\} \in \{a, b, c\}$
e) $\{0\} = 0$
f) $A \subseteq \mathcal{U}$

The next exercises will give you a chance to observe some patterns and draw some conclusions as a result of your observations:

16. Find all the subsets of the set $\{p, n\}$.

17. Find all the subsets of the set $\{p, n, d\}$.

18. Find all the subsets of the set: $\{p, n, d, q\}$.

19. On the basis of the pattern you detect in 16 through 18, how many subsets would you conjecture could be found for the set $\{p, n, d, q, h\}$? What is the effect of adding one more element to a set in the series?

■ **20.** Suppose that there are k elements in a set, how many different subsets could be found? Give an argument to prove or disprove your answer.

■ **21.** How many different amounts of money could you give someone if you had exactly one each of the following bills: one-dollar, two-dollar, and five-dollar? (You can give someone no money at all.) What does this problem have to do with Exercise 17?

2.2 SET OPERATIONS AND THEIR PROPERTIES

Operations

We shall see that the various operations for sets provide a logical basis for the arithmetic operations we deal with later. Many of the explanations and demonstrations used for teaching number concepts to children involve models that are based upon the properties of operations for sets.

UNION. The operation of the union of two sets is of special interest for the primary teacher who uses this concept over and over in teaching youngsters about addition in arithmetic. To help pupils understand the abstract notion involved in the addition $2 + 3$, a teacher might say: "I have two pieces of candy in this hand and three pieces in my other hand. If I put them all into this empty bag, how many are in the bag?" In this case a physical model for sets and their union provides a means for developing an abstraction.

The *union* of two sets is denoted by $A \cup B$ and is formed by uniting in one set the elements of the sets A and B.

DEFINITION. The union of two sets A and B is the set consisting of the elements of set A along with the elements of set B:

$$A \cup B = \{x \mid x \in A \quad \textit{or} \quad x \in B\}.$$

For any element x to be in the set $A \cup B$ it must satisfy at least one of the conditions, since the critical word in the definition is the connective *or*.

Example 1 Let $A = \{0, 5, 7, 13, 27\}$ and $B = \{0, 5, 7, 38, 45\}$. Find $A \cup B$.

Solution $A \cup B = \{0, 5, 7, 13, 27\} \cup \{0, 5, 7, 38, 45\}$
$= \{0, 5, 7, 13, 27, 38, 45\}$.

2.2 AIMS

You should be able to:

a) Define and apply the operations for union, intersection, complement, and Cartesian product.
b) Draw appropriate diagrams that illustrate unions, intersections, complements, and Cartesian products.
c) State and recognize the properties for the operations above.

5. Let $M = \{1, 3, 5, 9\}$, $N = \{2, 4, 5, 9, 10\}$ Write roster notation for $M \cup N$.

If an element appears on the rosters for both A and B, it is named only once on the roster for $A \cup B$.

Information about sets can be displayed in a drawing called a *Venn diagram*,* in which the interior of a closed curve such as a circle represents a set.

Example 2 Represent $A \cup B$ with a diagram.

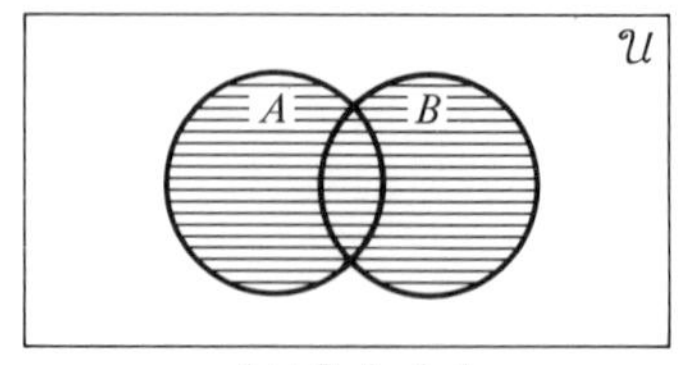

$A \cup B$ shaded

Two regions are used to represent sets A and B inside a region that represents the universal set $\mathscr{U}$. Since all of the elements of A belong to the union, A is shaded; similarly for B. Thus the entire shaded portion represents $A \cup B$. The diagram for $A \cup B$ shows the sets A and B overlapping, not because it is known that they actually do, but to allow for that possibility.

Try Exercise 5

INTERSECTION. Primary children who play the "game" of sorting attribute blocks frequently use the intersection operation although they may not verbalize it in such language. A child might be asked to sort out blocks on the basis of a single characteristic, such as "all of the triangular shapes" ($\{x \mid x$ is a triangular shape$\}$). The same child might similarly sort or classify "all those that are blue" ($\{x \mid x$ is a blue block$\}$). Finally the child might be asked to classify on the basis of both characteristics and select the blocks that are both triangular and blue ($\{x \mid x$ is triangular in shape and x is blue$\}$). Not all the triangular shapes are blue and not all the blue blocks are triangular in shape, but there will be some blocks that have both properties. We describe this formally.

The intersection operation used with any two sets A and B results in a new set A *intersect* B ($A \cap B$) consisting of the elements common to the sets A and B.

DEFINITION. The *intersection* of two sets A and B is the set of all those elements that are members of the set A and also members of the set B:

$$A \cap B = \{x \mid x \in A \quad \text{and} \quad x \in B\}.$$

If an element is a member of $A \cap B$, it must satisfy two conditions; the important word in the definition is the connective *and*.

Example 3 Let $A = \{0, 1, 3, 5, 7\}$ and $B = \{2, 3, 4, 5, 6\}$. Find $A \cap B$.

Solution $A \cap B = \{0, 1, 3, 5, 7\} \cap \{2, 3, 4, 5, 6\}$
$= \{3, 5\}$

* John Venn (1834–1923) was an English logician, a teacher of logic, and author of *The Logic of Chance* (1866), *Symbolic Logic* (1881), and *The Principles of Empirical Logic* (1889). His diagrams differed slightly from those shown here, since they did not include the surrounding rectangle for the universe $\mathscr{U}$.

A set diagram for intersection is as follows: the portion shaded twice represents the set of all those elements that are both members of set A and at the same time members of set B.

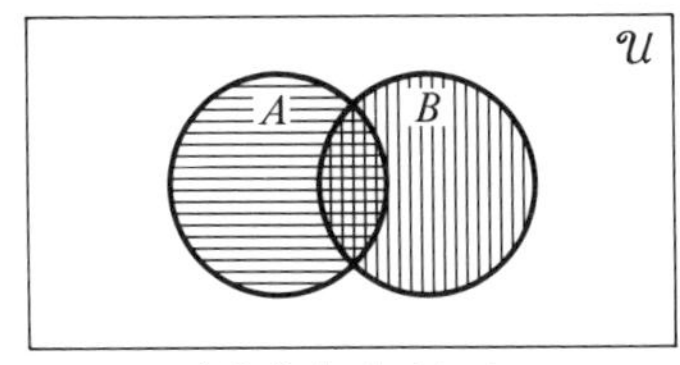

$A \cap B$ shaded twice

DISJOINT SETS. It may happen that the set $A \cap B$ is actually the empty set.

Example 4 Let $A = \{2, 4, 6, 8\}$ and $B = \{1, 3, 5, 7\}$. Find $A \cap B$.

Solution $A \cap B = \{\ \}$ since no elements satisfy both conditions for membership.

If the intersection $A \cap B$ is the empty set, then the sets A and B are said to be *disjoint*.

Try Exercise 6

6.

a) Let $M = \{1, 3, 5, 9\}$, $N = \{2, 4, 5, 9, 10\}$. Write roster notation for $M \cap N$.

b) Let $C = \{a, b, c\}$, $D = \{d, e, f, g\}$. Write roster notation for $C \cap D$.

COMPLEMENTATION. The complement of a set A is the set containing all those elements of the universal set that are not in A.

> **DEFINITION. The *complement* of a set A is $\bar{A}$:**
>
> $$\bar{A} = \{x \mid x \notin A \text{ and } x \in U\}.$$

The elements of $\bar{A}$ satisfy the negation of the statement that determines the membership for set A.

Example 5 Let $\mathscr{U} = \{3, 4, 5, 7, 9, 10, 12\}$ and $A = \{4, 7, 10\}$. Find the complement of A.

Solution $\bar{A} = \{3, 5, 9, 12\}$.

If a circle of a Venn diagram represents set A, then all of the universe outside of A must be shaded to show $\bar{A}$.

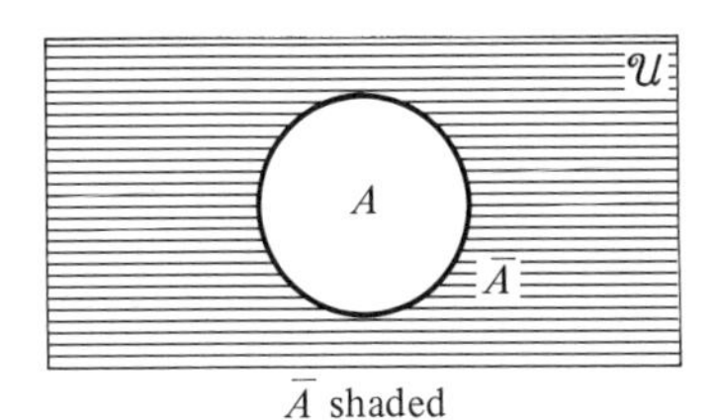

$\bar{A}$ shaded

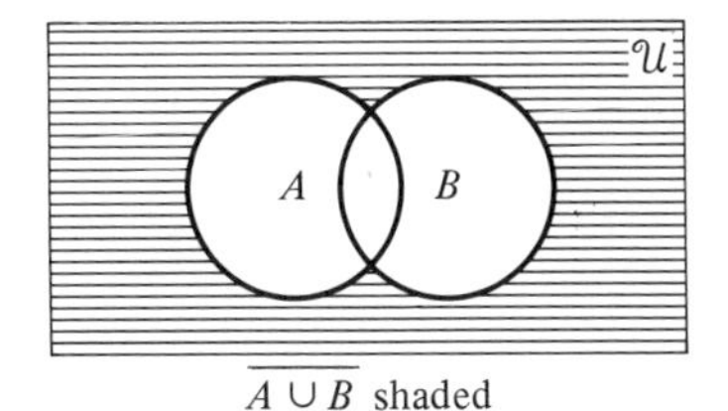

$\overline{A \cup B}$ shaded

7. Let $\mathscr{U} = \{1, 2, 3, 4, 5, 6, 7, 8\}$,
$N = \{1, 3, 5\}$.
Write roster notation for $\bar{N}$.

Try Exercise 7

CARTESIAN PRODUCT.* Both the union and intersection operations result in a new set whose elements are already elements of A or B. We now encounter an operation that produces a set whose members are not the original elements.

Suppose a photographer has two camera bodies, one loaded with black-and-white film and the other loaded with color-slide film along with a collection of four lenses that can be used on either of the cameras, say, 28 mm, 40 mm, 75 mm, and 150 mm. In how many ways can this photographer decide to take a picture? The body with the black-and-white film (B) could be used with any one of the four lenses, and the possibilities for the black-and-white film can be described with the pairs: (B, 28), (B, 40), (B, 75), and (B, 150). Similarly the camera with the color film (C) could be used with any one of the four lenses, so (C, 28), (C, 40), (C, 75), and (C, 150) are four additional possibilities, each of which would give a different photograph taken from the same spot. Each possibility is a pair consisting of, first, the choice of film and, second, the choice of the lens to use with the film; the set of all such possibilities is an example of a *Cartesian product* although it is unlikely the photographer would describe the situation in these words.

The *Cartesian product*, or *cross product*, of two sets A and B denoted by $A \times B$ and read "A cross B" is the set whose elements are ordered pairs of elements.

DEFINITION. The *Cartesian product* $A \times B$ of two sets A and B is the set of ordered pairs (a, b) such that the first component of each pair is an element of A and the second component is an element of B:

$$A \times B = \{(a, b) \mid a \in A \text{ and } b \in B\}.$$

Example 6 Find the Cartesian product $A \times B$, where

$$A = \{1, 3, 5\} \quad \text{and} \quad B = \{2, 4\}.$$

Solution $A \times B$ is the set of all possible ordered pairs with the first component from A and the second from B:

$$A \times B = \begin{Bmatrix} (1, 2) & (3, 2) & (5, 2) \\ (1, 4) & (3, 4) & (5, 4) \end{Bmatrix}.$$

Two ordered pairs (a, b) and (c, d) are the same if and only if $a = c$ and $b = d$. That is, the pair $(1, 2) = (1, d)$ if and only if $2 = d$, and the pairs (3, 4) and (4, 3) are not the same since $3 \neq 4$. Note that the set $B \times A$ would contain different ordered pairs:

$$B \times A = \begin{Bmatrix} (2, 5) & (4, 5) \\ (2, 3) & (4, 3) \\ (2, 1) & (4, 1) \end{Bmatrix}.$$

* Derived from Cartesius, the Latinization of Descartes. René Descartes (1596–1650) was the French philosopher and mathematician.

This single example (counterexample) proves that in general we cannot expect that $A \times B$ is the same as $B \times A$.

The cross products can be displayed on the usual coordinate graphs. The horizontal axis shows the elements used as first components and the vertical axis those that are used as second components. The two axes intersect in the origin. The graphs for Example 6 are:

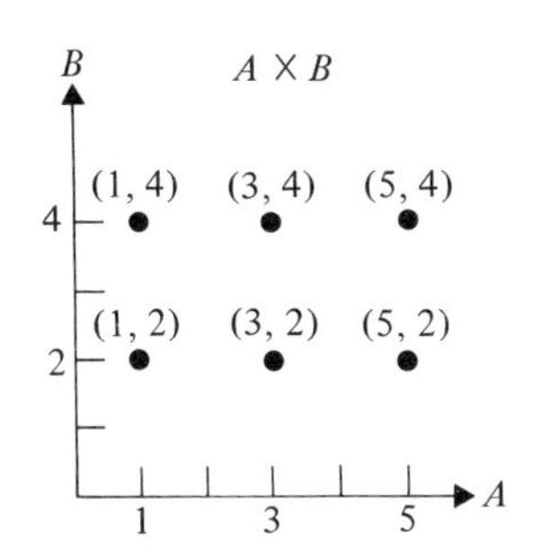

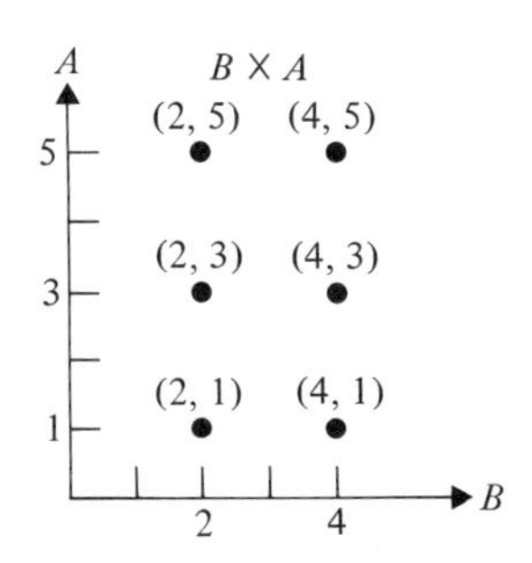

Try Exercise 8

8. Let $M = \{4, 5\}$, $N = \{1, 2, 3\}$.
a) Find $M \times N$
b) Find $N \times M$
c) Is $M \times N = N \times M$?
d) Make a graph of $M \times N$, of $N \times M$.

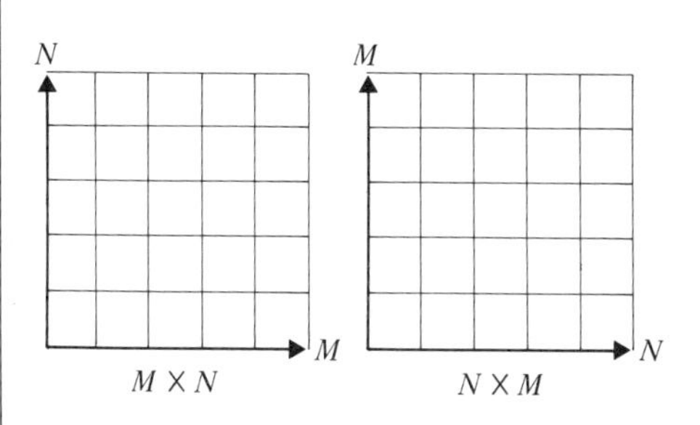

Properties of Set Operations

COMMUTATIVITY OF $\cup$ AND $\cap$. It is important that children realize quite early that the sum $2 + 4$ is identical to the sum $4 + 2$; it does not matter in which order 2 pieces of candy and 4 pieces of candy are put into an empty bag, since the total will be 6 pieces in either case. This commutative property for addition in arithmetic follows from the corresponding property for sets.

Example 7 Prove that the union operation for sets is commutative, that is, $A \cup B = B \cup A$ for all sets A and B.

Proof The set $A \cup B$ consists of those elements that are in set A or in set B. The set $B \cup A$ consists of those elements that are in set B or in set A. These statements both mean the same thing because they are of the form $p \vee q$ and $q \vee p$, respectively, and we have already established that these are equivalent statements (Exercise 7b of Exercise Set 1.5). Therefore the elements of $A \cup B$ are exactly the same as those of $B \cup A$.

A similar proof can be developed to establish that the intersection operation for sets is also commutative. The proof that $A \cap B = B \cap A$ is left for the exercises.

Try Exercise 9

9. Let $A = \{b, d, e, a\}$, $B = \{e, g, f, b\}$
a) Find $A \cap B$.
b) Find $B \cap A$.
c) How are $A \cap B$ and $B \cap A$ related?

A cross product in a third-grade textbook

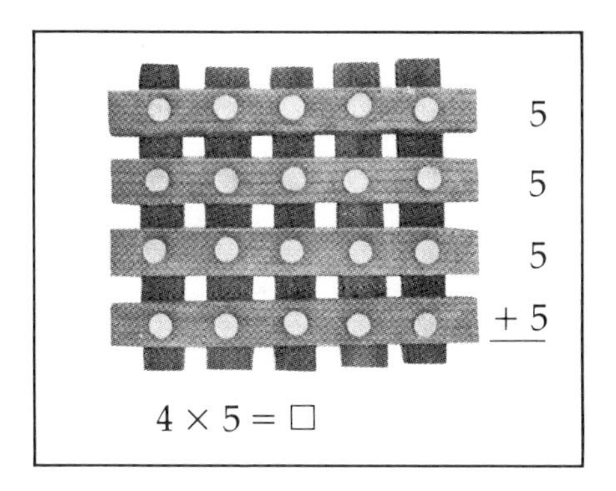

ASSOCIATIVITY FOR $\cup$ AND $\cap$. Since intersection is defined for two sets, what meaning is attached to $A \cap (B \cap C)$? The operation for the two sets inside the parentheses is carried out first; next, the intersection of A with the new set $(B \cap C)$ is found.

10. Use the same sets A, B, and C as in Example 8 of the text.
a) Find $A \cup B$, then find $(A \cup B) \cup C$.
b) Find $(B \cup C)$, then find $A \cup (B \cup C)$.

Example 8 Let
$$A = \{c, p, n, t, s\},$$
$$B = \{c, p, q, r, s\},$$
$$C = \{p, q, t, f, c\}.$$

Find a) $A \cap (B \cap C)$; b) $(A \cap B) \cap C$

Solution

a) $(B \cap C) = \{c, p, q\}$
$A \cap \{c, p, q\} = \{c, p\}$

b) $(A \cap B) = \{c, p, s\}$
$\{c, p, s\} \cap C = \{c, p\}$

In this example it did not seem to matter where the parentheses were placed: $A \cap (B \cap C) = (A \cap B) \cap C = \{c, p\}$. It can be proved that this is true no matter what sets are involved; this property is the associative law for intersection.

Example 9 Prove the following associative law for intersection of sets:

For any sets A, B, and C: $A \cap (B \cap C) = (A \cap B) \cap C$.

Proof We apply the definition of intersection repeatedly to $A \cap (B \cap C)$. The set $B \cap C$ has all the elements that make true the statement "$x \in B$ and $x \in C$," and the set $A \cap (B \cap C)$ has all the elements that make the statement $x \in A$ and ($x \in B$ and $x \in C$) true.

For the set $(A \cap B) \cap C$ we find that the elements of set $A \cap B$ satisfy the properties for membership in set A and also for set B. The elements of set $(A \cap B) \cap C$, besides satisfying the properties for $A \cap B$, also satisfy the properties for C.

The standard description for the sets $A \cap (B \cap C)$ and $(A \cap B) \cap C$ are statements of the form $p \wedge (q \wedge r)$ and $(p \wedge q) \wedge r$, respectively, that have been shown to be equivalent (Exercise 7c of Exercise Set 1.5). Since the standard descriptions for these sets mean the same thing, the elements that satisfy these descriptions will be the same elements, and the sets $A \cap (B \cap C)$ and $(A \cap B) \cap C$ must be the same sets.

In the proof above, all that was needed was an examination of the standard description of the sets that arose in the repeated application of the definition for intersection. A similar argument can be developed to prove that the associative law for union holds for any sets A, B, and C:

$$A \cup (B \cup C) = (A \cup B) \cup C.$$

Try Exercise 10

SUMMARY. If the standard description of a set involves an open sentence, then every element of the universe that makes the sentence true is a member of the set, and each one that makes it false is a member of the complement of the set. There is obviously a close relationship between the properties of the statements of logic and the laws for the set operations because the latter are based upon the corresponding properties already established for statements. The following important laws for set operations are proved either in the text or in an exercise set:

Commutative laws: $A \cup B = B \cup A$,
$A \cap B = B \cap A$.

Associative laws: $A \cup (B \cup C) = (A \cup B) \cup C$,
$A \cap (B \cap C) = (A \cap B) \cap C$.

Distributive laws: $A \cap (B \cup C) = (A \cap B) \cup (A \cap C)$,
$A \cup (B \cap C) = (A \cup B) \cap (A \cup C)$.

De Morgan's laws: $\overline{A \cup B} = \bar{A} \cap \bar{B}$,
$\overline{A \cap B} = \bar{A} \cup \bar{B}$.

Complementation: $A = \bar{\bar{A}}$.

EXERCISE SET 2.2

Unless otherwise directed, use the following sets for the exercises of this set:

$$\mathscr{U} = \{0, 1, 2, 5, 7, 10\}, \quad A = \{0, 1, 2, 5, 10\}, \quad B = \{0, 1, 5\}, \quad C = \{0, 1, 2, 7\}.$$

1. Use roster notation and compare the following:
a) $A \cap B$ with $B \cap A$
b) $A \cap C$ with $C \cap A$

2. Use roster notation and compare the following:
a) $A \cup (B \cup C)$ with $(A \cup B) \cup C$
b) $A \cap (B \cap C)$ with $(A \cap B) \cap C$

3. Use roster notation and compare the following:
a) $A \cup (B \cap C)$ with $(A \cup B) \cap (A \cup C)$
b) $C \cup (B \cap A)$ with $(C \cup B) \cap (C \cup A)$
c) $A \cap (B \cup C)$ with $(A \cap B) \cup (A \cap C)$
d) $C \cap (B \cup A)$ with $(C \cap B) \cup (C \cap A)$

4. Use two Venn diagrams for any sets A, B, and C to compare $A \cup (B \cap C)$ with $(A \cup B) \cap (A \cup C)$.

5. Use two Venn diagrams for any sets A, B, and C to compare $A \cap (B \cup C)$ with $(A \cap B) \cup (A \cap C)$.

6. How are the following sets related?
a) $A, \bar{A}, \bar{\bar{A}}$ b) $C, \bar{C}, \bar{\bar{C}}$

7. Use roster notation to compare the following sets:
a) $\overline{B \cup C}$ with $\bar{B} \cap \bar{C}$ b) $\overline{A \cap B}$ with $\bar{A} \cup \bar{B}$

8. On the basis of the inductive evidence of Exercises 6 and 7, what might you conjecture is true for all sets?

For any sets A, B, and C in 9–11 make a Venn diagram for the following sets:

9. a) $\bar{B} \cap A$ b) $(\bar{B} \cap A) \cup C$

10. a) $\overline{A \cap B}$ b) $\bar{A} \cup \bar{B}$

11. a) $\overline{A \cup B}$ b) $\bar{A} \cap \bar{B}$

12. Some of the following statements are always true for any sets A and B. State T for those that are always true and provide a counterexample for those that are not always true.

a) $A \subseteq (A \cup B)$ b) $A \cap B \subseteq B$
c) $A \cup A = A$ d) $A \cap A = \varnothing$
e) $B \cup \bar{B} \subset \mathscr{U}$ f) $B \cap \varnothing = \varnothing$
g) $B \cap \bar{B} = \varnothing$ h) $\overline{A \cup B} = \bar{A} \cup \bar{B}$

13. If the following statements are true, then it is possible to deduce what must be true about the sets A and B. State what special properties hold for A and B in each case.

a) $\bar{A} = \bar{B}$ b) $A \cup B = A$ c) $A \times B = B \times A$
d) $B \cup \bar{B} = B$ e) $A \cap B = B$ f) $A \times B = \varnothing$
g) $\varnothing \cup B = \varnothing$ h) $\bar{B} \cup \bar{A} = B$

14. Use roster notation for the following:

a) $A \times B$ b) $B \times B$ c) $\varnothing \times A$
d) $B \times \varnothing$ e) $(B \cap C) \times A$ f) $A \times (B \cap C)$

15. Illustrate that $B \times (A \cup C) = (B \times A) \cup (B \times C)$.

16. Three women, Nancy, Roxanne, and Burnette, form a set of possible golf partners in a tournament. Represent this set as $F = \{N, R, B\}$. Three men, Jack, George, and Ed, form another set of partners we call $M = \{J, G, E\}$. Find the set of all possible (female, male) partners for golf that can be formed from these two sets.

17. Prove that $A = \bar{\bar{A}}$ for any set A.

18. Prove that $A \cap B = B \cap A$ for any sets A and B.

■ **19.** Prove that $A \cup (B \cap C) = (A \cup B) \cap (A \cup C)$ for any sets A, B, and C.

2.3 AIMS

You should be able to:

a) Define and give examples of relations and functions and distinguish between them.
b) Represent relations graphically.

2.3 RELATIONS AND FUNCTIONS

RELATIONS. One of the most frequent uses of a Cartesian product is in a relation or function. Recall the example on p. 46: suppose the photographer decided that only the 75 mm and 150 mm lenses were suitable for a particular purpose; the set of possible pairs (film type, lens) under this restriction is a subset of the original cross product. Such a subset is called a *relation.*

A *relation* in mathematics is any subset of the set of ordered pairs of a Cartesian product. Usually a relationship of some sort exists for the components of these pairs; frequently the relationship is described by means of a rule. The set of all the first components of the ordered pairs of a relation is called the *domain* of the relation, and the set of all the second components is called the *range* of the relation. In the elementary school the domain of a relation is often referred to as the *input* and the range of a relation as the *output.*

Example 1 R_1 is the relation $\{(2, 4)(6, 12)(3, 6)(2, 5)\}$. What is the domain for R_1? the range?

Solution The domain is the set $\{2, 6, 3\}$, the range is the set $\{4, 12, 6, 5\}$.

Example 2 Given $A = \{4, 9, 10\}$ and $B = \{2, 10, 12\}$, find $A \times B$; then find the relation R_2 such that the first component is smaller than the second component.

Solution

$$A \times B = \left\{ \begin{matrix} (4, 12) & (9, 12) & (10, 12) \\ (4, 10) & (9, 10) & (10, 10) \\ (4, 2) & (9, 2) & (10, 2) \end{matrix} \right\},$$

(The shaded entries are labeled R_2.)

$$R_2 = \{(a, b) \mid a \text{ is less than } b\}:$$

$$R_2 = \{(4, 10) \quad (4, 12) \quad (9, 10) \quad (9, 12) \quad (10, 12)\}.$$

The relations of Examples 1 and 2 can also be represented graphically on the usual coordinate axes. Each point corresponds to one of the ordered pairs in the relation.

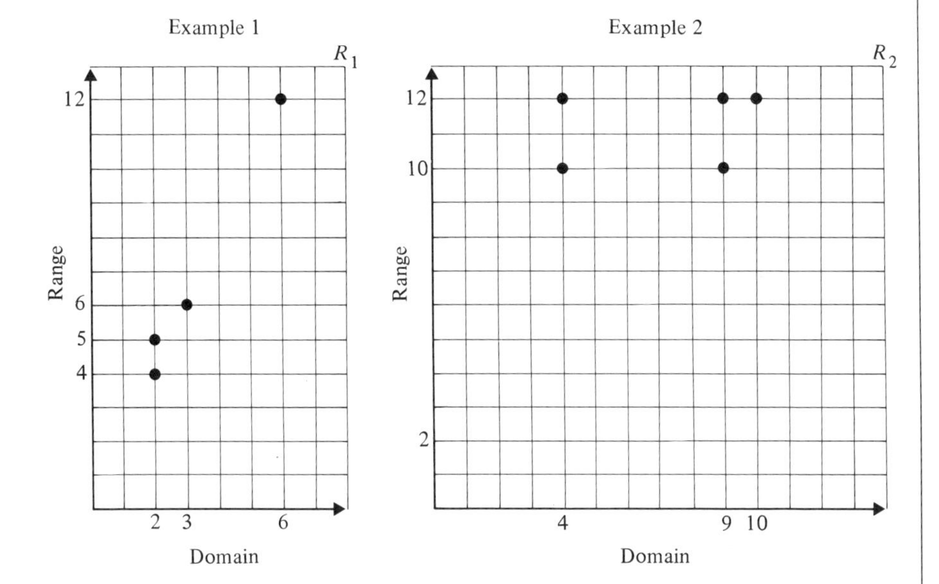

Try Exercise 11

FUNCTIONS. A very important type of relation called a function is often introduced to elementary pupils by means of a "function machine" that accepts a number (input), performs an operation according to some rule, and produces the result of the operation as output. If the machine is given a certain input, it is expected that the function rule will produce the same output each time. The set of ordered pairs (input, output) from such a machine is a relation, but because of the restriction on having a unique output, the relation is a special kind called a *function.* Every function is a relation, but not all relations are functions.

11. $A = \{1, 2, 3, 4\}$ and $B = \{1, 2, 3, 4\}$. Find the relation R, a subset of $A \times B$ such that

$$R = \{(a, b) \mid b \text{ is less than } a\}.$$

A function machine in an elementary textbook

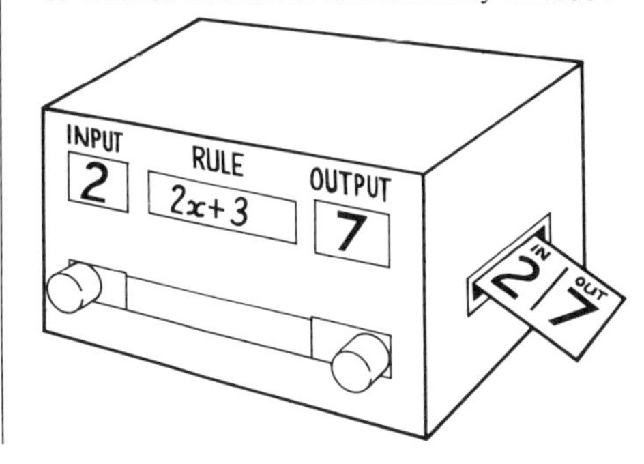

12. The domain of a relation R is

$\{0, 1, 2, 3, 4, 5\}$ and $R = \{(a, b) \mid b = 6a\}$.

a) Use roster notation to describe R.
b) What is the range for R?
c) Is R a function?

DEFINITION. A relation R is a *function* if and only if no two of its ordered pairs have the same first components and different second components. That is, for any a in the domain of R, if $(a, b) \in R$ and $(a, c) \in R$, then $b = c$.

For any given member of the domain of a function there is one and only one member of the range paired with it.

In the relation R_1 of Example 1 we find both pairs (2, 4) and (2, 5), so if 2 is the input, the output is not unique and R_1 is thus not a function. Similarly in Example 2, the pairs (4, 10) and (4, 12) are both members of R_2, so for the input 4 the output is not unique ($10 \neq 12$) and R_2 is also not a function.

If the member of the domain (input) for a function named f is x, then the output is denoted by $f(x)$ and is called the *function value* at x. The emphasis in the elementary school is likely to be on the rule used by the function machine instead of the set of ordered pairs as such; the rule can be used to find the set of ordered pairs, however.

Example 3 A function f has the output rule: $f(x) = 2x + 3$. Find the functional value (output) for each member of the domain $\{2, 4, 6, 8\}$.

Solution

$f(2) = 2 \times 2 + 3 = 7$, $\quad f(4) = 2 \times 4 + 3 = 11$,

$f(6) = 2 \times 6 + 3 = 15$, $\quad f(8) = 2 \times 8 + 3 = 19$.

The function is the set $\{(2, 7), (4, 11), (6, 15), (8, 19)\}$.

Try Exercise 12

It is easy to recognize the graph of a function on the usual Cartesian graph, since any vertical line will meet the graph of a function in no more than one point.

Example 4 Determine which of the following are functions:

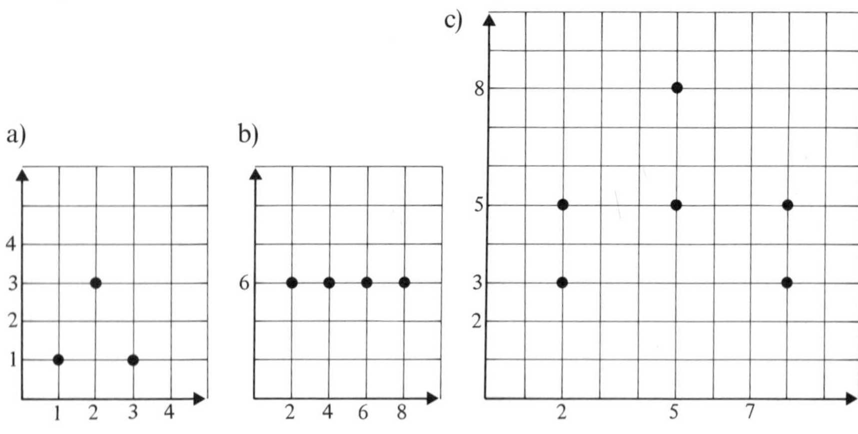

d) $\{(1,2), (1,4), (3,2), (3,4), (3,2), (3,5)\}$

e)

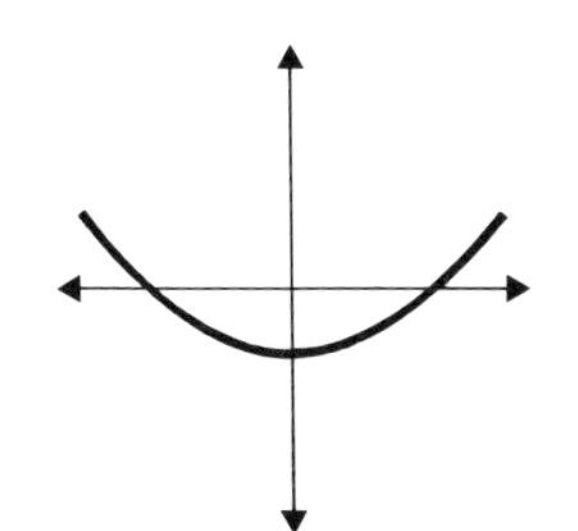

f)

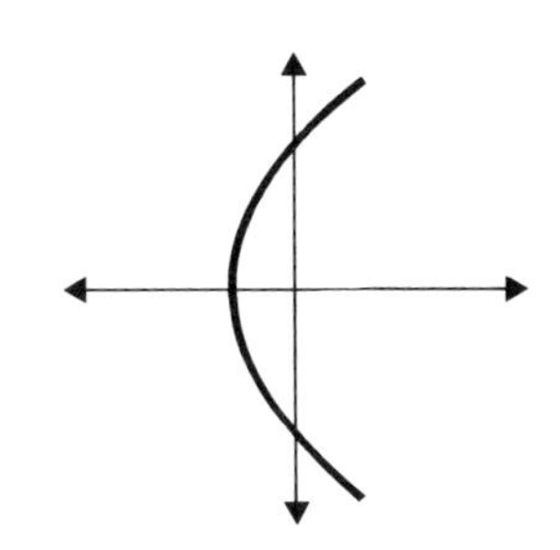

Solution The relations in (a), (b) and (e) are also functions, since each element of the domain is associated with exactly one element of the range; the remaining are not functions.

Try Exercise 13

If a relation is a subset of a cross product of two identical sets ($A \times A$), the relation is said to be *on A* instead of *from A to A*. In Example 4a, the relation is on the set $\{1, 2, 3, 4\}$, and the one in 4b is on the set $\{2, 4, 6, 8\}$.

13. Some of the following relations are functions, some are not. Pick out those that are functions.

a) $R = \{(3, 2), (3, 3), (3, 4)\}$
b) $R = \{(2, 3), (5, 3), (6, 3)\}$
c)

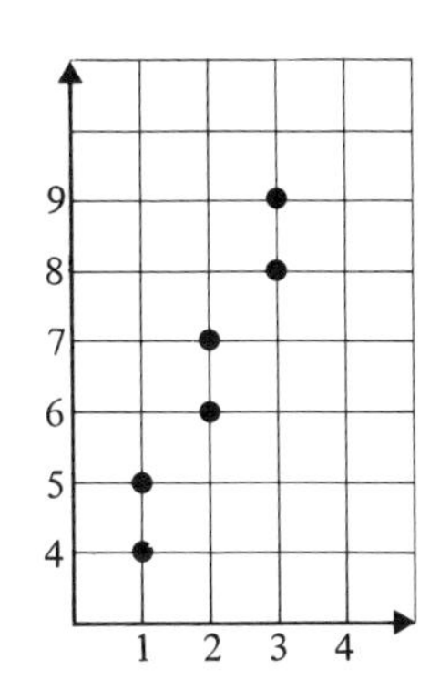

d)

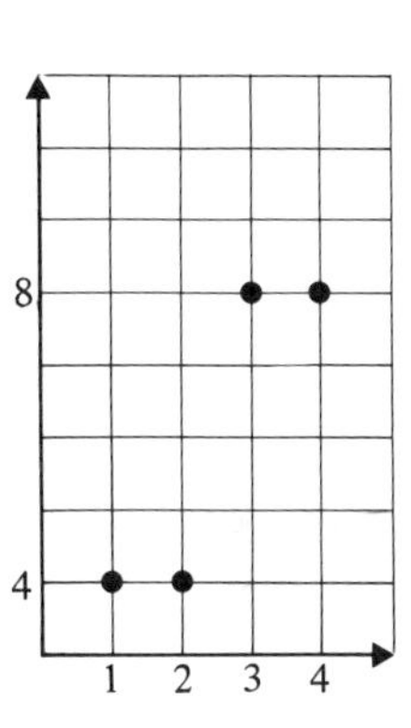

EXERCISE SET 2.3

1.
a) Use the roster method to describe the relation $R_1 = \{(a, b) \mid b = a + 1\}$ on the set $\{0, 1, 2, 3, 4, 5, 6\}$.
b) State the domain and range for R_1.
c) Is R_1 a function?

2.
a) Use the roster method to describe the relation "is less than" on the set $A = \{0, 1, 2, 3, 4, 5, 6\}$.
b) What are the domain and range of this relation?
c) Is the relation a function?

3. Make a graph on the usual coordinate system for the relation of Exercise 1.

4. Repeat Exercise 3 for Exercise 2.

5. Which of the following relations are also functions?
a) $\{(1, 1), (1, 2), (2, 1), (2, 2)\}$
b) $\{(7, 8), (8, 9), (9, 10), (10, 11)\}$
c) $\{(8, 7), (9, 7), (10, 8), (11, 8), (12, 9), (13, 9)\}$
d) $\{(7, 8), (7, 9), (8, 10), (8, 11), (9, 12), (9, 13)\}$

6. Find the domain for each relation in Exercise 5.

7. Find the range for each relation in Exercise 5.

8. Which of the following relations on the set of whole numbers are also functions?

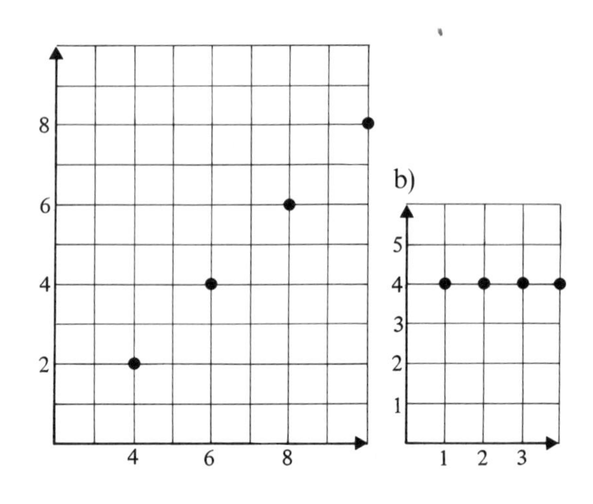

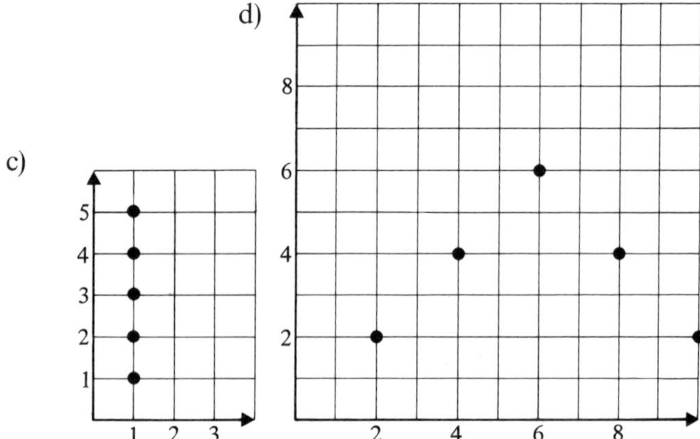

9. Each of the following rules in conjunction with the given domain determines a relation. Give the set of ordered pairs of the relation and state if the relation is a function.
a) The rule: $f(x) = 2x - 3$; the domain: $\{3, 5, 7, 9, 11\}$
b) The rule: $f(x) = x$ or $x + 1$; the domain: $\{0, 1, 2, 3, 4, 5\}$
c) The rule: $f(x) = x$; the domain: $\{2, 4, 6, 8\}$

In the mapping diagrams for the relations 10 and 11 each arrow corresponds to an ordered pair of the relation.
a) Describe the relation with the roster method.
b) State if the relation is also a function.

10.

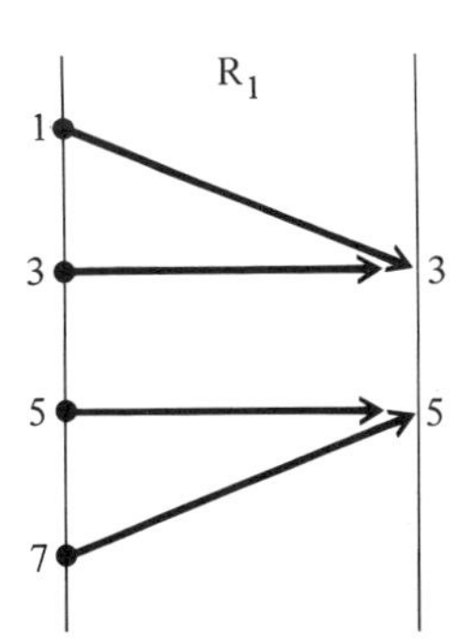

11.

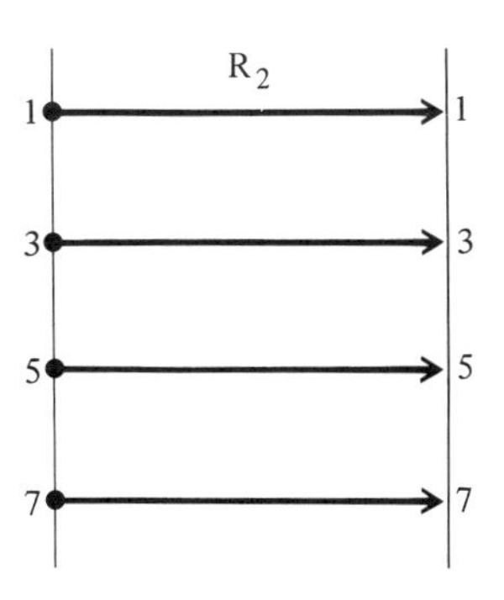

12. In a mapping diagram for a relation there is an easy way to determine if a relation is a function or not. What is the obvious clue?

13. Which elements of the set $A = \{1, 3, 9\}$ are related to the elements of $B = \{2, 4, 6\}$ by the relation "less than"?

2.4 PROPERTIES OF RELATIONS

2.4 AIMS

You should be able to:

a) Recognize and define the reflexive, symmetric, and transitive properties of a relation.
b) Determine if a relation on a set is an equivalence relation.
c) Prove that two sets are equivalent by exhibiting a one-to-one correspondence between the members of the sets.

Relations abound in the everyday world. You might be the *brother* or *sister of* someone, you are the *child of* your parents, the *grandchild of* your grandparents, in the *same school as* a neighbor, be *younger than* a friend, *taller than* your mother, and so on.

Relations are also everywhere in mathematics, and we have already discussed some relations for sets such as:

$A = B$ the relation is *equals*,
$A \subseteq B$ the relation is *subset of*,
$A \subset B$ the relation is *proper subset of*.

It does not seem possible that such a miscellany of relations both in and out of mathematics can be sensibly compared or that it is desirable to make the comparison in the first place.

There are three important properties that can be investigated for any relation, but they are of special interest in mathematics. In order to make the description of these properties apply to any relation and not just to a particular one, we shall use $a\text{R}b$ to mean *a is related to b*, while R could represent *equals*, or *subset of*, and so on. The following properties allow us to compare relations that seem on the face of it to have just about nothing in common.

THE REFLEXIVE PROPERTY

DEFINITION. A relation R on a set S is *reflexive* if and only if every element a of S is related to a, that is,

$$\forall a, \quad a\text{R}a.$$

This means that if the reflexive property holds, every element is related to itself. For example, if S is the set of your math-class members and the relation R is *the same age as*, then R is obviously a reflexive relation. You are always the same age as yourself.

Try Exercise 14

14. Is the relation "the same weight as" on the members of your mathematics class a reflexive relation?

THE SYMMETRIC PROPERTY

DEFINITION. A relation R on a set S is *symmetric* if and only if for all elements a and b of S it is true that if $a\text{R}b$, then $b\text{R}a$; that is,

$$\forall a \text{ and } b,\ a\text{R}b \rightarrow b\text{R}a$$

The symmetric property means that if the ordered pair (a, b) is a member of the relation, then the ordered pair (b, a) is also a member of the relation. If the set S is the set of your class members and the relation R is *as tall as*, then if person A is as tall as person B, it follows that person B is as tall as person A, and the relation is a symmetric one. This property does not say that for every element a of set S there must be an element b that is related to it, just that if there is such an element, then b is also related to a.

Try Exercise 15

15. Is the relation "weighs more than" on the members of your mathematics class a symmetric relation?

THE TRANSITIVE PROPERTY

DEFINITION. A relation R on a set S is *transitive* if and only if for all elements a, b, and c it is true that if $a\text{R}b$ and $b\text{R}c$, then $a\text{R}c$; that is,

$$\forall a,\ b, \text{ and } c,\ (a\text{R}b \wedge b\text{R}c) \rightarrow a\text{R}c.$$

This means that if a is related to b and it also happens that b is related to c, then in a transitive relation it will also be true that a is also related to c. If S is the set of your friends and if Ann is *shorter than* Bess, and Bess is *shorter than* Collette, then this is a transitive relation, since it is also true that Ann is *shorter than* Collette.

Try Exercise 16

16. Is the relation "intersects with" for the streets of a particular city a transitive relation?

Example 1 Suppose that S is $\{\{a, b\}, \{a\}, \{b\}, \varnothing\}$ and the relation R is *the subset of*. Is this relation:
a) reflexive? b) symmetric? c) transitive?

Solution

a) Since any set is a subset of itself, the relation is reflexive.
b) $\{a\} \subseteq \{a, b\}$ but $\{a, b\} \not\subseteq \{a\}$. This counterexample shows the relation is not symmetric.

c) We have $\varnothing \subseteq \{a\}$ and $\{a\} \subseteq \{a, b\}$; find that $\varnothing \subseteq \{a, b\}$ is true. In fact, no matter which sets we choose from S, if $A \subseteq B$ and $B \subseteq C$, then we find $A \subseteq C$, so the relation is transitive.

Try Exercise 17

A relation may have none of the three properties just described or just one, or two, or all three. When the relation has all three properties it is given the special name of *equivalence relation.*

DEFINITION. Any relation R on a set S is an *equivalence relation* if and only if it has the reflexive, symmetric, and transitive properties.

EQUALS RELATION. An important equivalence relation is the equals relation for numbers; if a and b are any numbers and a *equals* b, it means that "a" and "b" are both names for the same number. We can then replace or substitute "a" by "b" because it amounts to renaming.

It is this substitution property of equals that allows us to say that if $a = b$ and $c = d$, then $a + c = b + d$, since we have merely renamed a and c. Similarly, if $a = b$, then, by this basic renaming property of *equals*, $a \cdot c = b \cdot c$, since we have only renamed a.

ONE-TO-ONE CORRESPONDENCE. In the process of acquiring the abstract concepts of number, children are often asked to match, or pair, the elements

One-to-one matching in elementary mathematics

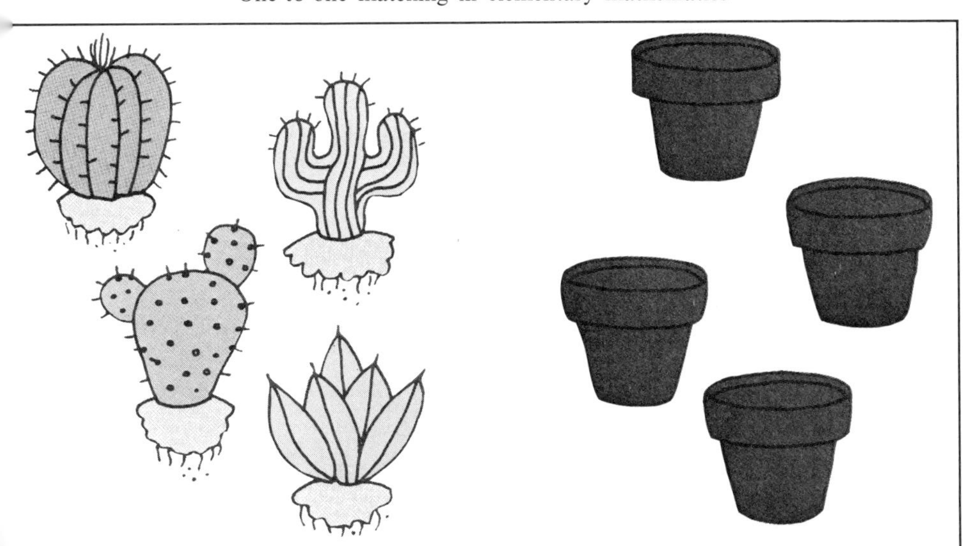

17. Replace each blank in the following statements with a set of S to make each statement true.

a) If $\varnothing \subseteq \{b\}$ and $\{b\} \subseteq$ ____, then ____ $\subseteq$ ____.

b) If $\{a\} \subseteq \{a, b\}$ and ____ $\subseteq$ ____, then ____ $\subseteq$ ____.

18.
a) Find three more one-to-one correspondences for the sets of Example 2.
b) Exhibit two different one-to-one correspondences between the elements of the sets

$$\{5, 13, 11, 6\} \text{ and } \{e, d, h, g\}.$$

of two sets for the purpose of comparison. In the preceding exercise, the first graders are expected to draw a line from each member of one set to a single member of the second set.

When we have exactly one cactus plant for each pot and exactly one pot for each plant, then we have established a one-to-one correspondence for the set of plants and the set of pots.

DEFINITION. The elements of two sets A and B are in *one-to-one correspondence* if and only if for each element in A there is one and only one element in B paired with it, and for each element in B there is one and only one element in A paired with it.

Example 2 Show that the elements of the set $\{3, 7, 2\}$ and the set $\{e, b, d\}$ can be put in one-to-one correspondence.

Solution Here are three of such possible correspondences:

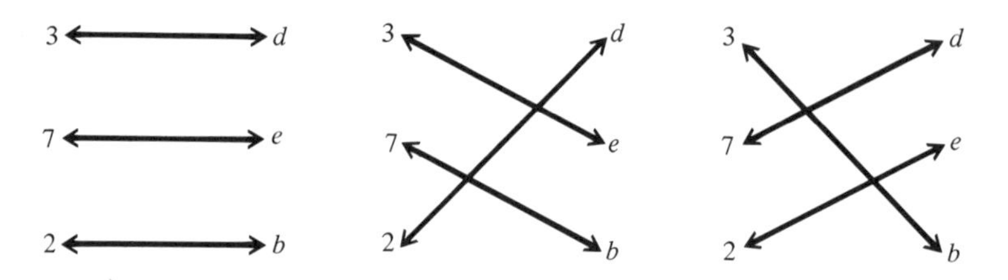

Try Exercise 18

The three properties discussed for relations hold true for the *one-to-one correspondence* relation for sets. We shall prove this point.

THE REFLEXIVE PROPERTY Any set A has a one-to-one correspondence with set A, since each element of A can be paired with itself.

$$\begin{array}{ll} \text{Set } A & \{a \quad b \quad c \quad \ldots\} \\ & \;\;\updownarrow \quad \updownarrow \quad \updownarrow \\ \text{Set } A & \{a \quad b \quad c \quad \ldots\} \end{array}$$

THE SYMMETRIC PROPERTY If a one-to-one correspondence exists for any sets A and B, then for any element a in A there is a unique element b in B and for any element b in B there is a unique element in A. Thus the pairs of the correspondence that show A is related to B can be reversed and used for a correspondence to show B is related to A.

THE TRANSITIVE PROPERTY If A has a one-to-one correspondence with B and B in turn has one with C, then these two correspondences can be used to show that A also has a one-to-one correspondence with C.

Each a in A is paired with a unique b in B, and each b in B is paired with a unique element in A. Also, each element of B is paired with a unique element of C, and each element of C is paired with a unique element of B.

Thus from a pair (a, b) of the first correspondence and the pair (b, c) from the second, a new pair (a, c) can be formed to show each element in A is paired with a unique element in C. This argument also works in reverse, since for each element c in C there is a corresponding unique b in B, which in turn has a unique a in A; thus every element in C has been paired with a unique element in A. Thus a one-to-one correspondence is established for sets A and C. This argument is easy to follow in an example.

Example 3 $A = \{1, 2, 3, 4\}$, $B = \{a, b, c, d\}$, $C = \{\bigstar, \bigcirc, \triangle, \square\}$. Show that a one-to-one correspondence for sets A and B and B and C can be used to produce a one-to-one correspondence for A and C.

Solution

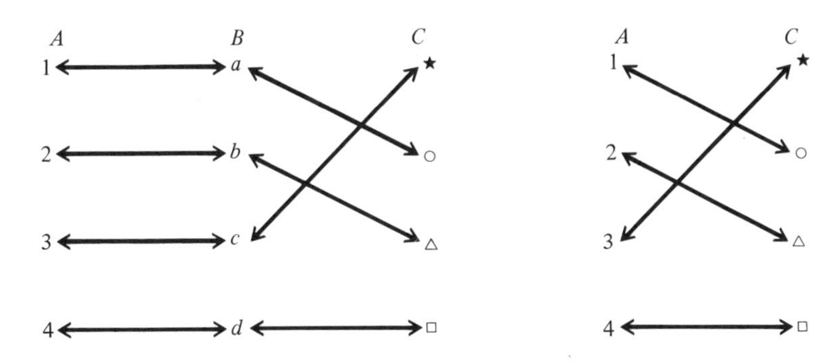

The fact that the relation "has a one-to-one correspondence with" has all three properties means it can be correctly called an equivalence relation, and the usual terminology reflects this. Any set A is defined to be *equivalent* to set B if and only if there is a one-to-one correspondence between their members; this is symbolized by $(A \sim B)$. Thus two nonempty sets are shown to be equivalent by exhibiting a one-to-one correspondence, and conversely, the empty set is equivalent only to itself.

The three properties for one-to-one correspondence can be restated in terms of the *equivalence relation* for sets.

For any sets A, B, and C the following properties hold:

1. Reflexive: $A \sim A$.
2. Symmetric: If $A \sim B$, then $B \sim A$.
3. Transitive: If $A \sim B$ and $B \sim C$, then $A \sim C$.

Try Exercise 19

19. Is the relation of *equality* for sets an equivalence relation?

20. Show that the set of whole numbers is equivalent to the set of natural numbers:

$$W = \{0, 1, 2, 3, 4, \ldots\}$$
$$N = \{1, 2, 3, 4, 5, \ldots\}$$

INFINITE SETS. The equivalence relation is useful for clarifying other concepts of mathematics. For example, we can now give a precise meaning to the word infinite.

DEFINITION. A set is *infinite* if and only if it is equivalent to a proper subset of itself. It follows that a set is *finite* if and only if it is impossible to find a proper subset that is equivalent to it.

Example 4 Show that the set $\{k, p, m\}$ is a finite set.

Solution All of the proper subsets can be listed:

$$\varnothing, \{k\}, \{p\}, \{m\}, \{k, p\}, \{k, m\}, \{p, m\}$$

None of these is equivalent to the set $\{k, p, m\}$.

Example 5 Show that the set $W = \{0, 1, 2, 3, 4, 5, \ldots\}$ is an infinite set.

Solution A proper set of the whole numbers W is the set of even whole numbers:

$$E = \{0, 2, 4, 6, \ldots\}.$$

Many correspondences are possible between the elements of the set of whole numbers and the set E. Here is one:

$$\begin{array}{rl} W = & \{0,\ 1,\ 2,\ 3,\ 4,\ \ 5, \ldots,\ \ n, \ldots\}, \\ & \ \ \updownarrow\ \ \updownarrow\ \ \updownarrow\ \ \updownarrow\ \ \updownarrow\ \ \ \updownarrow \qquad\quad \updownarrow \\ E = & \{0,\ 2,\ 4,\ 6,\ 8,\ 10, \ldots, 2n, \ldots\} \end{array}$$

The pairing rule is $n \leftrightarrow 2n$. Because W is equivalent to the set E (a proper subset), W is an infinite set.

In order to show a one-to-one correspondence for infinite sets, it is necessary to state a rule for the correspondence or pairing since not all of the pairs can be exhibited.

Try Exercise 20

EXERCISE SET 2.4

1. If the set S is the set of all persons, state which of the properties (reflexive, symmetric, transitive) of a relation are true for the following:

a) sister of
b) grandaughter of
c) father of
d) weighs more than
e) cousin of
f) is younger than

2. Each of the following is a relation on the set of all subsets of a nonempty set. Which properties (reflexive, symmetric, transitive) are true for each relation?

a) $\subset$
b) disjoint from
c) complement of
d) $=$
e) is equivalent to
f) $\subseteq$

3. Give an example of a relation that has the properties indicated in each of the following:
a) reflexive, symmetric, not transitive
b) reflexive, transitive, not symmetric
c) symmetric, transitive, not reflexive
d) neither reflexive nor symmetric, but transitive
e) neither reflexive nor transitive, but symmetric

4. Exhibit two different one-to-one correspondences between the elements of the sets $\{1, 2, 3, 4, 5\}$ and $\{a, b, c, d, e\}$.

5. Exhibit two different one-to-one correspondences between the elements of the sets $\{1, 2, 3, 4, 5, 6, 7, 8\}$ and $\{a, b, c, d, e, f, g, h\}$.

6. Prove that the "equals" relation on any set of numbers is an equivalence relation.

7. Prove that the relation "is logically equivalent to" is an equivalence relation on the set of statements of logic.

8. Exhibit a one-to-one correspondence between the elements of $\{11, 12, 13, 14, \ldots\}$ and those of the set of counting numbers $\{1, 2, 3, 4, 5, 6, \ldots\}$.

9. Exhibit a one-to-one correspondence between the elements of the set $\{0, 2, 4, 6, 8, \ldots\}$ (even whole numbers) and those of the set $\{1, 3, 5, 7, \ldots\}$ (odd whole numbers).

10. If set A is equivalent to set B, does it follow that the set A is equal to B? Why?

11. If set A is equal to set B, does it follow that the set A is equivalent to set B? Why?

12. Is every one-to-one correspondence also a function? Why?

13. Is every function also a one-to-one correspondence? Why?

14. Sometimes it is informative to study a definition to see if it could be shortened. Suppose the latter part of the definition for one-to-one correspondence: ". . . and for each element in B there is one and only one element in A paired with it," were omitted. Show that this omission changes the meaning completely.

■ **15.** Use the roster description for the relation "is an element of," a subset of the cross product $A \times B$, where

$$A = \{a, b, c\},$$
$$B = \{\varnothing, \{a\}, \{b\}, \{c\}, \{a, b\}, \{a, c\}, \{b, c\}, \{a, b, c\}\}.$$

16. Prove that for any sets A and B, the set $A \times B$ is equivalent to the set $B \times A$.

■ **17.** Prove that for any sets A, B, and C, the set $A \times (B \times C)$ is equivalent to $(A \times B) \times C$.

2.5 PROBLEM SOLVING

Techniques used by successful problem solvers in mathematics can be learned and deliberately practiced. The next example is suitable for elementary pupils, yet it provides an example of a useful attack and an opportunity to increase your own problem-solving skills.

2.5 AIMS

You should be able to:
a) Apply the five-step problem solving process to appropriate problems and proofs.
b) Use restatement of a problem and "looking back" as techniques in problem solving.

21. Give another restatement of the problem.

Example 1 Suppose that there are six circles in a triangular arrangement. Use just one of the numbers 1, 2, 3, 4, 5, 6 for each of the circles, so that the sum corresponding to each side of the triangle is 9. (Each of the three sums is 9.)

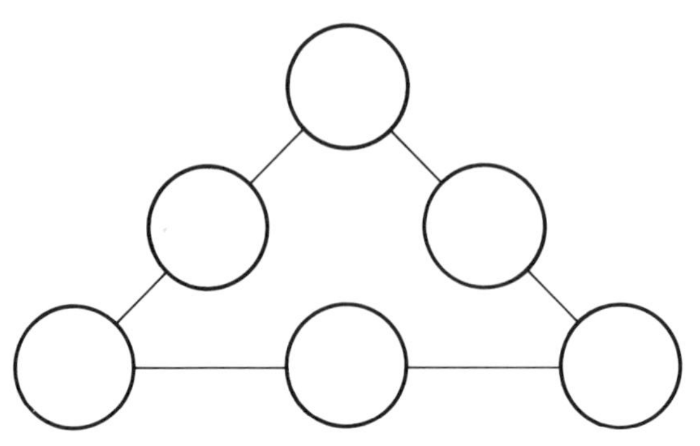

The five-step process will continue to be used.

1. What are the relevant *facts*?
 a) The numbers 1, 2, 3, 4, 5, 6 are all to be used just once in the diagram.
 b) The diagram must have an entry in each circle.

2. What is the *problem*? The sum of the entries along each side of the triangular figure must be 9. Can the problem be restated? Think of the entries as a, b, c, d, e, f. The problem now is to find a, b, c, d, e, f, so that

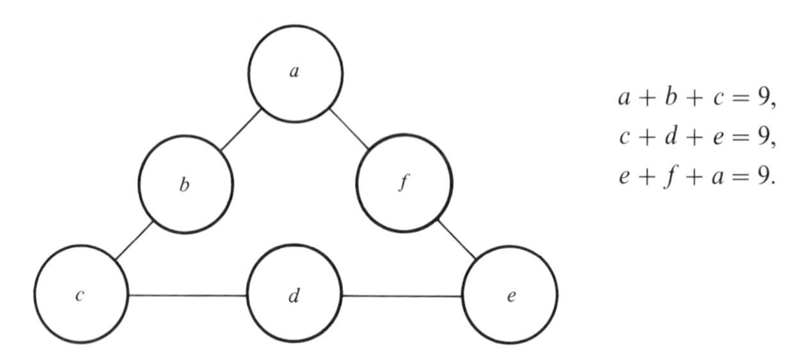

$$a + b + c = 9,$$
$$c + d + e = 9,$$
$$e + f + a = 9.$$

You may regard this restatement as more difficult to think about than the original problem.

Try Exercise 21

3. What *ideas* do you have for the solution? (If you already have a solution you do not need any ideas!) Here are some just in case you need them:
 a) Use 1, 2, 3 in the vertex (corner) positions.
 b) Use 4, 5, 6 in the vertex positions.

THE FIVE-STEP PROCESS

1. What are the relevant *facts*?
2. What actually is the *problem*?
3. What are my *ideas*?
4. How do I get a *solution*?
5. *Check* and *review* the solution.

c) Put labels for 1, 2, 3, 4, 5, 6 on poker chips and move them about in a trial-and-error procedure.

Try Exercise 22

22. Solve the problem.

CHECKING AND REVIEWING THE SOLUTION. If you wish to be a more successful problem solver, the last step should be more than a simple checking of the result. After it has been checked and it is clear that the solution does satsify the problem conditions, take time to:
Look back and ask yourself:

1. Is there an easier way to do this?
2. Is this related to other problems I have solved?
3. Is there an entirely different way to think about this problem?

Look forward and speculate:

1. Can I generalize this problem?
2. Can this problem be varied and lead to a new one?
3. Can this problem help me solve a future problem?

In the case of Example 1, a possible variation would be to change the required sum along each of the sides to 10, 11, 12, or 13 instead of 9. You can think of still other variations.

If you are not in the habit of doing this sort of thing for yourself as you solve problems in mathematics, start now to think about each one as you finish it. Such rethinking and reviewing will help you find ideas for solving future problems or perhaps a new problem to solve will be suggested to you.

PROOFS AS PROBLEMS. Some problems in mathematics demand that a proof be found of some statement. Such a proof can be thought of as a logical argument that will convince a listener that a statement is true. The five-step process can be used to solve such a problem also.

Example 2 Prove that for any sets A, B, and C the distributive law for intersection over union holds.

1. What are the relevant *facts*?

a) Any three sets A, B, and C are used.
b) Two operations are used to form the set $A \cap (B \cup C)$.
c) A different sequence of operations is used to form the set $(A \cap B) \cup (A \cap C)$.
d) The operations $\cup$ and $\cap$ are precisely defined. (This is not stated in the problem but must be kept in mind.)

2. What is the *problem*? Show (prove) that for any sets A, B, and C

$$A \cap (B \cup C) = (A \cap B) \cup (A \cap C).$$

”

However, after you solve it, or even better, while you are solving it, analyze what you are doing. It will greatly deepen your understanding of problem-solving methods, and you might discover new methods or a new application of an old method.

”

Wayne A. Wickelgren in *How to Solve Problems*. W. H. Freeman and Company, San Francisco, California (1974).

23.

a) What were the important steps of the proof?

b) Write the law for statements that was applied in the proof.

3. Do you have any *ideas*?

a) Look at some particular examples of the statement to be proved (see Exercise Set 2.2).

b) Make a drawing (see Exercise Set 2.2).

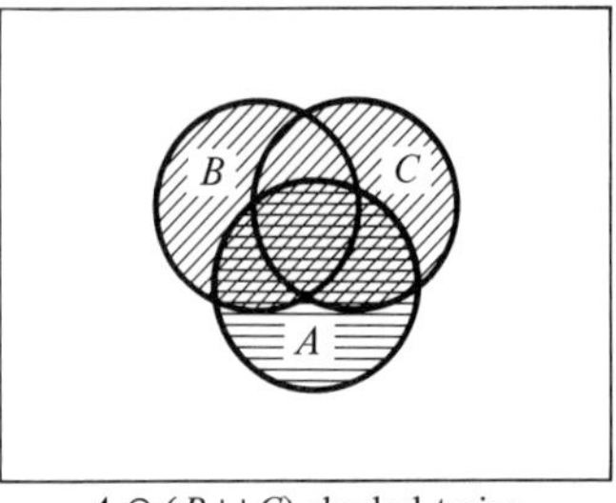

$A \cap (B \cup C)$ shaded twice

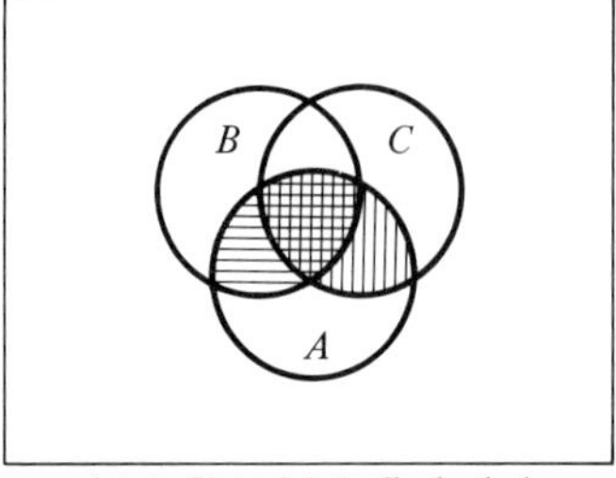

$(A \cap B) \cup (A \cap C)$ shaded

c) Describe the elements in $A \cap (B \cup C)$.
Describe the elements in $(A \cap B) \cup (A \cap C)$.

d) Compare the descriptions in (c).

e) Combine the ideas in (c) and (d).

4. The *solution*. We use idea (e) and start with $A \cap (B \cup C)$.

What are the elements of $B \cup C$? By the definition of $\cup$, those that are either in B or in C, that is, $x \in B$ or $x \in C$.

What elements are in $A \cap (B \cup C)$? Those that are in A and also in either B or C. What makes the statement $x \in A$ and ($x \in B$ or $x \in C$) true? The elements must either be in both A and B or they are in both A and C.

We now consider the set $(A \cap B) \cup (A \cap C)$.

What are the elements of $A \cap B$? By definition of $\cap$, those that are in both A and B, that is, $x \in A$ and $x \in B$.

What are the elements of $A \cap C$? Those that are in A and also in C, that is, $x \in A$ and $x \in C$.

What are the elements of $(A \cap B) \cup (A \cap C)$? Those that are in both A and B or else that are in both A and C.

A comparison of the description of the elements shows that the two sets are identical, that is,

$$A \cap (B \cup C) = (A \cap B) \cup (A \cap C).$$

The major portion of this analysis would not ordinarily appear in print; just a condensation of the argument presented in the solution step would be given.

Try Exercise 23

ANOTHER PROBLEM INVOLVING SETS. The same kind of analysis applied in the proof of Example 2 can be used for problems that are not proofs of a general statement.

Example 3 A team of college students was hired to take a survey of households in a city about frozen dessert purchases for the past month. One student reported that of those interviewed 103 households had purchased ice cream, 66 had purchased ice milk, 35 had purchased frozen yogurt, 9 had made no purchase of any of these items. Also, 20 had purchased all three types of frozen products, 36 had purchased ice cream and ice milk, 23 had purchased ice milk and frozen yogurt, and 24 had purchased ice cream and frozen yogurt. The supervisor looked at this information and said: "I also want to know how many households were surveyed, how many households purchased only ice cream, how many only ice milk, and how many only frozen yogurt." From the information given can these additional questions be answered?

Try Exercise 24

24.
a) Give the facts for this problem.
b) State the problem that is to be solved.

Now for the *ideas*. Two of those from Example 2 can be used.
a) Make a diagram.
b) Describe the elements in the sets.
c) How many households were in the described sets?

Problem solving

25. Use the information supplied about:

a) the 23 households that purchased ice milk and frozen yogurt.
b) the 24 households that purchased ice cream and frozen yogurt, and then
c) answer all the questions the supervisor wanted to know.

These ideas can be combined in looking for a solution.

a) Make a diagram. A Venn diagram is appropriate, since there are three distinct categories of households interviewed.

Let C be those who purchased ice cream,
M be those who purchased ice milk,
Y be those who purchased yogurt.

The diagram for this situation is:

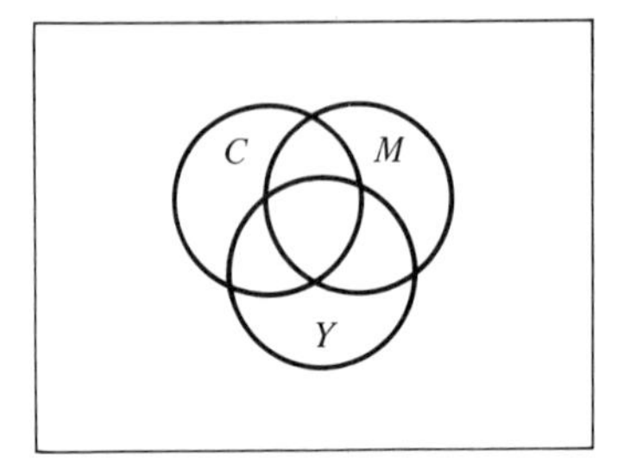

The supervisor knows that 20 households purchased all three items. Which set on the diagram must have 20 members? The one whose elements are in C and in M and in Y. This information can be added to the diagram.

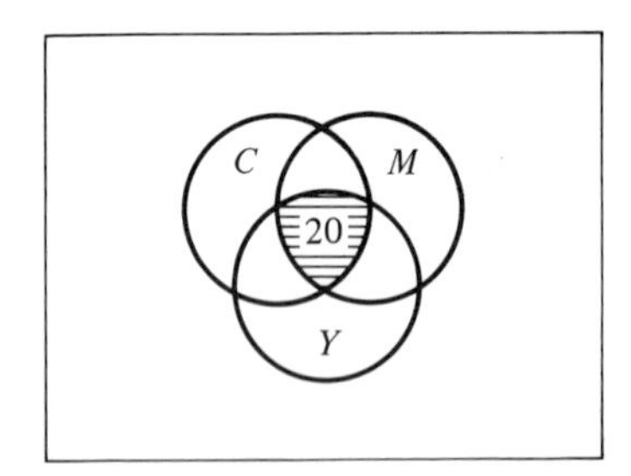

What other information does the supervisor have? 36 households purchased ice cream and ice milk. Which set on the drawing should have a total of 36 members? The set $C \cap M$. In this set, 20 households are already accounted for. Therefore there must be 16 that are in both C and in M but are not in Y. This information can be added to the diagram.

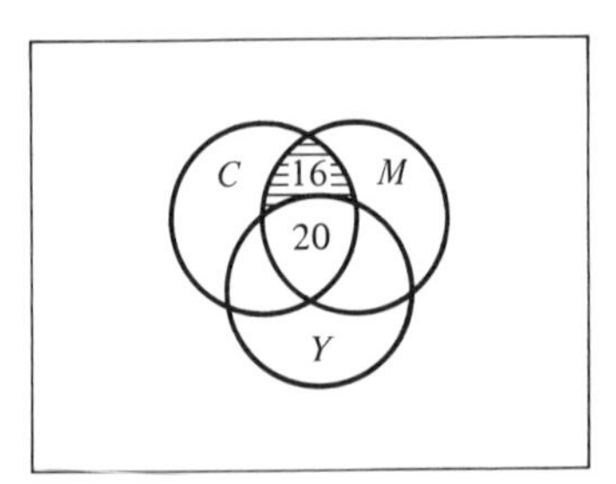

Try Exercise 25

EXERCISE SET 2.5

Use the five-step problem-solving process for all of the following. Be sure to do some *looking back* over your solutions. Problems 1 through 3 are variations of Example 1 of the text (p. 62).

1. This time the sum along each side of the triangle is to be 10.

2. If the same entries, 1, 2, 3, 4, 5, and 6 are used, what is the smallest total that could be required for one side of the triangle and the problem would still have a solution?

3. Write a variation of your own for Example 1. Solve your own problem or find out that it cannot be solved.

4. Start with 10 points, no three on the same line. How many line segments are needed to connect each point with every other point? A useful scheme is to start with fewer than 10 points to see what happens. *Hint*: It sometimes helps to recall a related problem (see Example 3 of Section 1.1).

■ **5.** A gourmet club that meets monthly has a total membership of 60 people. Fifty of them were at the first meeting of the year, while only 26 were at the second.

a) What is the smallest number of members that could have been at both?
b) What is the largest number of members that could have been at both?
c) The club secretary says that 20 people attended both parties. Are there any people who did not attend either one?

■ **6.** A school principal was planning the transportation to carry the members of the drama club, the band, and the gymnastics team to the fairgrounds. There are 45 members in the band, 26 in the drama club, and 26 on the gymnastics team. However, five students are in all three groups. There are two more pupils in both the band and the drama club than there are in both the band and the gymnastics team. There are two fewer students in both the drama club and the gymnastics team than there are in both the band and the gymnastics team. Five of the students are in neither the drama club nor in the band. If each school bus holds 33 students, how many buses will be needed to transport the students of the combined groups?

7. If you were unable to provide proofs in Problems 17–19 of Exercise set 2.2, try them again.

■ **8.** How can you get exactly 6 liters of water from a river by using two containers with capacities of 9 liters and 4 liters? The containers do not have markings of any kind on them.

9. Here is a fun–challenge problem for your deductive abilities. Determine what is going on here.

Two college students were working on the committee that had charge of planning a weekend at the college camp. In an effort to make some final arrangements, they were reviewing some lists of people who had said they might attend. Various club members contributed helpful bits of information about the people on one list:

a) If Art goes, then so will Betty.
b) Either Chuck or Diane, maybe both of them, will go.
c) Betty or Elaine will go, but if Betty goes, Elaine will not, and if Elaine goes, then Betty will not.

d) Chuck and Elaine have decided that if one goes, so will the other, and if one does not go, the other will do the same.
e) If Diane goes, then Art and Chuck will go too.

The poor bewildered students who got all this information said: "How do you expect us to figure out what is going on here?"

REVIEW TEST

1. Write roster and then the standard description for the following:
a) The set of natural numbers that are either less than 5 or greater than 10.
b) The set of whole numbers greater than 6 and less than 12.

2. Use the set of whole numbers W for a universal set.

$$\text{Let } A = \{x \mid x \text{ is in } W \text{ and } x \text{ is less than } 20\}$$
$$B = \{x \mid x \text{ is in } W \text{ and } x \text{ is greater than } 7\}.$$

Find the following: a) $A \cap B$ b) $A \cup B$

3. Show that the set of whole numbers is equivalent to the set of counting numbers.

4. If $A = \{5, 6, 7, 8, 9, 10\}$, find the relation $R_1 = \{(a, b) \mid a = 3 + b\}$ on the set A.

5. Some of the following statements are always true and some are not. Mark those that are true with T and those that are false with F.
a) An infinite set is one that can be put in one-to-one correspondence with a proper subset of itself.
b) If the intersection of two sets is { }, then the sets are disjoint.
c) Any set of ordered pairs is a one-to-one correspondence.
d) $\{\varnothing\}$ is the empty set.
e) $\{1, 2, 3\} = \{1, 3, 2\}$ f) $\{5, 6\} \in \{4, 5, 6, 7, 8\}$
g) Every one-to-one correspondence can be thought of as a function.
h) $A \cap \varnothing = \varnothing$
i) $A \times (B \cup C) = (A \times B) \cup (A \times C)$
j) $A \cap \bar{A} = A$ k) $A \cup A = A$
l) Every function is also a one-to-one correspondence.

6. Make a Venn diagram to illustrate the theorem: For any sets A and B, $\overline{(A \cup B)} = \bar{A} \cap \bar{B}$.

7. Prove the commutative law for the union operation for sets.

8. Label each of the following with one or more of these possible descriptions: relation, function, one-to-one correspondence.
a) $\{(2, 4), (2, 6), (3, 6), (3, 9), (4, 8), (4, 12)\}$
b) $\{(B, C), (C, D), (E, F), (A, C)\}$

c) Set A: 1, 2, 3 Set B: 7, 8, 12

d) 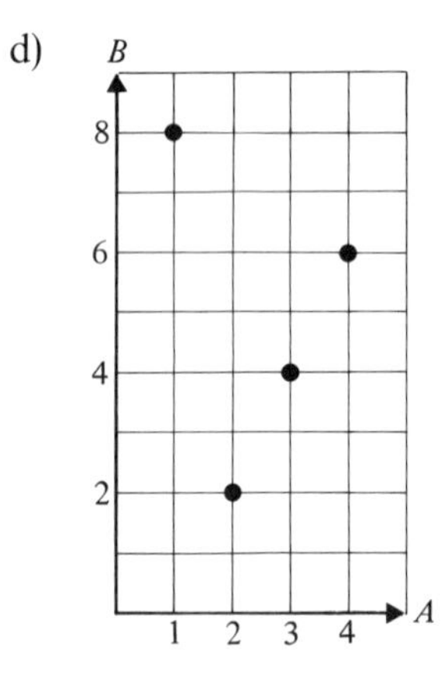

e) B: 20, 10; A: 1 2 3

f) $\{(A, A), (B, B), (C, C), (D, D)\}$

9. Prove that the relation *equals* for sets is an equivalence relation.

10. What does the opening page of this chapter have to do with the equivalence relation for sets?

■ **11.** Prove that for any sets A, B, and C, $A \cup (B \cap C) = (A \cup B) \cap (A \cup C)$.

■ **12.** Prove de Morgan's laws:
a) $\overline{A \cup B} = \bar{A} \cap \bar{B}$ b) $\overline{A \cap B} = \bar{A} \cup \bar{B}$

■ **13.** Prove that for any sets A, B, and C, $A \times (B \cup C) = (A \times B) \cup (A \times C)$.

FOR FURTHER READING

1. Zena Steinberg describes the work done with a class of first graders in "Will the set of children . . . ?" in *The Arithmetic Teacher*, 18:2, February 1971, pp. 105–108. She concludes: "This study of sets has proved to be most effective both as a foundation for computation and as a means of facilitating problem solving and logical thinking."

2. "The effect of sequence in the acquisition of three set relations: an experiment with preschoolers" by Edward A. Uprichard appears in *The Arithmetic Teacher*, 17:7, November 1970, pp. 597–604. This is a description of an experimental study to determine in which order it would be best to have preschoolers acquire the set concepts of "equivalence," "greater than," and "less than."

3. For a discussion of how set concepts can be both used and misused in K–2 see "What sets are not" by Herbert E. Vaughan in *The Arithmetic Teacher*, 17:1, January 1970, pp. 55–60. The author states he will "bring out some points that may sharpen your personal feeling for what a set is, suggest why the notion of set is an important one, and give some hints as to how to use the word in your classroom."

TEXT

Wheeler, Ruric E. *Modern Mathematics: An Elementary Approach*, Third Edition. Brooks/Cole Publishing Company, Monterey, California (1973, 1970, 1966).

chapter 3 numeration systems

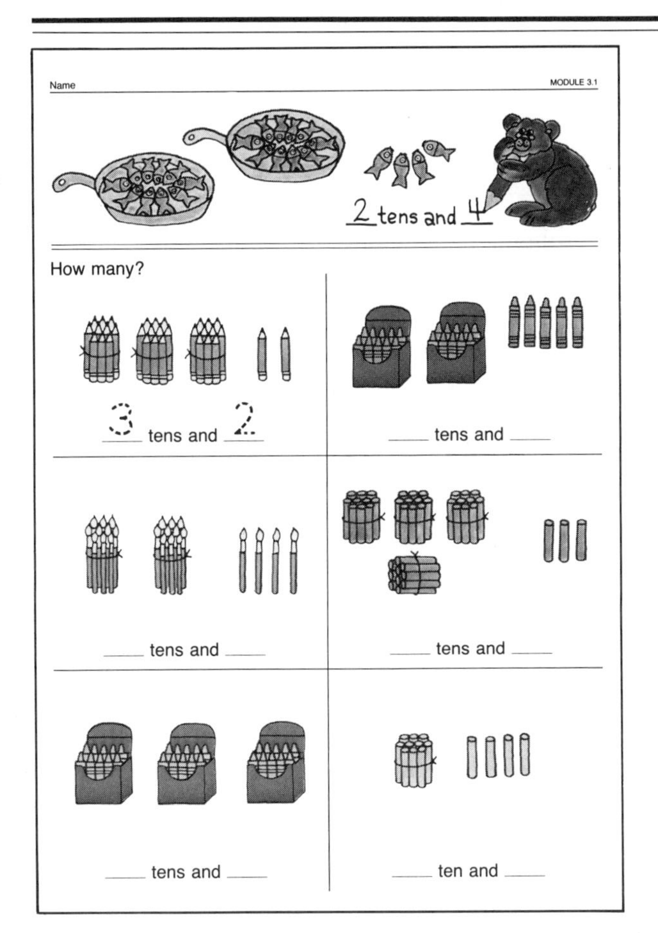

The idea of grouping by ten (the *basis* of our numeration system) is being developed here. Lack of understanding of this fundamental idea is a frequent cause of a child's difficulty with mathematics.

3.1 AIMS

You should be able to:

a) Convert from ancient Egyptian notation to standard decimal notation and the reverse, using the table of the text.
b) Convert from Roman notation to standard decimal and the reverse.
c) Illustrate and explain the terms additive, nonpositional, and subtractive, when applied to a numeration system.

3.1 ANCIENT NUMERATION SYSTEMS

The story of the development of the abstract place-value system used by just about the entire world covers a period of nearly 5000 years, so only a brief part of this history can be recounted here. Unorganized counting can be done by simply placing counters or marks in a one-to-one correspondence with the objects to be counted, and the earliest systems for counting used this nongrouped approach.

TALLY STICKS. Tally sticks with cut-out notches rather than written numerals were an old and universally used system for permanently recording numbers. Some tally sticks used only one type of notch as a counter, but in more sophisticated versions different sizes and positions meant different things, and the idea of grouping for convenience was introduced.

Quite often tally sticks show "crossings" in the forms of X's or V's, and these represent a significant step forward in the development of a recording system for numbers. Such tally-stick schemes are still practiced today in keeping scores when it is customary to let 𝍸 𝍸 𝍸 represent three groups of five each.

Not just unsophisticated peasants used tally sticks; this system was used from the 12th century until as late as 1826 by the English Royal Treasury to keep records of taxes owed and paid.

THE EGYPTIAN SYSTEM OF NUMERATION. The hieroglyphic* system of the ancient Egyptians used grouping by tens, and a stylized but recognizable picture symbol had the same meaning regardless of its position when written.

Picture of a:	*The symbol*	*Represents*	*Grouping*
Stroke	𓏺	1	𓏺𓏺𓏺𓏺𓏺𓏺𓏺𓏺𓏺𓏺 = 𓎆
Heel bone	𓎆	10	𓎆𓎆𓎆𓎆𓎆𓎆𓎆𓎆𓎆𓎆 = 𓍢
Scroll	𓍢	100	𓍢𓍢𓍢𓍢𓍢𓍢𓍢𓍢𓍢𓍢 = 𓆼
Lotus flower	𓆼	1 000	
Bent finger	𓂭	10 000	
Tadpole	𓆐	100 000	
Astonished man	𓁨	1 000 000	

The Egyptians usually grouped their numerals (written symbols for numbers) in ascending or descending order either horizontally or vertically, but in any case the meaning of the numeral is obtained by simply adding the values of the

* The Egyptian system here is the hieroglyphic system that was already perfected by 3200 BC. A later system, the hieratic (about 2000 BC), was a corruption of the hieroglyphic and was used by priests, merchants, and government officials. A still later system, the demotic (the 1st century BC), was very conventionalized, and it is hard to recognize the hieroglyphs from which the system developed.

separate parts. Such a system is called an *additive system*; performing addition in such a system is not much more difficult than reading a numeral.

Examples

1. = 10 000 + 300 + 20 + 6 = 10 326

2. = 100 000 + 30 000 + 4000 + 400 + 80 + 6 = 134 486

3. = 200 000 + 30 000 + 2000 + 400 + 10 + 3 = 232 413

Try Exercises 1 and 2

THE ROMAN SYSTEM OF NUMERATION. The Roman system of numeration was, at the start, also a completely additive system that used the principle of grouping into fives as well as tens. The Roman numerals I, X, and V for 1, 5, and 10, respectively, possibly arose from the notching schemes used on tally sticks.

The letters shown below are the ones used in Europe during the Middle Ages; they reflect the gradual changes made to the old abstract but letter-like Roman symbols.*

Symbol	*Represents*	*Grouping*	
I	1	IIIII ≐ V	$5 \times 1 = 5$
V	5	VV = X	$5 \times 2 = 10$
X	10	XXXXX = L	$5 \times 10 = 50$
L	50	LL = C	$50 \times 2 = 100$
C	100	CCCCC = D	$5 \times 100 = 500$
D	500	DD = M	$500 \times 2 = 1000$
M	1000		

In the old Roman system, when several letters are written together, the numbers they represent are simply added.

Examples

4. XXIII = 20 + 3 = 23
5. CLXXXXVIII = 100 + 50 + 40 + 5 + 3 = 198
6. MDCXXXII = 1000 + 500 + 100 + 30 + 2 = 1632

Try Exercises 3 and 4

1. Convert to standard decimal notation:

2. Convert 256 374 to Egyptian notation:

Convert to standard decimal notation:

3. XXXII

4. MDCCXXIII

* The Romans never wrote 1000 as M. They used (I), which bears a certain resemblance to the letter M to which the old symbol evolved.

Convert to standard decimal notation.

5. XLIX

6. MCMLXXXIV

Later a *subtractive principle* was incorporated: if the symbols I, X, or C precede one for either of the next two higher numbers, the smaller number is subtracted from the larger and these subtraction combinations were thus allowed:

$$\text{IV} = 5 - 1 = 4 \qquad \text{IX} = 10 - 1 = 9$$
$$\text{XL} = 50 - 10 = 40 \qquad \text{XC} = 100 - 10 = 90$$
$$\text{CD} = 500 - 100 = 400 \qquad \text{CM} = 1000 - 100 = 900$$

Examples

7. $\text{LIX} = 50 + (10 - 1) = 59$
8. $\text{CMXLIV} = (1000 - 100) + (50 - 10) + (5 - 1) = 944$
9. $\text{MCDXCII} = 1000 + (500 - 100) + (100 - 10) + 2 = 1492$

Try Exercises 5 and 6

Convert to Roman notation:

7. 579

8. 7449

Because of the special symbols for certain multiples of five as well as for ten, and because of the subtractive principle just cited, the number of symbols needed was slightly reduced.

Examples

10. $74 = 50 + 20 + 4 = 50 + 20 + (5 - 1)$
$= \text{LXXIIII} \quad \text{or LXXIV}$
11. $329 = 300 + 20 + 9 = 300 + 20 + (10 - 1)$
$= \text{CCCXXVIIII or CCCXXIX}$
12. $2438 = 2000 + 400 + 30 + 8 = 2000 + (500 - 100) + 30 + 8$
$= \text{MMCCCCXXXVIII or MMCDXXXVIII}$

Try Exercises 7 and 8

Calculation in writing with the Roman system was awkward, but that did not bother its users, since computation was done very quickly on an abacus, or counting board with counters. The only writing that was done was in recording the final answer.

A Roman hand-abacus had paired grooves for ones, tens, hundreds, thousands, and so on. On the lower grooves each of the "stones" represented one, but those in the upper grooves represented five of those just below it.

The markers that are to be counted are shoved toward the middle close to the groove labels. The schematic drawing of an abacus shows 3516 (markers to be counted are shaded):

$$3000 + 500 + 10 + 5 + 1 = 3000 + 500 + 10 + 6 = 3516.$$

Addition on such a board is accomplished by moving the appropriate markers. To add 10 to the amount 3516 on the board sketched below, another marker in the lower 10-groove would be pushed toward the center and 3526 would be read off the board.

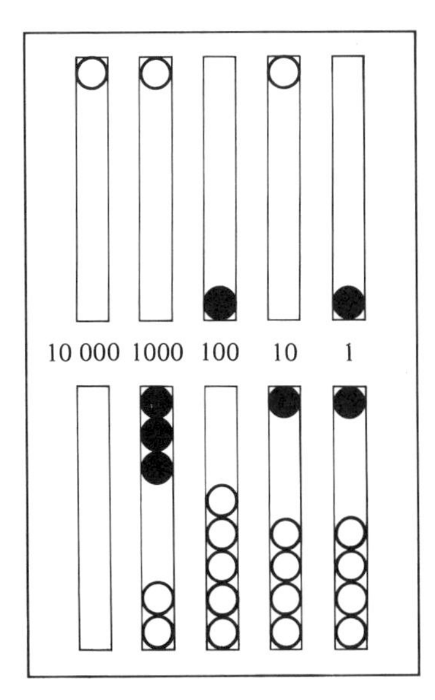

Try Exercises 9 and 10

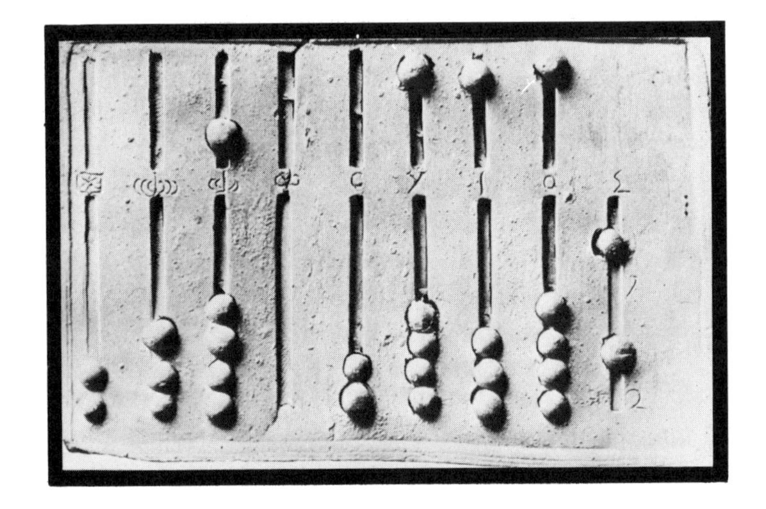

Roman hand abacus. Plaster cast of the specimen in the Cabinet des Medailles, Paris. Almost full size. Between the two rows of grooves are the Roman symbols for

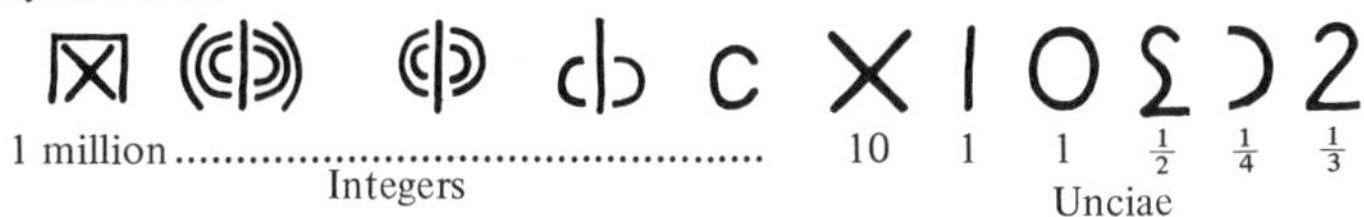

(Reproduced by permission from Karl Menninger's book *Zahlwort und Ziffer*, Vandenhoeck & Ruprecht, Verlagsbuchhandlung, Göttingen, 1958.)

9. Draw a schematic Roman abacus that shows each of the following:

a) 576

b) 2372

10. Use sketches of a Roman abacus to show

576 + 20.

ZERO. An abacus does not need a zero for computation, since it is obvious in which columns the pebbles or markers are used and hence are being counted. Only those in the "counting" position are recorded in the corresponding numeral.

No additive system, even with a subtractive feature, requires a symbol for zero, since the meaning of a symbol does not change with its position. For example, in the Roman numeral there is no need to indicate in a special way that there are no tens in the symbol for five hundred four; it is obvious from the numeral DIIII that 500 + 4 is intended.

In the case of the Romans there is a significant difference between computation and the writing of numbers. The Roman abacus certainly used the concepts of grouping and place value, but their numeration system failed to reflect the latter refinement. The Romans did not develop a symbol for zero, since it was not needed for either their computation or in the writing of their numerals.

EUROPE. The Europeans before and during the Middle Ages never did develop a true place-value numeration system but merely modified somewhat the old Roman system. By the 1500's, CCCM was being used to represent 300 000 but the final step toward our modern system was never made. Even though the clumsy written numerals were retained by the Europeans, they used exactly the same computational procedures that are taught in the elementary school today. The difference is that they used these procedures in conjunction with a counting board instead of with pencil and paper. In our modern system computation and the numeration scheme have become so closely linked that until the recent common use of calculators we have tended to forget that these are not the same thing.

CHINA. The Chinese had an early form of place-value system that involved multiplication of each place value named. A comparison of three systems indicates how the Chinese system worked:

Roman: CCCXXIIII
Chinese: 3C 2X 4
Modern: 324

The Chinese symbols for C and X represented the groupings, so 3C meant 3 of that grouping; this scheme is called a *named place-value system.* It was left to the Indians to introduce the abstract place-value system.

HINDU-ARABIC. With the invention of zero, some unsung genius freed computation from a counting board. Indian researchers believe that place-value notation started about 200 BC and possibly was stimulated by the use of counting boards covered by a layer of sand. Because such boards lacked the fixed columns of the Roman abacus and entries were made by making and erasing marks in the sand, a symbol of some sort was needed to indicate that a grouping was not being used.

> 1, 2, 3, 4, 5, 6, 7, 8, 9, and 0—these ten symbols which today all peoples use to record numbers, symbolize the world-wide victory of an idea. There are few things on earth that are universal, and the universal customs which man has successfully established are fewer still. . . . Indian numerals are indeed universal.

Karl Menninger in *Number Words and Number Symbols: A Cultural History of Numbers*, translated by Paul Broneer from the revised German edition (1958). M.I.T. Press, Massachusetts Institute of Technology, Cambridge, Massachusetts, U.S.A., and London, England, 1969.

The final step of this long history took place around 600 AD in India, when a completely developed system of numerals appeared that used just nine digits and zero. The present-day *abstract place-value* system of numeration (discussed in the next section) in which the meaning of every symbol of the numeral is determined by its position in the numeral, had arrived.

EXERCISE SET 3.1

Convert to standard decimal notation:

1. XXVII

2. XXXVIII

3. XLVI

4. XLIV

5. MCCXXIV

6. MCCXLIX

7. MCDXLIV

8. MMMCMXLIX

Convert to Roman notation:

9. 649

10. 432

11. 5 342

12. 6 579

13. 15 249

14. 7 509

Convert to standard decimal notation:

15. 𓂭𓍢𓎆𓎆𓎆𓏺𓏺

16. 𓂭𓆼𓍢𓎆𓎆𓏺𓏺𓏺

17. 𓏺𓏺𓎆𓎆𓍢𓂭𓂭𓂭𓆐
𓏺𓏺𓎆

18. 𓏺𓏺𓏺𓎆𓎆𓍢𓍢𓆼𓆼𓆼𓂭𓆐
𓏺𓏺 𓎆𓎆

Convert to Egyptian notation:

19. 345 267

20. 253 106

21. 503 126

22. 413 004

23. Compare the Roman and Egyptian numeration systems. Find similarities and differences.

24. Find the sums:
a) XXIII + XXVII b) LVII + CLVIII

25. Make schematic diagrams of the Roman abacus and find the sums of Exercise 24.

26. Suppose the Romans had actually used written computational methods. How might they have multiplied 4 × 49?

27. If the Romans had used written procedures instead of counting boards, what are the difficulties they would have had?

28. Are there any advantages to the Roman computation system? to the Roman numeration system?

29. Why is a symbol for zero not needed in an additive system of numeration?

3.2 AIMS

You should be able to:

a) Explain, in terms of place value and expanded notation, the meaning of standard notation for whole numbers.
b) Use the grouping principle to find decimal notation for the number of elements in a set.
c) Construct a decimal place-value chart.

3.2 DECIMAL NOTATION

The Hindu-Arabic numeration system was brought to Europe as early as the 10th and 11th centuries but it found no enthusiastic converts. Similarly, al-Khwarizmi's textbook on computation and place value introduced in the 12th century was largely ignored. In the 16th century the new scheme for recording numbers drifted across the Alps to northern Europe from the commercial houses of Italy and finally became accepted and gained widespread use.

PLACE VALUE. In an abstract place-value, or positional, system of numeration, each symbol has its value determined not only by the place it occupies in the numeral but also by the grouping used in the system. Thus in the decimal system the value of the two 4's in 44 depends upon both the position and the grouping by tens. The word *decimal* comes from the Latin *decem*, which means ten.

In the decimal system then, each digit represents a number of ones, tens, hundreds, and so on, with the values assigned to the place as shown in a *place-value chart.*

11. Explain the meaning of the following numeral: 87 924.

Place-Value Chart (base ten)

Ten thousands	*Thousands*	*Hundreds*	*Tens*	*Ones*
10 000 or 10^4	1000 or 10^3	100 or 10^2	10 or 10^1	1 or 10^0

Example 1 Explain the meaning of the following numeral:*

38 475.

Answer 3 ten thousands + 8 thousands + 4 hundreds + 7 tens + 5 ones.

12. Extend the place-value chart two places to the left of that in the text.

Try Exercises 11 and 12

GROUPING. The principle of grouping underlies a place-value numeration system, and the following example illustrates how the grouping of objects corresponds to place value.

Example 2 Form the groupings by circling and determine the number of objects in the following set.

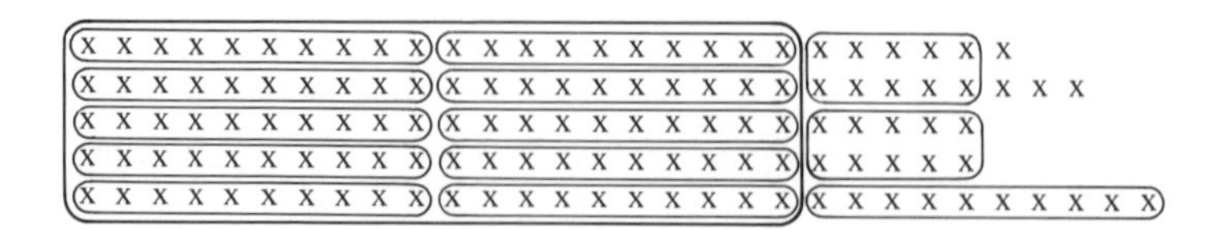

* You may be used to seeing 38 475 written with a comma: 38,475. In the metric system the standard numerals for whole numbers are written without a comma, and a space is left instead. Four-digit numerals can be written with or without a space as in 1 645 or 1645.

Solution Sets of ten are circled first; ten of those are then circled to make a set of one hundred. In this case there is one hundred, three tens, and four ones, so the standard decimal notation is 134.

Try Exercise 13

EXPANDED NOTATION. If the actual value of each place is stated in order to clarify the meaning, then the notation is described as *expanded.*

Example 3 Write expanded notation for 34 265.

Solution There are three possible variations:

a) $30\,000 + 4000 + 200 + 60 + 5$
b) $3 \times 10\,000 + 4 \times 1000 + 2 \times 100 + 6 \times 10 + 5 \times 1$
c) $3 \times 10^4 + 4 \times 10^3 + 2 \times 10^2 + 6 \times 10^1 + 5 \times 10^0$

Try Exercise 14

If the place-value chart shows grouping into periods, it can be of help when it is necessary to write out the spoken names for numbers. The periods conform to the scheme for spoken names because they first give the number of ones, then the number of thousands, and then the number of millions, and so on.

Place-Value Chart with Periods

10^8	10^7	10^6	10^5	10^4	10^3	10^2	10	1
millions			thousands			ones		

Example 4 Write the words that are spoken to name 64 327.

The spoken names reflect the groups of three (the periods) of the place-value chart:

sixty-four thousand three hundred twenty-seven.

Three hundred twenty-seven is the number of ones. Sixty-four is the number of thousands. It is not customary to say for 64 000: *6 ten thousands 4 thousands.*

There are other conventions in writing words to name numbers:

a) For numbers between ten and 100 such as 27, the words are hyphenated as in twenty-seven.
b) The word *and* is not used for whole numbers.

Example 5 Write words to name the number 46 214 325. The digits as written are already separated into groups of three. From the chart it can be seen that there are 46 millions, 214 thousands, and 325 ones. The name is thus:

forty-six million two hundred fourteen thousand three hundred twenty-five.

13. Form the groupings by circling sets of ten; determine the number of objects in this set.

x x x x x x x x x x x x x x x x
x x x x x x x x x x x x x x x
x x x x x x x x x x x x x x x x x
x x x x x x x x x x x x x x
x x x x x x x x x x x x x x x x
x x x x x x x x x x x x x x x
x x x x x x x x x x x x x x x x x x
x x x x x x x x x x x x x x x x
x x x x x x x x x x x x x x x x x
x x x x x x x x x x x x x x x x x
x x x x x x x x x x x x x x x

14. Write the three variations possible for the expanded notation for 573 428.

15. Write words to name the number
376 927 385

16. Extend the place-value chart in the text six places (two more periods) to the left.

This is never read as *4 ten millions 6 millions 2 hundred thousands 1 ten thousand* and so on. The use of the periods simplifies the reading aloud and the writing of the words for a number.

Try Exercises 15 and 16

EXERCISE SET 3.2

1. Explain in terms of place value the meaning of 756 425.

2. Repeat Exercise 1 for 4 258 316.

3. Form groupings by circling and write decimal notation for the number of objects in this set:

x x x x x x x x x x x x x x x x x x x x x
x x x x x x x x x x x x x x x
x x x x x x x x x x x x x x x x x x x x x x x x x x
x x x x x x x x x x x x x x x x x x x
x x x x x x x x x x x x x x x x x
x x x x x x x x x x x x x x x x x x x x x

4. a) Repeat Exercise 3 for the number of objects tallied:

b) Why was this exercise easier to do than Exercise 3?

卌 卌 卌 卌 卌 卌 卌 卌 卌 |||
卌 卌 卌 卌 卌 卌 卌 卌
卌 卌 卌 卌 卌 卌 卌 卌

Write words to name each of the following numbers:

5. 38

6. 95

7. 879

8. 463

9. 1492

10. 3275

11. 94 006

12. 48 404

13. 75 920 483

14. 12 846 924

Write three variations of expanded notation for each of the following numbers:

15. 12 418

16. 58 386

17. 492 148

18. 814 672

19. Study the opening page of this chapter. How is the concept of place value and that of grouping presented in this lesson?

3.3 NONDECIMAL NOTATION

BASE FIVE. Abstract place-value notation need not be based upon groupings of ten, and others have been used at various times in man's history. The place values for base five are powers of five: 1, 5, 5^2 (twenty-five), 5^3 (one hundred twenty-five), 5^4 (six hundred twenty-five), and so on. To indicate that a numeral refers to groupings of five, a subscript is used as in 342_{five}. A system with a base of five needs only five digits that could be totally new symbols, but we find it convenient to use the same symbols for one, two, three, four, and zero of the decimal system.

Example 1 Explain the meaning of the numeral 4231_{five}.

Answer is:

4 one hundred twenty-fives + 2 twenty-fives + 3 fives + one.

Try Exercise 17

The following example illustrates how the actual groupings based on five correspond to the place values in base five.

Chip trading in base ten

3.3 AIMS

You should be able to:

a) Make a place-value chart for a base other than ten.
b) Write expanded notation to explain nondecimal notation.
c) Convert whole-number nondecimal notation to decimal.
d) Convert whole-number decimal notation to nondecimal in three different ways.

17. Write expanded base-five notation for 3214_{five}.

18. Show the groupings by circling and determine base-five notation for the number of objects in this set.

x x x x x x x x x x x x x x x
x x x x x x x x x x x x x
x x x x x x x x x x x x x x
x x x x x x x x x x x x x x x x x x
x x x x x x x x x x x x x x x x x
x x x x x x x x x x x x x x x x x x
x x x x x x x x x x x
x x x x x x x x x x x

19. Extend the base-five place-value chart two more places to the left.

Example 2 Show the groupings by circling and determine the base-five notation for the number of objects in this set.

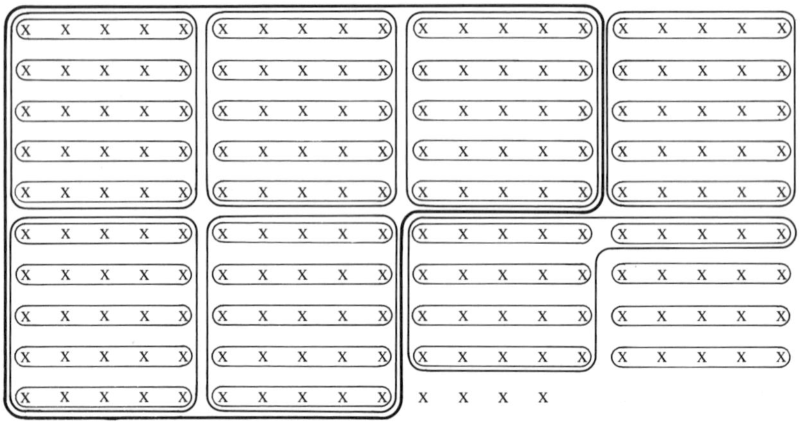

First groupings of five are circled, then five of these are circled to make a set of five fives, or twenty-five. If there are five twenty-fives, these are circled to make a set of one hundred twenty-five. In this case there is one set of one hundred twenty-five, there are two sets of twenty-five, three sets of five, and four left over. Base-five notation is therefore

$$1234_{\text{five}}.$$

Try Exercise 18

Here is a place-value chart for base five written in base-ten notation:

Place Values for Base Five

5^4 or 625	5^3 or 125	5^2 or 25	5^1 or 5	5^0 or 1

Since base-five notation for five is 10, written in base-five notation the same chart would be:

10^4 or 10 000	10^3 or 1 000	10^2 or 100	10^1 or 10	10^0 or 1

Try Exercise 19

Expanded notation can also be used to explain the meaning of nondecimal numerals.

Example 3 Write expanded notation for 3123_{five}.

Solution

$$\begin{aligned} 3123_{\text{five}} &= 3 \times 5^3 + 1 \times 5^2 + 2 \times 5 + 3 \\ &= 3 \times 125 + 1 \times 25 + 2 \times 5 + 3. \end{aligned}$$

Keep in mind that the expanded notation at the right in the example above is base ten, since in base five it would be:

$$\begin{aligned} 3123_{\text{five}} &= 3 \times 10^3 + 1 \times 10^2 + 2 \times 10 + 3 \\ &= 3 \times 1000 + 1 \times 100 + 2 \times 10 + 3. \end{aligned}$$

Try Exercise 20

20. Write expanded notation for 32132_{five}. First use base-ten notation for the expansion, then base five.

CONVERTING FROM NONDECIMAL TO DECIMAL NOTATION. Conversion from nondecimal to decimal notation follows directly from the expanded notation.

21. Convert 4132_{five} to decimal notation.

Example 4 Convert 3123_{five} to decimal notation.

Solution The decimal expansion is:

$$\begin{aligned} 3123_{\text{five}} &= 3 \times 125 + 1 \times 25 + 2 \times 5 + 3 \\ &= 375 + 25 + 10 + 3 \\ &= 413_{\text{ten}}. \end{aligned}$$

The multiplication and the addition indicated in the expanded notation are carried out to find the total, which is then written as a standard decimal numeral.

Try Exercises 21 and 22

22. Convert 23132_{five} to decimal notation.

CONVERTING FROM DECIMAL TO NONDECIMAL NOTATION. Three ways for converting from decimal to nondecimal notation follow.

A. GROUPING OBJECTS (illustrated with base three). This method uses the actual or schematic grouping of objects, as in the preceding exercises where X's were circled.

Example 5 Determine the base-three notation for 29.

A sketch is used to represent the objects being counted. First, all possible groupings of three are circled, then nine, and then twenty-seven.

(X X X) (X X X) (X X X) X
(X X X) (X X X) (X X X) X
(X X X) (X X X) (X X X)

23. Use the method of grouping to find base-three notation for 38.

Only three symbols are needed for base-three notation, and it is convenient to use 0, 1, and 2. In this case there are 2 ones, no threes, no nines, and one twenty-seven. The base-three notation is 1002_{three}.

Try Exercise 23

B. USING A PLACE-VALUE CHART (illustrated with base four). This method involves successive subtraction made in conjunction with a place-value chart.

Example 6 Determine base-four notation for 182.

Solution

Place Values for Base Four

4^4 or 256	4^3 or 64	4^2 or 16	4^1 or 4	4^0 or 4

(The chart is in decimal notation.) The number 182 is smaller than 256, so there is no 4^4 digit, but it is greater than 64, so there is a 4^3 digit. Subtract 64's:

$$\begin{array}{r} 182 \\ 64 \\ \hline 118 \\ 64 \\ \hline 54 \end{array}$$

There are two 64's with 54 left over. The 64's digit is 2: [2][ ][ ][ ].

The table is consulted to determine the next lower place value that can be subtracted:

$$\begin{array}{r} 54 \\ 16 \\ \hline 38 \\ 16 \\ \hline 22 \\ 16 \\ \hline 6 \end{array}$$

There are three 16's in 54, with 6 left over. Thus the 16's digit is 3: [2][3][ ][ ].

The next place value is 4 and 4's can be subtracted:

$$\begin{array}{r} 6 \\ 4 \\ \hline 2 \end{array}$$

There is one 4, with 2 left over. The 4's digit is 1: | 2 | 3 | 1 | | .
The left-over 2 is the ones digit: | 2 | 3 | 1 | 2 | .
Base-four notation for 182 is then 2312_{four}.

Try Exercise 24

24. Use the place-value chart to find base-four notation for 270.

C. SUCCESSIVE DIVISION (illustrated with base five). If all possible groups of five are circled, then the number of items not circled gives the ones digit for base-five notation. This number can also be determined by dividing the given number by five and finding the remainder.

Example 7 Find base-five notation for 23.

Solution First divide by five:

$$\begin{array}{r} 4 \\ 5\overline{)23} \\ \underline{20} \\ 3 \end{array}$$

There are four fives, with a remainder of three. The base-five notation is thus 43_{five}.

If any number is divided by five, the quotient will be the number of fives. The remainder will be the number of ones left over, hence it gives the ones digit of the base-five notation. If the quotient had turned out to be greater than 4, it could be divided again to find the number of 5^2 digits, or 25's. This time the remainder will be the number of 5's and the quotient will be the number of 25's.

Example 8 Find base-five notation for 89.

Solution Divide by five:

$$\begin{array}{r} 17 \\ 5\overline{)89} \\ \underline{5} \\ 39 \\ \underline{35} \\ 4 \end{array}$$

There are 17 fives in 89, and a remainder of 4. The ones digit is 4.

Divide again:

$$\begin{array}{r} 3 \\ 5\overline{)17} \\ \underline{15} \\ 2 \end{array}$$

25. Use successive division to find base-five notation for 194.

The 17 fives can be grouped into 3 twenty-fives plus two groups of five. The fives digit is 2.

Divide again:

$$\begin{array}{r} 0 \\ 5\overline{)3} \\ 0 \\ \hline 3 \end{array}$$

The 3 twenty-fives cannot be grouped into a group of 125. The twenty-fives digit is 3. Base-five notation for 89 is 324_{five}.

This procedure can be described as follows.

To find base-*b* notation for a whole number, divide the number by *b*; the remainder is the ones digit. Divide the quotient by *b*; the remainder is the *b*'s digit. Divide the new quotient by *b*; the remainder is the b^2 digit. Continue dividing in this manner until a quotient of 0 is obtained.

26. Extend the place-value chart for base two three places to the left.

The division method is an abstraction of the circling of the groupings in method A above. The first division by five corresponds to the initial grouping of objects into fives. The second division by five corresponds to the second grouping of fives into twenty-fives. The third division in Example 8 showed that it was not possible to group again into a group of 125.

Try Exercise 25

BINARY NOTATION. Any whole number greater than 1 can be used as a numeration base. (A base of 1 is a return to the tally system.) Base-two notation, known as *binary notation*, is of special interest because of the role it plays in modern computers. The place values are as follows:

Place Values for Base Two

2^3 or 8	2^2 or 4	2^1 or 2	2^0 or 1

(Chart is in base-ten notation.)

Try Exercise 26

The same methods used for the bases already discussed apply to base two.

Example 9 Write decimal notation for 101011_{two}:

Solution

$$\begin{aligned} 101011_{two} &= 1 \times 2^5 + 0 \times 2^4 + 1 \times 2^3 + 0 \times 2^2 + 1 \times 2 + 1 \\ &= 32 + 0 + 8 + 0 + 2 + 1 \\ &= 43_{ten}. \end{aligned}$$

Try Exercise 27

It is instructive to see consecutive numbers written in another base; only two symbols, 0 and 1, are needed for base two.

Example 10 Write the binary notation for the first nine natural numbers. They are:

one	1	six	110
two	10	seven	111
three	11	eight	1000
four	100	nine	1001
five	101		

A real advantage of the binary system is found in electronic calculators* and computers, since electronic switches in these devices can be either *on* or *off* and thus can be made to correspond to the numbers 1 and 0, respectively. The large number of digits in binary notation does not handicap these calculators because of their amazing operating speeds.

DUODECIMAL NOTATION. Base-twelve, or *duodecimal*, notation is interesting because two additional symbols must be introduced for ten and eleven; a convenient choice is t for ten and e for eleven.

Place Values for Base Twelve

12^2 or 144	12^1 or 12	12^0 or 1

Try Exercise 28

Example 11 Write expanded notation for $53e1t_{twelve}$.

Solution

$$53e1t_{twelve} = 5 \times 12^4 + 3 \times 12^3 + e \times 12^2 + 1 \times 12 + t.$$

Example 12 Write decimal notation for $53e1t_{twelve}$.

27. Write decimal notation for 1101101_{two}

28 Expand the place-value chart for base twelve two places to the left.

* Small electronic calculators use a binary-coded decimal system. Each *digit* of a numeral is represented by a *four-bit binary code.* The number 79 would be coded 0111 1001. For more information about this see "The Small Electronic Calculator" by Eugene W. McWhorter in *Scientific American*, 234:3, March 1976.

29. Write decimal notation for $t3e4_{\text{twelve}}$.

Solution Use the expanded notation from Example 11; convert e and t to base ten; then simplify:

$$\begin{aligned} 53e1t_{\text{twelve}} &= 5 \times 12^4 + 3 \times 12^3 + 11 \times 12^2 + 1 \times 12 + 10 \\ &= 5 \times 20736 + 3 \times 1728 + 11 \times 144 + 1 \times 12 + 10 \\ &= 110\,470_{\text{ten}} \end{aligned}$$

Try Exercise 29

EXERCISE SET 3.3

1. Make a place-value chart for base six. Show five places.

2. Make a place-value chart for base eight. Show five places.

Write expanded notation for the following, once in base ten, and once in the indicated base:

3. 13245_{six}

4. 46247_{eight}

Convert to decimal notation:

5. 13245_{six}

6. 46247_{eight}

7. 357_{eight}

8. 2315_{six}

Use the method of circling groupings to convert the following to the base indicated:

9. 54 to base four

10. 51 to base six

11. Convert to the indicated base using a place-value chart:
a) 41 to base six b) 103 to base eight

12. Use division to convert to the indicated base:
a) 108 to base six b) 412 to base eight

13. Write expanded notation for the following. Use base ten for the expansion, then convert to decimal notation.
a) 1111011_{two} b) 1001101_{two}

14. Write expanded notation for the following. Use base ten for the expansion, then convert to decimal notation:
a) $3e7te3_{\text{twelve}}$ b) $5t8e_{\text{twelve}}$

15. Convert to binary notation:
a) 39 b) 41

16. Convert to duodecimal notation:
a) 174 b) 335

17. Use the method of division for a flowchart that shows how to convert 89 to base-five notation (see Example 8 of the text).

18. Use the place-value chart and subtraction for a flowchart that shows how to find base-four notation for 182 (see Example 6 of the text).

■ 19. Make a flowchart that shows how to convert any standard decimal numeral to one written in any base b.

20. Develop a procedure to be used with your calculator to convert any standard decimal numeral to one written in a different, arbitrary base. Aim for a method that requires the minimum number of intermediate results to be recorded on paper.

21. It is possible to use still another method to convert from decimal to nondecimal notation: just count up to the number wanted. For example, in base five: 1, 2, 3, 4, 10, 11, 12, 13, 14, and so on. Count up to 36_{ten} in this fashion to find the base-five numeral for it.

■ 22 All the changes in notation in the text thus far have involved going from base ten to base b, or the reverse. Investigate the problem of going directly from, say, base two to base eight, without using base ten as an intermediate step. Use two different bases and see what you can find out for that case.

3.4 PROBLEM SOLVING

Although it is important to have an efficient notation for our number system, that is only part of the story, since there are many other occasions in mathematics when it is necessary to use other types of symbols.

TRANSLATION. Often one of the first steps in solving a mathematical problem is to translate the conditions of the problem into mathematical language or symbolism. Except for quite simple problems, a literal word-for-word translation is generally not possible.

It is not a good idea to depend upon the presence of a key word to determine how a translation to symbols should be made because even a simple word such as *remained* can be used in more than one way.

Example 1 There were 27 meters of plaid cloth in a bolt of fabric. Three meters were cut off for a dress, and 9 meters more were cut off for shirts than were cut off for the dress. How much material *remained* in the piece on the bolt?

Example 2 After 7 logs were burned in the fireplace, 13 logs *remained* in the bundle of firewood. How many logs were in the bundle to start?

Suppose that a child had been told that the word *remained* in a problem always meant "you should subtract."

It is far better to do some five-step* problem solving in such problems.

In Example 2, what are the relevant *facts*?
 a) 7 logs were burned
 b) 13 logs remained

Exactly what is the *problem*? Some possible statements of the problem are:
 a) How many logs were there to start?
 b) How many logs were there before any were burned?
 c) If 7 logs were replaced and combined with the 13 logs, what would you have?

Restatement of the problem as in (c) makes it easy to write out a sentence that describes the situation.

Here is an *idea for solution*:

7 logs burned + 13 logs left equal the total at the start,
hence $7 + 13 = \square$, where $\square$ represents the total at the start.

In this case the problem has been translated to a very brief statement in mathematical language, but it is also possible to translate such a problem into a concrete, or *physical*, model.

Try Exercise 30

3.4 AIMS

You should be able to:
a) Use the five-step process as an aid in the translation of the conditions of a problem into mathematical language.
b) Use a physical model to aid in the solution of problems.

30. How might you use physical objects to model the log problem for third graders?

* THE FIVE-STEP PROCESS
1. What are the relevant *facts*?
2. What actually is the *problem*?
3. What are my *ideas*?
4. How do I get a *solution*?
5. *Check* and *review* the solution.

A PHYSICAL MODEL CAN LEAD TO A SOLUTION. Not only young children benefit from seeing and thinking in terms of a physical model or representation of a problem.

31. State what the problem is. Can you restate it in slightly different words?

Example 3 There are 6 more children working in the math corner of the room than are seated in the reading area. The math corner seems crowded, so the teacher suggests that 5 children move from the math area to the reading area. A student noticed that after the move there were twice as many in the reading area as remained in the math corner. How many children were in the math corner before the teacher suggested the move?

What are the *relevant facts*?

At the start:
 a) There are an unknown number of children in the math corner (M).
 b) There are 6 more in the reading area (R).
Later:
 a) The teacher had 5 children move from the math to reading area.
 b) There are twice as many in the reading area as in the math area.

Try Exercise 31

32. If six white chips represent 6 more children, how many children are represented by 1 white chip?

What are some *ideas* for getting a solution?
 a) Translate the conditions to mathematical language.
 b) Write an equation and solve it.
 c) Interpret the conditions in a physical model.
 d) Use systematic trial and error.

To introduce a new technique we use poker chips of two different colors to interpret the conditions of the problem in a physical model. (Any small objects will do.)

At the start:

? children are at R:
a blue chip represents the
number of children at R

There are six more at M:
the number of children
is represented by

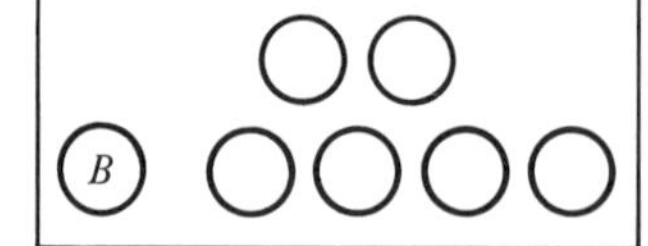

Try Exercise 32

The children are rearranged by the teacher: 5 go from the math area to the reading area.

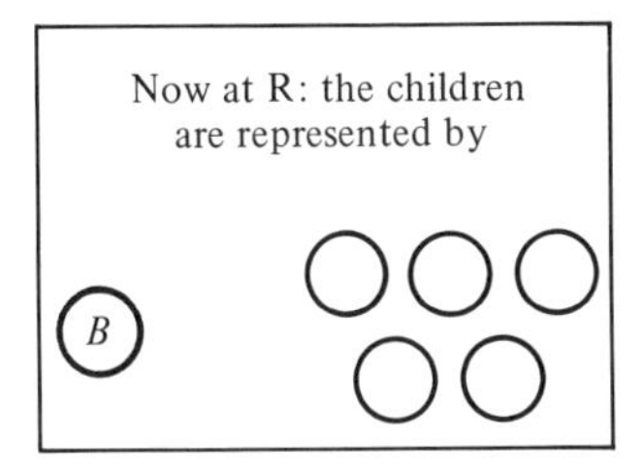

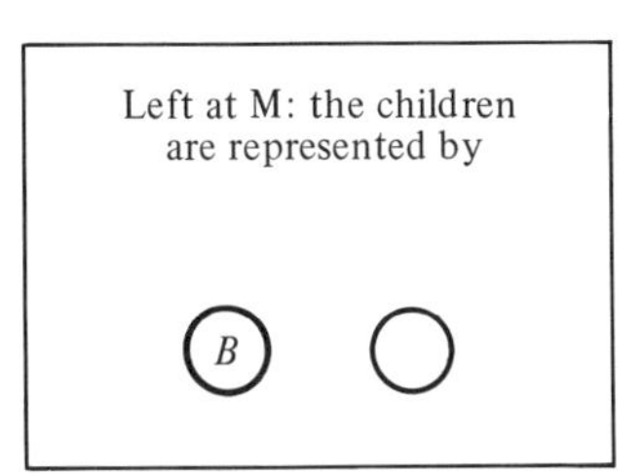

There are twice as many at R, as there are at M. If we knew how many children were represented by a blue chip, the solution would be immediate, so we try the idea (d) of systematic trial and error. If a blue chip represents 1 child, then the second model shows that there are 6 children in the reading area R. If a blue chip represents 2 children, then there are 7 children in the reading area, and so on.

Try Exercise 33

This information can be summarized in the following chart:

Value represented by the blue chip B	1	2	3	4
Number at the reading area R	6	7		
Number at the math area M	2	3		

When the number for area R is twice the number for area M, then the corresponding value for B is the solution.

Try Exercise 34

There are other ways to do this problem, but this method is usable by someone who must understand only counting and the idea that a chip can stand for something. The blue chip, as the unknown or mystery chip, becomes the variable x in the translation to an equation.

33.
a) If the blue chip represents 1, how many children are left in the math area?
b) Same as (a) for the blue chip representing 2 children.

34. Find the solution to the problem by completing the chart as far as necessary. Use the information on the chart to find what you want to know.

EXERCISE SET 3.4

Use the five-step solving process for all of these exercises. The first four exercises are very similar to the problem solved in the text. Use an actual physical model or make a sketch of one for solving these in particular.

1. Sandra ate three pieces of candy on Monday, five the next day, and two more on each succeeding day. How many did she eat on Saturday? How many altogether from Monday to Saturday?

2. A 43-year-old man has a son who is 16. How long ago was the father exactly four times as old as his son?

3. A woman is 26 years older than her daughter. In five years, she will be twice as old as her daughter. How old is her daughter now?

Separate 79 into two parts, so that the second will be greater than the first by 19.

5. Here is a "mind reading" trick. You have four cards, as shown. You ask a person to choose a number from 1 to 15 and tell you on which cards it appears. You quickly add the first numbers of those cards, and the sum is the person's number. Why does this work? *Hint*: Write binary notation for the first 15 natural numbers.

Card 1	Card 2	Card 3	Card 4
1 3 5	2 3 6	4 5 6	8 9 10
7 9 11	7 10 11	7 12 13	11 12 13
13 15	14 15	14 15	14 15

6. How can this trick be modified so that the person could choose a larger number?

7. To number the pages of a book, a printer used 1890 digits. From this information determine how many pages the book had.

8. Is there a base b for which 33_b will be even? Explain.

9. Is there a base b for which 46_b will be odd? Explain.

10. Find a base b for which 40_b will be square.

11. Find a base b for which 21_b is a multiple of 5.

12. There is a quick way to convert binary notation to decimal notation. Study the pattern in the two examples and find out how to use this quick method.

a) 1 0 0 1 0 1 1_{two} $= 75_{ten}$
   1 2 4 9 18 37 75

b) 1 0 1 0 0 0 1 1_{two} $= 163_{ten}$
   1 2 5 10 20 40 81 163

13. Use the quick method of Exercise 12 to find decimal notation for the following:

a) 10101_{two} b) 110110_{two} c) 111111_{two}

■**14.** Show that this quick method is valid (will give the proper result every time).

■**15.** Will your proof in Exercise 14 hold for other bases, not just for base two? These examples may help you decide:

a) 1 2 0 4 3_{five} $= 898_{ten}$
   1 7 35 179 898

b) 3 2 0 1_{six} $= 721_{ten}$
   3 20 120 721

■**16.** At a roadside stand, produce need be weighed only to the nearest ounce. The farmer running the stand has a balance and only four weights. With these, any whole number of ounces from 1 to 40 inclusive can be weighed. What weights does the farmer have? *Hint*: Think of other bases.

REVIEW TEST

1. Describe and illustrate each of the following:
a) additive numeration system,
b) positional system,
c) subtractive principle in a numeration system,
d) grouping,
e) named place-value system of numeration.

2. State some disadvantages of an additive numeration system.

3. Explain why the Roman numeration system was not a hindrance to their computation.

4. Why did the Romans not find it necessary to invent a symbol for zero?

5. Explain in terms of place value the meaning of 1 327 568.

6. Write the words to name 1 327 568.

7. Make a place-value chart for base seven. Show five places.

8. Use a place-value chart to convert to the indicated base:
a) 113 (base seven) b) 113 (base five)

9. Convert to the indicated base; use division:
a) 73 (base five)
b) 897 (base eight)

10. Convert to decimal notation:
a) 1100101_{two} b) $6t3e2_{\text{twelve}}$

11.
a) Write binary notation for the numbers from twelve to twenty-four inclusive.
b) Repeat (a) for base twelve.

12. Find a base b for which 14_b is a multiple of three.

13. Two sisters are three years apart in age. Eighteen years ago the older sister was twice the age of the younger. How old are the sisters now?

FOR FURTHER READING

1. For an interesting suggestion about making an improvement in current mathematical symbolism, read "Some thoughts on the use of computer symbols in mathematics" by Margaret Brown in the *Mathematical Gazette*, Vol. LVIII, No. 404, June 1974.

2. A discussion of the difficulties children encounter in relating their manipulation with physical objects to the symbols of arithmetic is found in "Symbolism and the world of objects" by Stanley M. Jencks and Donald M. Peck in *The Arithmetic Teacher*, 22:5, May 1975.

3. Some interesting facts about numeration systems that use bases of three, five, nineteen, and all the way up to forty-seven are in the article "The original counting systems of Papua and Guinea" by Edward P. Wolfers in *The Arithmetic Teacher*, 18:2, February 1971.

TEXTS

Bell, Max S., Karen C. Fuson, and Richard A. Lesh, *Algebraic and Arithmetic Structures.* The Free Press, New York, New York (1976).

May, Lola J., *Teaching Mathematics in the Elementary School.* The Free Press, New York, New York (1970). (A methods text.)

chapter 4 whole numbers

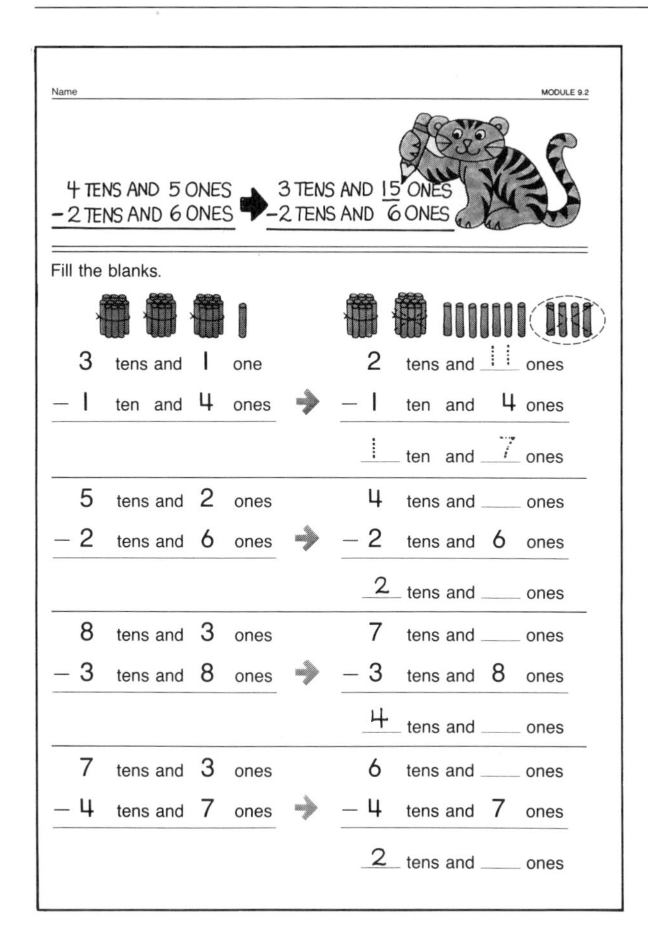
Name MODULE 9.2

4 TENS AND 5 ONES → 3 TENS AND 15 ONES
−2 TENS AND 6 ONES → −2 TENS AND 6 ONES

Fill the blanks.

3 tens and 1 one → 2 tens and 11 ones
− 1 ten and 4 ones → − 1 ten and 4 ones
1 ten and 7 ones

5 tens and 2 ones → 4 tens and ___ ones
− 2 tens and 6 ones → − 2 tens and 6 ones
2 tens and ___ ones

8 tens and 3 ones → 7 tens and ___ ones
− 3 tens and 8 ones → − 3 tens and 8 ones
4 tens and ___ ones

7 tens and 3 ones → 6 tens and ___ ones
− 4 tens and 7 ones → − 4 tens and 7 ones
2 tens and ___ ones

In this lesson, second graders are being introduced to the complexities of regrouping in the subtraction algorithm for whole numbers. The emphasis here is on understanding the process, not the final algorithm.

4.1 AIMS

You should be able to:

a) Find the cardinality of a finite set.
b) Describe in a precise way the whole numbers.

* *Cardinal* is derived originally from the Latin cardin-, cardo, hinge—something on which other things hinge.

4.1 CARDINALITY AND CARDINAL NUMBERS*

The preceding chapters form the basis for a study of the abstract properties of whole numbers and the operations defined for them, a system that must be distinguished from the numeration scheme for recording whole numbers discussed in the last chapter. Although the sophistications of sets, relations, and functions did not exist when the whole numbers were created out of necessity for the utilitarian purposes of daily life, these more recent concepts add not only to one's understanding of the whole-number system but also to the understanding of the subsequent systems derived from it.

Just as whole numbers were those first developed and used by early man, so do children find the answer to "How many do I have?" useful early in their lives. We have seen that a simple tally stick was often used to record man's first ideas of number and counting; such sticks provided a one-to-one correspondence between the notches and the articles being counted. Similarly, we use the concept of one-to-one correspondence to establish the meaning of whole numbers.

CARDINALITY. Intuitively, if two sets are equivalent, we say that there are as many elements in one set as in the other. Every set that is equivalent to the set $\{1, 2, 3, 4\}$ has at least one property in common with all other such sets, since all of them have the same number of elements—four. The equivalence relation is the basis for what we call counting.

When children are able to count correctly they often recite the names for the counting (natural) numbers in standard order: "one, two, three, four" and show the actual correspondence between the recited names and the four objects being counted.

In order to be as precise as possible in considering the whole numbers, we first define a sequence of standard sets as the following ordered finite sets:

$$
\begin{array}{ll}
W_0 = \{\ \} & W_4 = \{1, 2, 3, 4\} \\
W_1 = \{1\} & \vdots \\
W_2 = \{1, 2\} & W_m = \{1, 2, 3, 4, 5, \ldots, m\} \\
W_3 = \{1, 2, 3\} & \vdots
\end{array}
$$

Each set is an ordered subset of the counting numbers and is a proper subset of those that follow. No two of these sets are equivalent, and there is a unique number that corresponds to each set.

The *cardinal number* of a set A of this sequence is m if and only if there is a one-to-one correspondence between the members of A and the set $\{1, 2, 3, 4, \ldots, m\}$. The cardinal number for A is denoted by $n(A)$, which is read "n of A" or "the number of elements in A." The empty set has the cardinal number 0, the set $\{1\}$ and any set equivalent to it has cardinal number 1, the set $\{1, 2\}$ and any set equivalent to it has cardinal number 2, and so on. The set W of *whole numbers* $\{0, 1, 2, 3, 4, \ldots\}$ is the set of all the cardinal numbers of this sequence of standard finite sets. The cardinality function assigns a cardinal number to each set that is equivalent to one of the standard sets. For such sets the range of the cardinality function is the set of whole numbers.

Example 1 Find the cardinal number of each of the following:

$$A = \{a, b, d\}, \qquad B = \{k, m, n\}, \qquad C = \{p, q, r, s\}.$$

Solution

$n(A) = n(B) = 3$, since both sets A and B can be shown to be equivalent to the set $\{1, 2, 3\}$;
$n(C) = 4$ since $C \sim \{1, 2, 3, 4\}$.

Try Exercise 1

Example 2 Prove that if two finite sets are equivalent, then they have the same cardinality. That is, for any finite sets A and B, if $A \sim B$, then $n(A) = n(B)$.

Proof The set B has some cardinality; suppose that $n(B)$ is m and thus $B \sim \{1, 2, 3, 4, \ldots, m\}$. But since $A \sim B$, by the transitive property of the equivalence relation for sets we have $A \sim \{1, 2, 3, 4, \ldots, m\}$ and $n(A) = m$.

Try Exercise 2

Cardinal numbers

1. Use a one-to-one correspondence with a standard set to find the cardinality of the following sets:
a) $\{2, 4, 6, 8\} = A$
b) $\{1, 3, 5, 7, 9, 11, 13\} = B$
c) $\{m, t, q, p, r, w\} = C$

2. If A and B are any two sets such that $A = B$, what can be said about $n(A)$ and $n(B)$?

ORDER. A second aspect of whole numbers is not concerned with equivalence of sets but focuses upon the order instead.

In the usual ordering, 0 is first, 1 is next and is second, 2 is next and is third, and so on. This ordering is often pictured on a number line that has some point representing 0 and equally spaced points to the right of 0 representing the remaining whole numbers in succession or order.

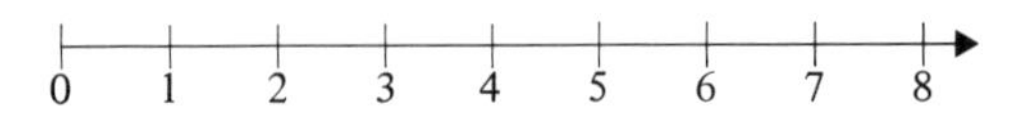

Any whole number to the right of an arbitrary number n of this line is greater than the number n, and any number to the left of n is less than n.

Some psychologists believe that children develop the cardinal concept of number (equivalence of sets) before they develop the ordinal concept (the order of the number line). Recent evidence shows that it is more than likely that these two aspects develop together or that possibly the ordinal may even develop earlier.

A lesson on the order of whole numbers from 1 to 9 in first grade

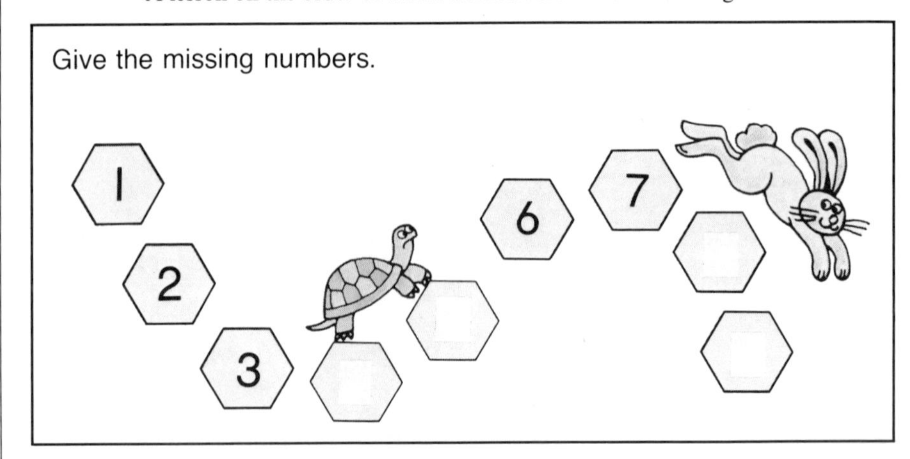

EXERCISE SET 4.1

1. Let $A = \{a, b, c\}$, $B = \{5, 6\}$, and $C = \{b, c, d, e\}$. Find the following:

a) $n(A)$
b) $n(A \cup C)$
c) $n(A \cup B)$
d) $n(A \times (B \cup C))$
e) $n((A \times B) \cup (A \times C))$
f) $n(A \cap B)$

2. Some of the following statements are always true for any sets A and B. State *true* for those that are true and give a counterexample for those that are not:

a) If $n(A) = n(B)$, then $A = B$.
b) If $A = B$, then $n(A) = n(B)$.
c) $n(A \cup B) = n(A) + n(B)$.
d) If $n(A) = n(B)$, then $A \sim B$.

3. Is the relation "has the same cardinality as" an equivalence relation for sets?

4. Prove that if two finite sets have the same cardinality, then the sets are equivalent. That is, for sets A and B, if $n(A) = n(B)$, then $A \sim B$.

■ **5.** Assume that every member of your mathematics class has at least one friend in the class. Prove that in this case there must be at least two people with the same number of friends in the class.

■ **6.** The concept of cardinality can be extended to infinite sets: two infinite sets have the same cardinality if and only if they have a one-to-one correspondence. Any set equivalent to the set of counting numbers $\{1, 2, 3, 4, \ldots\}$ has cardinality $\aleph_0$ (read "aleph null"), a *transfinite cardinal number.*

a) What is the cardinality of the set of whole numbers? the even whole numbers? the odd whole numbers?

b) Explain why the following statements of transfinite arithmetic are plausible:

$$\aleph_0 + \aleph_0 = \aleph_0, \qquad 2\aleph_0 = \aleph_0, \qquad 10 + \aleph_0 = \aleph_0.$$

4.2 OPERATIONS FOR WHOLE NUMBERS

4.2 AIMS

You should be able to:

a) State and apply the definitions for addition and multiplication of whole numbers.

b) Apply, recognize, and prove the commutative, associative, and distributive laws for addition and multiplication of whole numbers.

c) State, apply, and prove the properties of the identity for addition and the identity for multiplication.

Addition

The whole numbers have been defined in terms of sets, and it is not surprising that the basic operations for whole numbers can be derived from appropriate set operations. A child who has five cents and is given ten additional cents can find the total of fifteen by joining the two sets and simply counting (finding the cardinality). This model of addition is used in the formal definition for the sum of two whole numbers.

WHOLE-NUMBER ADDITION. If A and B are two disjoint finite sets with $n(A) = a$ and $n(B) = b$, then the *sum* of a and b is defined to be the unique number $n(A \cup B)$. More briefly:

$$\text{If } A \cap B = \varnothing, \text{ then } a + b = n(A) + n(B) = n(A \cup B).$$

The a and b of the definition are given the name of *addend*, or *term*.

The operation of addition can be thought of as a function defined on a domain consisting of ordered pairs of whole numbers with the set of whole numbers as the range.

The sets used in the definition must be disjoint. If $A = \{d, e, f\}$ and $B = \{e, f, g, h, j\}$, then $n(A) = 3$ and $n(B) = 5$, but $n(A \cup B) = 6$, which is not the result we would get by first counting a set of three objects united with a completely different set of five objects.

3. Use the definition of addition to show that $5 + 2 = 7$.

Try Exercise 3

From a second-grade textbook

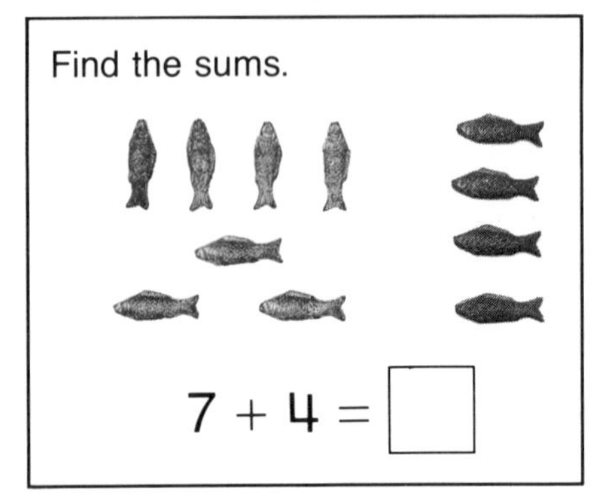

It does not matter which disjoint sets are used in the definition; as long as the respective cardinalities remain the same, the sum will remain the same.

COMMUTATIVE PROPERTY OF ADDITION. The commutative property for addition of whole numbers follows from the commutative property for set union:

> **For any pair of whole numbers a and b, $a + b = b + a$.**

It is easy to show that this is true. If A and B are disjoint sets such that $n(A) = a$ and $n(B) = b$, then by the definition of addition, the following statements are true:

$$n(A \cup B) = n(A) + n(B) = a + b,$$
$$n(B \cup A) = n(B) + n(A) = b + a.$$

But $n(A \cup B) = n(B \cup A)$, since $A \cup B$ and $B \cup A$ are exactly the same set. Therefore, by the transitive property of the equals relations for numbers, $a + b = b + a$.

This means that the order in which two numbers are added does not affect the answer.

ASSOCIATIVE PROPERTY FOR ADDITION. Similarly, the associative property for set union leads to a corresponding property for the addition operation for whole numbers:

> **For any whole numbers a, b, and c, $a + (b + c) = (a + b) + c$.**

This solves the problem of how to add three or more whole numbers because the definition provides for only two numbers at a time. The associative law ensures that:

$$2 + (5 + 3) = 2 + 8 = 10,$$
$$(2 + 5) + 3 = 7 + 3 = 10.$$

Example 1 Show that the associative law holds for the addition of whole numbers.

To do this we let A, B, and C be three mutually disjoint sets (no pair of sets has a common element) with $n(A) = a$, $n(B) = b$, and $n(C) = c$. By using the definition of whole-number addition twice we have:

$$\begin{aligned} n(A \cup (B \cup C)) &= n(A) + n(B \cup C) && [A \text{ is disjoint from } B \cup C] \\ &= a + (n(B) + n(C)) \\ &= a + (b + c) && [\text{by the assumptions above}]. \end{aligned}$$

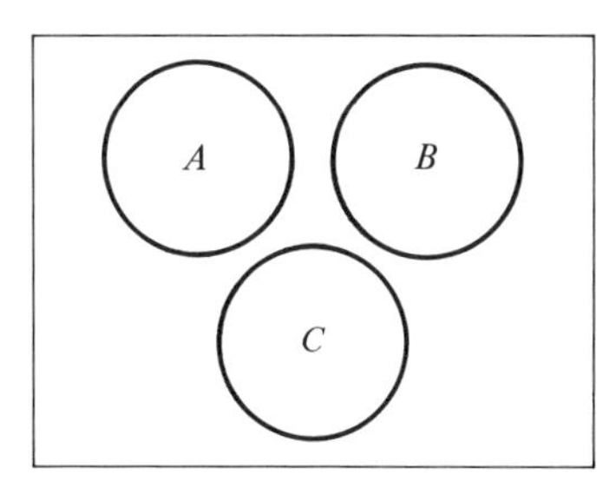

Similarly, using the definition of addition we have:

$$\begin{aligned} n((A \cup B) \cup C) &= (n(A \cup B)) + n(C) \qquad [A \cup B \text{ is disjoint from } C] \\ &= (n(A) + n(B)) + c \\ &= (a + b) + c. \end{aligned}$$

By the associative property for the union of sets we know that

$$A \cup (B \cup C) = (A \cup B) \cup C.$$

The cardinality of equal sets is the same:

$$n(A \cup (B \cup C)) = n((A \cup B) \cup C).$$

By the transitive and substitution properties of equality, we have:

$$a + (b + c) = (a + b) + c.$$

Try Exercise 4

4.

a) If $3 + (4 + 7) = 3 + 11 = 14$, then what law assures you that $(3 + 4) + 7 = 14$?

b) If $x + 2 = 5$, why is it true that $2 + x = 5$?

Multiplication

Recall the example of the photographer in Chapter 2 with two types of film and four lenses who was able to use eight different combinations for photographing a scene. A Cartesian product similar to that of this illustration is often used in the elementary school to develop the concept of whole-number multiplication. The next example illustrates a particular case.

Example 2 Let $A = \{a, b\}$ and $B = \{e, f, g, h\}$. Find the cardinality of the set $A \times B$.

Solution

The Cartesian product $A \times B$ is $\left\{\begin{matrix} (a, h) & (b, h) \\ (a, g) & (b, g) \\ (a, f) & (b, f) \\ (a, e) & (b, e) \end{matrix}\right\}$

Thus, $n(A) = 2$, $n(B) = 4$, $n(A \times B) = 8$. This Cartesian product models $2 \cdot 4 = 8$.

Try Exercise 5

5. Let $A = \{a, b, c\}$ and $B = \{1, 2, 3, 4\}$.

a) Find $A \times B$.

b) What is $n(A \times B)$?

c) What whole-number product is determined?

From a third-grade textbook

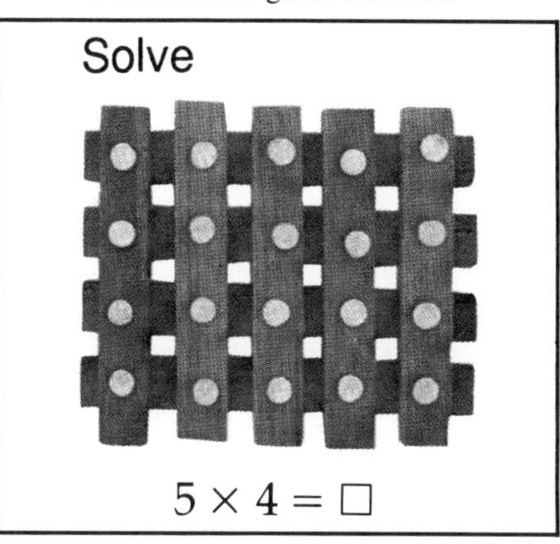

6. Use the definition to prove that the product $2 \cdot 3 = 6$.
From a third-grade textbook

This model leads naturally to the following definition:

WHOLE-NUMBER MULTIPLICATION. If A and B are any two sets with $n(A) = a$ and $n(B) = b$, then the *product* of the whole numbers a and b is the unique whole number $n(A \times B)$. More briefly:

$$n(A) \cdot n(B) = a \cdot b = n(A \times B)^*$$

The product can be written as $a \cdot b$, or $(a)(b)$, or (if letters are being used) as just ab. The multiplication operation is a function that assigns the whole number $n(A \times B)$ to the whole-number pair (a, b). The two whole numbers a and b are called *factors* and the product is called a *multiple* of either of the factors.

Try Exercise 6

Just as for addition, it does not matter which sets are used to determine a product. If a set A' has $n(A') = a$ and a set B' has $n(B') = b$, then $A \times B$ is equivalent to $A' \times B'$ and $n(A \times B)$ is the same as $n(A' \times B')$, since equivalent sets have the same cardinality.

COMMUTATIVE PROPERTY FOR MULTIPLICATION. The commutative property for multiplication of whole numbers is easily proved, since it depends upon a corresponding set property; the set $A \times B$ is equivalent to $B \times A$.

7. Let $A = \{1, 2\}$ and $B = \{a, b, c\}$.
a) Find roster notation for $A \times B$ and $B \times A$.
b) Prove that set $A \times B$ is equivalent to set $B \times A$.

Example 3 Show that the commutative property holds for multiplication of whole numbers. That is, for any whole numbers a and b,

$$a \cdot b = b \cdot a.$$

Solution We start with any two sets A and B such that $n(A) = a$ and $n(B) = b$. The sets $A \times B$ and $B \times A$ are equivalent sets (Problem 16 of Exercise Set 2.4). The definition of multiplication gives:

$$n(A \times B) = n(A) \cdot n(B) = a \cdot b,$$
$$n(B \times A) = n(B) \cdot n(A) = b \cdot a.$$

But because of the fact that $A \times B \sim B \times A$, we have

$$n(A \times B) = n(B \times A)$$

since equivalent sets have the same cardinality. By the transitive property of equals, we have

$$a \cdot b = b \cdot a.$$

Try Exercise 7

ASSOCIATIVE PROPERTY FOR MULTIPLICATION. Since multiplication has been defined for two whole numbers, $a \cdot b \cdot c$ could be interpreted to mean either $a \cdot (b \cdot c)$ or $(a \cdot b) \cdot c$. Because the operation is associative, it does not

* In the definition, the dot $(\cdot)$ is used for multiplication of whole numbers and the $(\times)$ is reserved for the Cartesian product of the two sets.

matter which interpretation is used; the answer is the same. This property follows from the fact that the set $A \times (B \times C)$ is equivalent to the set $(A \times B) \times C$. An example illustrates this.

Example 4 If $A = \{1, 2\}$, $B = \{3, 4\}$, and $C = \{c, d, e\}$, show that $A \times (B \times C)$ is equivalent to $(A \times B) \times C$.

Solution We find $A \times (B \times C)$:

$$A \times \begin{Bmatrix} (3, e) & (4, e) \\ (3, d) & (4, d) \\ (3, c) & (4, c) \end{Bmatrix} = A \times (B \times C)$$

$$= \begin{Bmatrix} (1, (3, e)) & (1, (3, d)) & (1, (3, c)) & (1, (4, e)) & (1, (4, d)) & (1, (4, c)) \\ (2, (3, e)) & (2, (3, d)) & (2, (3, c)) & (2, (4, e)) & (2, (4, d)) & (2, (4, c)) \end{Bmatrix}.$$

Hence,

$$n(A) = 2,$$
$$n(B \times C) = n(B) \cdot n(C) = 2 \cdot 3 = 6,$$
$$n(A \times (B \times C)) = n(A) \cdot n(B \times C) = 2 \cdot 6 = 12.$$

Try Exercise 8

In this case the general proof is not much more difficult than working out a particular case. It is easy to show that a member of the set $A \times (B \times C)$ such as $(a, (b, c))$ has a corresponding element $((a, b), c)$ in the set $(A \times B) \times C$. Details are left for the exercises.

Try Exercises 9 and 10

Repeated use of the associative and the commutative properties or laws can help in performing the basic operations for three or more whole numbers. They are especially useful in doing calculations mentally.

Example 5 Add $7 + 5 + 3 + 25$.

Solution This can be done from left to right:

$$(7 + 5) + 3 + 25 = (12 + 3) + 25 = 15 + 25 = 40.$$

It is easier when the terms are grouped as follows:

$$7 + (5 + 3) + 25 = 7 + (3 + 5) + 25 = (7 + 3) + (5 + 25)$$
$$= 10 + 30 = 40.$$

Example 6 Multiply $5 \cdot 7 \cdot 6 \cdot 2$.

Solution This can be done from left to right:

$$(5 \cdot 7) \cdot 6 \cdot 2 = (35 \cdot 6) \cdot 2 = 210 \cdot 2 = 420.$$

8. Use roster notation to describe $(A \times B) \times C$. What is $n(A \times B)$? $n(C)$? $n((A \times B) \times C)$?

9. State the law(s) illustrated in each of the following:
a) $3 + 7 = 7 + 3$
b) $(3 \cdot 5) \cdot 6 = 6 \cdot (3 \cdot 5)$
c) $5 \cdot (3 + 2) = (2 + 3) \cdot 5$

10. State the law(s) illustrated in each of the following:
a) $(x + y) + (z + 2) = (z + 2) + (x + y)$
b) $(3 \cdot x) \cdot 6 = (3 \cdot 6) \cdot x$
c) $(x + y) + 5 = x + (5 + y)$

It is slightly easier when the factors are regrouped as follows:

$$5 \cdot 7 \cdot 6 \cdot 2 = (5 \cdot 2) \cdot (7 \cdot 6) = 10 \cdot (7 \cdot 6) = 10 \cdot 42 = 420.$$

Try Exercises 11 and 12

11. Show that applications of the commutative and associative laws can make the following addition easier:

$$6 + 27 + 4 + 15 + 3 + 5.$$

12 Repeat Problem 11 for the following product:

$$25 \cdot 6 \cdot 4 \cdot 2 \cdot 7 \cdot 5.$$

THE DISTRIBUTIVE PROPERTY. The distributive law combines the addition and multiplication operations, as in the following diagram that illustrates the statement:

$$2 \cdot (3 + 5) = (2 \cdot 3) + (2 \cdot 5).$$

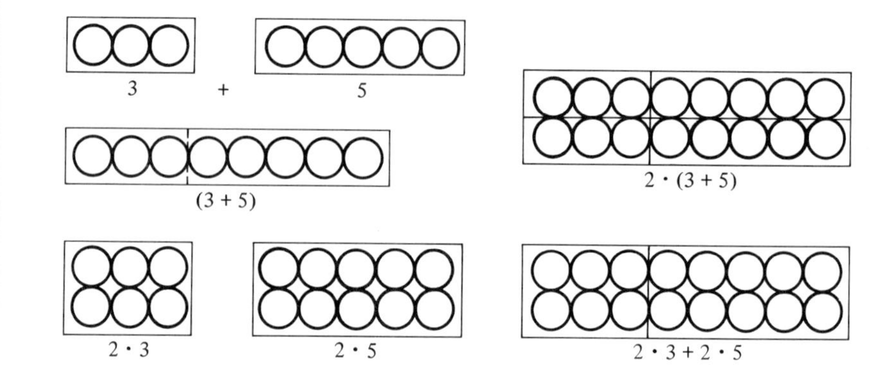

Try Exercise 13

13. Draw a diagram to illustrate the following example of the distributive law:

$$3 \cdot (2 + 4) = 3 \cdot 2 + 3 \cdot 4.$$

Because of the distributive law, such a diagram can be made no matter what whole numbers are involved:

THE DISTRIBUTIVE LAW. For any whole numbers a, b, and c,

$$a \cdot (b + c) = (a \cdot b) + (a \cdot c).$$

The proof follows from corresponding properties of sets.

Consideration of the general case is actually easier than working through some specific example. We use any sets A, B, and C such that B and C are disjoint and $n(A) = a$, $n(B) = b$, and $n(C) = c$. We already know* that

$$A \times (B \cup C) = (A \times B) \cup (A \times C).$$

Since equal sets have the same cardinality, we get

$$n(A \times (B \cup C)) = n((A \times B) \cup (A \times C)).$$

From the definitions for multiplication and then for addition we obtain:

$$n(A \times (B \cup C)) = n(A) \cdot n(B \cup C) = a \cdot (b + c).$$

* See Problem 13 of Review Test in Chapter 2.

By the definition of addition and the fact that $A \times B$ and $A \times C$ are disjoint we have:

$$n((A \times B) \cup (A \times C)) = n(A \times B) + n(A \times C) = a \cdot b + a \cdot c.$$

The fact that equality is transitive means that

$$a \cdot (b + c) = (a \cdot b + a \cdot c).$$

When the distributive law is written in reverse order, we get

$$(a \cdot b + a \cdot c) = a \cdot (b + c);$$

the number a is said to be *factored* out of the sum $a \cdot b + a \cdot c$.

Try Exercise 14

14. Use factoring to rewrite the following sums as products:
a) $2 \cdot 9 + 2 \cdot 7$
b) $3 \cdot 4 + 3 \cdot 11$

The important properties for the two basic operations for any whole numbers a, b, and c are summarized:

The commutative laws:

$$a + b = b + a; \qquad a \cdot b = b \cdot a$$

The associative laws:

$$a + (b + c) = (a + b) + c; \qquad a \cdot (b \cdot c) = (a \cdot b) \cdot c$$

The distributive law:

$$a \cdot (b + c) = (a \cdot b) + (a \cdot c).$$

15. Since $a + 0 = a$, what property insures that $0 + a = a$ also?

IDENTITY 0 FOR ADDITION. The number zero is the identity for addition because the operation of addition, using it as an addend, results in no change in the sum:

$$0 + a = a + 0 = a.$$

This property follows from the definition of addition and the properties of the empty set. Let A be any set with $n(A) = a$; then $A \cup \varnothing = A$. Since equal sets have the same cardinal number, it is also true that

$$n(A \cup \varnothing) = n(A);$$

but $n(A)$ is given as a, so $n(A \cup \varnothing) = a$.

On the other hand, since $A \cap \varnothing$ is also $\varnothing$, then, from the definition of addition, we get

$$n(A \cup \varnothing) = n(A) + n(\varnothing) = a + 0.$$

Thus we have $n(A \cup \varnothing) = a$ and $n(A \cup \varnothing) = a + 0$. By the transitive property of equality, $a = a + 0$.

Try Exercise 15

16. Use an appropriate set A and the definition of multiplication to find the product $3 \cdot 0$.

17. The definition for multiplication has been used in the example to prove $1 \cdot 3 = 3$. Why is it also true that $3 \cdot 1 = 3$?

IDENTITY PROPERTIES

For all whole numbers a:

$$a + 0 = 0 + a = a,$$
$$a \cdot 0 = 0 \cdot a = 0,$$
$$a \cdot 1 = 1 \cdot a = a.$$

The number zero in multiplication also has an important property, which is illustrated in the next example.

Example 7 Use the set $A = \{a, b\}$ and the empty set to find the product $2 \cdot 0$.

Solution By the definition of multiplication:

$$n(A \times \varnothing) = n(A) \cdot n(\varnothing) = 2 \cdot 0.$$

But observe that $A \times \varnothing = \varnothing$, so that $n(A \times \varnothing) = n(\varnothing) = 0$. We conclude that $2 \cdot 0 = 0$.

Any set A could have been used in this example, since for any set A, $A \times \varnothing = \varnothing$. The arguments and reasoning are exactly the same, except that $n(A)$ is now a instead of 2, as it was in the particular case used above.

Try Exercise 16

No other number behaves the way 0 does, because the only set that behaves in the unique way required in these proofs is the empty set and $n(\varnothing)$ is 0.

IDENTITY 1 FOR MULTIPLICATION. If any whole number a is multiplied by the number 1, the product is the original factor a:

$$1 \cdot a = a \cdot 1 = a.$$

This identity property follows from the definition of multiplication and the properties of sets.

Example 8 Use the set $B = \{d\}$ and $A = \{a, b, c\}$ to find the product $1 \cdot 3$.

Solution We use the definition and find the Cartesian product

$$B \times A = \{(d, c), (d, b), (d, a)\}.$$

Then,

$$n(B) \cdot n(A) = 1 \cdot 3,$$

but $n(B) \cdot n(A) = n(B \times A) = 3$. Hence $1 \cdot 3 = 3$.

Try Exercise 17

A general proof for the identity property follows exactly the procedure used in the example above, and we outline the method.

Example 9 Show that for any whole number a, its product with the whole number 1 is a.

Solution We would now let B be any set such that $n(B) = 1$ just as before, and A is any set such that $n(A) = a$. Then, just as in Example 8, we need to show that the set A is equivalent to the set $B \times A$. This is easily done and the conclusion follows from

$$n(B \times A) = n(A) = a \qquad \text{and} \qquad n(B \times A) = n(B) \cdot n(A) = 1 \cdot a.$$

This means that $1 \cdot a = a$.

EXERCISE SET 4.2

In Problems 1 through 7 use the sets $A = \{a, b, c\}$, $B = \{1, 2\}$, $C = \{p, q, r, s, t\}$.

1. Use the definition to show that $3 + 2 = 5$.

2. Use the definition for addition to show that $3 + 5 = 8$.

3. Use A, B, C, and the definition for addition to show that

$$3 + (2 + 5) = (3 + 2) + 5.$$

4. Use the definition to show that $2 \cdot 5 = 10$.

5. Use the definition to show that $5 \cdot 3 = 15$.

6. It is also possible to define multiplication for whole numbers in terms of repeated addition. Show how this might be done for the product $5 \cdot 3$.

7. Repeat Problem 6 for the product $2 \cdot 5$.

8. Describe how small objects or counters could be used to demonstrate the commutative law for the addition of whole numbers. How might a number line be used for the same purpose?

9. Describe how counters could be used to demonstrate the commutative law for multiplication of whole numbers. How might the number line be used for the same purpose?

10. Use the associative and commutative laws to make the following calculations easier than proceeding from left to right:
a) $4 + 8 + 7 + 2 + 3 + 3 + 3 + 3 + 31$
b) $6 \cdot 2 \cdot 8 \cdot 4 \cdot 5 \cdot 25$

11. Prove that the associative property holds for whole-number multiplication.

12. Make a diagram for the following:
a) $4 \cdot (5 + 6) = (4 \cdot 5) + (4 \cdot 6)$
b) $3 \cdot (4 + 1) = (3 \cdot 4) + (3 \cdot 1)$

13. Use the distributive law to
a) factor out 3: $18 + 36$
b) factor out 7: $49 + 14$

14. Show that for all whole numbers a, b, and c:

$$(a + b) \cdot c = (a \cdot c) + (b \cdot c)$$

15. Use the definition of multiplication to show:
a) $0 \cdot 5 = 0$ b) $1 \cdot 5 = 5$

16. You have an opportunity to make a dessert from one of the following ice creams: almond, coffee, vanilla, maple walnut, and strawberry, and one of the following toppings: fudge, caramel. Use a diagram of a Cartesian product to determine the number of different dessert choices available to you. What does one point of the product represent?

■ **17.** Show that for any whole numbers a and b, if $a \cdot b = 0$, then $a = 0$ or $b = 0$.

18. Use the definition of equals and that for whole-number addition to show that for any whole numbers a, b, and c, if $a = b$, then $a + c = b + c$.

19. Show that for any whole numbers a, b, and c, if $a = b$, then $a \cdot c = b \cdot c$.

■ **20.** Why is it *not* possible to prove, using the corresponding set properties, that the following is true?

$$a + (b \cdot c) = (a + b) \cdot (a + c)$$

Are there any values of a, b, and c for which the statement does happen to be true?

4.3 AIMS

You should be able to:

a) Define a binary operation.
b) State and apply the definitions for subtraction and division.
c) Determine for an operation on a set if the following properties hold: closure, commutative, and associative.
d) Explain why division by zero is impossible.

4.3 SUBTRACTION AND DIVISION

What is an Operation?

Addition of whole numbers is an operation such that for any two whole numbers a number called their sum can be found:

$$3 + 2 = 5, \qquad 7 + 4 = 11, \qquad 8 + 7 = 15.$$

This is a relation in which a pair of numbers is associated with a third number. To emphasize this respect of the relation, it helps to write the following:

$$(3, 2) \to 5, \qquad (7, 4) \to 11, \qquad (8, 7) \to 15.$$

Multiplication is also a relation in which an ordered pair of whole numbers is associated with a single number:

$$(5, 2) \to 10, \qquad (9, 4) \to 36, \qquad (7, 6) \to 42.$$

In both addition and multiplication, a pair of numbers is used for the operation and they are both described as binary operations.

DEFINITION. Any relation that associates with each ordered pair of elements from a set a unique element from the same set is called a *binary operation*.

CLOSURE. Any two whole numbers can be added, and the answer is always a unique whole number, so the set of numbers is closed under the addition operation.

DEFINITION. A set is said to be *closed* under an operation if and only if for every ordered pair of elements in the set the result of performing the operation with them *exists* and is a *member of the set*.

Try Exercise 18

18. Is the set $\{1, 2, 3, 4, 5, 6\}$ closed under the usual addition for whole numbers?

The definition for binary operation requires that the set be closed under the operation, since an element is associated with each pair.

Operations for whole numbers other than the usual addition and multiplication can be invented for the purposes of illustration. An operation symbol ★ (star) is used for the following invented operation:

$$\begin{array}{lcl} (3, 2) \to 7 & \text{or} & 3 \star 2 = 7 \\ (2, 3) \to 8 & \text{or} & 2 \star 3 = 8 \\ (4, 3) \to 10 & \text{or} & 4 \star 3 = 10 \\ (5, 8) \to 21 & \text{or} & 5 \star 8 = 21 \\ (5, 10) \to 25 & \text{or} & 5 \star 10 = 25 \\ (6, 2) \to 10 & \text{or} & 6 \star 2 = 10 \end{array}$$

(Read "3 star 2 equals 7," etc.)

Try Exercises 19–21

19. Find the following:
a) $1 \star 2$ b) $3 \star 4$ c) $8 \star 5$

Such invented operations are interesting to investigate, since it is possible to create some that do not have the familiar commutative and associative properties of the whole numbers. The operation given above is not commutative because there is a counterexample in the list of pairs.

20. What pattern did you find in the associated pairs in the text?

Subtraction

Ordinarily addition and multiplication are considered to be the fundamental operations for whole numbers. It is helpful in teaching children about addition to discuss the related subtraction at the same time. If a set of three blocks and one of four different blocks are used to model the sum $3 + 4 = 7$, then the removal of the original set of 4 blocks will model $7 - 4 = 3$. The definition for subtraction can be based on sets, but it is also possible to apply its inverse relation to addition.

DEFINITION. For any whole numbers a and b the *difference* $a - b$ is the number c if and only if $a = c + b$. More briefly:

$$a - b = c \leftrightarrow a = c + b.$$

21. Define algebraically the operation called *star* or ★.

The difference $a - b$ can also be read "a minus b" or "a subtract b."

Example 1 Find the following differences:

a) $9 - 4$ b) $12 - 5$ c) $4 - 9$

Solution

a) $9 - 4 = 5$ since $5 + 4 = 9$.
b) $12 - 5 = 7$ since $12 = 7 + 5$.
c) The difference $4 - 9$ is not defined for whole numbers, since there is no whole number c such that $4 = c + 9$.

We see in Example 1 that the set of whole numbers is not closed under subtraction, nor is it commutative, since $9 - 4 \neq 4 - 9$. This lack of closure eventually motivated the creation of the numbers called integers, a system discussed later in Chapter 8. For this reason also, subtraction is not, in the precise sense of the definition, a binary operation for whole numbers. Traditionally it has always been referred to in the elementary school as one of the four arithmetic operations, and children are not given those cases that do not have whole-number differences when they first study subtraction.

By definition of subtraction, it follows that when 0 is subtracted from any number the result is the original number:

$$a - 0 = a \leftrightarrow a = a + 0.$$

22. Since $3 + 0 = 3$, by the definition of subtraction what is $3 - 0$?

Try Exercise 22

Division

Many of the comments made in connection with subtraction hold for division. The definition depends upon the relation between division and multiplication.

DEFINITION. For any whole numbers a, b, and c ($b \neq 0$), the *quotient* $a \div b$ is the number c if and only if $a = c \cdot b$. More briefly:

$$a \div b = c \leftrightarrow a = c \cdot b.$$

In the definition, a is given the special name of *dividend*, b is called the *divisor*, and c is called the *quotient*.

23. Use the definition for division to find the quotient for the whole-number pair (24, 6). Can a quotient in the set of whole numbers be found for the pair (27, 5)?

Example 2 $18 \div 6 = c$ if and only if $18 = c \cdot 6$. Since 3 makes the last equation true, then

$$18 \div 6 = 3.$$

Example 3 Find the quotient for 15 and 6.
$15 \div 6 = c$ if and only if $15 = c \cdot 6$ is true. There is no whole number c that will make the last equation true, and $15 \div 6$ does not exist in the system of whole numbers.

Strictly speaking, neither subtraction nor division should be referred to as an operation for the whole numbers. When elementary pupils first study division of whole numbers they are given only such pairs (a, b) for which $a \div b$ is a whole number.

Try Exercise 23

DIVISION BY 1. It follows from the definition of division that if a number is divided by 1, the result is that number. For any whole number n, $n \div 1$ is the number which, when multiplied by 1, gives n. This must be the original number n.

DIVISION BY 0 IS NOT DEFINED. If 0 is multiplied by any number, the result is 0, but dividing a number by 0 is not defined. Suppose it were possible to divide 5 by 0. By the definition of division, $5 \div 0$ is the number which, when multiplied by 0, equals 5. But there is no such number.

What about $0 \div 0$? This is the number c which, when multiplied by 0, gives 0. Any number would work this time, so, if this were allowed, the division operation would not result in a unique quotient. Thus division by 0 is excluded.

Try Exercise 24

24. Only one of the following is equal to zero. Which is it?

a) $7 \div 0$
b) $0 \div 9$
c) $0 \div 0$

A glance toward the classroom

Related sentences are developed as early as the second grade in elementary school. For example,

$$8 - 5 = \square, \qquad \square + 5 = 8, \qquad 8 = \square + 5.$$

EXERCISE SET 4.3

1. An operation $\triangle$ is defined for whole numbers as follows:

$$a \triangle b = 3 \cdot a + 2 \cdot b.$$

a) Compare $2 \triangle 3$ with $3 \triangle 2$. Is the operation commutative?
b) Compare $2 \triangle (3 \triangle 4)$ with $(2 \triangle 3) \triangle 4$. Is the operation associative?
c) Explain why it is easy to see that the operation is closed for the set of whole numbers.

■ **2.** Invent a binary operation of your own for the whole numbers. Make sure that there is closure for the set of whole numbers for your operation. See if you can find one that is commutative and not associative or vice versa.

■ **3.** An operation $\square$ is defined for whole numbers as follows:

$$a \square b = a^b.$$

(To find $a \quad b$, the first number a is raised to the bth power.)
a) Show that the set of whole numbers is closed under $\square$.
b) Decide if the operation $\square$ is commutative.
c) Decide if the operation $\square$ is associative.

4. Compare the following pairs of numbers:
a) $6 - (3 - 2)$ with $(6 - 3) - 2$,
b) $9 - (5 - 4)$ with $(9 - 5) - 4$.

5. Is subtraction associative? That is, for all whole numbers a, b, and c for which the differences exist, is it true that

$$a - (b - c) = (a - b) - c? \text{ Why?}$$

6. Make a diagram to illustrate:
a) $3 \cdot (5 - 3) = 3 \cdot 5 - 3 \cdot 3$
b) $4 \cdot (6 - 5) = 4 \cdot 6 - 4 \cdot 5$

7. From the diagrams of Problem 6, make a conjecture about the truth of the following: For any whole numbers a, b, and c with $b - c$ defined, $a \cdot (b - c) = a \cdot b - a \cdot c$. ■ Prove your conjecture.

8. Do the following:
a) Compare $6 - (3 + 2)$ with $(6 - 3) - 2$.
b) Compare $15 - (5 + 3)$ with $(15 - 5) - 3$.
■ c) If the difference $a - (b + c)$ is defined, is it always true that $a - (b + c) = (a - b) - c$?

9. Use the definition of subtraction to find the differences:
a) $19 - 7$ b) $12 - 9$
c) $11 - 11$ d) $21 - 21$

10. Show that the difference $n - n$ for any whole number n is 0.

11. If $a = b$ and the difference $a - c$ is a whole number, is it true that $a - c = b - c$? Justify your answer.

12. Compare the following pairs of differences:
a) $(5 - 3)$ with $(5 + 2) - (3 + 2)$
b) $(8 - 4)$ with $(8 + 10) - (4 + 10)$
c) $(20 - 9)$ with $(20 + 100) - (9 + 100)$

■ **13.** Use the definition of subtraction and the meaning of equals to prove that if $a - b = k$, where k is a whole number, then for any whole number c

$$(a + c) - (b + c) = k.$$

■ **14.** Prove the *cancellation law for addition.* For any whole numbers a, b, and c, if $a + c = b + c$, then $a = b$.

15. Use the definition of division to find the quotient:
a) $14 \div 7$ b) $15 \div 15$ c) $21 \div 3$ d) $13 \div 13$

16. Show that for any whole number $n (n \neq 0)$ it is true that $n \div n = 1$.

17. Compare the following pairs of numbers:
a) $(6 \div 3) + (18 \div 3)$ with $(6 + 18) \div 3$,
b) $(16 \div 4) + (24 \div 4)$ with $(16 + 24) \div 4$,
c) $(36 \div 6) + (12 \div 6)$ with $(36 + 12) \div 6$.

18. On the basis of Problem 17, state a property for whole-number division that you believe holds true.

4.4 AIMS

You should be able to:

a) State and apply the definitions for the "less than" and the "greater than" relations for whole numbers.
b) Determine which of the three properties (reflexive, symmetric, transitive) hold for relations defined on the set of whole numbers.

4.4 ORDER

We have used the words "less than" earlier in this chapter on an intuitive basis in terms of a number line. However, the relation can be defined precisely in terms of a whole-number operation.

DEFINITION. For any whole numbers a and b, a is less than b ($a < b$) if and only if there is a whole number c not equal to zero such that $a + c = b$. More briefly:

$$a < b \leftrightarrow a + c = b,\ c \neq 0.$$

This definition is equivalent to stating that a is less than b if and only if the difference $b - a$ is a nonzero whole number, since, by the definition for subtraction, $a + c = b$ is equivalent to the statement $c = b - a$.

Example 1

a) $11 < 19$ since $11 + 8 = 19$;
b) $19 \nless 11$ since $19 + c = 11$ has no whole-number solution;
c) $10 \nless 10$ since $10 + 0 = 10$ and the whole number c in the definition cannot be 0.

The definition for the relation "greater than" follows immediately:

DEFINITION. **For any whole numbers a and b, b is greater than a ($b > a$) if and only if a is less than b. More briefly:**

$$b > a \leftrightarrow a < b.$$

Example 2

a) $19 > 11$ since $11 < 19$;
b) $11 \ngtr 19$ since $19 \nless 11$;
c) $10 \ngtr 10$ since $10 \nless 10$.

Try Exercise 25

As before, we are interested in establishing the properties of these relations defined on the set of whole numbers. Since we have seen that $10 \nless 10$, the reflexive property does not hold, and since $11 < 19$ but $19 \nless 11$, neither does the symmetric. The order of the whole numbers on the number line indicates that the "less than" relation does have the transitive property.

Order as displayed in playthings

25. Write one of the symbols

$$<, >, \nless, \text{ or } \ngtr$$

to make a true statement out of each of the following:

a) 3 ____ 0
b) 7 ____ 7
c) 15 ____ 17
d) 21 ____ 3

26. State the conclusion that results from the transitive property for $<$ in each of the following:

a) $x < y$ and $y < y + 2$, ____;
b) $2 < y + 2$ and $y + 2 < 8$, ____;
c) $4 < 3 \cdot 4$ and $3 \cdot 4 < 6 \cdot 4$, ____.

Example 3 Show that the "less than" relation for whole numbers has the transitive property:

For any whole numbers a, b, and c, if $a < b$ and $b < c$, then $a < c$.

Solution We start with the definition of "less than":

If $a < b$, then there is a nonzero whole number m such that $a + m = b$. Also, if $b < c$, there is a nonzero whole number n such that $b + n = c$. We replace b in this second statement by $a + m$:

$$b + n = (a + m) + n = c.$$

By the associative property for addition:

$$a + (m + n) = c.$$

Since m and n are both nonzero whole numbers, so is the sum $m + n$, and, by the definition of "less than," we get

$$a < c.$$

The even shorter proof for the transitivity of the "greater than" relation is left for the exercises.

Try Exercise 26

In an exercise set earlier in this chapter, the cancellation law for addition was established (Problem 14 of Exercise Set 4.3):

For any whole numbers a, b, and c, if $a + c = b + c$, then $a = b$.

This law can be used to show that a similar result holds for the "less than" relation. We observe that if $10 + 5 < 25 + 5$, then $10 < 25$ is also true; this holds in general.

Example 4 Show that for any whole numbers a, b, and c, if $a + c < b + c$, then $a < b$.

Solution Since we know that $a + c < b + c$, by the definition of "less than" there is a nonzero whole number n such that

$$b + c = (a + c) + n.$$

By the associative and commutative properties for addition:

$$b + c = a + (c + n) = a + (n + c).$$

Again, by the associative property for addition:

$$b + c = (a + n) + c.$$

Now, from the cancellation law for addition we know:

$$b = a + n.$$

Finally, $a < b$ by the definition of the "less than" relation.

Try Exercise 27

27. Use the result in Example 4 to complete the following statements:
a) $37 < 52$, therefore $30 <$ ___;
b) if $x + 2 < 5$, then $x <$ ___;
c) if $2x + 4 < x + 8$, then $x <$ ___.

If the statements for the "less than" ($<$) relation and the "equals" ($=$) relation are combined with the connective *or*, a new relation "less than or equal to" ($\leqslant$) is created. Its properties follow from the compound statement by which it is described.

EXERCISE SET 4.4

1. Use the definitions for $<$ and $>$ to determine which will complete the following statements to make them true.
a) 0 ___ 13 b) 13 ___ 0 c) $5 \cdot 2$ ___ $6 \cdot 2$
d) $(6 + 2)$ ___ $(8 + 2)$ e) $x + 5$ ___ $x + 10$
f) if $x < y$, then $x + 2$ ___ $y + 2$

2. In each of the following, find the set of whole numbers that make the inequality true when used as a replacement for the variable x:
a) $x + 2 < 5$ b) $0 < x - 7$
c) $x + 3 < 2 \cdot x - 5$ d) $2 \cdot x - 2 < 3 \cdot x - 3$

3. State which properties (reflexive, symmetric, transitive) hold for the following relations on the whole numbers:
a) is greater than ($>$)
b) is not equal to ($\neq$)
c) is greater than or equal to ($\geqslant$)
d) is not less than ($\nless$)

4. Show that the "greater than" relation for whole numbers is a transitive relation.

5.
a) If $5 < 6$, how are $5 + 8$ and $6 + 8$ related?
b) If $17 < 31$, how are $17 + 4$ and $31 + 4$ related?
c) If $0 < 15$, how are $0 + 7$ and $15 + 7$ related?

6. Prove that for any whole numbers a, b, and c, if $a < b$, then $a + c < b + c$.

7. Prove the corresponding theorem for the "greater than" relation (see Problem 6).

8. Show that for any whole numbers a, b, and c, if $a + c > b + c$, then $a > b$.

9. Is it true that for any whole numbers a, b, c, and d, if $a < b$ and $c < d$, then $a + c < b + d$? Why?

■**10.** Armando, Ben, Carla, and Donna have discovered some interesting facts about their respective weights. Armando and Ben together weigh the same as Carla and Donna together. Armando and Carla together are heavier than Ben and Donna together. Carla weighs less than Donna. Place the four friends in order according to weight.

4.5 ALGORITHMS

4.5 AIMS

You should be able to:

Justify the ordinary algorithms for addition, subtraction, multiplication, and division for whole numbers on the basis of the properties of the whole numbers and the place-value notation system.

One of the benefits of a place-value system is that it is necessary to learn only a certain few combinations for each of the operations. If the basic addition and multiplication facts are memorized, then any two whole numbers less than 10 can be added or multiplied. The memorized facts are used with a systematic procedure called an *algorithm** for performing operations with larger numbers. The justification of such an algorithm for computation in arithmetic rests upon the properties of the whole numbers and the characteristics of the numeration system. A "rearranging" principle that involves both the associative law for addition and the commutative law for addition is used repeatedly in the algorithms for addition and is as follows:

For any whole numbers a, b, c, and d, $(a + b) + (c + d) = (a + c) + (b + d)$.

We start with $(a + b) + (c + d)$. By using the laws for addition and by treating $(a + b)$ as a single term we have:

$$\begin{aligned}
(a + b) + (c + d) &= ((a + b) + c) + d && \text{[by the associative law]}\\
&= (a + (b + c)) + d && \text{[by the associative law]}\\
&= (a + (c + b)) + d && \text{[by the commutative law]}\\
&= ((a + c) + b) + d && \text{[by the associative law]}\\
&= (a + c) + (b + d) && \text{[by the associative law]}
\end{aligned}$$

This last step is exactly what we set out to prove.

This "rearranging" principle can be extended to include an arbitrary number of terms and will be used repeatedly in the discussion that follows.

Addition

The ordinary algorithms for addition use the place-value system properties and the "rearranging" principle proved above.

Example 1 Justify the usual procedure for adding 87 and 19 in the algorithm:

$$\begin{array}{r} 87 \\ 19 \\ \hline 106 \end{array}$$

Solution We use expanded notation for the sum $87 + 19$:

$$\left.\begin{aligned} 87 &= 80 + 7 \\ 19 &= 10 + 9 \end{aligned}\right\} \text{by place-value numeration}$$

Hence

$$\begin{aligned}
87 + 19 &= (80 + 7) + (10 + 9) && \text{[the addends have been renamed]}\\
&= (80 + 10) + (7 + 9) && \text{[by the "rearranging" principle]}\\
&= (80 + 10) + (16) && \text{[by the basic fact]}\\
&= (80 + 10) + (10 + 6) && \text{[by the place-value notation]}\\
&= 10 \cdot (8 + 1) + (10 + 6) && \text{[by the distributive law]}\\
&= 10 \cdot (9) + (10 + 6) && \text{[by the basic fact]}
\end{aligned}$$

* An alternative word used specifically for the algorithms of the four arithmetic operations is *algorism*. The more general term of *algorithm* can refer to any procedure which is highly systematic. Procedures for computers are referred to as algorithms, so the term is being broadened. Both terms are derived from the name of the Arabic mathematician Al-Khwarizmi (ca. 825).

$= (10 \cdot (9) + 10) + 6$ [by the associative law]
$= 10 \cdot (9 + 1) + 6$ [by the distributive law]
$= 10 \cdot 10 + 6$ [by the basic fact]
$= 10^2 + 6 = 106$ [by the place-value notation]

Try Exercise 28

28. State the property that justifies each of the following statements:
a) $10 \cdot 9 + (10 + 6) = (10 \cdot 9 + 10) + 6$
b) $10 \cdot (9 + 1) + 6 = (10 \cdot 9 + 10) + 6$

This lengthy and detailed explanation is not the algorithm. Computation would not proceed very rapidly if all of these explanatory steps were included each time.

INTERMEDIATE ADDITION ALGORITHM. Children learn an intermediate algorithm in which a great deal of the explanation can be written along the side if desired.

Example 2 Use the partial-sum method to find 87 + 19.

Algorithm	*Explanation*	*Reason*
87 19 16	1. Add the ones $(7 + 9 = 16)$	1. [a basic fact and the "rearranging" principle]
90	2. Add the tens $(8 \cdot 10 + 1 \cdot 10 = 9 \cdot 10)$	2. [distributive law and a basic fact]
6	3. Add the ones $(6 + 0 = 6)$	3. [same as 1]
100	4. Add the tens $(9 \cdot 10 + 1 \cdot 10 = 10^2)$	4. [same as 2]
$100 + 6 = 106$		5. [by the place-value notation]

Try Exercise 29

29. Use the method of partial sums to find the following:

$$\begin{array}{r} 87 \\ 92 \\ +76 \\ \hline \end{array}$$

ORDINARY ADDITION ALGORITHM. In this way the ordinary algorithm is gradually developed to the final form illustrated in the next example.

Example 3 Use the ordinary right-to-left algorithm to add.

Solution The algorithm:

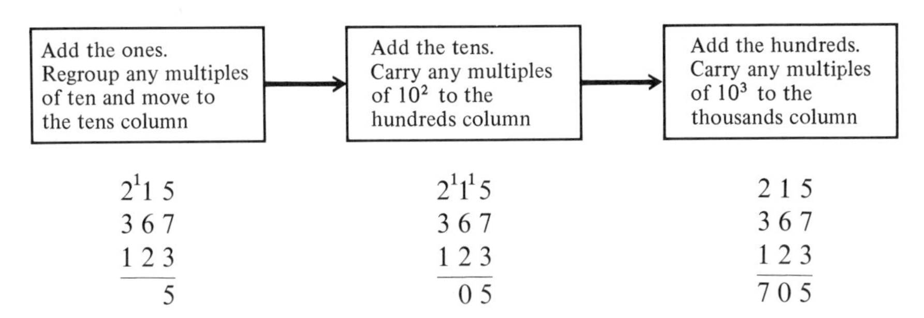

Here is an algorithm for a "dust," or chalkboard, for which erasure is easy. The problem is not recopied, as is done here to show the process, but just erased.

536	539	619	1019
+483	48	4	

↖ ↑ ↗
Erasures

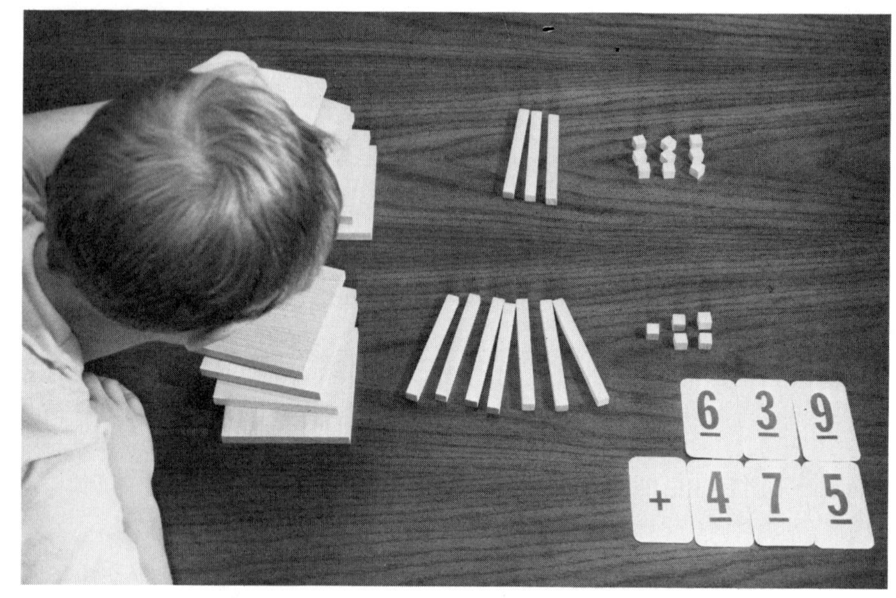

Base ten blocks used for the development of the addition algorithm

Expanded notation can be used to justify this procedure. Here is the same addition with expanded notation:

$$\begin{array}{rcrcr}
1 \cdot 100 & & 1 \cdot 10 & & \\
2 \cdot 100 & + & 1 \cdot 10 & + & 5 \\
3 \cdot 100 & + & 6 \cdot 10 & + & 7 \\
1 \cdot 100 & + & 2 \cdot 10 & + & 3 \\
\hline
7 \cdot 100 & & \boxed{10 \cdot 10} & & 15 = \boxed{10} + 5 \\
7 & & 0 & & 5
\end{array}$$

The nine terms of the original expanded notation can be rearranged because of the principle proved above. The ones are added first:

$$5 + 7 + 3 = 15;$$

the 15 is renamed 10 + 5 (regrouped*) so that a single digit remains in the ones column. The tens are then added:

$$1 \cdot 10 + 1 \cdot 10 + 6 \cdot 10 + 2 \cdot 10 = 10 \cdot 10$$

by repeated application of the distributive law. The $10 \cdot 10 = 10^2$ is regrouped to the hundreds column.

* This regrouping is often referred to as *carrying*. In the example above, a ten is carried (after the ones are added) to the tens column. This terminology is possibly a remnant in the language of the fact that on a counting board the counters were literally carried or transferred to another column.

Multiplication

The usual multiplication algorithm can be justified in a corresponding way. The product $3 \cdot 316$ is first written in expanded notation:

1. $3 \cdot 316 = 3 \cdot (300 + 10 + 6)$.
   The distributive law is applied, with $(300 + 10)$ treated as a single term:
2. $3 \cdot 316 = 3 \cdot (300 + 10) + 3 \cdot 6$.
   The distributive law is applied again together with the associative property for multiplication:

$3 \cdot 316 = 3 \cdot 300 + 3 \cdot 10 + 3 \cdot 6$

3.	$= 9 \cdot 100 + 3 \cdot 10 + 18$	3. ______________
	$= 9 \cdot 100 + 3 \cdot 10 + (10 + 8)$	[by the place-value notation]
	$= 9 \cdot 100 + ((3 \cdot 10 + 1 \cdot 10) + 8)$	[by the associative property for addition]
4.	$= 9 \cdot 100 + (3 + 1) \cdot 10 + 8$	4. ______________
	$= 9 \cdot 100 + 4 \cdot 10 + 8$	[by renaming $(3 + 1)$, a basic fact]
	$= 948$	[by the place-value notation]

The *intermediate algorithm* for this product is:

Algorithm	*Expanded notation*	*Explanation*
316	300 + 10 + 6	
×3	×3	
18	18	[this is 3 × 6]
30	30	[this is 3 × 10]
900	900	[this is 3 × 300]
948	948	[by the addition algorithm]

Try Exercise 30

The usual multiplication algorithm depends upon the rearrangement of terms in a sum and the distributive law, which can be used for any number of terms. Consider the product P given below:

$$P = (a + b + c) \cdot (d + e).$$

The sum $(a + b + c)$ is a whole number; we treat it as one term and use the distributive law:

$P = (a + b + c) \cdot d + (a + b + c) \cdot e$

$= d \cdot (a + b + c) + e \cdot (a + b + c)$ [by the commutative law for multiplication]

$= d \cdot (a + b) + d \cdot c + e \cdot (a + b) + e \cdot c$ [by the distributive law and by treating $(a + b)$ as one term]

$= d \cdot a + d \cdot b + d \cdot c + e \cdot a + e \cdot b + e \cdot c$ [by the distributive law]

The last six products can be rearranged in any order in the sum, as we have already discussed. Any number of terms can appear in either factor of the

30. In the justification of the multiplication of 3 × 316, the reasons in the steps marked 3 and 4 are not supplied. State the laws or properties that make these steps correct.

31. Use expanded notation to explain the multiplication algorithm for
a) 325 × 4
b) 324 × 35

product P above; this merely increases the number of times the distributive law is applied. This repeated application underlies all of the multiplication algorithms.

Example 4 Use $(427) \cdot (36)$ to justify two forms of the multiplication algorithm.

Final form	*Intermediate algorithm*		*Explanation*	
427	427		(400 + 20 + 7)	[expanded notation]
× 36	36		(30 + 6)	
2562	42	sum is 2562	42	[this is 6 × 7]
12810	120		120	[this is 6 × 20]
15372	2400		2400	[this is 6 × 400]
	210	sum is 12810	210	[this is 30 × 7]
	600		600	[this is 30 × 20]
	12000		12000	[this is 30 × 400]
	15372		15372	[by addition]

This last example is a model of the general product considered above. From the statement containing $(a + b + c) \cdot (d + e)$ we know that

$$(400 + 20 + 7) \cdot (30 + 6)$$

can be expressed as

$$30 \cdot 400 + 30 \cdot 20 + 30 \cdot 7 + 6 \cdot 400 + 6 \cdot 20 + 6 \cdot 7.$$

These are precisely the products displayed in the intermediate algorithm and the explanation in expanded notation.

Try Exercise 31

Subtraction

The standard subtraction algorithm depends on the whole-number properties, place-value notation, and the definition of subtraction. The difference $43 - 21$ is 22 if and only if $43 = 22 + 21$, but the definition is not an effective method of actually finding this difference. What is actually done in the algorithm is the following:

$$\begin{aligned} 43 - 21 &= (40 + 3) - (20 + 1) \\ &= (40 - 20) + (3 - 1) \\ &= 20 + 2 = 22. \end{aligned}$$

Since the associative law does not hold for subtraction, this procedure appears to violate this principle. If we think of 43 objects with 40 in one pile and 3 in another, then removing first 20 of them and then another 1 is intuitively correct.

We need the following result to justify the rearranging done in the algorithm:

If a, b, c, and d are such that all the subtractions in the following are possible, then

$$(a + b) - (c + d) = (a - c) + (b - d).$$

To show this, we let the difference $(a + b) - (c + d) = m$,
the difference $a - c = p$,
the difference $b - d = n$,

where m, n, and p are all whole numbers.

We will be justified in using the algorithm if we can show that

$$m = p + n.$$

By the definition of subtraction, we can make three statements:

$$(a + b) - (c + d) = m \leftrightarrow (a + b) = m + (c + d),$$
$$a - c = p \leftrightarrow \quad a = p + c,$$
$$b - d = n \leftrightarrow \quad b = n + d.$$

We use the last two equations to find $a + b = (p + c) + (n + d)$.

$a + b = p + ((c + n) + d)$ 1. [by the associative law for addition]
$= p + (n + (c + d))$ 2. ________
$= p + n + (c + d)$ 3. ________

However, from $a + b = m + (c + d)$ we can write:

$$m + (c + d) = p + n + (c + d).$$

By the cancellation law for addition, $m = p + n$, which is what we set out to show. Thus we are justified in the rearranging done in finding the difference $(43 - 21)$. This type of reasoning can be extended to cover whole numbers whose numerals have more than two digits.

Try Exercise 32

32. Give the justification for the steps labeled 2 and 3 in the text.

Example 5 Justify the subtraction algorithm for:

$$\begin{array}{r} 325 \\ -186 \\ \hline \end{array}$$

Solution

$$325 - 186 = (300 + 20 + 5) - (100 + 80 + 6).$$

If we attempt to use an extension of the statement above we have:

$$325 - 186 = (300 - 100) + (20 - 80) + (5 - 6).$$

The difference $20 - 80$ is not a whole number and neither is $5 - 6$. Some

Regrouping in second grade

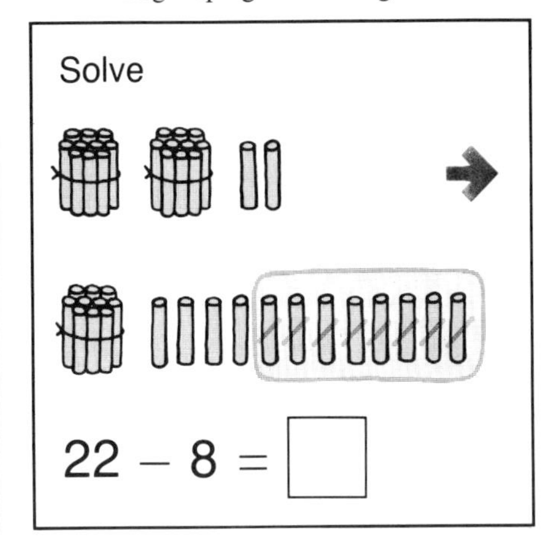

33. Justify the subtraction algorithm for

$$\begin{array}{r} 7246 \\ -6519 \\ \hline \end{array}$$

regrouping* is needed, which we do by renaming and applying the associative and commutative properties for addition:

$$\begin{aligned}(300 + 20 + 5) &= (200 + 100) + (10 + 10) + 5 \\ &= 200 + (100 + 10) + (10 + 5) \\ &= 200 + 110 + 15.\end{aligned}$$

Now we can proceed as above:

$$\begin{aligned}325 - 186 &= (200 + 110 + 15) - (100 + 80 + 6) \\ &= (200 - 100) + (110 - 80) + (15 - 6) \\ &= 100 + 30 + 9 = 139.\end{aligned}$$

In the more conventional form:

$$\begin{array}{r} 325 \\ -186 \\ \hline \end{array} \qquad \begin{array}{r} 300 + 20 + 5 \\ -(100 + 80 + 6) \\ \hline \end{array} \qquad \begin{array}{r} 200 + 110 + 15 \\ -(100 + 80 + 6) \\ \hline 100 + 30 + 9 \end{array} = 139$$

The regrouping can also be shown in the abbreviated algorithm:

$$\begin{array}{r} {\scriptstyle 2\,11} \\ 3\,\not{2}\,5 \\ -1\,8\,6 \\ \hline 1\,3\,9 \end{array}$$

Try Exercise 33

Division

The definition of division in terms of multiplication can be applied directly to find $15 \div 5$. Since $3 \cdot 5 = 15$, it is true that $15 \div 5 = 3$. A physical model of this division helps to understand the algorithms that follow. The array shows 3×5, or three sets of five each for a total of 15.

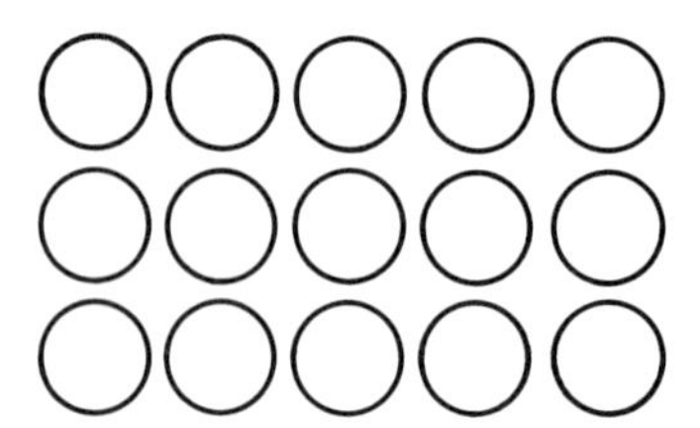

Multiplication of whole numbers can be regarded as repeated addition, so for division it makes sense to begin with 15 objects, remove sets of 5, and record the corresponding subtractions:

* This type of regrouping is often referred to as *borrowing.* This is not a very accurate description but it makes more sense when interpreted in terms of an abacus, or a counting board, where actual objects are manipulated.

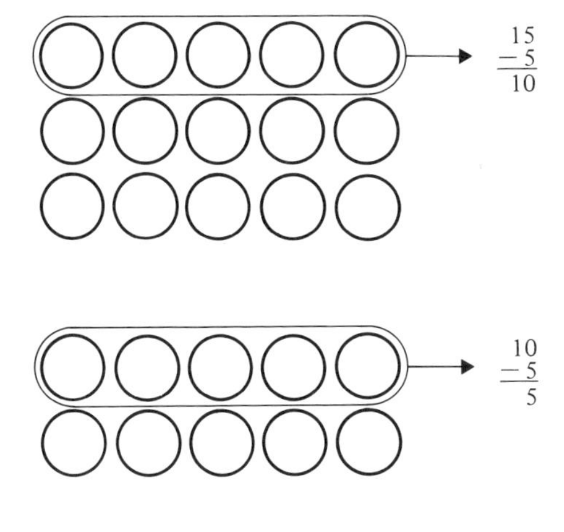

Exactly three sets of five can be removed, so $3 \times 5 = 15$ and $15 \div 5 = 3$.

Try Exercise 34

34. Draw X's. Show the removal of sets and the corresponding subtractions to find $20 \div 4$.

Not all divisions are possible in whole numbers; for example, $13 \div 5$ is not a whole number. Repeated subtractions can still be done, and in this case 13 is 2 fives with 3 left over, that is,

$$13 = 2 \cdot 5 + 3.$$

The definition of the word quotient is broadened to apply to this case; 2 is still called the *quotient*, but now there is a *remainder* of 3 that remains after all possible fives have been removed.

For any whole number D and a divisor d, $d \neq 0$, there are always unique whole numbers Q and R such that the following is true:

$$D = d \cdot Q + R, \text{ where } 0 \leqslant R < d.$$

This statement, which provides the basis of the various division algorithms, will not be proved here.

The repeated subtraction of 5 described above is a primitive algorithm for the division of whole numbers. It can be refined slightly as in the following example in which 1506 is to be divided by 32. Instead of subtracting just 32 each time, it is more efficient to subtract multiples of 32.

35.
a) By guessing, multiplying, and subtracting as in the text, divide 1738 by 42.
b) Check your answer.

Example 6 The basic subtraction algorithm:

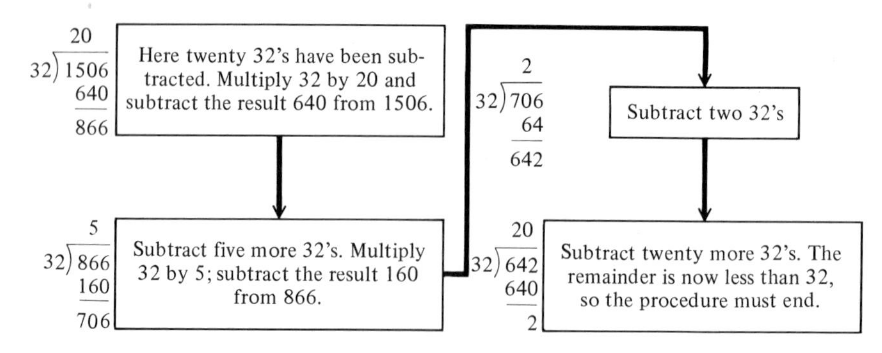

A total of (20 + 5 + 2 + 20) thirty-twos have been subtracted to give a quotient of 47 and a remainder of 2, an answer which can be checked by using the fact that $32 \cdot 47 + 2$ must equal 1506.

Try Exercise 35

The successive subtractions done above can be condensed in an algorithm that is not yet the conventional highly abbreviated form.

Example 7 The "stacking" algorithm:

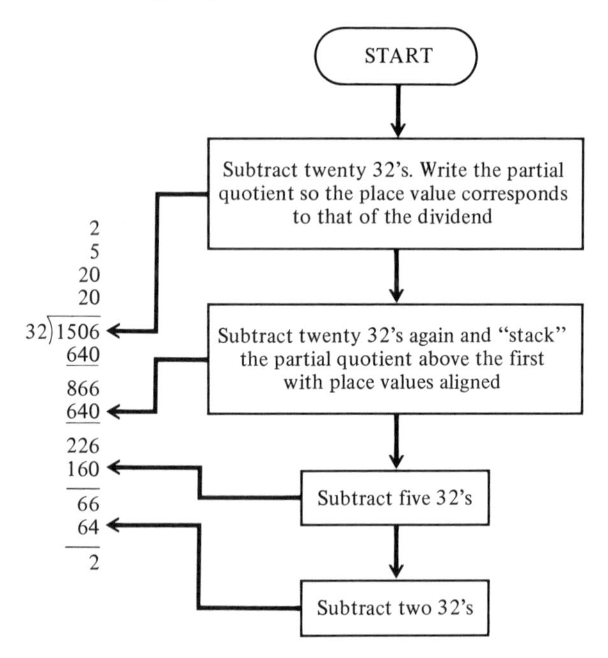

The final quotient is the sum of the partial quotients:

$$Q = (20 + 20 + 5 + 2) = 47.$$

Again the remainder is 2, and $1506 = 32 \cdot 47 + 2$.

Try Exercise 36

The algorithm described so far is "Guess, multiply, subtract until the remainder is less than the divisor; add the guesses to get the quotient." The following flowchart describes the algorithm more precisely.

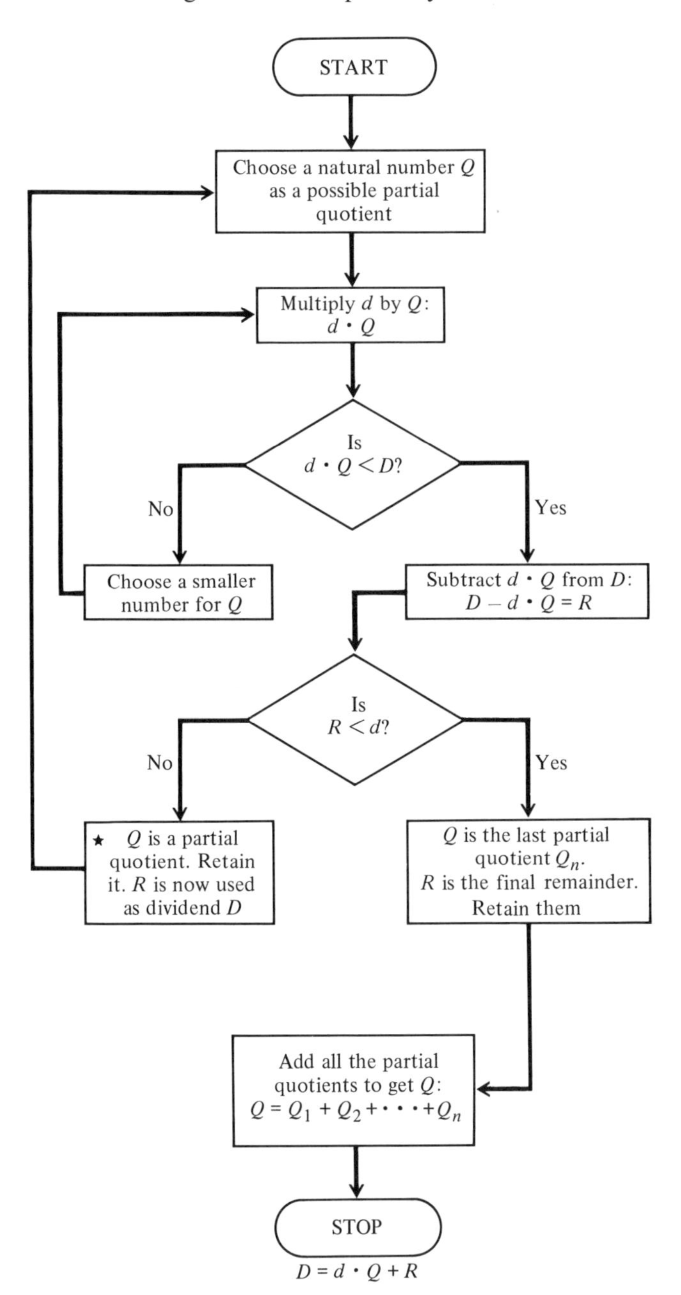

36. Use the more condensed "stacking algorithm" to divide 2453 by 47. Check your answer.

Calculator usage in fourth grade

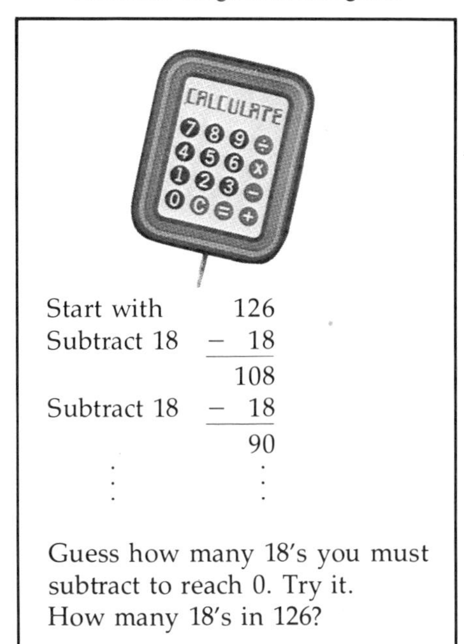

37. The columns below show the flowchart applied to $32\overline{)1506}$. The flowchart process is not yet completed. Complete it for this division.

Flowchart	*First Q*	*Second Q*
Q	20	20
Find $d \cdot Q$	$32 \cdot 20 = 640$	$32 \cdot 20 = 640$
Is $d \cdot Q < D$?	$640 < 1306$?	$640 < 866$?
	Yes	Yes
$D - d \cdot Q = R$	$1506 - 640 = 866$	$866 - 640 = 226$
$R < d$?	$866 < 32$?	$226 < 32$?
	No	No
	(you have reached ★ on the flowchart)	(and so on)

38. Divide 14 839 by 43. Use the method of the example in the text.

Try Exercise 37

The refinements to the algorithm consist mainly of looking for systematic ways of choosing partial quotients. The following example illustrates how partial quotients can be chosen to obtain the digits of the answer directly.

Example 8 The nearly-final algorithm:

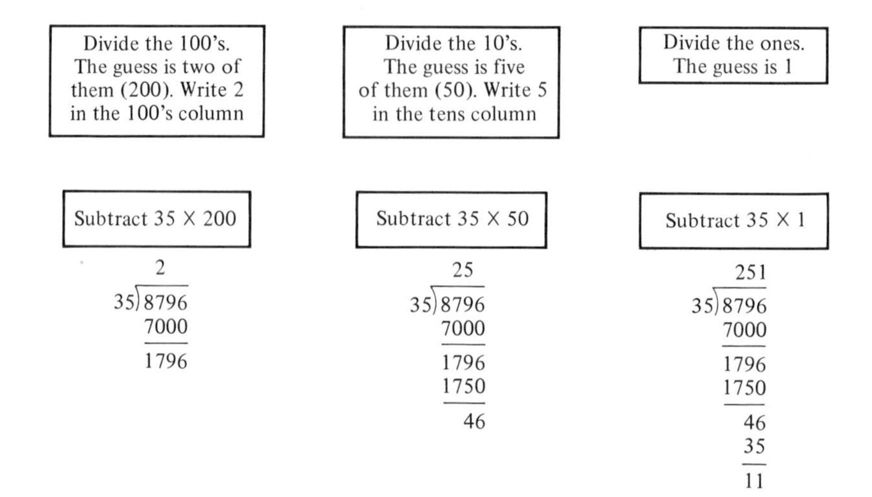

The partial quotients are already added to give the final quotient:

$$200 + 50 + 1 = 251$$

and the remainder R = 11. This can be checked by showing:

$$8796 = 35 \cdot 251 + 11.$$

There are further refinements with respect to how many zeros are written, and hints can be given that will help in making the "guesses." These details will not be pursued here.

Try Exercise 38

EXERCISE SET 4.5

1. Use expanded notation and prove that the ordinary addition algorithm gives correct results for the following:

a) $\begin{array}{r} 286 \\ +729 \\ \hline \end{array}$ b) $\begin{array}{r} 345 \\ +887 \\ \hline \end{array}$

2. Use the partial-sums algorithm and write an accompanying explanation for each of the following:

a) $\begin{array}{r} 125 \\ 276 \\ +132 \\ \hline \end{array}$ b) $\begin{array}{r} 237 \\ 386 \\ +241 \\ \hline \end{array}$

3. Write a flowchart for an elementary student for the algorithm that uses partial sums to add $37 + 56 + 92$.

5. Invent an algorithm for addition that proceeds from left to right.

4. Write a flowchart for use by an elementary student for the final compact addition algorithm for $363 + 217 + 125$.

6. Use the algorithm of Problem 5 for the following:

a) $\begin{array}{r} 7621 \\ 1973 \\ 2059 \\ +8722 \\ \hline \end{array}$ b) $\begin{array}{r} 1108 \\ 199 \\ 4417 \\ +2365 \\ \hline \end{array}$ c) $\begin{array}{r} 1234567 \\ +7654321 \\ \hline \end{array}$

7. Use expanded notation and justify the intermediate algorithm for multiplication in each of the following:
a) 786×4 b) 1305×26

8. Justify the usual compact multiplication algorithm for
a) 539×6 b) 401×24

9. Make a flowchart for the compact algorithm for finding the product 539×6.

10. Devise, describe, and justify a multiplication algorithm in which one proceeds from left to right.

11. Use the algorithm of Problem 10 to find
a) 359×6 b) 563×4
c) 5079×63 d) 1023×45

12. Apply the theorem proved in the text to show that the ordinary subtraction gives correct results for
a) $62 - 38$ b) $53 - 27$

13. Make a sketch that models each of the following differences (see the opening page of this chapter for ideas):
a) $62 - 38$ b) $53 - 27$ c) $43 - 29$

14. Make a flowchart for the usual algorithm for finding the difference $504 - 295$.

15. Make a flowchart for an elementary student to use with a calculator to find the difference $507 - 419$.

16. Devise and justify a left-to-right subtraction algorithm.

17. Use the algorithm of Problem 16 to subtract:

a) $\begin{array}{r} 60322 \\ -59999 \\ \hline \end{array}$ b) $\begin{array}{r} 4567 \\ -\ 678 \\ \hline \end{array}$

18. Draw X's; show removal of sets and the corresponding subtractions to find
a) $24 \div 4$ b) $36 \div 9$

19. Use the flowchart of the text to find $1527 \div 35$.

20. Use the flowchart of the text to find $3717 \div 17$.

21. Demonstrate the "stacking" algorithm and compare it with the usual compact algorithm for each of the following:
a) $1527 \div 35$ b) $3717 \div 17$

■ **22.** You have been in the habit of using a calculator to do your computations. One day you find that the [÷] key is not operating although the [+], [−], and [×] keys still do. How would you divide 2538 by 17 under these circumstances, still using the calculator? Make a flowchart for your procedure.

■ **23.** Suppose in the example above you find that only the [+] and [−] keys work. How would you divide 3754 by 279, still using the calculator as much as possible? Make a flowchart for your procedure.

24. An elementary-school child shows this problem in "borrowing" to you and asks you to figure out what is wrong:

$$\begin{array}{rr} & 9+8+7+6+5+4+3+2+1 = 45 \\ \text{Subtract} & 1+2+3+4+5+6+7+8+9 = 45 \\ \hline & 8+6+4+1+9+7+5+3+2 = \ \ 0 \\ & 45 = \ \ 0 \end{array}$$

■**25.** Here is an algorithm for subtraction you may never have seen. Find the difference 463 − 248:

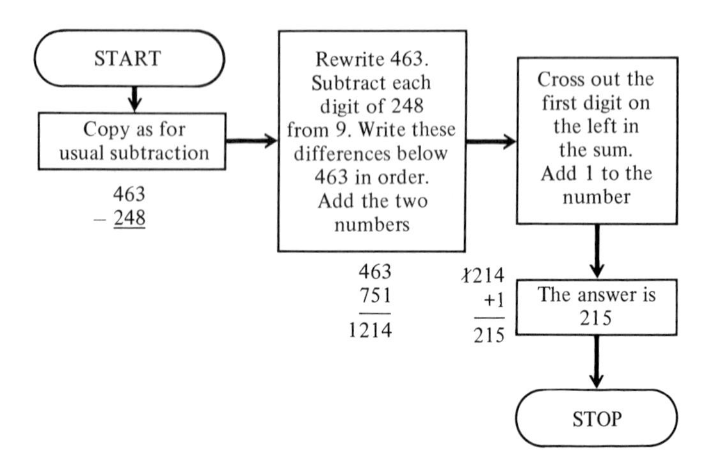

Try the algorithm on some other difference. Analyze why this algorithm works. Show that it is correct.

4.6 AIMS

You should be able to:

a) Adapt the ordinary algorithms of base-ten addition, multiplication, subtraction, and division for use with other bases.

4.6 COMPUTATION IN NONDECIMAL BASES

In order to do decimal arithmetic, you need to memorize all the addition combinations up to $9 + 9$ and all multiplication combinations up to $9 \cdot 9$ because these basic facts are used in the algorithms for computation with larger numbers. In a nondecimal base the number of facts that are needed is different but the same principles apply.

Base Five

Addition

+	1	2	3	4
1	2	3	4	10
2	3	4	10	11
3	4		11	12
4	10	11		

Multiplication

·	1	2	3	4
1	1	2	3	4
2	2	4	11	13
3	3		14	
4	4		22	31

The tables are read from left to right and answers are at the intersection of a row and column. Unless you plan to do a lot of computation in another base, it is not worthwhile to memorize these tables.

Try Exercise 39

39. Complete the addition and the multiplication tables (that are incomplete as given in the text) for base five.

ADDITION (BASE FIVE). The algorithms for calculations with decimal notation can easily be adapted for use in another base; base five is illustrated first. The ones can be added to start as in the ordinary algorithm for base ten.

Example 1 Add; use the intermediate algorithm (all notation is base five).

Solution

```
    1 2 2
  + 4 2 1
  -------
        3   [add the ones]
      4 0   [add the fives]
  1 0 0 0   [add the twenty-fives]
  -------
  1 0 4 3   [by place-value notation]
```

Just as with the base-ten algorithm, the addition can be condensed.

Example 2 Add 214 + 432 + 123 (notation is base five).

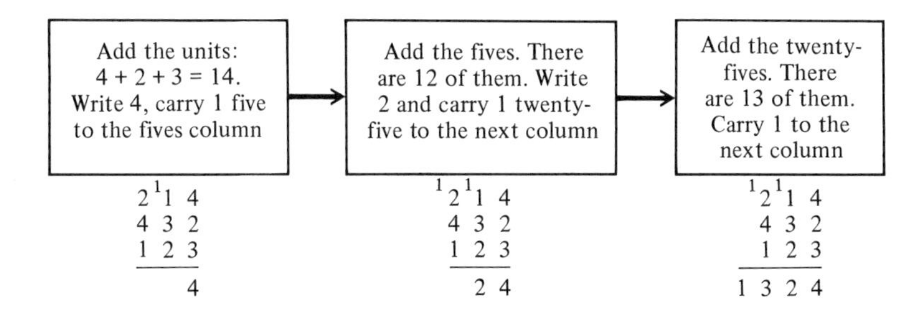

Try Exercises 40 and 41

40. Use the intermediate algorithm and add (notation is base five):

```
   4 1 2
  +3 2 4
  ------
```

41. Add (notation is base five):

```
   4 3 3
   2 1 2
  +1 2 4
  ------
```

SUBTRACTION (BASE FIVE)

Example 3 Subtract 321 − 123. Notation is base five. Then do the same calculation in base ten as a check.

Regroup; a five is needed in the ones place: 11 − 3 = 3 → Regroup; a twenty-five is needed in the fives place: 11 − 2 = 4 → Subtract the twenty-fives: 2 − 1 = 1

```
  3 ¹2̸ ¹1        ²3̸ ¹2̸ ¹1        ²3̸ 2 1
 − 1 2 3         − 1 2 3         − 1 2 3
 -------         -------         -------
       3             4 3           1 4 3
```

42. Subtract (notation is base five):

$$\begin{array}{r} 313 \\ -221 \\ \hline \end{array}$$

43. Multiply (notation is base five):

$$\begin{array}{r} 4132 \\ 3 \\ \hline \end{array}$$

44. Multiply (notation is base five):

$$\begin{array}{r} 234 \\ 43 \\ \hline \end{array}$$

This can be checked by doing the same calculation in base ten.

Base five	*Base ten*	
3 2 1	8 6	
−1 2 3	−3 8	$143_{five} = 48_{ten}$
1 4 3	4 8	

Try Exercise 42

MULTIPLICATION (BASE FIVE). Multiplication in base five uses the same algorithms as in base ten.

Example 4 Multiply 342 by 4 (notation is base five).

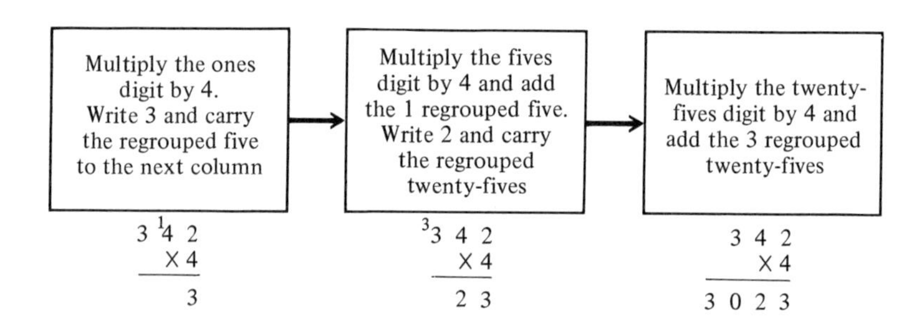

Example 5 Multiply 342 by 34; use an intermediate algorithm (notation is base five).

$$\begin{array}{rl} 342 & \\ \times\ \ 34 & \\ \hline 13 & [\text{this is } 4 \times 2] \\ 310 & [\text{this is } 4 \times 40] \\ 2200 & [\text{this is } 4 \times 300] \\ 110 & [\text{this is } 30 \times 2] \\ 2200 & [\text{this is } 30 \times 40] \\ 14000 & [\text{this is } 30 \times 300] \\ \hline 24333 & [\text{by addition}] \end{array}$$

Try Exercises 43 and 44

DIVISION (BASE FIVE). The division algorithm discussed in the previous section will likewise apply here. If the procedure can be described as "guess, multiply, subtract repeatedly, and finally add the guesses," it can be done in any place-value system in which it is, of course, possible to multiply, subtract, and add.

Example 6 Divide (all notation is base five).

```
            1000 + 100 + 2 + 2     The procedure is:
        34)43202
           34000 <------------     Guess 1000, multiply, subtract 34000
            4202
            3400 <------------     Guess 100, multiply, subtract
             302
             123 <------------     Guess 2, multiply, subtract
             124
             123 <------------     Guess 2, multiply, subtract.
               1                   The remainder is less then 34
```

The quotient is 1104_{five} and the remainder is 1. To check, it can be shown that $43202 = 1104 \times 34 + 1$:

```
     1104
  ×    34
     4431
    33220
 +      1
    43202
```

Try Exercise 45

Calculations in any other base use the same principles along with the basic tables for addition and multiplication. Base two is easy but cumbersome.

Base Two

Addition

+	0	1
0	0	1
1	1	10

Multiplication

·	0	1
0	0	0
1	0	1

45. Divide (notation is base five). Check your answer by multiplying quotient and divisor and then adding the remainder.

$$43\overline{)44212}$$

46. Add (notation is binary):

```
10101
10110
11011
10111
10001
11111
-----
```

Example 7 Add (notation is binary).

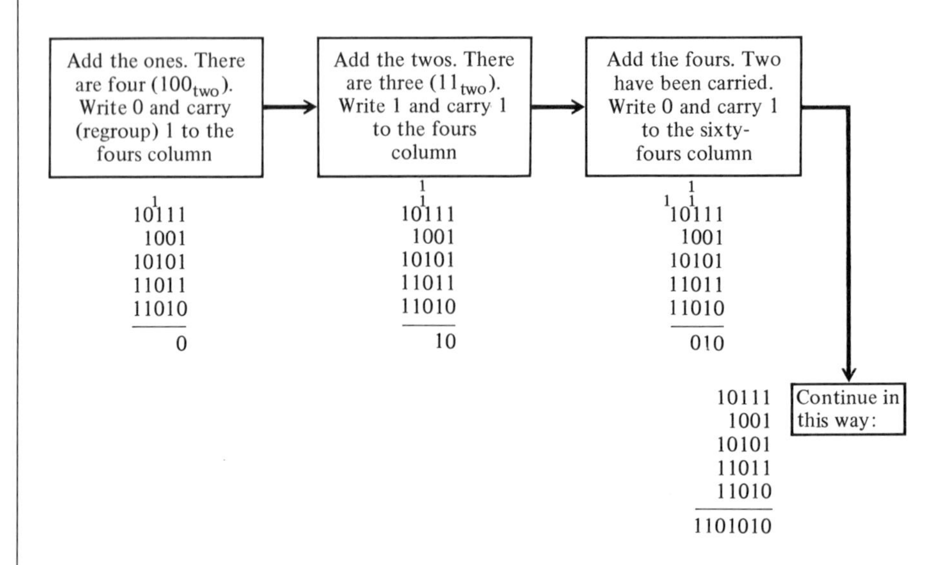

47. Multiply (notation is binary):

```
1101101
  11001
-------
```

Check by doing the calculation in base ten.

Try Exercise 46

Another glance at the multiplication table indicates that the algorithm for finding products in base two is simple.

All that is needed to perform calculations in any base, then, is the basic tables and the familiar algorithms that are used in the decimal system. These depend only upon the commutative, associative, and distributive properties of the operations for the system of whole numbers and a place-value system of notation.

Try Exercise 47

There is an interesting algorithm sometimes called the "Russian-peasant" method. The method requires that the user can add, can multiply by 2, and can divide by 2. To multiply two numbers, the peasant began by doubling the one number and dividing the other by 2. The process was continued until the second series reached 1. (The Russian peasant did not record the remainders, but they help to understand the process.) Then all the even numbers were crossed out in the series being divided, along with the corresponding ones in the other series. Finally the remaining numbers in the series being multiplied were added to get the result.

Example 8 Multiply 217 and 35 by the Russian-peasant method.

Multiply by 2 here:	Divide by 2 here:	Remainders after divisions:	
217	35	1	
434	17	1	
~~868~~ ←	~~8~~ ←	~~0~~ ←	Cross out even numbers
~~1736~~ ←	~~4~~ ←	~~0~~ ←	
~~3472~~ ←	~~2~~ ←	~~0~~ ←	
6944	1 Stop		
7595			

Add those not crossed out

EXERCISE SET 4.6

1. Prepare the basic-fact tables for addition and multiplication for base four.

In Problems 2–4 use your tables from Problem 1 and the intermediate algorithms to perform the following computations:

2. a) Add: $\begin{array}{r}322\\ 32\\ \hline\end{array}$ b) Add: $\begin{array}{r}3123\\ 213\\ \hline\end{array}$

3. a) Multiply: $\begin{array}{r}312\\ \times\ \ 3\\ \hline\end{array}$ b) Multiply: $\begin{array}{r}132\\ \times\ 23\\ \hline\end{array}$

4. a) Divide: $2\overline{)312}$ b) Divide: $12\overline{)213}$

5. Prepare the basic-fact tables for addition and multiplication for base six.

In Problems 6–8 use the tables from Problem 5 and any algorithms you choose to perform the following computations:

6. a) Add: $\begin{array}{r}305\\ +132\\ \hline\end{array}$ b) Add: $\begin{array}{r}3123\\ +\ 213\\ \hline\end{array}$

7. a) Subtract: $\begin{array}{r}403\\ -344\\ \hline\end{array}$ b) Subtract: $\begin{array}{r}5102\\ -3015\\ \hline\end{array}$

8. a) Multiply: $\begin{array}{r}513\\ \times\ \ 4\\ \hline\end{array}$ b) Multiply: $\begin{array}{r}454\\ \times\ 23\\ \hline\end{array}$

9. In what numeration system has the following subtraction been performed? One digit in the answer is wrong. Which digit is it?

$$\begin{array}{r}63532\\ -\ \ 4534\\ \hline 54665\end{array}$$

In Problems 10–11 is it possible to determine the numeration system in which the following have been written? What are some bases they could not be written in?

10. a) $\begin{array}{r} 11011 \\ \times \quad 10 \\ \hline 110110 \end{array}$ b) $\begin{array}{r} 123 \\ \times \ 13 \\ \hline 1599 \end{array}$

c) $\begin{array}{r} 156 \\ -\ 34 \\ \hline 122 \end{array}$ d) $\begin{array}{r} 348 \\ -231 \\ \hline 117 \end{array}$

11. a) $\begin{array}{r} 156 \\ \times\ 23 \\ \hline 4314 \end{array}$ b) $\begin{array}{r} 207 \\ \times\ 78 \\ \hline 17412 \end{array}$

c) $\begin{array}{r} 233 \\ -134 \\ \hline 44 \end{array}$ d) $\begin{array}{r} 621 \\ -172 \\ \hline 427 \end{array}$

12. Children are sometimes asked to make a table of multiples of a divisor to help them do division. This would also be helpful in base-five computation for finding the quotient in 23)12314.

a) Prepare such a table:

1 × 23	23
2 × 23	101
3 × 23	
4 × 23	
10 × 23	
100 × 23	
20 × 23	
200 × 23	

b) Use your table to find Q and R: $23\overline{)12314}$

13. Repeat Problem 12 with an appropriate table for base six and the division $12\overline{)4525}$.

14. Use the Russian-peasant algorithm to multiply the following:

a) 119 by 357 b) 321 by 117

■ **15.** The Russian peasant in all probability did not understand the binary system, but it is possible to use that system to explain why this algorithm works. Explain why this is a valid algorithm. *Hint*: What is the significance of the remainders? Observe that $100011_{\text{two}} = 35_{\text{ten}}$.

■ **16.** Make a flowchart for the Russian-peasant multiplication algorithm that will work for any pair of numbers.

4.7 AIMS

You should be able to:

a) Analyze a computation, with a view to simplifying it for mental computation.

b) Estimate the result of a calculation.

4.7 CALCULATING

In the Middle Ages people found counting boards very satisfactory and were most reluctant to give them up for the algorithms that required writing. Today people are so used to doing the algorithms with pencil and paper that they view the advances made in the way of electronic calculators with the same sort of suspicion.

All of these techniques are more or less "mechanical" in the sense that there is some sort of automation involved in the calculation.

MENTAL CALCULATION. At one time there was more emphasis in the schools on performing calculations "in one's head." Even in the electronic age it is often necessary to do some quick estimation for immediate use, and it is possible to apply the properties of the whole-number system in a creative way to develop shortcuts in computation. We show some ways of calculation without paper and pencil, but you can be imaginative and devise many others of your own.

Examples Addition

1. $\begin{array}{r} 38 \\ +54 \\ \hline \end{array}$ Add the tens first to get 80, then add the ones to get 12. The answer is 92.

   Reason: This is the intermediate stage of the addition algorithm used left to right.

2. $\begin{array}{r} 89 \\ +61 \\ \hline \end{array}$ Increase 89 to 90 by adding 1; decrease 61 to 60 by subtracting 1. Add; the answer is 150.

   Reason: By associativity for addition, since $89 + 61 = 89 + 1 + 60 = (89 + 1) + 60$.

3. $\begin{array}{r} 25 \\ +73 \\ \hline \end{array}$ Know that $25 + 75 = 100$. This sum is 2 less, so the answer is 98.

Try Exercise 48

Examples Subtraction

4. $\begin{array}{r} 57 \\ -23 \\ \hline \end{array}$ Subtract the tens to get 30, subtract the ones to get 4. The answer is 34.

   Reason: A left-to-right subtraction algorithm.

5. $\begin{array}{r} 82 \\ -47 \\ \hline \end{array}$ Increase both numbers by 3, to get a "round" number on the bottom. Then $85 - 50 = 35$.

   Reason: If $a - b = k$, then $(a + c) - (b + c) = k$ also (Problem 13 in Exercise Set 4.3).

Try Exercise 49

Some multiplication and division can easily be done mentally.

Examples Multiplication and division

6. 87×2 Double the 80 to get 160. Double the 7 to get 14. The answer is 174.

   Reason: The distributive law: $(80 + 7) \cdot 2 = 87 \cdot 2$.

7. 36×4 Double 36 to get 72. Double 72 to get 144.

   Reason: The associative law: $36 \times (2 \times 2) = (36 \times 2) \times 2$.

48. Do these calculations without pencil and paper:

a) $\begin{array}{r} 57 \\ +49 \\ \hline \end{array}$ b) $\begin{array}{r} 463 \\ +258 \\ \hline \end{array}$

c) $\begin{array}{r} 98 \\ +102 \\ \hline \end{array}$ d) $\begin{array}{r} 249 \\ +501 \\ \hline \end{array}$

e) $\begin{array}{r} 50 \\ +27 \\ \hline \end{array}$ f) $\begin{array}{r} 57 \\ +78 \\ \hline \end{array}$

49. Do these calculations without pencil and paper:

a) $\begin{array}{r} 87 \\ -34 \\ \hline \end{array}$ b) $\begin{array}{r} 71 \\ -36 \\ \hline \end{array}$

c) $\begin{array}{r} 473 \\ -157 \\ \hline \end{array}$ d) $\begin{array}{r} 453 \\ -275 \\ \hline \end{array}$

50. Do these calculations without pencil and paper:

a) 78×2 b) 47×4

c) 93×5 d) $4600 \div 5$

e) 52×25

51. Estimate:

a) 583×312

b)
```
  384
  214
  566
 +143
```

c) $756 \div 23$

8. 87×5 Multiply 87 by 10 to get 870. Then divide by 2 to get 435.
   Reason: $10x \div 2 = 5x$, $(10 \times 87) \div 2 = 5 \times 87$.
9. 43×25 Multiply 43 by 100 to get 4300. Divide the result by 4. To do this divide by 2, then divide 2150 by 2 again to get 1075.
   Reason: $100x \div 2 = 50x$, $50x \div 2 = 25x$.
10. $2700 \div 5$ Divide by 10 to get 270, then multiply by 2 to get 540.
   Reason: $100x \div 10 = 10x$, $2 \times 10x = 20x$; $100x \div\ = 20x$.

Try Exercise 50

ESTIMATION. An estimate is an approximate answer to either a problem or a calculation. As you calculate you should regularly estimate answers to detect gross errors. This is just as important with a calculator as with pencil and paper, because it is possible to push a wrong button or copy a result incorrectly.

To make such estimates we round the numbers involved and then proceed to do the computation. Whole numbers can be rounded to the nearest ten (the nearest multiple of ten). Thus 32 is rounded to 30, 37 is rounded to 40, and so on. If there are five units, then an even number of tens is rounded down and an odd number of tens is rounded up. This means 125 is rounded to 120 and 155 is rounded to 160. Analogous rules apply for rounding to the nearest hundred or the nearest thousand, and so on.

Examples

11. 478×296 Round to 500 and 300, then multiply:

$$500 \times 300 = 5 \cdot 3 \cdot 100^2 = 150\,000.$$

A calculated answer of 14 148 is incorrect.

12.
```
  269     Round to   300     Estimate of the sum is 1100.
  114                100
  323                300
 +415               +400
```

For an answer of 2121, this estimate says "Check your work."

13. $668 \div 32$ Round the divisor to 30. Then approximate 668 to some multiple of 30, say 600. Divide for an estimate of 20. An answer of 208 is clearly incorrect.
(As a check on the estimate, multiply 32 by 20 = 640, which is close.)

Try Exercise 51

CALCULATORS. Although small calculators are in schoolrooms and in the homes of elementary pupils, it is still necessary for these pupils to understand the basic concepts of whole numbers and the operations with them. Surprisingly, a calculator can sometimes help to provide this understanding, since with one it is possible to investigate interesting problems in arithmetic that would be too time-consuming to be practical otherwise. The potential for the schoolroom could be enormous.

Using estimation in sixth grade

The label on your cereal box says that the box contains 596 g. You estimate that you eat about 50 g of cereal each morning. About how many days will a full box of cereal last you?

In a classroom

Imagine a teacher who had thought of putting pupils to work on the following problem to give them some practice in subtraction. Here is the problem posed to the students:

Example 14 Choose any four digits you want from the set

$$\{1, 2, 3, 4, 5, 6, 7, 8, 9\};$$

at least two different digits must be chosen. Now form the largest number from these digits and the smallest one from the same digits. Subtract the smaller from the larger. Your answer will be another four-digit number. Use these digits and repeat the process.

Pretty soon a hand is raised.

Alicia: I finished. I can't keep going. I have the number 6174 and the next time I get 6174 again.

Teacher: Let me see your work.

Alicia: Here it is:

$$\begin{array}{r} 8631 \\ -1368 \\ \hline 7263 \end{array} \quad \begin{array}{r} 7632 \\ -2367 \\ \hline 5265 \end{array} \quad \begin{array}{r} 6552 \\ -2556 \\ \hline 3996 \end{array} \quad \begin{array}{r} 9963 \\ -3699 \\ \hline 6264 \end{array} \quad \begin{array}{r} 6642 \\ -2466 \\ \hline 4176 \end{array} \quad \begin{array}{r} 7641 \\ -1467 \\ \hline 6174 \end{array} \quad \begin{array}{r} 7641 \\ -1467 \\ \hline \end{array}$$

At this point another hand is raised.

Brian: I finished. I keep getting the same number each time.

Teacher: Let me see your work.

Brian: Here it is:

$$\begin{array}{r} 4221 \\ -1224 \\ \hline 2997 \end{array} \quad \begin{array}{r} 9972 \\ -2799 \\ \hline 7173 \end{array} \quad \begin{array}{r} 7731 \\ -1377 \\ \hline 6354 \end{array} \quad \begin{array}{r} 6543 \\ -3456 \\ \hline 3087 \end{array} \quad \begin{array}{r} 8730 \\ -0378 \\ \hline 8352 \end{array} \quad \begin{array}{r} 8532 \\ -2358 \\ \hline 6174 \end{array} \quad \begin{array}{r} 7641 \\ -1467 \\ \hline 6174 \end{array}$$

Teacher: 6174! . . . Well! That is interesting.

Hands start popping all over the classroom.

52. Try a few cases of the task in Example 14. What is causing the hands to pop up all over the classroom?

Using estimation in sixth grade

Try Exercise 52

Just as you have found in your investigation with your calculator, every child in the hypothetical classroom would have had 6174 turn up in the course of the repeated subtractions. In both of the examples for Alicia and Brian, 6174 turned up as the sixth difference.

This raises all sorts of interesting questions.

1. Even though you may have looked at 100 examples and you may suspect that 6174 always turns up, that does not prove it. Does 6174, in fact, have to turn up?
2. You may already know the answer to this. Does 6174 always occur on the sixth difference?
3. If it does require a different number of subtractions to reach 6174, is there a way to recognize from the outset how long it will take?

This is clearly a case of how the calculator has stirred up more interesting mathematical questions than you might have suspected. These will be left for you to pursue.

EXERCISE SET 4.7

Do the following calculations mentally:

1. a) $\begin{array}{r} 48 \\ +36 \\ \hline \end{array}$ b) $\begin{array}{r} 56 \\ +48 \\ \hline \end{array}$ c) $\begin{array}{r} 97 \\ +103 \\ \hline \end{array}$ d) $\begin{array}{r} 101 \\ +\ 91 \\ \hline \end{array}$

2. a) $\begin{array}{r} 97 \\ -36 \\ \hline \end{array}$ b) $\begin{array}{r} 62 \\ -35 \\ \hline \end{array}$ c) $\begin{array}{r} 84 \\ -22 \\ \hline \end{array}$ d) $\begin{array}{r} 582 \\ -137 \\ \hline \end{array}$

3. a) 58×2 b) 2145×2 c) $48 \div 2$ d) $3134 \div 2$.

4. a) 86×5 b) 68×5 c) 82×25 d) 68×25

5. Devise and justify a method of multiplying by 9 mentally.

6. Use your method developed in Problem 5 to multiply the following:
a) 63×9 b) 49×9 c) 17×9 d) 111×9

7. Devise and describe a method of multiplying by 101.

8. Use your method to multiply the following:
a) 123×101 b) 71×101

9. Find and justify a method for dividing by 5 mentally.

10. Use your method to find the following:
a) $3200 \div 5$ b) $635 \div 5$ c) $485 \div 5$

Estimate the results in the following:

11. a) 312×295 b) 488×124
c) 76×17 d) 83×21

12. a)
$$\begin{array}{r} 412 \\ 489 \\ 606 \\ +107 \\ \hline \end{array}$$
b)
$$\begin{array}{r} 38 \\ 145 \\ 76 \\ +503 \\ \hline \end{array}$$
c)
$$\begin{array}{r} 109 \\ 253 \\ 348 \\ +818 \\ \hline \end{array}$$
d)
$$\begin{array}{r} 112 \\ 808 \\ 919 \\ +531 \\ \hline \end{array}$$

13. a) $4078 - 1976$ b) $5163 - 4812$
c) $671 - 387$ d) $342 - 193$

14. a) $765 \div 24$ b) $828 \div 21$
c) $495 \div 9$ d) $616 \div 12$

15. It is unlikely you will ever meet this problem, but suppose you were offered a summer job consisting of handing out $100 bills. You are to hand one out every minute until you have given away a billion dollars. You want to know how long this job will last. What is your estimate? A week? a month? Find out.

All of the following are activities with a calculator that are suitable for elementary pupils. Try them to see what possible insights might be developed in the students.

16. How big is a thousand? Figure out how to *count* on your calculator. Make a flowchart that describes how your calculator can be used to count. Count to 1000.

17. What numbers must be multiplied to produce an answer between 495 and 500:

$$\square \times \square =$$

Make a guess, then check it on your calculator. How many guesses did you need before you got it?

18. Enter 347 on your calculator. You want a number on the display: △△8 (the first two digits are the same). What numbers can you add to make this happen? Keep a record of your tries.

19. Repeat Problem 18, but this time subtract a number.

20. Start with 37 on the display of your calculator. Add one number and land between 83 and 90. Did you get there on the first try?

21. Start with 143 on the display. Add one number and land between 215 and 225. Did you get there on the first try?

22. Start with 137 on the display and multiply by one number to land between 1500 and 1510. How many tries?

23. *Annihilation game.* 375 is on the display. What must you do to get a 0 in place of the 7? 3142 is on the display. What must you do to get 0 instead of the 1? 4163 is on the display. What must you do to get 0's instead of the 4 and the 6? Is there more than one way to do each of these?

4.8 AIMS

You should be able to:

a) Use the five-step problem-solving process as an aid in deducing additional facts in a problem.
b) Make a restatement of a problem in an effort to generate ideas and find a solution.

4.8 PROBLEM SOLVING

Even the simplest of everyday problems do not present themselves in the form of an algorithm or as a simple equation to be solved, so children must learn to translate problems into mathematical language.

In the first grade, problems are initially presented in the form of pictures such as: Three ducks are sitting in a pond. Two more ducks wander toward the pond and join the first three. The child is asked how many ducks are in the pond. The translation that appears below such a picture would be:

$$3 + 2 = \square.$$

Later, youngsters will need to be able to translate from the language of the problem to a mathematical statement even if the problem is only a slightly disguised practice for a recently learned algorithm. Many children decide to multiply because that is what they did on the previous ten problems, not because they thought about it.

There are some interesting problems* that depend upon an understanding of the properties of the whole-number system and the base-ten notational system.

RESTATEMENT OF THE PROBLEM

Example 1 Each letter in the following problem represents one of the digits 0, 1, 2, 3, 4, 5, 6, 7, 8, or 9. Determine what the addition problem is:

$$\begin{array}{r} \text{U} \\ \underline{\text{ZZZ}} \\ \text{UPPP} \end{array}$$

The *facts* are: U, Z, P; each represents a different digit;

$$\text{U} + \text{ZZZ} = \text{UPPP}.$$

The *problem*: Find U, Z, and P.

A restatement of the problem: What 3-digit number, all digits alike, can be added to a 1-digit number to yield a 4-digit number whose last three digits are alike?

Some *ideas*:

1. The 3-digit number is greater than 990. Why?
2. The 1-digit number can not be 0. Why?
3. The 3-digit number is 999. Why?
4. The first digit of UPPP must be 1. Why?

Getting a *solution*: the additional facts deduced in the idea stage have led directly to a solution:

$$999 + 1 = 1000.$$

* As a reminder:

THE FIVE-STEP PROCESS
1. What are the *relevant* facts?
2. What actually is the *problem*?
3. What are my *ideas*?
4. How do I get a *solution*?
5. *Check* and *review* the solution.

Looking back: Is there *another* solution to this problem that was missed by this train of ideas?

DEDUCTION OF NEW FACTS. In the next example we will concentrate on how additional facts can be deduced from those given in the original problem. The *new facts* may actually change the way you look at a problem and help you to reach a solution.

Example 2 Each letter in the following problem represents one of the digits 1, 2, 3, 4, 5, 6, 7, 8, or 9. Determine what this addition problem is:

$$GO = ON + ON + ON + ON.$$

The *facts*: Each letter G, O, N, represents a different digit; no letter represents zero;

$$GO = ON + ON + ON + ON.$$

The *problem*: Find G, O, and N. A restatement of the problem: What 2-digit number, when multiplied by 4, yields a 2-digit number which has the ones digit the same as the tens digit of the number multiplied? (This may or may not help!)

Some *ideas*:

1. $GO < 100$, hence $ON < 25$ (a new fact).
2. ON is one of $\{11, 12, \ldots, 19, 21, \ldots, 24\}$, since 0 is not one of the allowed digits (a new fact).
3. Take each of the numbers listed in idea 2, multiply by 4 and see if it is a solution.
4. The letter O is either 1 or 2 (from idea 2). The problem now is changed to:

$$G2 = 2N + 2N + 2N + 2N$$

or

$$G1 = 1N + 1N + 1N + 1N.$$

5. $N + N + N + N = 4 \times N$ is an even number. Hence the letter O is 2.

A solution: **Try Exercise 53**

This example could have been solved by using idea 3. By taking the time to deduce a few more facts, we reduced the problem to one that could be done mentally.

Looking *back* over your solution: Is it possible there is another solution that you missed? What would be the effect of allowing O to be one of the digits?

Problems such as these develop in pupils a firmer grasp of the notational system as well as overall problem-solving skills.

53.

a) In idea 5, why is it impossible for the letter O to be 1?
b) Use any or all of the ideas to solve the problem. By this time it should be easy.

EXERCISE SET 4.8

Use the five-step process in solving all of these problems. Practice the orderly deduction of new facts to help you in problem solving.

In Problems 1–5 each letter stands for a single digit; different letters stand for different digits. Find the digits.

1. A + A + A = HA

2.
```
  SEND
 +MORE
 -----
 MONEY
```

3.
```
   TEN
   TEN
+FORTY
------
 SIXTY
```

4.
```
 ABCDEF
×     3
-------
 BCDEFA
```

5.
```
 ABCABC
×     7
-------
 DEADEA
```

In Problems 6 and 7 the X's stand for any digits. Some of the digits are supplied. Find the missing ones.

6. Multiplication:

```
    4XX
    XXX
    ---
    XXX
   X3X
  XXXX
 ------
 1XX9X7
```

7. An exact division:

```
      9XX
3X)X4X9X
   X1X
   ---
    XXX
    X1X
    ---
     2X5
     2X5
     ---
```

8. Elementary pupils enjoy number tricks. Here is one that will baffle them: "Choose any number and add 6. Multiply the result by 2. Now add 4 to what you have. Divide by 2. Subtract the number you started with." The trick is, the answer is always 8. Analyze this trick to find out how it works. *Hint*: Use a physical model.

9. Another number trick: "Choose any number; add 1. Now double the result and add 3 to what you get. Multiply this answer by 4, then add 4. Divide the result by 8 and finally subtract 3." The answer is the number chosen at the start. Analyze this trick.

10. The following diagram is a calendar for the month of August 1979. Any month of any year will do. Block off *any* nine numerals in a square formation on the calendar. It is possible to find the sum of the numbers in such a square by simple addition. There is a remarkably simple formula that will give the sum as a product. Find the formula. Prove it is true no matter which square is blocked off.

August 1979

S	M	T	W	T	F	S
			1	2	3	4
5	6	7	8	9	10	11
12	13	14	15	16	17	18
19	20	21	22	23	24	25
26	27	28	29	30	31	

11. A certain whole number is divisible by 8. The same number, when divided by 7, gives a remainder of 3. Find an expression for the quotient. Find such a number.

12. Students in the drama class sold 845 tickets for a play they were producing. The price of the tickets was \$2, \$3, and \$4, depending upon location in the auditorium. They sold the least of the \$4 seats, four times as many of \$3 as the \$4 seats, and twice as many of the \$2 seats as of the \$3 seats. How many tickets of each price did they sell?

13. Discover a pattern and find the next four numbers in the following sequence: 563, 1127, 1691, 2255, . . .

14. Study the patterns below and conjecture what the missing numbers should be:

$94 \times 94 = 8836$
$93 \times 95 = 8835$

$71 \times 71 = 5041$
$70 \times 72 = 5040$

$67 \times 67 = 4489$
$66 \times 68 =$

$85 \times 85 = 7225$
$84 \times 86 =$

Prove that this is not just a coincidence; it works every time.

The remaining problems are similar to some found in elementary textbooks. Solve them as if you were using the five-step process in explaining the problems to elementary-school pupils.

15. Suppose that you walk 1 km in 10 minutes; how far could you walk in 1 hour? On your bike you can go 3 km in 10 minutes. Your friend lives 6 km from your house. How much time will you save on a round trip to his house and back to yours if you take your bike instead of walking?

16. The twins Beth and Bill are planning a picnic for their birthday. They figure each person will eat two hot dogs. There are to be 14 people altogether at the picnic. How many hot dogs should they buy? The rolls for the hot dogs come in packages of 8. How many packages of rolls do they need?

17. The X's in the following addition problem stand for any digits. Find the missing ones. (This problem has a rather interesting feature!) Is there more than one solution?

```
 XXXX5
 XXXX6
 XXXX4
 XXX83
 -----
 56438
```

18. There are 450 children in the school who are going to be taken to the museum for a special program. Each bus holds 36 children. How many busloads will be needed to transport all of them?

REVIEW TEST

1. Choose appropriate sets A and B and use the definition of addition to show that $2 + 3 = 5$.

2. Use the same sets as in Problem 1 to show that $2 \cdot 3 = 6$.

3. Make a suitable diagram to illustrate the distributive law for the example: $2 \cdot (3 + 4) = 2 \cdot 3 + 2 \cdot 4$.

4. Prove that if a and b are any whole numbers, and $a = b$, then

$$a + 2 = b + 2.$$

5. Each of the following statements can be justified by one or more of the laws for the arithmetic operations for whole numbers. Identify all the properties that are used in each case. The variables represent whole numbers.

a) $(y + 2) \cdot (x + 4) = (4 + x) \cdot (2 + y)$
b) $5 \cdot (y + x) = 5 \cdot x + 5 \cdot y$
c) $7 + 8 + 3 = (15 + 3) = 7 + 11$
d) $49 + 77 = 7 \cdot (7 + 11) = 7 \cdot (8 + 10)$

6. An operation $\star$ is defined on the set of whole numbers as follows:

$$a \star b = 2 \cdot a \cdot b.$$

a) Is the set of whole numbers closed under this operation?
b) Is the operation $\star$ commutative? associative?
c) Is there an identity element for the $\star$ operation?

7. Use expanded notation to explain and justify the intermediate forms of the algorithm for:

a) Addition:
$$\begin{array}{r} 736 \\ 241 \\ +637 \\ \hline \end{array}$$

b) Multiplication:
$$\begin{array}{r} 342 \\ \times\ 93 \\ \hline \end{array}$$

8. Explain and justify the "stacking" division algorithm for

$$59\overline{)3937}$$

9. Prepare the basic tables for addition and multiplication for base eight.

10. Use the table of Problem 9 and the intermediate algorithms to perform the following (notation is base eight):

a) Addition:
$$\begin{array}{r} 137 \\ 256 \\ +402 \\ \hline \end{array}$$

b) Multiplication:
$$\begin{array}{r} 172 \\ \times\ 35 \\ \hline \end{array}$$

11. Devise and justify a shortcut for

a) division by 5
b) multiplication by 5

12. Estimate the result in each of the following:

a) $214 \cdot 396$
b) $315 + 510 + 703 + 111$
c) $563 - 125$
d) $395 \div 9$

13. Deduce three facts in the following that would help you to solve the problem. Each letter stands for a different single digit. ■ Find a solution:

$$\begin{array}{r} \text{AHAHA} \\ +\text{TEHE} \\ \hline \text{TEHAW} \end{array}$$

14. Show that for any whole number a, $a \cdot 0 = 0 \cdot a = 0$.

15. Show that for any whole numbers a, b, c and d, if $a > b$ and $c > d$, then $a + c > b + d$.

16. Some of the following statements are always true. Mark such statements T and provide a counterexample for the remaining ones.

a) If A and B are any two sets and $A = B$, then $n(A) = n(B)$.
b) For any sets A and B, $n(A) + n(B) = n(A \cup B)$.
c) For finite sets, the relation "has the same cardinality as" is a symmetric relation.
d) The set A is not the same set as $\bar{A}$, and $n(A) \neq n(\bar{A})$.
e) The set of whole numbers is closed under subtraction.
f) Division on the set of whole numbers has the associative property.
g) For any whole numbers x and y, $3x + 9y = 3 \cdot (x + 3y)$ by the distributive law.
h) If $n(A) = 5$, and $n(B) = 7$, then $n(A \cup B) = 12$.

■ **17.** Prove that if $n(A \times B) = k$, and set A' is equivalent to set A and set B is equivalent to set B', then $n(A' \times B') = k$.

FOR FURTHER READING

1. For ideas on motivation see "LET'S DO IT! More indoor games to motivate computational skills" by James V. Bruni and Helene J. Silverman, *The Arithmetic Teacher*, 24:5, May 1977. All sorts of games are described, involving dominoes, dice, the hundreds board, and cards. Also many ideas are offered for practice in matching, counting, and combinations.

2. Different types of subtraction algorithms are discussed in "Research on a 'new' method of subtraction" by Ruth C. Hoppe. This appears in *The Arithmetic Teacher*, 22:4, April 1975. She asks: "Why do most schools and all textbooks in the United States perpetuate a method of subtracting multidigit numbers that is awkward for everyday use . . . ?"

3. More details about division algorithms can be found in "A study of teaching division through the use of two algorithms" by L. Scott in *School Science and Mathematics*, 63, 1963, pp. 739–752.

TEXTS

Copeland, Richard W., *How Children Learn Mathematics: Teaching Implications of Piaget's Research*, Second Edition. Macmillan Publishing Co., Inc., New York, New York (1974). (A methods text.)

Schminke, C. C., Norbert Maertens, and William R. Arnold, *Teaching the Child Mathematics*. The Dryden Press, Hinsdale, Illinois (1973). (A methods text.)

chapter 5 number theory

Prime and composite numbers

Numbers that have exactly two different factors are called **prime numbers.**

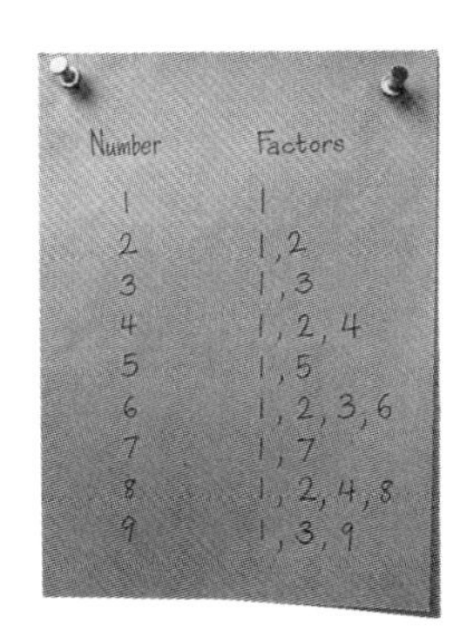

1. Name the prime numbers less than ten. 1, 2, 3, 5, 7
2. Name the prime numbers between 10 and 20. 11, 13, 17, 19

Composite numbers are larger than 1 and have more than 2 factors.
Example: 6 has four factors: 1, 2, 3, and 6.
6 is a composite number.

Tell whether each number is a prime number or a composite number.

3. 4 C 4. 12 C 5. 13 P 6. 18 C 7. 19 P 8. 21 C

Factor trees show prime factors of a number. The top row must have all prime numbers.

Copy and complete these factor trees.

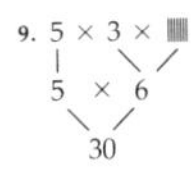

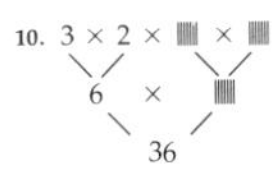

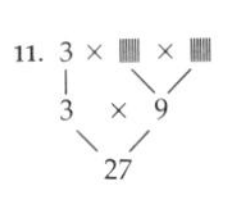

The fifth graders who work with the factor trees of this lesson have probably been introduced to the concepts of prime and composite numbers at least a year earlier. This work is needed before fractions are developed.

5.1 WHY NUMBER THEORY?

So far we have developed the basic properties of the system of whole numbers and what we need to do now is look more closely at some of the interesting results that follow from the basic properties. Number theory, the study of the properties of natural numbers, sounds awesome until you realize that its subject matter is ordinary arithmetic. Pierre de Fermat,* who founded this particular branch of mathematics, was in the habit of describing his mathematical discoveries in his letters to his friends who seemed to find such correspondence interesting.

The more you understand about how the whole numbers behave, the more you will be able to understand not only this system but those that will be developed later in this book.

Very often the puzzle-type problems you encounter depend for their easy solution upon number theory. Many such recreational problems can be understood and solved by an elementary-school pupil who finds them both a mathematical challenge and fun to do. An elementary pupil could easily discover that the square matrix of the next example turns out to have a rather unusual property.

Example 1 What is unusual about this square?

2	4	6	22
3	6	9	33
5	10	15	55
7	14	21	77

To find out more about this square, use the flowchart on the facing page.

As you have discovered, every time four numbers are selected in the manner set out by the flowchart, the product of these four turns out to be 13 860!

1. What is so special about this matrix that this happens?
2. How is such a square constructed?
3. Can a square be made to produce any preselected number as the product?
4. Can larger squares of this type be constructed?

Examination of the patterns in the matrix, along with some information about prime numbers, can answer such questions and will be pursued in subsequent exercises.

* Pierre de Fermat (1601–1665) was an outstanding French mathematician who "fathered the modern theory of numbers" along with many other things.

The Age of Louis XIV by Will and Ariel Durant. Simon and Schuster, New York, New York (1963).

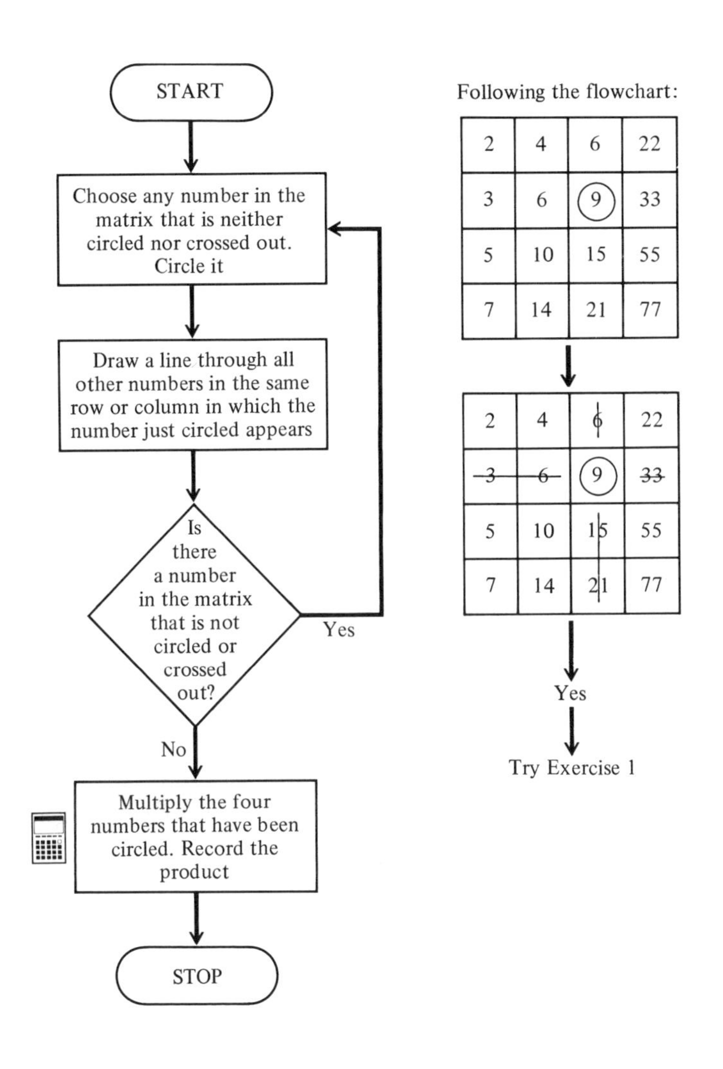

1.

a) Use the flowchart until you reach STOP in the example started in the text.

b) Use the flowchart three more times but select a different entry to start each time.

EXERCISE SET 5.1

1. Discover all you can about this magic square before you read any further. Save your work for later in the chapter.

5.2 DIVISIBILITY AND THE "DIVIDES" RELATION

We consider first another relation on the set of whole numbers that does a great deal toward simplifying and understanding the subject matter of this chapter. It is just as important as the "equals", or "greater than", or "less than", relations we have already discussed; it is called the "divides" relation.

5.2 AIMS

You should be able to:
Define the "divides" relation and apply the definition in proving some properties of the relation.

DIVIDES **For any whole numbers a and b, $a \neq 0$, a *divides* b (symbolized $a|b$) if and only if there is a whole number c such that $a \cdot c = b$.**

The statement $a|b$ is usually read "a divides b" but it can also be read "a is a factor of b" or "b is a multiple of a" since these sentences all mean the same thing. The statement $a \nmid b$ is read "a does not divide b."

Example 1

a) $8|32$ because $8 \cdot 4 = 32$.

b) If $x \neq 0$, then $x|2x$ because $x \cdot 2 = 2x$.

c) $1|64$ because $1 \cdot 64 = 64$.

d) $6|0$ because $6 \cdot 0 = 0$.

Try Exercise 2

2.
a) If $p \neq 0$, is $p|p$ a true statement? Why?
b) Is $7|37$ a true statement? Why?

There are some properties of the "divides" relation that will prove to be especially useful for what is done later in the chapter:

For any whole numbers a, b, and c, $a \neq 0$, if $a|b$ and $a|c$, then $a|(b + c)$.

The proof follows from the definition of "divides" and the basic properties of whole numbers:

If $a|b$ and $a|c$, then $a \cdot m = b$ and $a \cdot n = c$, by the definition of "divides."

We want to know something about $b + c$, and from the renaming property of "equals" we can write:

$$b + c = a \cdot m + a \cdot n.$$

By the distributive law of multiplication over addition we have:

$$b + c = a \cdot (m + n).$$

Since m and n are both whole numbers, so is their sum, and we can state $a|(b + c)$ by the definition of "divides."

The following examples illustrate how this theorem can be applied.

Example 2

a) $6|12$ and $6|24$, therefore, by the theorem, $6|(12 + 24)$ or $6|36$.

b) $5|35$ and $5|70$, therefore $5|105$.

c) $2|10m$ and $2|100m$, therefore $2|(110m)$.

Try Exercise 3

3. Use the theorem just proved in the text to complete the following (true) statements:
a) Since $7|14$ and $7|21$, then we conclude that ____.
b) Since $4|200$ and $4|3000$, then we conclude that ____.
c) Since $3|99$ and $3|9$, then we conclude that ____.

It is also simple to show that for any whole numbers a, b, and c with $a \neq 0$, if $a|b$, then it is also true that $a|b \cdot c$. This theorem allows us to conclude that if $2 | 10$, then it follows that $2 | 10 \cdot 10$ and $2 | 10 \cdot 100$. The proof is left for the exercises.

A third useful property is illustrated in the following: We observe that $5 | 15$, and $5 | 75$, which could have been stated as $5 | (15 + 60)$. Now we see that it is also true that $5 | 60$. Does this happen in every case? The example suggests that:

For any whole numbers a, b, and c, $a \neq 0$, if $a | b$ and $a | (b + c)$, then $a | c$ also.

The proof is as follows: Since $a | b$ and $a | (b + c)$, from the definition of "divides," we can write:

$$a \cdot m = b \quad \text{and} \quad a \cdot n = b + c.$$

We need to find out if c can be written $a \cdot p$, where p is a whole number. By the definition of subtraction we can state that

$$a \cdot n - b = c,$$

but since $b = a \cdot m$, we can write instead that:

$$a \cdot n - a \cdot m = c.$$

By the distributive law: $a \cdot (n - m) = c$. Since $n - m$ is a whole number, by the definition of "divides" we have that $a | c$.

The following examples illustrate the application of this theorem.

Example 3

a) Since $37 | 111$ and $37 | 185$, we conclude $37 | 74$.

b) If there is a number p such that $p | 180$ and $p | 105$, then it must be the case that $p | 75$ because $180 = 105 + 75$.

c) $12 \nmid 151$ since, although $151 = 130 + 21$ and $13 | 130$, it is not true that $13 | 21$.

Try Exercise 4

The following is a summary of the properties of the "divides" relation discussed in the text. For any whole numbers a, b, and c ($a \neq 0$):

1. If $a | b$ and $a | c$, then $a | (b + c)$.
2. If $a | b$, then $a | b \cdot c$.
3. If $a | b$ and $a | (b + c)$, then $a | c$.
4. If $a | b$, then $a \leqslant b$.
5. If $a | b$ and $b | c$, then $a | c$ (the transitive law).

Statements 1 and 3 have been proved in the text, the rest have been left for the exercises.

4. Use the appropriate theorems in:

a) If 2 divides 24, then which is true?

$$2 | 24 \cdot 3 \quad \text{or} \quad 2 \nmid 24 \cdot 3$$

b) Since $7 | 70$ and $7 \nmid 23$, which is true?

$$7 | (70 + 23) \quad \text{or} \quad 7 \nmid (70 + 23)$$

c) Since $7 | 252$ and $7 | 49$, which is true?

$$7 | 203 \quad \text{or} \quad 7 \nmid 203$$

EXERCISE SET 5.2

1. State T for those of the following statements that are true and F for those that are false:

a) $14|7$ b) $7|14$
c) $0|37$ d) $37|0$
e) $10|(10 \cdot m + 2)$ f) $2|2 \cdot n$
g) $2|(2 \cdot n + 1)$ h) $3p|p$

2. For each of the following open sentences find two whole numbers that make the statement true:

a) $a|12$ b) $b|15$
c) $6|m$ d) $7|m$
e) $p|27$ f) $q|64$
g) $6|(p + 12)$ h) $7|(p + 2)$

3. Find three different divisors for each of the following:

a) 14 b) 126 c) 66
d) 115 e) 171 f) 58

4. In each of the following, either prove the statement for all whole numbers or give a counterexample:

a) If $p \neq 0$, then $p|p$.
b) If $a \neq 0$ and $a | b$, then $(a + c) | (b + c)$.
c) If $a \neq 0$, $c \neq 0$ and $a \cdot c|b \cdot c$, then $a|b$.
d) If $p \neq 1$ and $p|N$, then $p \nmid N + 1$.

5. The symbols $a \div b$ and $a|b$ do not mean the same thing. Discuss the difference between them.

6. Illustrate and then prove the following statement:
For any whole numbers a, b, and c ($a \neq 0$),
if $a | b$, then $a | b \cdot c$.

7. Prove the transitive property for the "divides" relation:
For any whole numbers a, b, and c ($a \neq 0$ and $b \neq 0$),
if $a|b$ and $b|c$, then $a|c$.

8. Show that for all natural numbers a and b, if $a|b$, then $a \leqslant b$.

5.3 AIMS

You should be able to:

a) Define a prime and a composite number.
b) Use the sieve of Eratosthenes to find all the primes less than some specified number, say, 200.
c) Find the prime factorization of a given number.

5.3 PRIMES AND COMPOSITES

If six checkers are used to form rectangular arrays, then the two possible arrays that use all the checkers each time show 6 to have the factors 1, 2, 3, and 6.

2 X 3

On the other hand, the only array possible for 13 checkers points out the fact that 13 is a prime number and has exactly two different factors: 13 and 1.

1 X 13

A natural number is a *prime* number if and only if it has exactly two different factors. It is *composite* if and only if it has more than two different factors.

Because 1 does not satisfy the definition for either a prime or a composite, the set of *natural numbers* is the union of the following disjoint sets: number 1, the primes, the composites.

It is not always simple to determine whether a large number is prime; for example, is 111 111 111 111 a prime number or not? No one has succeeded in finding a formula that will consistently produce prime numbers, although some famous mathematicians have tried.

SIEVE OF ERATOSTHENES. The sieve of *Eratosthenes** is a simple (but not very rapid) way of finding prime numbers; it starts with the following diagram:

1	2	3	4	5	6	7	8	9	10
11	12	13	14	15	16	17	18	19	20
21	22	23	24	25	26	27	28	29	30
31	32	33	34	35	36	37	38	39	40
41	42	43	44	45	46	47	48	49	50
51	52	53	54	55	56	57	58	59	60
61	62	63	64	65	66	67	68	69	70
71	72	73	74	75	76	77	78	79	80
81	82	83	84	85	86	87	88	89	90
91	92	93	94	95	96	97	98	99	100

The number 1 is not a prime, so it is marked out. The number 2 is prime, so it is circled but its multiples such as 4, 6, 8, and so on are composite and these are next to be marked out. At this stage 3 has not been marked out, since it is not a multiple of 2; thus it is the next prime.

Try Exercise 5

The smallest number after 3 that has not been eliminated as a multiple of either 2 or 3 is 5, which means it is the next prime. The multiples of 5 are marked out next. Because 6 is already eliminated along with all of its multiples, 7 is the next prime found with this process, and its multiples must be marked out next.

Try Exercise 6

Because 8, 9, and 10 have already been removed as multiples of 2, 3, or 5, the next prime is 11. The multiples $2 \cdot 11$, $3 \cdot 11$, $4 \cdot 11$, $5 \cdot 11$, and $7 \cdot 11$ have already been removed as multiples of 2, 3, 5, or 7, and we see that the only multiple of 11 that has not been removed is $11 \cdot 11 = 121$, which is greater than 100. There is no necessity to continue the process beyond 10. The numbers that are left are the primes less than 100.

Try Exercise 7

5.

a) Mark out the multiples of 2 that are greater than 2 (mark out every second number).

b) Mark out the multiples of 3 that are greater than 3 (mark out every third number).

6.

a) Mark out the multiples of 5 that are greater than 5 (mark out every fifth number).

b) Mark out the multiples of 7 that are greater than 7.

1	2	3	4	5	6	7	8	9	10
11	12	13	14	15	16	17	18	19	20
21	22	23	24	25	26	27	28	29	30
31	32	33	34	35	36	37	38	39	40
41	42	43	44	45	46	47	48	49	50
51	52	53	54	55	56	57	58	59	60
61	62	63	64	65	66	67	68	69	70
71	72	73	74	75	76	77	78	79	80
81	82	83	84	85	86	87	88	89	90
91	92	93	94	95	96	97	98	99	100

7.

a) Check your chart and confirm that what has been stated about the multiples of 11 is true.

b) Circle all the numerals not marked out. These represent the primes less than 100.

* Eratosthenes was a Greek mathematician and astronomer (ca. 276–195 B.C.).

IS THERE A LARGEST PRIME? It might appear, as you study your chart for finding the primes below 100, that the primes seem to be getting fewer and fewer and might possibly disappear altogether. At some point, when the numbers are large enough, do the primes simply end?

Suppose we look at a scheme that seems to produce primes. If the scheme always worked, then we could continue to construct primes forever and thus answer the question.

Example 1 If successive primes are multiplied and 1 is added to the product, is the result another prime?

Solution We try this scheme by using the first primes that have remained in our "sieve."

$$2 + 1 = 3 \quad \text{(a prime)}$$
$$2 \cdot 3 + 1 = 7 \quad \text{(a prime)}$$
$$2 \cdot 3 \cdot 5 + 1 = 31 \quad \text{(a prime)}$$

So far the scheme works pretty well and we continue:

$$2 \cdot 3 \cdot 5 \cdot 7 + 1 = 211 \quad \text{(a prime)}$$
$$2 \cdot 3 \cdot 5 \cdot 7 \cdot 11 + 1 = 2311 \quad \text{(a prime)}$$
$$2 \cdot 3 \cdot 5 \cdot 7 \cdot 11 \cdot 13 + 1 = 30\,031 = 59 \cdot 509 \quad \text{(a composite!)}$$

We must conclude that this scheme does not yield a prime each time, since we have produced a counterexample.

While the result of the last example was disappointing in that we cannot continue to grind out primes each time, we can still utilize the pattern in solving the original problem. We notice that the produced primes were larger than the primes involved in the computation and that the composite found had a prime factor that was larger than any of the primes used to build it. We can apply these ideas in a step-by-step argument very similar to that given by Euclid* to prove that there is no largest prime.

Proof

1. Assume there is a largest prime and call it p.
2. Find the product of all the primes from 2 to p and add 1 to the product:

$$(2 \cdot 3 \cdot 5 \cdot 7 \cdot 11 \cdot \cdots \cdot p) + 1; \quad \text{call this } N + 1.$$

3. If $N + 1$ is prime, it is larger than any primes on the list from 2 to p.
4. If $N + 1$ is composite, it must be divisible only by a prime larger than p (the prime p and those less than p do not divide $N + 1$).
5. In either case, a prime larger than p exists, which contradicts the original assumption. The conclusion must be that there is no largest prime.†

* Euclid (ca. 300 B.C.) was one of the mathematical giants of ancient Greece.

† In 1977, Hugh C. Williams of the University of Manitoba discovered a new prime. All of its digits are 1 and there are 317 of them. It is called R_{317} for short to avoid writing it in the form

$$\underbrace{111\cdots1}_{317}.$$

At the present time there are only four of these primes known: R_2, R_{19}, R_{23}, and R_{317}. This last one was found nearly 50 years after the previous one.

Reported in *Scientific American*, 238:2, pp. 89–90, February 1978.

THE UNIQUE FACTORIZATION THEOREM. A *prime factorization* of a number is a product in which all the factors are prime. Such representation is particularly useful in work with rational numbers expressed by fractions, which will be discussed in Chapter 6.

Example 2 Find the prime factorization of 24.

Solution

$$\begin{aligned}\text{Start with } 24 &= 2 \cdot 12 \quad \text{(now factor the composite 12)}\\ &= 2 \cdot 2 \cdot 6 \quad \text{(now factor the composite 6)}\\ &= 2 \cdot 2 \cdot 2 \cdot 3 = 2^3 \cdot 3 \quad \text{(prime factorization).}\end{aligned}$$

Example 3 Find the prime factorization of 48.

Solution

$$\begin{aligned}\text{Start with } 48 &= 4 \cdot 12 \quad \text{(both 4 and 12 are composite)}\\ &= 2 \cdot 2 \cdot 6 \cdot 2 \quad \text{(6 is still a composite)}\\ &= 2 \cdot 2 \cdot 2 \cdot 2 \cdot 3 = 2^4 \cdot 3 \quad \text{(prime factorization).}\end{aligned}$$

Try Exercise 8

8. Find the prime factorization of 72.

"Discover any new prime numbers lately?"

(Reprinted from *The American Scientist*, Vol. 64, November-December 1976, by permission of Sidney Harris.)

Prime factorizations can be diagrammed in a tree-like figure in which the factoring continues until a prime is reached at the end of each branch. Even if the tree diagrams are not started with a factorization that includes a prime factor, the ends of the branches will eventually show the same information.

Example 4 Use a tree diagram to find the prime factorization of 32.

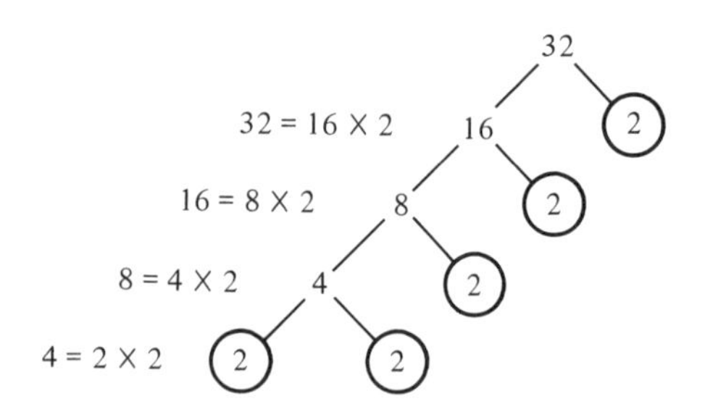

The possible factor pairs: 1×32, 2×16, 4×8.

The prime factorization is $2 \cdot 2 \cdot 2 \cdot 2 \cdot 2$ or 2^5. The set of all possible factors is $\{1, 2, 4, 8, 16, 32\}$.

Using poker chips to factor 24

Example 5 Find the prime factorization of 42. How many different prime factors do you find?

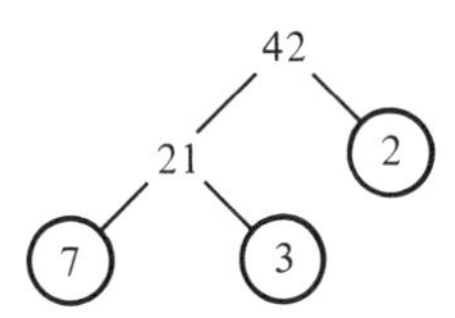

Solution The possible factor pairs: 1×42, 2×21, 3×14, 6×7.

The prime factorization is $7 \cdot 3 \cdot 2$; there are three different prime factors: 2, 3, and 7, and the set of all possible factors of 42 is $\{1, 2, 3, 6, 7, 14, 21, 42\}$.

Try Exercise 9

There must be an infinite number of composite numbers as well as an infinite number of primes, since each prime can be used with one or more of the other primes to produce a composite number. Different factor diagrams for the same composite always yield the same prime factors, no matter how they are started. This fact is an important theorem of number theory and is given here without proof.

THE UNIQUE FACTORIZATION THEOREM. Every natural number greater than 1 has one and only one prime factorization, that is, it is unique except for the order of the factors.

Example 6 Is $77 \cdot 10 = 35 \cdot 22$?

Solution Rather than multiply and compare, it is easier to factor:

$$77 \cdot 10 = 7 \cdot 11 \cdot 5 \cdot 2,$$
$$35 \cdot 22 = 5 \cdot 7 \cdot 2 \cdot 11.$$

The prime factorizations are the same; hence the equation given in the example is true.

Try Exercise 10

In a search for the primes of a factorization it is efficient to use a systematic approach and first check for the smallest prime 2, then 3, and so on. A calculator can be very useful, since this procedure can become tedious if the number is large. There is no need to consider primes whose square is greater than the number being factored.

9.

a) Finish an alternative tree diagram for Example 4:

b) Finish an alternative tree diagram for Example 5:

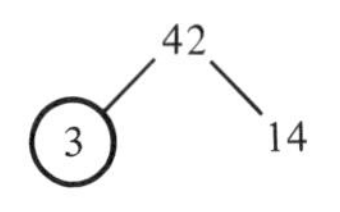

10. Without multiplying, determine whether

$$5^2 \cdot 6^3 \cdot 7 \cdot 11 = 5 \cdot 6^3 \cdot 7 \cdot 11 \cdot 25$$

is a true statement.

Example 7 Is 247 a prime number?

Solution First we see which primes need to be tested and then test them in order.

Since $13^2 = 169, 17^2 = 289$, and $169 < 247 < 289$, only the primes 2, 3, 5, 7, 11, 13 need be tested. If 13 is not a factor, then 17 cannot be a factor either, since it would need to be multiplied by a prime number greater than 13 and 17^2 is already larger than 247. We find that 2, 3, 5 and 7 are not factors. However, 13 is a factor, since $247 = 13 \cdot 19$.

EXERCISE SET 5.3

1. Find all the different prime factors of each of the following:
a) 24 b) 120 c) 720 d) 5040

2. Complete the following tree diagrams:

a) 

b) 

3. Youngsters can learn about factoring by working with arrays.
a) In how many ways can 81 checkers be placed in a rectangular array? Why?
b) In how many ways can 72 checkers be placed in a rectangular array? Why?

4. Study the sieve made earlier. Look for patterns. Are there numerals with two lines through them? three lines? Do they form a pattern?

5. Use the sieve method to find all the prime numbers between 100 and 200. The process is finished when multiples of what number are eliminated? (Keep the list of primes for your use.)

6. Devise an algorithm for finding the prime factorization of any number less than 1000.

7. Make a flowchart for the algorithm of Problem 6. Use the flowchart to find the prime factorization of 825.

8. Find the prime factorization of the following numbers:
a) 60 b) 126 c) 171 d) 153

9. Decide which of the following numbers are primes:
a) 211 b) 201 c) 319
d) 307 e) 2311 f) 407

10. The number of different possible factors of a number depends upon the prime factorization. Find the prime factorization, then find all possible factors for each of the following:
a) 48 b) 51 c) 64
d) 81 e) 154 f) 252

■ **11.** Do you see any patterns in Problem 10 that would allow the number of divisors of a number to be predicted from the prime factorization? Make a conjecture and prove it.

12. Use prime factorization to determine which of the following are true statements:
a) $2^2 \cdot 5 \cdot 7 = 20 \cdot 7$ b) $39 \cdot 63 \cdot 9 = 13 \cdot 81 \cdot 7$
c) $28 \cdot 28 = 3 \cdot 7 \cdot 4 \cdot 7 \cdot 4$ d) $36 \cdot 36 = 4 \cdot 18 \cdot 18$

13. Suppose that m is a whole number and n is another whole number (these need not be prime numbers). A certain whole number can be expressed as $m \cdot m$. Is it possible that this same number can be expressed as $3 \cdot n \cdot n$? Explain your answer.

14. Suppose that m is a whole number and n is another whole number (again these need not be prime numbers). A certain whole number can be expressed as $m \cdot m$. Is it possible that this same number can be expressed as $4 \cdot n \cdot n$? Explain.

In Problems 15–18 we return to the mystic square in Example 1 of Section 5.1.

15. Assume that the square is a portion of a multiplication table. Use your knowledge about factors and primes to help you find the numbers that belong on the margins of the table. For example, what two numbers were multiplied to give the product 77?

×				
	2	4	6	22
	3	6	9	33
	5	10	15	55
	7	14	21	77

16. Explain why the product in the square of Example 1 is always 13 860.

17. How could you make such a square for the final product 10 920?

■ **18.** For what numbers is it impossible to make such a square? Is it possible to make such a square for which the magic number is the year of your birth?

19. Since $1^2 + 1 = 2$, a prime number, it is tempting to think that this formula $x^2 + 1$ will generate prime numbers. How many prime numbers expressible as $x^2 + 1$ exist that are less than 500?

20. The formula $x^3 + 1$ produces a prime number when $x = 1$. Find all the prime numbers less than 1000 generated by this formula. Before you start, estimate the number of cases you will have to investigate.

21. One mathematician hoped that $x^2 + x + 41$ would always yield a prime number. Investigate this rule and see ■ why.

5.4 DIVISIBILITY TESTS

5.4 AIMS

You should be able to:

a) Determine whether any number is divisible by 2, 4, 5, 3, 9, or 10 without dividing the number itself.

b) Prove that the divisibility tests are valid.

The "divides" relation properties can be used to establish the well-known tests that allow us to recognize certain factors of a number without actually carrying out the division.

DIVISIBILITY TEST FOR 2 AND 5. We need only consider the units digit of a number to determine if the number is divisible by 2.

Example 1 Show that 2 divides 2468.

Solution 2468 can be written as $(2 \cdot 10^3 + 4 \cdot 10^2 + 6 \cdot 10) + 8$. First we consider

$$2460 = 2 \cdot 10^3 + 4 \cdot 10^2 + 6 \cdot 10.$$

11. Use expanded notation to show that 175 is not divisible by 2.

Since $2 \mid 10$, a property of the "divides" relation assures us that $2 \mid 2 \cdot 10^3$, $2 \mid 4 \cdot 10^2$ and $2 \mid 6 \cdot 10$, which means by another property of the "divides" relation that 2 divides the sum of these numbers, so we have $2 \mid 2460$ or $2 \mid 10 \cdot 246$.

Now we look at the 8 and observe that $2 \mid 8$. Since $2 \mid 2460$ and $2 \mid 8$, we conclude that $2 \mid (2460 + 8)$, which we wanted to prove.

This result holds in general, since any whole number whatsoever can be expressed as $10 \cdot m + n$, where n is the units digit. This is obvious from our decimal notation system. We know that $2 \mid 10 \cdot m$, since $2 \mid 10$; if in addition $2 \mid n$, then $2 \mid (10 \cdot m + n)$ by a property of the "divides" relation. On the other hand, if $2 \nmid n$, then $2 \nmid (10 \cdot m + n)$.

This entire line of reasoning can be carried out in the same way for 5, since $5 \mid 10$ also. Thus the proof is valid for 5 as well. We have just proved that:

For any whole number *a*, 2 *divides* *a* if and only if 2 divides the units digit of *a*; also, 5 divides *a* if and only if 5 divides the units digit of *a*.

From this statement we can see that a number divisible by 2 must have 0, 2, 4, 6, or 8 as a last digit, while a number divisible by 5 has either 0 or 5 for the last digit.

12. Use expanded notation to show that 45 624 is divisible by 4.

Try Exercise 11

DIVISIBILITY BY 4. It is necessary to look at the last two digits of the numeral for a whole number in order to determine if it is divisible by 4. If the number represented by the last two digits is divisible by 4, so is the number itself. We look at an example and then do the proof.

Example 2 Use expanded notation to show that 4636 is divisible by 4.

Solution $4636 = 4 \cdot 10^3 + 6 \cdot 10^2 + 36$. Since $4 \mid 10^2$, it must also hold that $4 \mid 10^3$ and $4 \mid (4 \cdot 10^3 + 6 \cdot 10^2)$; that is, $4 \mid 4600$. Also $4 \mid 36$, so $4 \mid (4630 + 36)$, which is what was to be shown.

This reasoning holds for a general proof: Because of the decimal notation system, any whole number can be represented as $100 \cdot m + n$, where n is the number formed by the last two digits of the numeral. Since $4 \mid 100$ and hence 4 will divide $100 \cdot m$, we can reason (as we did in the proof above) that 4 will divide $100 \cdot m + n$ if and only if 4 divides n.

Try Exercise 12

DIVISIBILITY BY 8. The familiar rule for divisibility by 8 can be established with similar reasoning and the proof is left as an exercise.

A whole number is *divisible by* 8 if and only if 8 divides the number represented by the last three digits of its numeral.

DIVISIBILITY BY 9 AND 3. The simple test for deciding whether a number is divisible by 9 is suggested by looking at the following pattern:

Number	*Sum of digits*	*Number*	*Sum of digits*
18	9	63	9
27	9	891	18
45	9	1782	18

In each case in the table, 9 divides both the number and the sum of the digits.

Example 3 Show that 2475 is divisible by 9 by examining the expanded notation.

Idea If the pattern of the table holds, we want to separate the sum of the digits from the other terms in the sum:

$$2475 = 2 \cdot 1000 + 4 \cdot 100 + 7 \cdot 10 + 5.$$

We want to get $(2 + 4 + 7 + 5)$ isolated in some way and we try:

$$2475 = 2 \cdot (999 + 1) + 4 \cdot (99 + 1) + 7 \cdot (9 + 1) + 5.$$

By the usual laws for whole numbers this can be rearranged as:

$$2475 = (2 \cdot 999 + 4 \cdot 99 + 7 \cdot 9) + (2 + 4 + 7 + 5).$$

Now 9 divides 9, 99, and 999, so $9 \mid (2 \cdot 999 + 4 \cdot 99 + 7 \cdot 9)$. Therefore 9 divides 2475 if and only if 9 divides $(2 + 4 + 7 + 5)$, which is the sum of the digits.

The procedure of the example can be used to prove the general case. Any power of ten can be written as:

$$10^n = \underbrace{9999999\cdots}_{n \text{ digits}} + 1.$$

We know that $9 \mid 9$, $9 \mid 99$, and $9 \mid 999$, and so on, so that any whole number can be written as $9 \cdot m + S$, where S represents the sum of the digits. Since 9 divides $9 \cdot m$, the number itself is divisible by 9 if and only if the sum of the digits S is divisible by 9. Since $3 \mid 9$, the same reasoning holds for 3, so the proof is correct for 3 also and we have proved the following:

A whole number is *divisible by* 9 if and only if the sum of its digits is divisible by 9. It is *divisible by* 3 if and only if the sum of its digits is divisible by 3.

Try Exercise 13

13.

a) Use the method of the proof and the reasoning there to prove that 467 is not divisible by 9.

b) Use the method of the proof to prove that while 669 is not divisible by 9, it is divisible by 3.

EXERCISE SET 5.4

1. Prove the divisibility test for 8 stated in the text.

2. State and prove a divisibility test for 10.

In Problems 3–10, test for divisibility by 2, 3, 4, 5, 9, and 10:

3. 135

4. 225

5. 422

6. 504

7. 240

8. 480

9. 2160

10. 4260

11. If your test shows that a number is divisible by 2 and by 5, can you conclude without further testing that it is divisible by 10?

12. If a number is divisible by 2 and also by 4, can you conclude it is also divisible by 8?

13. State and prove a divisibility test for 6.

14. Under what circumstances is the following statement true?

For whole numbers a, b, and c,
if $a \mid b$ and $c \mid b$, then $a \cdot c \mid b$.

■ 15. Find a test for divisibility by 11. Prove that it is valid.

16. Test the following for divisibility by 2:

a) 660 b) 801 c) 1984 d) 1776
e) 82 071 f) 5661 g) 67 342 h) 83 763

17. Check the numbers in Problem 16 for divisibility by 3, by 6, by 11, by 22, and by 33.

■ 18. Use base-seven notation. Find and prove a divisibility test for 2.

■ 19. Use base-four notation. Find and prove a divisibility test for 2.

20. Suppose a number is not divisible by 2; need you continue and test for 4 and for 8? Explain your answer.

■ 21. The divisibility test for 9 is the basis of a once widely-used procedure for checking whole-number computation called "casting out nines," which uses remainders after division by 9. If a whole number is divided by 9, the remainder is the same as if the sum of the digits is divided by 9. For example, to check the addition:

$$\begin{array}{r l} 4312 & \rightarrow 4+3+1+2 \rightarrow 10 \rightarrow 1+0 \rightarrow 1 \\ +7614 & \rightarrow 7+6+1+4 \rightarrow 18 \rightarrow 1+8 \rightarrow 9 \rightarrow 0 \end{array} \Big\} \; 1+0 = 1$$
$$11926 \rightarrow 1+1+9+2+6 \rightarrow 19 \rightarrow 1+9 \rightarrow 10 \rightarrow 1+0 = 1$$

a) Show that when 9 divides a whole number, the remainder is the same as the sum of the digits.
b) Explain why this casting-out-nines procedure provides a check for addition. Is it a foolproof check?
c) Perform the addition and check by casting out nines for: 4326 + 8536.
d) Show that the casting-out-nines check holds in the following case: 8109 − 7824.
e) Investigate the problem of casting out nines as a check for the multiplication of two whole numbers.

5.5 GREATEST COMMON DIVISOR, LEAST COMMON MULTIPLE

GREATEST COMMON DIVISOR (GCD). In the solution of equations and in other computation it often facilitates matters to be able to simplify by recognizing a common divisor of two or more numbers. This is true of the computations we will be doing in the next chapter.

Two whole numbers, chosen at random, may or may not have any divisors other than 1 that are the same. In the case of 18 and 42, for example, the set of divisors of 18 is $\{1, 2, 3, 6, 9, 18\}$ and the set for 42 is $\{1, 2, 3, 6, 7, 14, 21, 42\}$. The intersection of these sets consists of their common divisors and is $\{1, 2, 3, 6\}$, which means that the greatest common divisor of 18 and 42 is 6, a fact that will be useful in the next chapter when we wish to simplify 12/42.

DEFINITION. The *greatest common divisor* of any two whole numbers *a* and *b* is the greatest divisor *d* such that *d* divides *a* and *d* divides *b*. It is denoted GCD (*a*, *b*).

The greatest common divisor is sometimes called the greatest common factor.

Try Exercise 14

If the greatest common divisor is found by use of the intersection of the two sets of divisors it can be very time-consuming, so we reduce the work by the use of prime factorizations.

Example 1 Find the GCD of 60 and 270.

Solution We first find the prime factorizations:

$$60 = 2 \cdot 2 \cdot 3 \cdot 5, \qquad 270 = 2 \cdot 3^3 \cdot 5.$$

In the factorization for 60 and for 270 there is 2, so that 2 is a common divisor; however, $2 \cdot 2$ is not a common divisor, since it does not appear in the factorization of 270. Similarly, 3 and 5 are both common divisors, and the GCD is

$$2 \cdot 3 \cdot 5 = 30.$$

This same method can be used for three or more numbers.

Example 2 Find the greatest common divisor of 35, 105, and 42.

Solution We find the prime factorizations:

$$35 = 5 \cdot 7, \qquad 105 = 3 \cdot 5 \cdot 7, \qquad 42 = 2 \cdot 3 \cdot 7.$$

The only common factor is 7 and it is the GCD.

Try Exercises 15 and 16

Example 3 Find the greatest common divisor of 40 and 27.

5.5 AIMS

You should be able to:

a) Find the GCD of two or more whole numbers.
b) Find the LCM of two or more natural numbers.
c) Determine whether two numbers are relatively prime.

14.

a) What is GCD of 25 and 35?
b) What is GCD of 4 and 9?
c) What is GCD of 17 and 0?

Use prime factorization to find the GCD of:

15. 80 and 155.

16. 105, 180, and 210.

17. Which of the following pairs are relatively prime?
a) 23 and 47
b) 69 and 64
c) 22 and 14
d) 33 and 121

Solution We factor: $40 = 2 \cdot 2 \cdot 2 \cdot 5$, $27 = 3 \cdot 3 \cdot 3$. The prime factorizations in this case show no prime factors in common, but 1 divides both 40 and 27, so it is the GCD.

The integers 40 and 27 of this example are said to be *relatively prime*, although each of the numbers is a composite number.

Any two whole numbers *a* and *b* are called *relatively prime* if and only if their greatest common divisor is 1.

Try Exercise 17

EUCLID'S ALGORITHM. If the numbers are large, finding the greatest common divisor by means of the factorization method becomes increasingly difficult, so a method that uses the division algorithm over and over is preferable.

Example 4 Use the division algorithm repeatedly to find the greatest common divisor of 180 and 105.

Step 1 The division algorithm, with 105 as the divisor, gives:

$$180 = 105 \cdot 1 + 75, \qquad 0 \leqslant 75 < 105.$$

Now, any divisor of 180 and 105 will have to be a divisor of 75, since by a property of the "divides" relation:

$$\text{if } a \mid (b + c) \text{ and } a \mid b, \text{ then } a \mid c;$$

hence,

$$\text{if } a \mid 180 \text{ and } a \mid 105, \text{ then } a \mid 75.$$

Step 2 The process is repeated to find the GCD of 105 and 75:

$$105 = 75 \cdot 1 + 30, \qquad 0 \leqslant 30 < 75.$$

Any divisor of 105 and 75 will have to be a divisor of 30 by the reasoning in Step 1.

Step 3 The process is repeated to find the GCD (75, 30):

$$75 = 30 \cdot 2 + 15, \qquad 0 \leqslant 15 < 30.$$

Any divisor of 75 and 30 will have to be a divisor of 15.

Step 4 The process is repeated to find the GCD(30, 15):

$$30 = 15 \cdot 2 + 0, \qquad 0 \leqslant 0 < 15.$$

Finally we see that the greatest common divisor of 30 and 15 must also be a divisor of 0, and since the GCD(15, 0) is 15, we now know that 15 is the greatest common divisor we have been trying to find.

This method, called *Euclid's algorithm*, has shown the following in Example 4:

$$\text{GCD}(180, 105) = \text{GCD}(105, 75) = \text{GCD}(75, 30) = \text{GCD}(30, 15) = \text{GCD}(15, 0).$$

Since the GCD(15, 0) is 15, we know that the GCD(180, 105) is also 15.

Euclid's algorithm applied to the two whole numbers a and b can be represented as follows:

$$a = b \cdot Q_1 + R_1, \qquad 0 \leqslant R_1 < b.$$

Any divisor of a and b must also be a divisor of R_1:

$$\begin{aligned} b &= R_1 \cdot Q_2 + R_2, & 0 &\leqslant R_2 < R_1, \\ R_1 &= R_2 \cdot Q_3 + R_3, & 0 &\leqslant R_3 < R_2, \\ &\vdots & &\vdots \\ R_{n-2} &= R_{n-1} \cdot Q_n + R_n, & 0 &\leqslant R_n < R_{n-1}, \\ R_{n-1} &= R_n \cdot Q_{n+1} + 0. \end{aligned}$$

A zero finally shows up as the remainder because each succeeding R is less than the preceding one, and from the last line we know that the GCD(R_n, 0) is R_n, which is the remainder in the second last line and is the greatest common divisor we are looking for.

Try Exercise 18

18. Use Euclid's algorithm to find GCD (12, 66)

Will this method work if the two numbers in question have no prime factors in common?

Example 5 Use Euclid's algorithm to find the GCD(16, 27).

Step 1 $27 = 16 \cdot 1 + 1$, $0 \leqslant 11 < 15$. A divisor of 27 and 16 must also be a divisor of 11.

Step 2 The process is repeated: $16 = 11 \cdot 1 + 5$, $0 \leqslant 5 < 11$.

Step 3 $11 = 5 \cdot 2 + 1$, $0 \leqslant 1 < 5$.

Step 4 $5 = 1 \cdot 5 + 0$, $0 \leqslant 0 < 1$.

The GCD(1, 0) is 1, so the GCD(16, 27) is also 1. We have shown that 27 and 16 are relatively prime.

Try Exercise 19

19. Use Euclid's algorithm to determine which of the following pairs are relatively prime:

a) 175 and 24
b) 210 and 121
c) 1678 and 531
d) 96 and 352

LEAST COMMON MULTIPLE. In addition to the divisors of a number we also must think about the multiples of a number, since the idea of a common multiple is needed in the computation done with fractions in the next chapter. Because every divisor of a nonzero whole number is less than or equal to the number itself, the set of divisors for such numbers is a finite set, which is not the case for the set of multiples.

We know from the definition of the "divides" relation that if a number a divides a number m, then m is a multiple of a. Thus $a \cdot n$ is a multiple of a for any whole number n.

Example 6

a) Find the set of multiples of 7.
b) Find the set of multiples of 14.

Solution

a) The multiples are: $\{m \mid m = 7 \cdot n \text{ and } n \text{ is a whole number}\}$. The set is $\{0, 7, 14, 21, 28, \ldots\}$.
b) The multiples are $\{m \mid m = 14 \cdot n \text{ and } n \text{ is a whole number}\}$. The set is $\{0, 14, 28, 42, \ldots\}$.

Try Exercise 20

20.
a) Find the multiples of 3.
b) Find the multiples of 9.
c) What relation holds for these two sets?

The intersection of the sets of multiples found in Example 6 contains many numbers, but we are interested in the smallest such nonzero number. For the common multiples of 7 and 14 this smallest number is 14.

DEFINITION. The *least common multiple* of the whole numbers *a* and *b* is the smallest nonzero whole number *m* such that *a* divides *m* and *b* divides *m*. It is denoted LCM (*a*, *b*).

The least common multiple for two or more numbers can be found by considering the sets of multiples for the numbers and finding their intersection.

Example 7 Find the least common multiple (LCM) of 24 and 36.

Solution The set of nonzero multiples of 24 is:

$$M_{24} = \{24, 48, 72, 96, 120, 144, 168, 192, 216, 240, \ldots\}.$$

The set of nonzero multiples for 36 is:

$$M_{36} = \{36, 72, 108, 144, 180, 216, 252, \ldots\}.$$

The set $M_{24} \cap M_{36}$ is the set of nonzero multiples that are common to 24 and 36:

$$M_{24} \cap M_{36} = \{72, 144, 216, \ldots\}.$$

The smallest, or least, of these common multiples is 72.

Try Exercise 21

21.
a) Find the next three multiples of 24.
b) Find the next three multiples of 36.
c) Find the next common multiple of 24 and 36.

It is possible to find least common multiples in a more efficient way by using prime factorization.

Example 8 Find the LCM(24, 36) using prime factorization.

Solution

$$24 = 2 \cdot 2 \cdot 2 \cdot 3 = 2^3 \cdot 3,$$

so each multiple of 24 has $2^3 \cdot 3$ in its prime factorization. Also,

$$36 = 3 \cdot 3 \cdot 2 \cdot 2 = 2^2 \cdot 3^2,$$

so each multiple of 36 has $2^2 \cdot 3^2$ in its prime factorization.

The number $2^3 \cdot 3^2$ is a multiple of both numbers and is the smallest one possible, so it is the least common multiple.

To find the LCM of several numbers by this method, we find each prime factorization and use each distinct factor the greatest number of times that it occurs in any factorization. A theorem relating the GCD(a, b) and LCM(a, b) will be explored in Exercise Set 5.5 (see Problems 6 and 7).

Try Exercise 22

22.

a) Find the LCM (18, 24).

b) Find the LCM (3, 17).

EXERCISE SET 5.5

1. Use prime factorization to find the greatest common divisor of each set of numbers:

a) 126, 30, 105

b) 1540, 210, 1820

c) 750, 525, 825

2. Find the set of common divisors for the following pairs:

a) 64 and 32 b) 9 and 16 c) 49 and 51

d) 24 and 48 e) 25 and 55 f) 121 and 66

3. Use Euclid's algorithm to find the greatest common divisor for the following:

a) 5674 and 672 b) 6155 and 730

c) 18 927 and 663 d) 1984 and 1088

e) 4769 and 296 f) 37 982 and 3761

4. Show that if p is a prime number and a is any number, then there are just two possibilities for the greatest common divisor of a and p.

5. Find the LCM of each set of numbers:

a) 30, 105, 140

b) 10, 105, 14, 15

c) 8778, 3542

6. Complete the following table:

a	b	$a \cdot b$	GCD(a, b)	LCM(a, b)
60	126	7560		
74	102			
91	95			
105	180			
210	285			
266	151			

7. Study the table of Problem 6 for patterns and relationships in the last three columns. ■ Make a conjecture and prove it.

8. Find the GCD and the LCM of each pair:

a) 5, 7 b) 11, 13 c) 2, 23 d) 3, 61

9. What is the GCD of two prime numbers? The LCM of two prime numbers?

10. If two numbers are relatively prime, what is their GCD? LCM?

11. If two numbers have the same number for their LCM and GCD, what can be said about the two numbers?

12. Show that for any natural number R_n, the GCD(R_n, 0) is R_n.

13. If M_3 is the set of nonzero multiples of 3 between 1 and 500, and M_7 is the set of nonzero multiples of 7 between 1 and 500, describe the set $M_3 \cap M_7$ using the roster method.

14. Make a flowchart for finding the GCD(a, b) by any method you wish. Use the flowchart to find GCD(84, 3080).

15. Assume that $a \neq 0$ and $a | m$. Find the GCD(a, m).

■ **16.** Suppose that the greatest common divisor of a and b is d; also, that $a = p \cdot d$ and $b = q \cdot d$. Prove that p and q are relatively prime.

5.6 AIMS

You should be able to:

a) Use the algebraic description of even and odd numbers to determine some further properties of the whole numbers.

b) Recognize and develop simple relations between the triangular numbers, square numbers, and rectangular numbers.

5.6 SPECIAL KINDS OF NUMBERS

EVEN AND ODD. The whole numbers that are divisible by 2 are the *even* numbers, and those that are not are the *odd* numbers, a classification frequently useful in problem solving. Since 2 divides every even number, any even number is $2 \cdot n$ for some whole number n. But the odd numbers must have a remainder of 1 when divided by 2, hence are $2 \cdot n + 1$ for some whole number n. From the divisibility tests for 2 it follows that the last digit of an even number in standard notation is 0, 2, 4, 6, or 8.

Try Exercise 23

23.

a) Express the even number 48 as $2 \cdot n$.

b) Divide the odd number 75 by 2 and express it as $2 \cdot n + 1$.

TRIANGULAR NUMBERS. Looking for patterns in a sequence of numbers may help find a solution for a complex problem; if familiar numbers are present, a solution may be easy. The Chinese seem to have been the first to investigate such sequences by using a geometric representation for various numbers. The Pythagoreans thought that everything could be explained by numbers and although they made an effort to explain geometric shapes in this way, they were not completely successful.

Imagine checkers or other such counters arranged as follows:

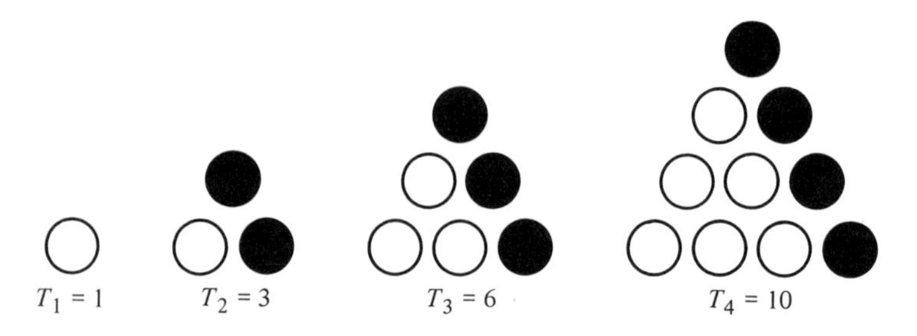

The sequence of numbers that goes along with this pattern consists of the *triangular numbers.*

Try Exercise 24

24. Use counters or diagrams to find the next three triangular numbers.

A study of the triangular arrangements and the shaded objects will show that the triangular numbers may be thought of as follows:

Triangular numbers

First number T_1	1
Second number T_2	$1 + 2$
Third number T_3	$1 + 2 + 3$
Fourth number T_4	$1 + 2 + 3 + 4$

Each of the numbers is built up from the previous one. The kth triangular number T_k is the sum of all the natural numbers up to and including k:

$$T_k = 1 + 2 + 3 + \cdots + k.$$

Try Exercise 25

25. The 13th triangular number is the sum of what natural numbers?

SQUARE NUMBERS. Numbers that go with the following sequence of geometric patterns are called *square numbers.*

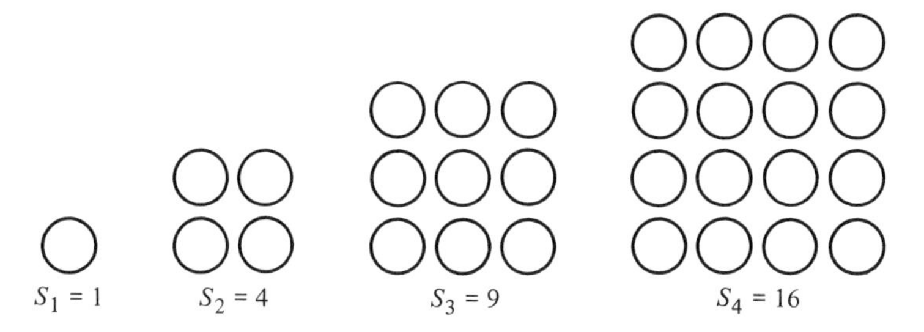

Try Exercise 26

26. Use counters or diagrams to find the next three square numbers. What is added in each case to get the next number in the sequence?

The square numbers can be thought of as the following:

Square numbers

First number	$S_1 = 1^2 = 1$
Second number	$S_2 = 2^2 = 1 + 3$
Third number	$S_3 = 3^2 = 1 + 3 + 5$
Fourth number	$S_4 = 4^2 = 1 + 3 + 5 + 7$

The kth square number S_k is the sum of the first k odd numbers:

$$S_k = 1 + 3 + 5 + \cdots + (2 \cdot k - 1).$$

Try Exercise 27

27. Find the following sum *without* adding:

$$1 + 3 + 5 + 7 + 9 + 11.$$

28. Use poker chips or diagrams to lay out the next three rectangular numbers. Verify that R_5, R_6, and R_7 all satisfy the formula for R_k.

RECTANGULAR NUMBERS. Numbers that go with the following geometric pattern are called *rectangular numbers*.

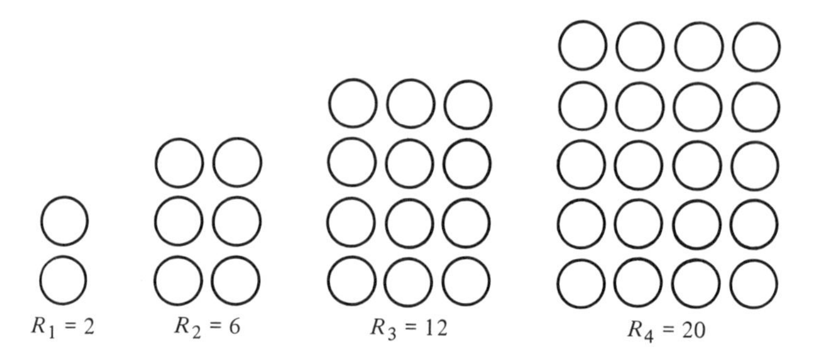

From the geometric arrangement it is easy to see the following:

Rectangular numbers

First number $R_1 = 2 = 1 \cdot 2$

Second number $R_2 = 6 = 2 \cdot 3$

Third number $R_4 = 12 = 3 \cdot 4$

Fourth number $R_4 = 20 = 4 \cdot 5$

The kth rectangular number is $R_k = k \cdot (k + 1)$.

Try Exercise 28

You might have observed that:

$$R_1 = 2 = 1 + 1 = T_1 + T_1,$$
$$R_2 = 6 = 3 + 3 = T_2 + T_2,$$
$$R_3 = 12 = 6 + 6 = T_3 + T_3,$$
$$R_4 = 20 = 10 + 10 = T_4 + T_4.$$

Each rectangular number can be regarded in two different ways, either as a product or as a sum of corresponding triangular numbers. In particular,

$$R_8 = 8 \cdot 9$$

and it is also the following sum:

$$R_8 = (1 + 2 + 3 + 4 + 5 + 6 + 7 + 8) + (1 + 2 + 3 + 4 + 5 + 6 + 7 + 8)$$
$$= T_8 + T_8$$

Hence,

$$R_8 = 8 \cdot 9 = 2 \cdot T_8 \qquad \text{and} \qquad T_8 = \frac{8 \cdot 9}{2} = 36.$$

The sum of the first eight natural numbers is thus 36.

"There is an irresistible fascination in finding numbers having specified properties and one soon falls under their spell and begins to understand why so many men were willing to devote so much time to this subject."

Recreations in the Theory of Numbers—The Queen of Mathematics Entertains by Albert H. Beiler. First published by Dover Publications, New York (1964), p. 1.

Similar reasoning will give the sum of the first k natural numbers. Since

$$R_k = k \cdot (k+1) \qquad \text{and} \qquad R_k = T_k + T_k,$$

it follows that

$$2 \cdot T_k = k \cdot (k+1) \qquad \text{and} \qquad T_k = \frac{k \cdot (k+1)}{2},$$

which is the sum of the first k natural numbers.

Try Exercise 29

Numbers obtained from geometric patterns, such as the square numbers and triangular numbers, are called *figurate numbers*. The geometric patterns need not be confined to a plane; the cubes: 1, 8, 27, 64, 125, . . . can easily be built with 3-dimensional models with blocks or checkers. Many relationships between the figurate numbers can also be found algebraically as, for example, the cubes which can be shown to be related to the triangular numbers:

$$\begin{array}{lll} T_1 = 1 & 1^2 = 1^3 & = 1 \\ T_2 = 3 & 3^2 = 1^3 + 2^3 & = 9 \\ T_3 = 6 & 6^2 = 1^3 + 2^3 + 3^3 & = 36 \\ T_4 = 10 & 10^2 = 1^3 + 2^3 + 3^3 + 4^3 & = 100 \end{array}$$

Pyramidal numbers can be constructed using the triangular or square numbers as bases in the way oranges might be piled in a grocery store; a square-based pyramid with 4 oranges on a side would require a total of 30 oranges.

Try Exercise 30

Working with triangular numbers

29. Find the sum of all the natural numbers from 1 to 18 inclusive.

30.
a) Verify that T_4^2 is the sum of the first four cubic numbers.
b) How many balls are required to build a triangular-based pyramid with 10 balls in the base?

In a classroom

Mark: I have been working with my calculator, and if you add three numbers in a row they always add to a multiple of three.

Teacher: How many of these did you try?

Mark: At least 30 and they all worked.

Carmen: Well, that doesn't mean there aren't some that *won't* work.

Teacher: That's true. But perhaps we could try to prove it. I think you are saying that the sum of any three consecutive numbers is always divisible by three.

Mark: Consecutive . . . that means each number is one more than the one before? Yes, I think that is what is true.

Teacher: Maybe we could prove this. What do we know?

Carmen: We think that if three numbers in a row are added the sum is a multiple of three.

Teacher: Do you have any ideas for working on this?

Carmen: You know, if you look at three numbers in a row—one of them is odd or two of them are odd. Just look at 3, 4, 5; that's odd, even, odd. Now look at 4, 5, 6; that's even, odd, even.

Mark: That's because the odd and even numbers go back and forth all the time.

Carmen: What happens if you add the numbers in a row with those two different patterns?

Teacher: Let's see if we can find out. Where shall we start? With the odd, even, odd? What do we know about even numbers?

Mark: They have a factor of 2. So every even number is 2 times something. You could write $2 \cdot \square$. (Goes to board.)

Teacher: Good. Now what is the number just before that number?

Carmen: Easy: $2 \cdot \square - 1$. (Writes at board.)

Teacher: The number just after the even number?

Carmen: $2 \cdot \square + 1$. Look, if we add these three numbers what happens? (Points to the sequence on the board.)
$2 \cdot \square - 1,\ 2 \cdot \square,\ 2 \cdot \square + 1.$
You get $6 \cdot \square$, which is divisible by 3.

Teacher: All you have to do now is look at the even, odd, even pattern and if that works out you will know that Mark was right.

In fact, discovery of correct results often follows the sequence: A guess on the basis of an experiment, an attempt at proof which fails but suggests a modified guess, which leads to another attempt at proof, and so on. One of the charms of mathematics is the pervading spirit of discovery.

Burton W. Jones in *The Theory of Numbers*. Rinehart and Company, Inc., New York (1955).

Try Exercise 31

Even elementary pupils can make discoveries which, if not new to the world, are exciting to them. The preceding interchange between a teacher and some students shows them thinking about properties of numbers and discovering some on their own.

The result just discovered by these students is not a profound one, but anyone would be pleased at finding it without any help.

31. Check the other sequence: even, odd, even for three consecutive whole numbers and find the proof for this second possible pattern.

EXERCISE SET 5.6

1.
a) Study particular examples to investigate the behavior of even and odd numbers under the operation of addition.
b) Use a physical model to demonstrate your results and find an informal proof for your conjectures.
c) Give an algebraic proof of your results that can be recorded in the appropriate table to the right.

2. Repeat Problem 1 for multiplication:

$+$	*Even*	*Odd*
Even		
Odd		

$\times$	*Even*	*Odd*
Even		
Odd		

3. Is the set of odd numbers closed under addition?

4. Is the set of odd numbers closed under multiplication?

5. What is the 25th rectangular number?

6. What is the 97th rectangular number?

7. Find the sum

$$1 + 1 + 2 + 2 + 3 + 3 + \cdots + 112 + 112.$$

8. Find the sum

$$1 + 1 + 2 + 2 + 3 + 3 + \cdots + 200 + 200.$$

9. What is the sum of the first 17 natural numbers?

10. What is the sum of the first 527 natural numbers?

11. What is the 8th triangular number?

12. What is the 35th triangular number?

13. Think of the square numbers in the following way: What relationship do you see between the square numbers and the triangular numbers?

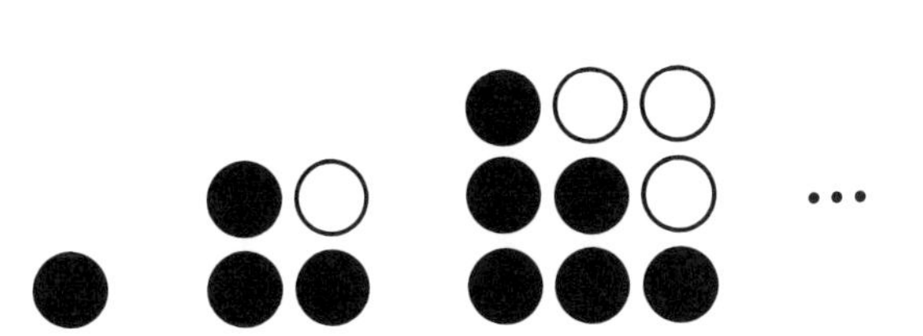

14. Slightly different diagrams can be made for any of the polygonal numbers. The first three pentagonal numbers (five equal sides) could be drawn as follows:

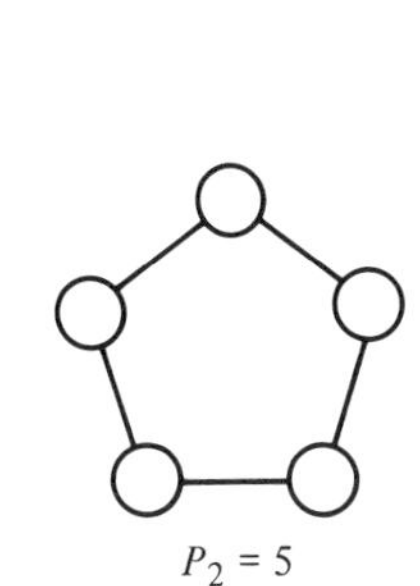
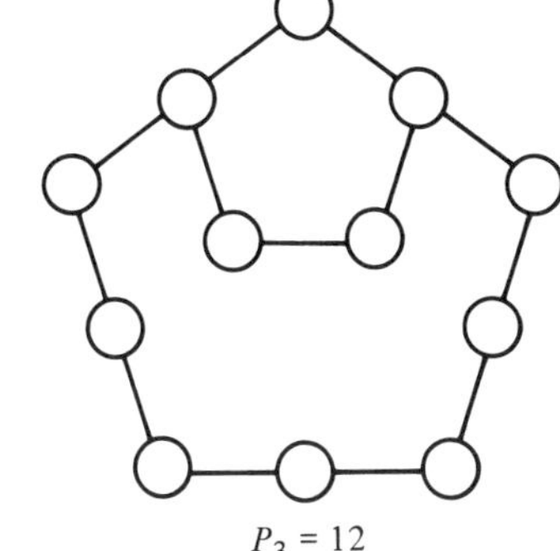

$P_1 = 1$ $P_2 = 5$ $P_3 = 12$

The next number in the sequence will be drawn with an additional circle on each of the sides and so on.

a) Use poker chips or make sketches of the next few pentagonal numbers.
b) What is the pattern developing in the sequence?
■ c) How might these numbers be related to the triangular and other figurate numbers?

15. Repeat Problem 14 for hexagonal numbers. These form configurations with six equal sides. The first is 1; the second is 6.

16. A store manager making a display of oranges remarked to a customer that the number of oranges piled into a ■ square-based pyramid could just as well have been arranged into a large square formation. The customer looked at the pile of oranges and said: "I know that there is such a number; still it does not seem possible that you counted correctly." How many oranges would have had to be in the pile in order to make it possible to dismantle the pyramid and then arrange the oranges in a square?

5.7 AIMS

You should be able to:

a) Apply the first four sections of this chapter to solving problems in number theory.
b) Continue to develop general problem-solving skills.

5.7 PROBLEM SOLVING

This time we shall deal with some problems in number theory.

Example 1 Find the sum of the first 100 even natural numbers.

What are the *facts*? 2, 4, 6, 8, . . . , 196, 198, 200 are the first 100 even numbers.

What is the *problem*? Find the sum: $2 + 4 + 6 + 8 + \cdots + 198 + 200$.

Some *ideas* that might occur to you include:

1. Add them with the usual addition.
2. Use a calculator to do it.
3. Each even number is 2 times some other number; rewrite the sum as:

$$2 \cdot 1 + 2 \cdot 2 + 2 \cdot 3 + \cdots + 2 \cdot 99 + 2 \cdot 100$$

4. Subtract the sum of the odd numbers in between from the sum of all the numbers to 200. (Now a new problem has been introduced—what is the sum of the odd numbers up to 199?)

32.

a) Write down any other information or ideas that you might use to solve this problem.
b) Solve the problem using idea 2.

Try Exercise 32

How to get a *solution*. There must be an easier way than ideas 1 or 2.

Use idea 3. Rewrite the sum by factoring as:

$$2 \cdot (1 + 2 + 3 + 4 + \cdots + 98 + 99 + 100). \qquad \text{(Why?)}$$

The sum of the first k natural numbers was found in the previous section; it is T_k. The sum of the first 100 natural numbers is $T_{100} = 100 \cdot 101 \div 2 = 5050$. The sum needed here is twice that, so $2 \cdot 5050 = 10\,100$ is the required sum.

Use idea 4 instead. A new problem was introduced that must be solved first. How can the required odd numbers be added? In the previous section it was shown that the kth square number is the sum of the first k odd numbers. How many do we have here? 100. We use the sum of the first 100 odd natural numbers: $S_{100} = 100^2$.

Now the sum of all the natural numbers to 200 is needed. The 200th triangular number gives this:

$$T_{200} = \frac{200 \cdot 201}{2} = 20\,100.$$

By subtraction: $20\,100 - 10\,000 = 10\,100$ (of course, it is the same answer as worked out with idea 3.)

If one of the ideas you thought about was the rectangular numbers, you might think of them as being formed in the following way:

```
                                  X X X X X
                       X X X X    X X X X|X
           X X X       X X X|X    X X X|X|X
X X        X X|X       X X|X|X    X X|X|X|X
R1 = 2     R2 = 2 + 4  R3         R4
```

Even for these simple problems there is not just one route to a solution. It is good to think of several different ideas that might work and then choose what seems easiest to carry out. In case you get a solution immediately, it never hurts to think about other ways to do something (compare idea 1 with idea 3 or 4 in Example 1).

As a reminder:

THE FIVE-STEP PROCESS

1. What are the relevant *facts*?
2. What actually is the *problem*?
3. What are my *ideas*?
4. How do I get a *solution*?
5. *Check* and *review* the solution.

From a sixth-grade textbook

Choose any number from 1 through 9.
Multiply your number by 429.
Then multiply the product by 259.

Can you guess what the answer would be for any other number from 1 through 9?

Check your guess.

EXERCISE SET 5.7

1. It has been shown that the sum of the first k natural numbers is $k(k + 1)/2$. Explain why this will always turn out to be a natural number even though the formula involves division.

2.
a) What is the sum of the first 32 natural numbers?
b) What is the sum of the first 16 odd numbers?
c) What is the sum of the first 16 even numbers?
Hint: One way to solve a problem is to see if it is anything like one you have already solved.

3. Solve Example 1 of the text using the idea of rectangular numbers proposed there. How do the solutions based on different ideas compare?

■ **4.** The GCD of the numbers 6, 24, and 60 is 6. Now think of *any* number of the form $k(k + 1)(k + 2)$, such as $7 \cdot 8 \cdot 9$. Do these numbers have any common factors (other than 1) with the set $\{6, 24, 60\}$? If so, what is the GCD?

■ **5.** A student was playing with a calculator and noticed that for

$$1 \cdot 2 \cdot 3 \cdot 4 \cdot 5 = 120, \qquad 2 \cdot 3 \cdot 4 \cdot 5 \cdot 6 = 720,$$
$$3 \cdot 4 \cdot 5 \cdot 6 \cdot 7 = 2520$$

the GCD(120, 720, 2520) was 120. Imagine the student's excitement upon looking at the following:

$$8 \cdot 9 \cdot 10 \cdot 11 \cdot 12, \qquad 9 \cdot 10 \cdot 11 \cdot 12 \cdot 13,$$
$$10 \cdot 11 \cdot 12 \cdot 13 \cdot 14, \qquad 11 \cdot 12 \cdot 13 \cdot 14 \cdot 15$$

and discovering that the GCD for these four was also 120. Was the student just lucky or can any numbers constructed in this way be added to the list and have the GCD for the entire list remain 120?

6. A young married couple is studying their new work schedule. The husband works for the fire department and gets every sixth night off from his work. The wife is a nurse who has every fourth night off. They are having this discussion on Saturday and see that he is off the next day (Sunday), while she is off the day after that (Monday). They are trying to figure out when they will have dinner together in the next two months. Help them out in their problem.

7. *Goldbach's Conjecture.* In 1742 Goldbach made the conjecture that every even number greater than 4 is the sum of two odd primes. Neither a proof nor a counterexample has ever been found for this statement. Find one or more ways of expressing the following as the sum of two odd primes:

a) 8 b) 10 c) 12 d) 20 e) 88 f) 128

8. Can any odd numbers be written as the sum of two odd primes?

■ **9.** Show that any three numbers, n, $n+2$, and $n+4$, where n is a prime number greater than 3, cannot all be prime numbers.

10. *Perfect numbers.* Every divisor of a number is called a *proper divisor* except for the number itself. A *perfect number* is one that is the sum of its proper divisors. For example, 6 is a perfect number because $6 = 1 + 2 + 3$. A formula for finding even perfect numbers was proved correct by Leonhard Euler (1707–1783), who showed that $2^{n-1} \cdot (2^n - 1)$ will produce such a number when $n > 1$ and $(2^n - 1)$ is a prime. Prove that the following are perfect numbers:

a) $28 = 2^2 \cdot (2^3 - 1) = 4 \cdot 7$ (the second perfect number)
b) $496 = 2^4 \cdot (2^5 - 1)$ (the third)
c) $8128 = 2^6 \cdot (2^7 - 1)$ (the fourth)

11. Show that the following are not perfect numbers:
a) $2^3 \cdot (2^4 - 1)$ b) $2^5 \cdot (2^6 - 1)$

The next two problems are similar to those posed to elementary students.

12. Find two primes whose product is 10 001. Each prime is less than 150.

13. Fill in the remaining lines of the table:

n	$n^2 - 1$
1	$1^2 - 1 = 0 = 8 \cdot 0$
3	$3^2 - 1 = 8 = 8 \cdot 1$
5	$5^2 - 1 = 24 = 8 \cdot 3$
7	$7^2 - 1 = 48 = 8 \cdot 6$
9	
11	
13	

a) Without calculating further, do you think that $71^2 - 1$ is a multiple of 8?

b) What sequence of numbers do you recognize in this table? (An elementary pupil might not see this.)

14. What conjecture can you make about the pattern in Problem 13? ■ Prove it.

15. Play a game of *Reverse*. Choose a six-digit number. It could be your birthday or some other number you particularly like. For example, May 27, 1962, or 5/27/62, or 052762 you would enter as 267250. You are to arrive at the reverse of this number using only the addition, subtraction, multiplication, or division operations with two-digit numbers only. It is to be done in not more than 10 steps. If you do it in fewer than 5, you are a gold-ribbon player.

REVIEW TEST

1. Find the prime factorization of 2590.

2. Use a tree diagram to find the prime factorization of 539.

3. Test 481 for divisibility by 37 directly by first expressing 481 as $37 \cdot n + r$.

4. In each of the following, find all those whole numbers x that make the statement true.

a) $x \mid 36$ b) $x \mid (3x + 1)$ c) $x \mid 7x$

5. Find the LCM(30, 121).

6. Find the GCD(24, 120, 360).

7. Use Euclid's algorithm to determine if 64 and 81 are relatively prime.

8. Test 1107 for divisibility by 2, 3, 4, 5, 6, 9, and 10.

9. Write a proof that shows the divisibility test for 3 is valid.

10. What is the smallest number that is divisible by any number from 2 to 10 inclusive?

11. Find the following counting numbers:

$\{x \mid x \text{ is less than 40 and has } \textit{exactly} \text{ 3 divisors}\}$.

What is another way to describe these numbers?

12.
a) Find a base b for which 14_b is a multiple of 3.
b) In base-b notation give the first ten multiples of 3.
■ c) Determine a divisibility rule for 3 in this base b.

13. The three numbers 3, 5, and 7 are consecutive odd numbers that are all prime. Prove that there are no other such triples.

14. Is the "divides" relation on the set of whole numbers an equivalence relation? Provide a proof or counter-examples for each property.

15. What is the sum of the first 28 natural numbers?

16. What is the sum of the first 14 odd natural numbers?

17. What is the sum of the first 14 even natural numbers?

18. Prove that 28 is a perfect number.

19. Show that $x^2 - 1$ is prime when $x = 2$, but will never be prime for any other value of x.

20. Fill in the remaining lines of the table:

n	$n^3 - n$
1	$1^3 - 1 = 1 - 1 = 6 \cdot 0$
2	$2^3 - 2 = 8 - 2 = 6 \cdot 1$
3	$3^3 - 3 = 27 - 3 = 6 \cdot 4$
4	
5	
6	

a) Without calculating further, do you think that $65^3 - 65$ is a multiple of 6?
■ b) Make a conjecture about the pattern in the table and prove it.

21. Karl Friedrich Gauss (1777–1855), a remarkable mathematician of whom it is said that he did mathematics before he could talk, was once given the task of adding the first 100 numbers. Gauss studied the pattern and came up with the answer very quickly. How might Gauss have solved the problem?

$$\begin{array}{cccccccc} 1 & 2 & 3 & 4 & \ldots & 97 & 98 & 99 & 100 \\ 100 & 99 & 98 & 97 & \ldots & 4 & 3 & 2 & 1 \end{array}$$

FOR FURTHER READING

1. Perfect numbers are perfectly useless but have a fascinating history. (The earth was created in 6 days, the moon has a 28 day trip around the earth.) More interesting material about perfect numbers can be found in *Mathematical Magic Show* by Martin Gardner. Alfred A. Knopf, New York (1977).

2. *The Numerology of Dr. Matrix* by Martin Gardner, Simon and Schuster, New York (1967) is charmingly written and easy to read. The book is loaded with all sorts of oddities about numbers that would fascinate elementary pupils.

TEXTS

Baur, Gregory R., and Linda Olsen George, *Helping Children Learn Mathematics: A Competency-Based Laboratory Approach.* Cummings Publishing Company, Inc., Menlo Park, California (1976). (A methods text.)

D'Augustine, Charles H., *Multiple Methods of Teaching Mathematics in the Elementary School.* Harper and Row, New York (1968). (A methods text.)

chapter 6

nonnegative rational numbers—fractions

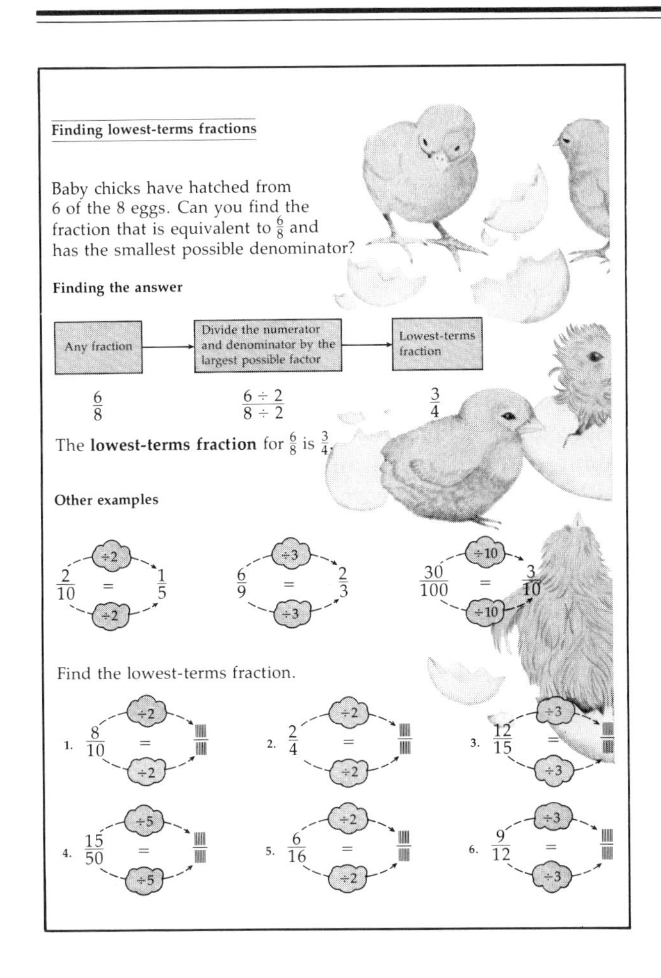
Finding lowest-terms fractions

Baby chicks have hatched from 6 of the 8 eggs. Can you find the fraction that is equivalent to $\frac{6}{8}$ and has the smallest possible denominator?

Finding the answer

Any fraction → Divide the numerator and denominator by the largest possible factor → Lowest-terms fraction

$\frac{6}{8}$ $\frac{6 \div 2}{8 \div 2}$ $\frac{3}{4}$

The **lowest-terms fraction** for $\frac{6}{8}$ is $\frac{3}{4}$.

Other examples

$\frac{2}{10} = \frac{1}{5}$ (÷2) $\frac{6}{9} = \frac{2}{3}$ (÷3) $\frac{30}{100} = \frac{3}{10}$ (÷10)

Find the lowest-terms fraction.

1. $\frac{8}{10}$ = ▒ (÷2) 2. $\frac{2}{4}$ = ▒ (÷2) 3. $\frac{12}{15}$ = ▒ (÷3)

4. $\frac{15}{50}$ = ▒ (÷5) 5. $\frac{6}{16}$ = ▒ (÷2) 6. $\frac{9}{12}$ = ▒ (÷3)

Factoring of whole numbers must be mastered by fourth graders before they can cope with finding the lowest-terms fractions.

6.1 AIMS

You should be able to:

a) Use a variety of models to illustrate problems that involve the concept of a rational number.
b) Define a nonnegative rational number.

6.1 WHAT ARE RATIONAL NUMBERS?

The whole numbers were sufficient for the modest arithmetic needs of early humans who got along with counting and performing simple arithmetic operations. It did not take very long before the ordinary demands of agriculture and commerce made it obvious that the whole numbers were inadequate.

If land, tools, supplies, equipment, and profits have to be apportioned or divided up, it is not usually the case that the amounts of any of these things can be divided "evenly" in terms of the units under discussion. In any event, as as long ago as 1700 B.C., the Egyptians were using other numbers in addition to whole numbers and had special symbols for one-half, one-third, one-fourth, and so on.

The Egyptians did not sit down one day and decide to guarantee that the closure property held for division. The people who invented fractions and helped develop rational numbers merely responded to the need for "fractional" parts, that arose in many different circumstances.

Interpretations which compare area or length arose out of practical problems, and they are still used in the development of the concept of rational numbers in elementary school. An important aspect, then, is in the description of a certain number of the equal parts of a whole; such parts may be areas or lengths.

The idea of a *ratio*, the term that gave rise to the word "rational," is also fundamental. It is possible to compare two sets on the basis of the ratio of the number of elements in one set to the number of elements in another. The statement: "For every 20 bushels of grain you produce, you are to provide 1 bushel for the granaries of the King," describes a ratio 20 to 1.

It is only a fairly recent development that anyone has thought in abstract terms about providing closure for division and "tidied up" the mathematics of fractions and rational numbers. Because for any whole numbers a and b, $a \neq 0$, every equation $ax = b$ does not have a whole-number solution, some additional numbers are needed. Mathematicians tend to be more interested in this abstract viewpoint.

Before defining nonnegative rational numbers we shall examine some interpretations that motivate the definition. We are following in our development the sequence used in a large number of elementary-school mathematics programs, in which nonnegative rational numbers and fractions are treated before negative numbers are.

A COUNTING MODEL. When children are first taught the basic concepts of fractions, a physical model or drawing of some whole object that can be cut or divided (fractured) into equal portions is used as a starting point.

Children are very familiar with the sharing of equal numbers of cookies or candies, and this type of sharing is the basis for a model in which a set of objects is partitioned into equivalent subsets and these subsets are regarded as the equal parts.

Example 1 Represent one-fourth using a dozen eggs as the unit.

Solution The original set of 12 eggs is partitioned into four equivalent sets. Each of these sets is one-fourth ($\frac{1}{4}$) of the dozen.

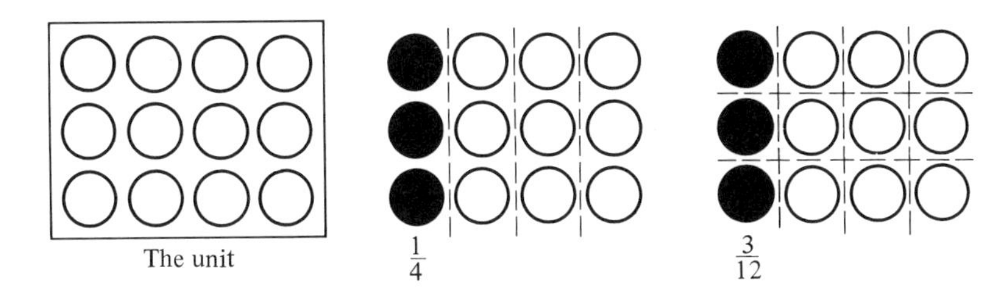

The same model could be used to represent three-twelfths ($\frac{3}{12}$) if the original partitioning had been done into 12 equivalent subsets, which suggests that $\frac{3}{12}$ and $\frac{1}{4}$ are the same.

Try Exercise 1

1. Use a dozen as the unit and illustrate $\frac{2}{3}$ by partitioning the whole into equivalent sets.

AN AREA MODEL. Youngsters do not have much occasion to consider problems about land area but do have a great deal of experience in sharing parts of a pizza or portions of a cake or candy bar, a practical problem for them corresponding to a problem that involves dividing land. Both candy and land are based upon a real-life situation in which some whole is portioned out into equal parts. Circular "pie" drawings or rectangular "bars" represent a whole pie or candy bar.

Example 2 Represent three-fourths ($\frac{3}{4}$) with a pie drawing and a bar drawing.

Solution In both sketches the objects are shown as cut into fourths, equal parts of the same size, and the parts shaded represent three of these fourths.

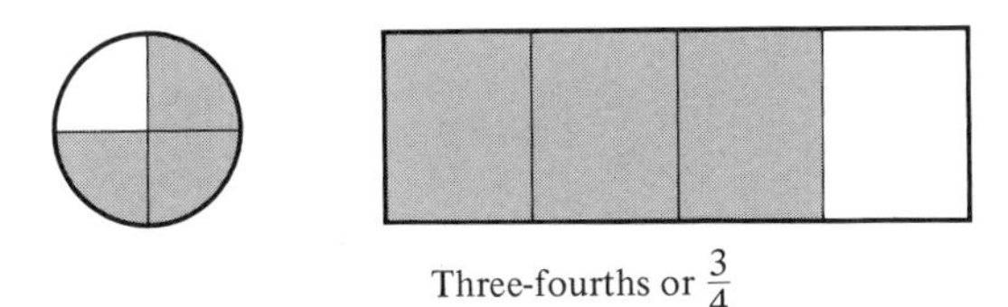

Three-fourths or $\frac{3}{4}$

These sketches interpret three-fourths as three parts of the four equal parts of the whole. In general $\frac{m}{n}$ can be represented by an area with n equal areas marked out (each piece is one-nth) and m of the equal portions shaded to represent the number.

2. Make a pie drawing and a bar drawing to represent three-eighths ($\frac{3}{8}$).

Try Exercise 2

THE RATIO MODEL. The models discussed so far have all started with some whole that was partitioned, or divided, into equal portions either by counting or measuring area. With a ratio interpretation, usually two distinct sets are compared; there is not a partitioning of a single unit that took place in the previous examples.

Pictured are two sets with 8 objects in the first and 12 in the second. The ratio of the number in set A to the number in set B is 8:12, or 8/12.

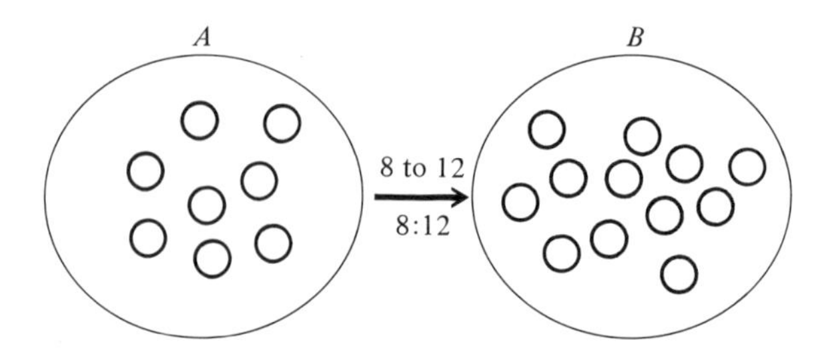

3. Find another ratio that compares the number of elements in A with the number in B.

The sets could also be described with a correspondence as follows:

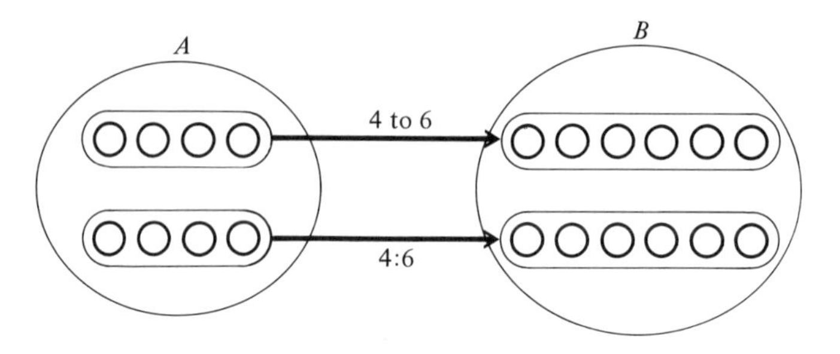

The relation between the sets has not changed but can be described as 4:6 or 4/6.

Try Exercise 3

This comparison can also be thought of as a function on the set A, in which 4 elements of A correspond to 6 elements of B. Since the ratio for the same sets A and B has been variously described as 8:12, or 4:6, or 2:3, it is obvious that the ratio is not unique for a particular pair of sets. Some ratios in a sixth-grade textbook are:

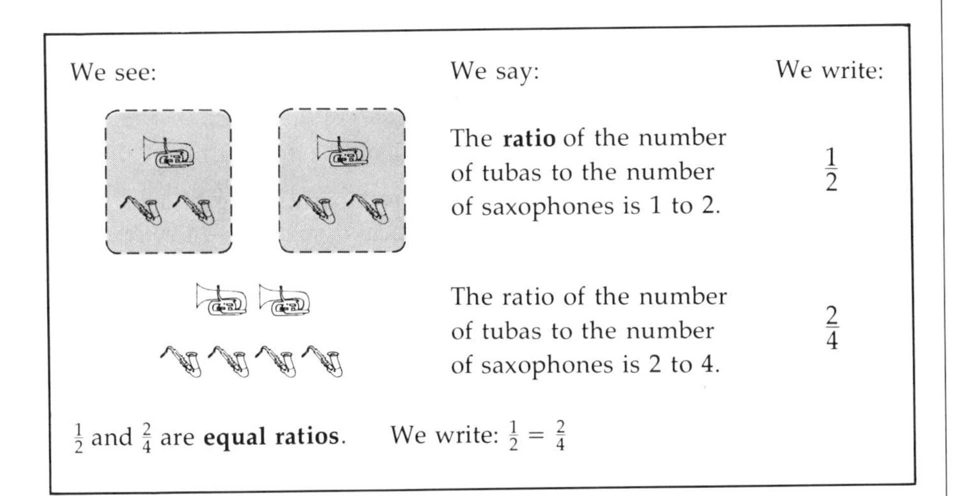

A NUMBER-LINE MODEL. Children progress from the following fourth-grade approach to an abstract number line in which the segment from a point labeled 0 to an arbitrary point labeled n is divided into a specified number of equal parts.

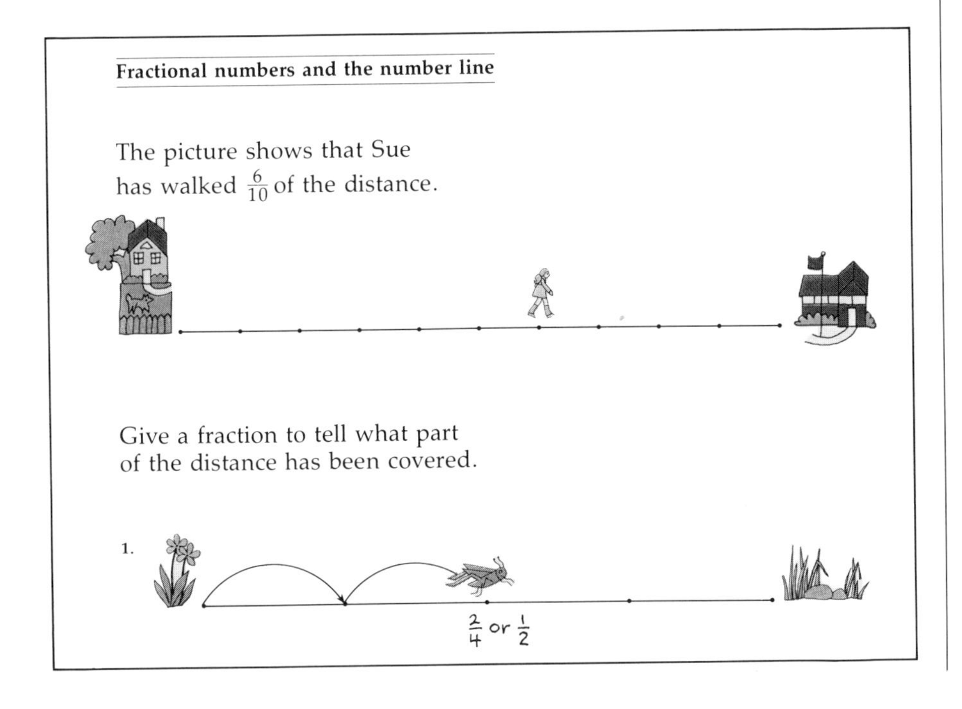

4. Show two interpretations for $\frac{4}{5}$ on a number line.

None of the models discussed indicate that the new numbers are actually the missing quotients needed to provide closure for division. We see that the number-line interpretation can be used in this way.

Example 3 Represent the quotient $3 \div 4 = 3/4$ on the number line.

Solution Just as $24 \div 4$ can be interpreted as a partitioning of a set of 24 objects into four sets of 6 each to find the quotient 6, so can $3 \div 4$ be regarded as a partitioning of three units of a number line into four equal parts to find the quotient 3/4. This point is exactly the same one that would be found under the interpretation in which three of the equal parts (fourths) of a single unit would be marked off. These are two different ways to look at 3/4.

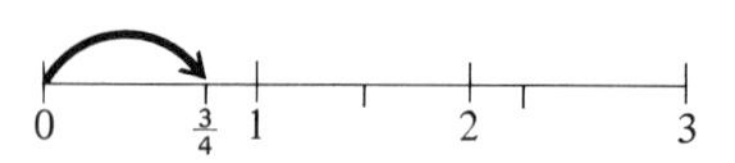

Try Exercise 4

All of the interpretations and models discussed above can be shown to be consistent with the idea that rational numbers were created to take care of a deficiency of the whole-number system in that it fails to have closure for division.

Certain quotients such as $3 \div 5$ (also symbolized $\frac{3}{5}$ or 3/5), $7 \div 9$ (7/9) and $12 \div 13$ (12/13) are not whole numbers. We wish to supply all the "missing quotients" for whole-number division, except those in which 0 is the divisor, since division by 0 remains undefined. The following definition for nonnegative rational numbers has the effect of creating a number for each quotient of all possible whole-number pairs (except for division by 0).

DEFINITION. A *nonnegative rational number* is any number that can be represented by $\frac{a}{b}$ such that a and b are any whole numbers and $b \neq 0$.

Thus all the following numbers have something in common:

$$4,\ 15/2,\ 16.311,\ 2/7,\ 0.6666\ldots,\ 0.$$

All of these can be named with a/b such that a and b are some whole number and $b \neq 0$ and hence are nonnegative rational numbers: for 15/2, $a = 15$ and $b = 2$; for $16.311 = \frac{16311}{1000}$, $a = 16311$, $b = 1000$; for $0.666\ldots = 2/3$, $a = 2$ and $b = 3$.

Not all the numbers of the definition are new to us.

Example 4 Show that any whole number is a nonnegative rational number.

Solution Any whole number a can be regarded as the quotient of a and 1 (that is, $a \div 1 = a/1$) and thus satisfies the definition above. For example, 4 is the quotient $4 \div 1$ and 0 is the quotient $0 \div 1$.

Nowhere in the definition of a nonnegative rational number does it state that the representation is unique, and the models based on practical problems which motivated the creation of these numbers have already indicated that more than one name is possible. We can think of the whole number 3 as any of the quotients $12 \div 4$, $15 \div 5$, $18 \div 6$, and so on.

EXERCISE SET 6.1

1. Not all of the following can be represented with a dozen as the unit: 1/3, 1/4, 5/11, 3/8, 2/9. Make diagrams of those that can be and explain what the difficulty is with the others.

2. Repeat Problem 1 for: 3/4, 5/8, 7/12, 1/6, 3/11.

3. Use the counting model with a unit of 15 objects to represent 2/15.

4. Use the counting model with a unit of 25 objects to represent 3/5.

5. Use a pie diagram to represent:
a) 5/6 b) 4/3

6. Use a bar diagram to represent:
a) 3/5 b) 7/6

7. Start with a set A of 24 objects and a set B of 36 objects. Diagram at least three different ratios using these sets.

8. Start with a set A of 18 objects and a set B of 36 objects. Diagram at least three different ratios using these sets.

9. Let set A be a set of 14 counters. Make a sketch to illustrate the ratio 2:3 using set A and another set B. How many counters are needed in set B? What other ratios can be illustrated with these two sets?

10. Let set A be a set of 9 counters. Make a sketch to illustrate the ratio 3:4 using A and another set B. How many are needed in set B? How many other ratios can be illustrated with this same pair of sets?

11. Use two number lines drawn to the same scale to represent
a) three-fifths and the quotient $3 \div 5$.
b) four-thirds and the quotient $4 \div 3$.

12. In the following pairs of diagrams, the first one represents the whole, the second a part of this whole. What number is represented in each case?

a)

b)
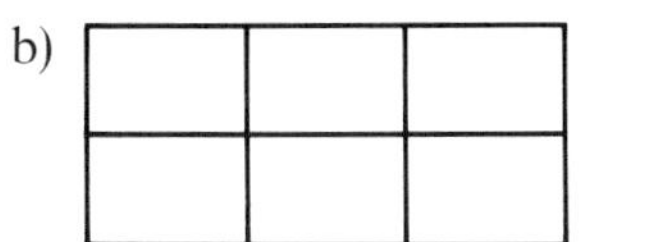
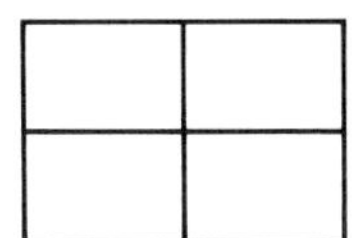

13. Bug A hops in equal-sized hops along a number line starting at 0 and stops after 20 hops. Bug B hops in equal-sized hops along the same number line starting at 0 and stops after 24 hops. By an amazing coincidence they both land on the same point. How can that be? How can you write a ratio to describe this situation?

14. Are the bugs of Problem 13 ever going to land on the same spot again if they continue to hop in the same way along the number line? Had they ever landed on the same point before the one described in Exercise 13? If so, what points were they?

6.2 AIMS

You should be able to:

a) Use bar diagrams or a number line to show equivalence of fractions.
b) Find an arbitrary number of fractions to represent a nonnegative rational number.
c) State and apply the definition of equality for nonnegative rational numbers.

6.2 NONNEGATIVE RATIONAL NUMBERS AND FRACTIONS

The fact that the symbols for a nonnegative rational number are not unique seems to cause more difficulty than it should. We are used to representing the whole number two with the symbol "2." However, we could just as well represent two by the difference 3 − 1, or 5 − 3, or 9 − 7; it is just not usually done. The situation for nonnegative rationals is somewhat similar.

The symbol $\dfrac{\text{“}a\text{”}}{b}$ that appears in the definition of a rational number is what we call a *fraction*; each fraction contains two numerals, one at the top and one at the bottom of the symbol. The *numerator* of the fraction is the whole number that corresponds to the symbol "a" on top, while the *denominator* is the whole number that corresponds to the symbol "b" on the bottom. Since a fraction is a symbol or a name, the fraction $\dfrac{\text{“}5\text{”}}{10}$ is not the same symbol as $\dfrac{\text{“}1\text{”}}{2}$, but we shall see that the two symbols both name the same rational number (some examples precede the definition of equality for rational numbers).

Example 1 Use bar diagrams of the same size and shape to represent 2/3, 4/6, and 6/9.

$\frac{2}{3}$

$\frac{4}{6}$

$\frac{6}{9}$

In the diagram representing 4/6, the bar was divided into twice as many parts as the diagram for 2/3, but then twice as many of the parts were shaded for 4/6 as for 2/3, so the total area did not change. For consistency with this area model, all three fractions of Example 1 should represent the same rational number.

Try Exercise 5

5. Use identical bar diagrams to represent 1/3, 2/6, and 4/12. Compare the areas.

Working with models for addition of fractions

Example 2 Use a number line to show that 1/2, 4/8, and 5/10 are all the same number.

Solution Three number lines drawn to the same scale help us avoid confusion.

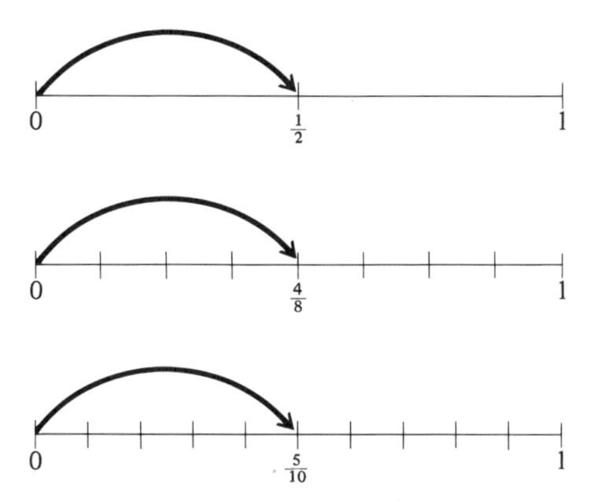

Under the assumption that any point of the number line corresponds to some number, then in this interpretation 1/2, 4/8, and 5/10 are all the same number.

6. Use a number line to show that 3/4, 6/8, and 9/12 correspond to the same point on a number line.

Try Exercise 6

THE EQUALS RELATION. We need a means of determining under what conditions two fractions name the same nonnegative rational number. That is, how do we know when two names represent the same number? When is the number a/b equal to the number c/d? The examples discussed so far make the following definition reasonable.

DEFINITION.* **The nonnegative rational number $\frac{a}{b}$ is *equal* to the nonnegative rational number $\frac{c}{d}$ if and only if $a \cdot d = b \cdot c$.**

Example 3 Use the definition for equals to show that

a) $1/2 = 4/8$ b) $4/8 = 5/10$ c) $1/2 = 5/10$

Solution

a) $1/2 = 4/8$ because $1 \cdot 8 = 2 \cdot 4$
b) $4/8 = 5/10$ because $4 \cdot 10 = 8 \cdot 5$
c) $1/2 = 5/10$ because $1 \cdot 10 = 2 \cdot 5$

We can also apply the definition for equality for nonnegative rational numbers and some appropriate whole-number properties to show that:

***Equality* for nonnegative rational numbers is an equivalence relation.**

1. It is *reflexive*: any nonnegative rational number equals itself.
   Since $a \cdot b = b \cdot a$ by the commutative law for whole numbers, we have $\frac{a}{b} = \frac{a}{b}$ by the definition of equality for nonnegative rational numbers.

2. It is *symmetric*: for any rationals $\frac{a}{b}$ and $\frac{c}{d}$, if $\frac{a}{b} = \frac{c}{d}$, then $\frac{c}{d} = \frac{a}{b}$.
   By the definition of equals for nonnegative rationals we have
   $$\frac{a}{b} = \frac{c}{d} \leftrightarrow a \cdot d = b \cdot c,$$
   but $a \cdot d = b \cdot c \rightarrow c \cdot b = d \cdot a$ by the commutative law for whole numbers.
   Therefore, $\frac{c}{d} = \frac{a}{b}$ by the definition of equality for nonnegative rational numbers.

3. It is *transitive*: if $\frac{a}{b} = \frac{c}{d}$ and $\frac{c}{d} = \frac{e}{f}$, then $\frac{a}{b} = \frac{e}{f}$.

* The criterion of this definition is sometimes referred to as the *cross-product rule* (which is confusing), or the *criss-cross product* (which is descriptive and less confusing).

The argument depends again on the definition of equality and whole-number properties and is left for the exercises.

Try Exercise 7

7. Use the definition to prove:

a) $\frac{5}{15} = \frac{8}{24}$

b) $\frac{7}{36} = \frac{14}{72}$

EQUIVALENT FRACTIONS. The definition of equality for nonnegative rational numbers leads to a method of generating as many fractions (names) for a rational number as we wish. The following is the key:

If $\frac{a}{b}$ is a nonnegative rational number, then for any whole number p, $p \neq 0$, $\frac{a \cdot p}{b \cdot p} = \frac{a}{b}$.

To show this, we depend upon the definition of equality for nonnegative rational numbers and whole-number properties.

We let a, b, and p be any whole numbers with $p \neq 0$ and $b \neq 0$. Then $a \cdot (b \cdot p) = (a \cdot b) \cdot p = b \cdot (a \cdot p)$ because of the associative and commutative laws for whole-number multiplication. Therefore we conclude by the definition of equality that

$$\frac{a}{b} = \frac{a \cdot p}{b \cdot p}.$$

Two fractions, such as "$\frac{a}{b}$" and "$\frac{a \cdot p}{b \cdot p}$" that name the same nonnegative rational number, are said to be *equivalent fractions*, so finding another name or representation in this way is finding an equivalent fraction.

Example 4 Find four fractions that are equivalent to the fraction "$\frac{3}{8}$".

Solution We use the result just proved:

$$\frac{3 \cdot 2}{8 \cdot 2} = \frac{6}{16}, \quad \frac{3 \cdot 3}{8 \cdot 3} = \frac{9}{24}, \quad \frac{3 \cdot 4}{8 \cdot 4} = \frac{12}{32}, \quad \frac{3 \cdot 5}{8 \cdot 5} = \frac{15}{40}.$$

Thus we can write for the number $\frac{3}{8}$ either $\frac{6}{16}$, $\frac{9}{24}$, $\frac{12}{32}$, or $\frac{15}{40}$. It is possible to find an infinite number of representations in this way since p can be any of the natural numbers.

Try Exercise 8

8. Find four different fractions that represent the number 6/7.

LOWEST-TERMS FRACTION. There is always a simplest fraction that can be used to represent a rational number.

DEFINITION. The name "$\frac{a}{b}$" for a nonnegative rational number, where a and b are whole numbers, $b \neq 0$, is the *lowest-terms fraction* if and only if GCD$(a, b) = 1$, that is, a and b are relatively prime.

9.

a) Read the opening page of this chapter if you have not already done so.

b) Represent $\frac{18}{81}$ in lowest terms.

Example 5 Find the lowest-terms fraction for the following:

a) $\frac{16}{32}$ b) $\frac{98}{100}$ c) $\frac{6}{10}$

Solution

a) $\frac{16}{32} = \frac{16 \cdot 1}{16 \cdot 2} = \frac{1}{2}$, $p = 16$, GCD(1,2) = 1;

b) $\frac{98}{100} = \frac{49 \cdot 2}{50 \cdot 2} = \frac{49}{50}$, $p = 2$, GCD(49, 50) = 1;

c) $\frac{6}{15} = \frac{2 \cdot 3}{5 \cdot 3} = \frac{2}{5}$, $p = 3$, GCD(5, 2) = 1.

Try Exercise 9

It is now a simple matter to see that every fraction for a nonnegative rational can be derived from the lowest-terms fraction.

Example 6 Show that if $\frac{a}{b}$ is expressed in the lowest-terms fraction, then every other nonnegative rational number equal to it must be $\frac{a \cdot p}{b \cdot p}$ for some natural number p.

We reason as follows. Suppose that $\frac{c}{d}$ is another nonnegative rational that is equal to $\frac{a}{b}$. If $\frac{a}{b} = \frac{c}{d}$, then $a \cdot d = b \cdot c$.

Now $a \mid a \cdot d$, so it must also be true that $a \mid b \cdot c$. Since the GCD(a, b) = 1, it means $a \mid c$, and $c = a \cdot q$, where q is a whole number, not zero.

Also $b \mid b \cdot c$, so it must also be true that $b \mid a \cdot d$. Since the GCD(a, b) = 1, it means that $b \mid d$ and $d = b \cdot p$, where p is a whole number.

Therefore, $\frac{a}{b} = \frac{c}{d} = \frac{a \cdot q}{b \cdot p}$, and it follows by the definition of equality for nonnegative rational numbers that

$$a \cdot (b \cdot p) = b \cdot (a \cdot q), \qquad (a \cdot b) \cdot p = (a \cdot b) \cdot q.$$

Since $a \cdot b$ divides both terms, we have $p = q$. This means that $\frac{c}{d}$ is really $\frac{a \cdot p}{b \cdot p}$, and we are done.

Fractions, fractional numbers, and rational numbers are not treated by all authors in exactly the same way, and there is lack of uniformity of terminology. Often the quotes that indicate a symbol is being discussed are omitted with the understanding that the context clarifies the discussion.

EXERCISE SET 6.2

1. Use bar diagrams drawn to the same scale to show 3/8, 6/16, and 12/32. What do the diagrams illustrate?

2. Repeat Problem 1 for 3/7, 6/14, and 12/28.

3. Find five different fractions for each of the following:

a) 3/11 b) 4/12 c) 2/13 d) 1/11
e) 0/5 f) 0/2 g) 16/4 h) 18/3

In Problems 4–7, not all of the numbers are equal. Find the one that does not belong in each case.

4. 3/5, 6/10, 9/14, 15/25, 18/30, 12/20

5. 2/9, 4/18, 6/27, 8/36, 9/45, 12/54

6. 15/2, 30/4, 45/5, 90/12, 120/16, 75/10

7. 20/44, 35/77, 30/66, 5/11, 40/88, 50/99

8. Use the counting model to demonstrate that the following are equal:

a) 1/2, 2/4, 3/6, and 6/12 b) 3/5, 6/10, and 15/25

9. Use the ratio model to show that the following are equal:

1/2, 3/6, and 6/12.

10. Repeat Problem 9 for 3/5, 6/10, and 15/25.

11. Determine if the "equals" relation holds for the following pairs:

a) $\frac{8542}{9834}$ and $\frac{3762}{4332}$ b) $\frac{9513}{22197}$ and $\frac{3723}{8687}$

12. Prove that the transitive property holds for the "equals" relation for nonnegative rational numbers.

■ **13.** The denominator of 4/7 is a prime number. Is it always possible to find for any nonnegative rational number a fraction with either the denominator or the numerator prime?

6.3 OPERATIONS: ADDITION AND SUBTRACTION

6.3 AIMS

You should be able to:

a) Apply the definitions for addition and subtraction.
b) Recognize and use the commutative and the associative laws for addition.
c) Convert from mixed notation to a fraction.
d) Prove some properties for the addition and subtraction operations.

Addition

DEFINITION. Any of the models could be adapted to motivate a definition for the addition of two nonnegative rational numbers. The set of examples from a sixth-grade textbook shown on p. 192 uses the pie model and a number line to formulate a rule for addition when the denominators are alike.

The general definition for addition of nonnegative rational numbers is consistent with these models.

ADDITION. For any nonnegative rational numbers $\frac{a}{b}$ and $\frac{c}{d}$, their sum is

$$\frac{a}{b}+\frac{c}{d}=\frac{a\cdot d+b\cdot c}{b\cdot d}.$$

10. Use the definition to perform the addition:

a) $\frac{3}{10} + \frac{4}{10}$

b) $\frac{2}{5} + \frac{1}{3}$

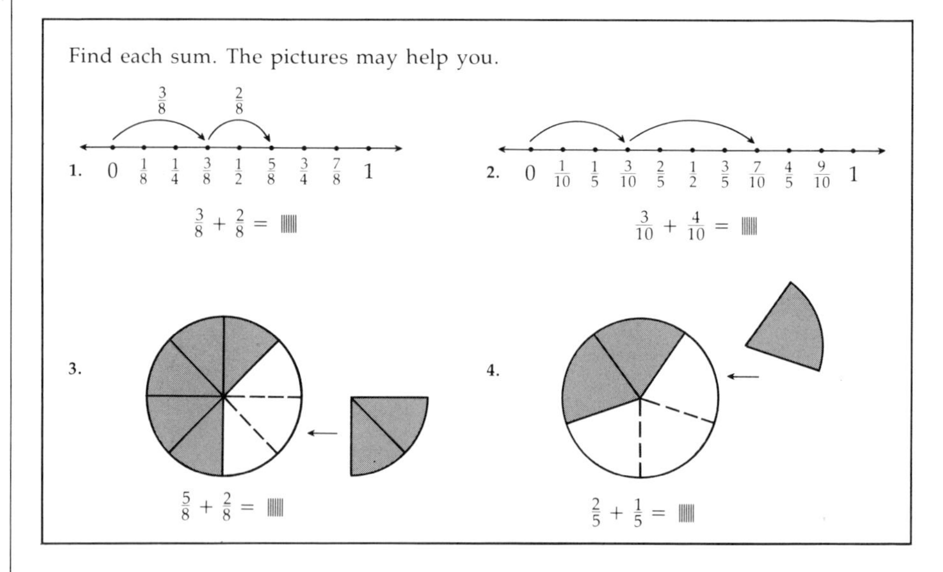

Try Exercise 10

Is the sum $\frac{a \cdot d + b \cdot c}{b \cdot d}$ always a rational number? Yes, because the numerator is the sum of two whole numbers ($a \cdot d$ and $b \cdot c$) and the denominator is the product of two nonzero whole numbers, hence it is also not zero.

Does it make any difference which representation is used for a rational number when finding a sum? The following example answers this.

Example 1 Show that no matter which fractions are used for two nonnegative rational numbers in a sum, the sums will be equal.

Solution Assume that $\frac{a}{b}$ and $\frac{c}{d}$ are any two nonnegative rational numbers expressed in the lowest-term fractions. We find the sum:

$$\frac{a}{b} + \frac{c}{d} = \frac{a \cdot d + b \cdot c}{b \cdot d}.$$

Also we know that any other rational number equal to $\frac{a}{b}$ must be $\frac{a \cdot p}{b \cdot p}$, where p is a natural number (see Example 6 in Section 6.2). Similarly, any other number equal to $\frac{c}{d}$ must be $\frac{c \cdot q}{d \cdot q}$, where q is a natural number. We use the definition to add these:

$$\frac{a \cdot p}{b \cdot p} + \frac{c \cdot q}{d \cdot q} = \frac{(a \cdot p) \cdot (d \cdot q) + (b \cdot p) \cdot (c \cdot q)}{(b \cdot p) \cdot (d \cdot q)} = \frac{p \cdot q \cdot (a \cdot d + b \cdot c)}{p \cdot q \cdot (b \cdot d)}$$

(by the usual and familiar whole-number properties for multiplication). But

$$\frac{a \cdot d + b \cdot c}{b \cdot d} = \frac{p \cdot q \cdot (a \cdot d + b \cdot c)}{p \cdot q \cdot (b \cdot d)}.$$

This means that it does not matter if we add $\frac{3}{4}$ and $\frac{1}{4}$ to get the sum $\frac{4}{4}$ or $\frac{6}{8}$ and $\frac{2}{8}$ to get the sum $\frac{8}{8}$, since the sums are in fact equal.

Try Exercise 11

11. Show that the sums are the same:

a) $\frac{1}{7} + \frac{4}{7}$ and $\frac{2}{14} + \frac{8}{14}$

b) $\frac{5}{6} + \frac{1}{3}$ and $\frac{10}{12} + \frac{2}{6}$

In adding nonnegative rational numbers represented by fractions with the same denominator, we find it is often simpler to merely add their numerators instead of using the definition for addition.

Example 2 Find the sum of $\frac{a}{c}$ and $\frac{b}{c}$.

$$\frac{a}{c} + \frac{b}{c} = \frac{a \cdot c + c \cdot b}{c \cdot c}$$ [by the definition of addition for nonnegative rationals]

$$= \frac{c \cdot (a + b)}{c \cdot c}$$ [by the commutative law for multiplication and the distributive law, both for whole numbers]

$$= \frac{(a + b)}{c}$$ [by the definition of equality for nonnegative rational numbers]

We have just proved the following *rule for addition*:

For any nonnegative rational numbers $\frac{a}{c}$ and $\frac{b}{c}$, their sum is given by $\frac{a+b}{c}$.

This rule makes it easy to add, since it is always possible to find fractions with a common denominator.

Example 3 Add $\frac{1}{3}$ and $\frac{2}{5}$.

Solution We find equivalent fractions with common denominators:

$$\frac{1}{3} = \frac{1 \cdot 5}{3 \cdot 5} = \frac{5}{15} \quad \text{and} \quad \frac{2}{5} = \frac{2 \cdot 3}{5 \cdot 3} = \frac{6}{15};$$

then

$$\frac{1}{3} + \frac{2}{5} = \frac{5}{15} + \frac{6}{15} = \frac{11}{15}.$$

Try Exercise 12

12. Add $\frac{3}{5}$ and $\frac{5}{7}$. Use the common-denominator method.

13. Find the least common denominator and the appropriate equivalent fractions to add $\frac{12}{70}$ and $\frac{7}{15}$.

It is more efficient, but not necessary, to use the least common multiple of the individual denominators as a common denominator.

Example 4 Find the least common denominator and then add:

$$\frac{3}{22} + \frac{2}{14}.$$

Solution The denominators are factored: $22 = 2 \cdot 11$, $14 = 2 \cdot 7$. Hence the LCM is $2 \cdot 7 \cdot 11$. The required equivalent fractions are found:

$$\frac{3}{22} = \frac{3 \cdot 7}{2 \cdot 11 \cdot 7} = \frac{21}{154}, \qquad \frac{2}{14} = \frac{2 \cdot 11}{2 \cdot 7 \cdot 11} = \frac{22}{154}.$$

Thus, the sum is:

$$\frac{3}{22} + \frac{2}{14} = \frac{21}{154} + \frac{22}{154} = \frac{21 + 22}{154} = \frac{43}{154}.$$

Try Exercise 13

THE ASSOCIATIVE PROPERTY. From the definition of addition it is easy to show that the familiar properties of the whole numbers also hold for the non-negative rationals. Thinking about the general case is no more difficult than working a specific example.

Example 5 Show that the associative law for addition holds for nonnegative rational numbers.

Since it is always possible to find common denominators, without loss of generality the law can be stated as:

> **For any nonnegative rational numbers $\frac{a}{d}$, $\frac{b}{d}$, and $\frac{c}{d}$,**
>
> $$\left(\frac{a}{d} + \frac{b}{d}\right) + \frac{c}{d} = \frac{a}{d} + \left(\frac{b}{d} + \frac{c}{d}\right).$$

The sums are calculated in two ways:

1. $\left(\frac{a}{d} + \frac{b}{d}\right) + \frac{c}{d} = \left(\frac{a + b}{d}\right) + \frac{c}{d}$ [by the definition of addition for non-negative rational numbers, used twice]

$$= \frac{(a + b) + c}{d}$$

2. $\frac{a}{d} + \left(\frac{b}{d} + \frac{c}{d}\right) = \frac{a}{d} + \left(\frac{b + c}{d}\right)$ [same as in (1)]

$$= \frac{a + (b + c)}{d}$$

The numerators are $(a + b) + c$ and $a + (b + c)$, which are sums in the system of whole numbers where associativity holds for addition; so they are the same. The two nonnegative rational sums are then the same, since both numerators and denominators are the same.

Try Exercise 14

14. Illustrate the associative property of addition:

a) Use 2/9, 5/9, 1/9

b) Use 1/3, 1/6, 1/12

THE COMMUTATIVE PROPERTY. Just as one can prove the associative property by applying the definition for addition of nonnegative rational numbers and the corresponding laws for whole numbers, so can the commutative law be shown to hold.

For any nonnegative rational numbers $\frac{a}{b}$ and $\frac{c}{d}$,

$$\frac{a}{b} + \frac{c}{d} = \frac{c}{d} + \frac{a}{b}.$$

The proof is left for the exercises.

15. Find the sum of 0/1 and 2/9.

THE IDENTITY FOR ADDITION. The nonnegative rational number $0/n$, $n \neq 0$, is the additive identity. In order to add zero to the number a/b, the representation $0/b$ is the most convenient one:

$$\frac{a}{b} + \frac{0}{b} = \frac{a + 0}{b} = \frac{a}{b}.$$

There is no other identity, since, if there were it could be called I' and then $I' + 0/1$ must equal $0/1$. On the other hand, $I' + 0/1$ must equal I'. This means that I' and $0/1$ are the same and there is exactly one identity.

Try Exercise 15

Subtraction

The definition of subtraction used for the whole numbers can be extended to include the nonnegative rational numbers.

16. Use the definition of subtraction to show that

a) $\frac{9}{17} - \frac{2}{17} = \frac{7}{17}$

b) $\frac{2}{3} - \frac{1}{2} = \frac{1}{6}$

SUBTRACTION. For any nonnegative rational numbers $\frac{a}{b}$ and $\frac{c}{d}$, their *difference* $\frac{a}{b} - \frac{c}{d}$ is the number $\frac{e}{f}$ if and only if the sum of $\frac{e}{f}$ and $\frac{c}{d}$ is $\frac{a}{b}$:

$$\frac{a}{b} - \frac{c}{d} = \frac{e}{f} \leftrightarrow \frac{a}{b} = \frac{e}{f} + \frac{c}{d}.$$

Try Exercise 16

Since fractions can always be found that have the same denominator for a given pair of rational numbers, it is useful to prove a *rule for subtraction* that can be applied for that case.

17. Find the difference by first finding a common denominator:

$$\frac{3}{8} - \frac{1}{3}$$

For any nonnegative rational numbers $\frac{a}{d}$ and $\frac{c}{d}$, the difference $\frac{a}{d} - \frac{c}{d}$ is $\frac{a-c}{d}$.

If we show that $\frac{a-c}{d} + \frac{c}{d} = \frac{a}{d}$ is true, then, by the definition of subtraction, we will have proved that the difference given is the correct one. We do this:

$$\frac{a-c}{d} + \frac{c}{d} = \frac{(a-c)+c}{d}$$ [by the definition of addition of nonnegative rationals]

$$= \frac{a}{d}$$ [by the definition of subtraction of whole numbers, $(a - c) + c = a$.]

It is easier to apply the rule just proved than to try to use the definition itself for subtraction.

Try Exercise 17

A sixth-grade textbook motivates the rule for subtraction we have just proved by means of suitable diagrams.

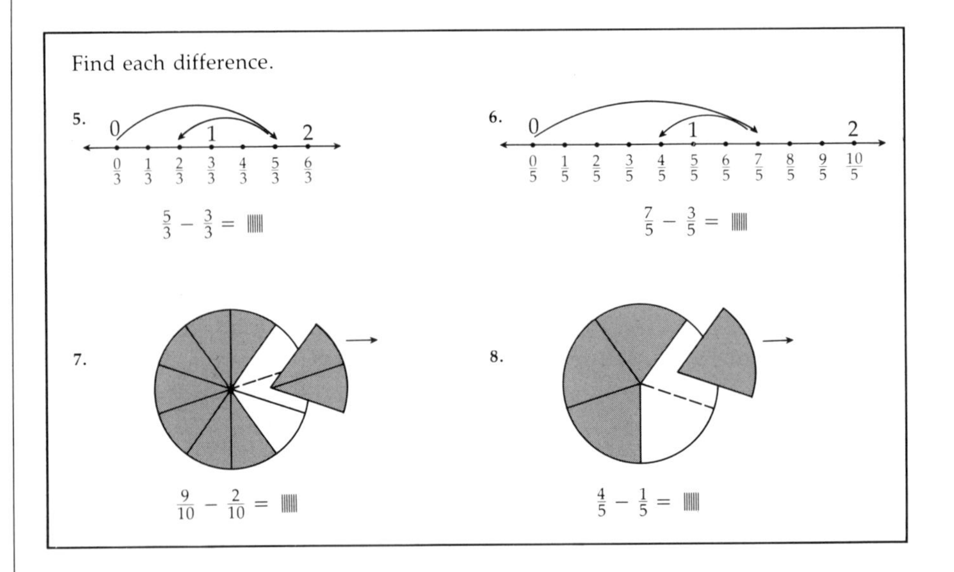

Subtraction for nonnegative rational numbers, just as for whole numbers, is not associative. Strictly speaking, it is also not an operation, since there are some pairs of nonnegative rational numbers for which there is no difference that is also a nonnegative rational.

Mixed Numerals

The meaning of so-called *mixed* notation becomes clear if the missing plus sign is written. For example,

$$2\frac{2}{3} \quad \text{means} \quad 2 + \frac{2}{3}.$$

This sum of two rational numbers can be found by supplying a common denominator:

$$2\frac{2}{3} = \frac{2}{1} + \frac{2}{3} = \frac{2 \cdot 3}{1 \cdot 3} + \frac{2}{3} = \frac{6}{3} + \frac{2}{3} = \frac{8}{3}.$$

Try Exercise 18

18.

a) Represent $5\frac{3}{8}$ as a fraction.

b) Read and work through the fourth-grade textbook excerpt on mixed numerals.

Since addition is commutative and associative, numbers can be grouped and ordered for convenience in adding. Ordinarily the whole numbers are added separately first.

Example 6 Add:

$$\begin{array}{r} 4\frac{3}{5} \\ 2\frac{1}{5} \\ 7\frac{4}{5} \\ \hline \end{array}$$

Solution By the associative and commutative law for addition of nonnegative rational numbers, the numbers 4, 2, and 7 are added first to get 13. The remaining terms are added to get $\frac{8}{5}$:

$$4 + \frac{3}{5} + 2 + \frac{1}{5} + 7 + \frac{4}{5}$$
$$= (4 + 2 + 7) + \left(\frac{3}{5} + \frac{1}{5} + \frac{4}{5}\right) = 13 + \frac{8}{5}.$$

However, $\frac{8}{5} = \frac{5}{5} + \frac{3}{5} = 1\frac{3}{5}$, so the sum is finally $14\frac{3}{5}$.

Try Exercise 19

19.

Add.
$$\begin{array}{r} 6\frac{2}{3} \\ 5\frac{1}{3} \\ 4\frac{2}{3} \end{array}$$

The area models can be used to good advantage in displaying representations of mixed notation. The following examples from a fourth-grade text show how this might be done.

Mixed numerals

$1\frac{1}{2}, 3\frac{3}{4}, 2\frac{1}{3}, \ldots$

Improper fractions

$\frac{9}{1}, \frac{8}{4}, \frac{10}{10}, \ldots$

Give a mixed numeral and an improper fraction for each problem.

Example:

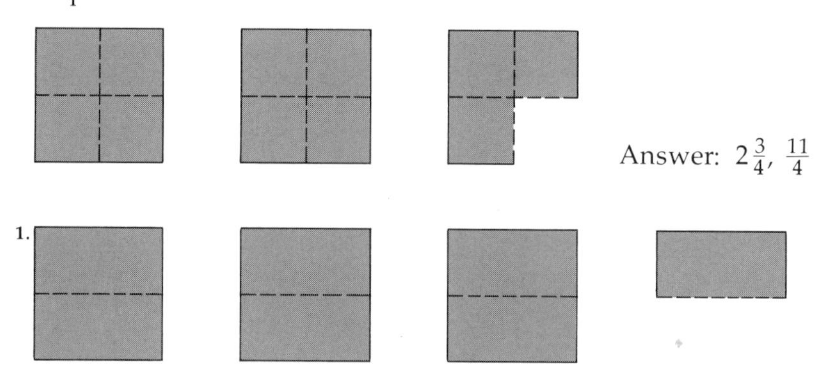

EXERCISE SET 6.3

1. Add:

a) $\dfrac{3}{10}+\dfrac{7}{10}$ b) $\dfrac{6}{11}+\dfrac{8}{11}$ c) $\dfrac{33}{64}+\dfrac{93}{64}$ d) $\dfrac{38}{100}+\dfrac{57}{100}$

2. Subtract:

a) $\dfrac{30}{32}-\dfrac{2}{32}$ b) $\dfrac{29}{10}-\dfrac{12}{10}$ c) $\dfrac{36}{16}-\dfrac{25}{16}$ d) $\dfrac{56}{100}-\dfrac{44}{100}$

3. Find the least common denominator and add:

a) $\dfrac{23}{42}+\dfrac{12}{105}$ b) $\dfrac{17}{90}+\dfrac{312}{420}$

4. Find the least common denominator and subtract:

a) $\dfrac{39}{45}-\dfrac{21}{70}$ b) $\dfrac{143}{150}-\dfrac{112}{315}$

5. Illustrate the commutative property of addition using

a) 3/5 and 7/8 b) 7/12 and 19/11

6. Illustrate the associative property of addition using

a) 3/15, 2/15, and 7/15 b) 3/19, 5/19, and 6/19

7. Convert to improper fractions:

a) $3\dfrac{2}{3}$ b) $5\dfrac{4}{7}$ c) $17\dfrac{5}{12}$ d) $23\dfrac{5}{16}$

8. Add: $4\dfrac{5}{12}+7\dfrac{9}{12}+8\dfrac{3}{12}+9\dfrac{7}{12}$

9. Add: $8\dfrac{3}{11}+7\dfrac{9}{11}+8\dfrac{5}{11}+13\dfrac{7}{11}$

10. Show that the set of nonnegative rational numbers is closed under addition and not closed under subtraction.

11. Prove the commutative law of addition for nonnegative rational numbers.

12. A rule for converting from a mixed numeral to an improper fraction is as follows: To convert $a\frac{b}{c}$ multiply a by c, write $\frac{a \cdot c + b}{c}$. Show that this rule is correct.

13. A rule for converting a fraction that represents a rational number greater than 1 is: To convert $\frac{a}{b}$, divide a by b, getting a quotient Q and a remainder R. The answer is $Q\frac{R}{b}$. Prove that this rule is correct.

Problems 14–17 are suitable for elementary pupils. Solve them as if you were going to explain them to such students. Try to find appropriate models or diagrams that could be used in the explanation.

14. You have cut up a pizza and given 1/3 of it to one friend and 1/4 to another. How much do you have left for yourself?

15. There is a number which, when added to half of itself, gives the sum $13\frac{1}{2}$. Find the number.

16. Kim mowed 1/3 of the lawn in the morning and 3/8 of it just after lunch. How much of the lawn was left to mow?

17. You are planning a garden and think you would like to have 1/4 of the garden in tomatoes, 1/3 corn, and 1/8 stringbeans. If you do as planned, how much of the garden space would be filled?

■ **18.** Prove that although subtraction is not associative, yet for any nonnegative rational numbers m, n, p, r, s, and t it is true that:

$$(m + n + p) - (r + s + t) = (m - r) + (n - s) + (p - t).$$

(Assume that all the indicated differences are nonnegative rational numbers.)

6.4 OPERATIONS: MULTIPLICATION AND DIVISION

Multiplication

DEFINITION. The area model suggests a sensible definition for the product of two nonnegative rational numbers.

Example 1 Use the area model to find $\frac{3}{4}$ of $\frac{1}{3}$. We begin with one unit of area, mark off 3 equal areas and shade 1/3 of the original figure.

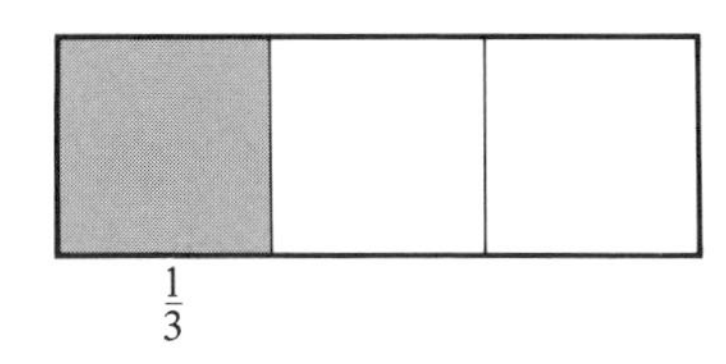

6.4 AIMS

You should be able to:

a) Multiply and divide nonnegative rational numbers represented by fractions.
b) Recognize, apply, and prove the associative and commutative properties for multiplication.
c) Prove that the distributive property for multiplication over addition holds and that 1 is the identity for multiplication.
d) Prove that division can be done by multiplying by a multiplicative inverse.

20. Use a rectangular diagram to show $\frac{2}{3}$ of $\frac{3}{4}$.

The shaded region is portioned into fourths and three of these are shaded:

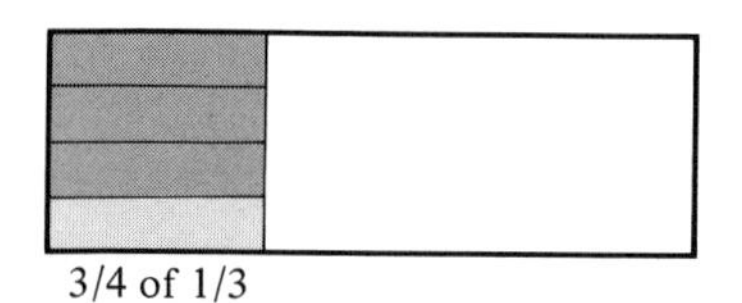

3/4 of 1/3

The area shaded twice is 3/12 of the original rectangle.

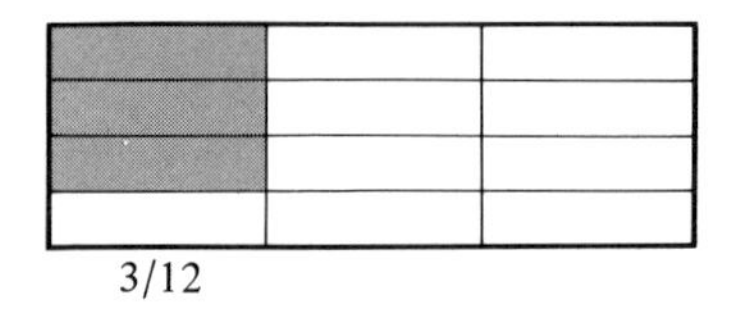

3/12

The denominator of 3/12 is $3 \cdot 4$, the total number of equal parts. The numerator is $3 \cdot 1$, the total number of parts shaded twice.

Try Exercise 20

To represent multiplication of a/b and c/d, imagine a rectangle* divided into d equal parts with c of them shaded. The shaded area is then divided into b equal parts and a of them are shaded. The original unit would be divided into $b \cdot d$ equal parts, and $a \cdot c$ of them would be shaded twice. This leads to the following definition.

* Observe that there is a great deal of resemblance between this model and the models for whole-number products.

Cross product
3 X 4

Counting
3 X 4

DEFINITION. If $\frac{a}{b}$ and $\frac{c}{d}$ are any two nonnegative rational numbers, the *product* of $\frac{a}{b}$ and $\frac{c}{d}$ is $\frac{a \cdot c}{b \cdot d}$. That is:

$$\frac{a}{b} \cdot \frac{c}{d} = \frac{a \cdot c}{b \cdot d}.$$

The product of any two nonnegative rational numbers is again a nonnegative rational, since $a \cdot c$ is the product of two whole numbers and hence a whole number and $b \cdot d$ is the product of two nonzero whole numbers and hence is a nonzero whole number.

It does not matter which fractions are used to represent the factors in a product. If we multiply $\frac{1}{2}$ and $\frac{1}{4}$ to get $\frac{1}{8}$, it is the same as if we used $\frac{2}{4}$ and $\frac{2}{8}$ to get the product $\frac{4}{32}$ since $\frac{1}{8}$ is equal to $\frac{4}{32}$.

Example 2 Show that equivalent fractions can be used to find the product of $\frac{a}{b}$ and $\frac{c}{d}$.

Solution We show that no matter what fractions represent the factors, the product is $\frac{a \cdot c}{b \cdot d}$. Suppose that we use $\frac{a \cdot p}{b \cdot p}$ for $\frac{a}{b}$, where p is a natural number. Similarly, we use $\frac{c \cdot q}{d \cdot q}$ for $\frac{c}{d}$, where q is a natural number. We use these in the product:

$$\frac{a \cdot p}{b \cdot p} \cdot \frac{c \cdot q}{d \cdot q} = \frac{(a \cdot p) \cdot (c \cdot q)}{(b \cdot p) \cdot (d \cdot q)} = \frac{p \cdot q \cdot a \cdot c}{p \cdot q \cdot b \cdot d}.$$

We see that this product is really $\frac{a \cdot c}{b \cdot d}$.

Youngsters sometimes think that somehow the answer might be different if they use an equivalent fraction in a product. This example shows they need not worry.

Try Exercise 21

ASSOCIATIVE PROPERTY. From the definition of multiplication it is not difficult to establish the familiar properties such as associativity and commutativity for multiplication of nonnegative rationals. The next example follows exactly the steps of the exercise completed above.

Example 3 Show that the associative law of multiplication holds for nonnegative rational numbers.

It must be shown that for any nonnegative rational numbers $\frac{a}{b}$, $\frac{c}{d}$, and $\frac{e}{f}$, the following is true: $\left(\frac{a}{b} \cdot \frac{c}{d}\right) \cdot \frac{e}{f} = \frac{a}{b} \cdot \left(\frac{c}{d} \cdot \frac{e}{f}\right)$.

1. $\left(\frac{a}{b} \cdot \frac{c}{d}\right) \cdot \frac{e}{f} = \left(\frac{a \cdot c}{b \cdot d}\right) \cdot \frac{e}{f}$ $\quad$ [by the definition of multiplication of $\frac{a}{b} \cdot \frac{c}{d}$]

   $= \frac{(a \cdot c) \cdot e}{(b \cdot d) \cdot f}$ $\quad$ [by the same definition]

2. $\frac{a}{b} \cdot \left(\frac{c}{d} \cdot \frac{e}{f}\right) = \frac{a}{b} \cdot \left(\frac{c \cdot e}{d \cdot f}\right)$ $\quad$ [by the definition of multiplication of $\frac{c}{d} \cdot \frac{e}{f}$]

   $= \frac{a \cdot (c \cdot e)}{b \cdot (d \cdot f)}$ $\quad$ [by the same definition]

These results are the same because:

the numerators are the same: $(a \cdot c) \cdot e = a \cdot (c \cdot e)$,
the denominators are the same: $(b \cdot d) \cdot f = b \cdot (d \cdot f)$,

and both of these are true, since the associative law holds for whole numbers.

21. Use the definition of multiplication.

a) Find $\frac{2}{3} \cdot \frac{4}{7}$, then $\left(\frac{2}{3} \cdot \frac{4}{7}\right) \cdot \frac{9}{11}$.

b) Find $\frac{4}{7} \cdot \frac{9}{11}$, then $\frac{2}{3} \cdot \left(\frac{4}{7} \cdot \frac{9}{11}\right)$.

c) Compare your answers in parts (a) and (b).

22. Use the definition of multiplication to verify that multiplication is commutative:

$$\frac{9}{5}\cdot\frac{7}{11}=\frac{7}{11}\cdot\frac{9}{5}$$

23. Find the lowest-terms fraction to represent $\frac{60}{700}$.

Thus we have shown that the two products of the nonnegative rational numbers are the same.

COMMUTATIVE PROPERTY. The commutative property follows just as easily from the definition of multiplication.

For any two nonnegative rational numbers $\frac{a}{b}$ and $\frac{c}{d}$,

$$\frac{a}{b}\cdot\frac{c}{d}=\frac{a\cdot c}{b\cdot d}=\frac{c\cdot a}{d\cdot b}=\frac{c}{d}\cdot\frac{a}{b}.$$

Details of the proof are left for the exercises.

Try Exercise 22

IDENTITY FOR MULTIPLICATION. The number 1 is the identity. We need only examine what happens when any nonnegative rational number is multiplied by 1/1.

By the definition of multiplication we have

$$\frac{a}{b}\cdot\frac{1}{1}=\frac{a\cdot 1}{b\cdot 1}=\frac{a}{b},$$

since $a\cdot 1=a$ and $b\cdot 1=b$ holds for whole numbers. This property does not depend upon the representation of 1 that is used.

ALTERNATIVE METHOD FOR SIMPLIFYING. The lowest-terms fraction for a nonnegative rational number can be found by first factoring the numerator and the denominator and then using the definition for multiplication to isolate factors of 1/1.

Example 4 Simplify $\frac{12}{630}$.

Solution Because 12 and 360 are not relatively prime, they are both factored:

$$\frac{12}{630}=\frac{2\cdot 2\cdot 3}{2\cdot 3\cdot 3\cdot 5\cdot 7}$$

$$=\frac{2}{2}\cdot\frac{3}{3}\cdot\frac{2}{3\cdot 5\cdot 7} \quad \text{[by the definition of multiplication]}$$

$$=\frac{2}{3\cdot 5\cdot 7}=\frac{2}{105} \quad \text{[by the identity property of multiplication]}$$

Try Exercise 23

MULTIPLICATIVE INVERSES. A property of the nonnegative rational numbers which the whole numbers do not have is that each such number, except for 0, has a multiplicative inverse.*

* If 0/1 had an inverse it would mean that 1/0 is an allowable quotient, but it is not.

DEFINITION. **For each nonnegative rational number a/b ($a/b \neq 0/b$) there is a nonnegative rational number b/a, called the *multiplicative inverse* of a/b, such that**

$$\frac{a}{b} \cdot \frac{b}{a} = 1.$$

It follows from the definition that a/b is also the multiplicative inverse of b/a; they are called *reciprocals* of each other. The multiplicative inverse is sometimes written as $(a/b)^{-1}$, which is read as "the multiplicative inverse of a/b."

Example 5 Show that 12/7 and 7/12 are multiplicatives inverses of each other.

Solution Their product must be the identity:

$$\frac{12}{7} \cdot \frac{7}{12} = \frac{7}{12} \cdot \frac{12}{7} = \frac{7 \cdot 12}{12 \cdot 7} = \frac{84}{84}.$$

Try Exercise 24

24. Find the multiplicative inverse of each of the following:

a) 8/13

b) 5

c) $a \cdot b, a \neq 0, b \neq 0$

CANCELLATION LAW. The fact that every nonnegative rational number, aside from zero, has a multiplicative inverse allows the following *cancellation law for multiplication* to be proved.

For all nonnegative rational numbers a/b, c/d, and e/f ($a/b \neq 0/1$),

$$\text{if } \frac{a}{b} \cdot \frac{c}{d} = \frac{a}{b} \cdot \frac{e}{f}, \text{ then } \frac{c}{d} = \frac{e}{f}.$$

The proof follows the steps and the reasoning used in the next example and is left for the exercises.

Example 6 Find x/y if $\frac{1}{2} \cdot \frac{3}{4} = \frac{1}{2} \cdot \frac{x}{y}$.

Solution Since 1/2 is not 0, it has a reciprocal 2/1. Multiply both terms by 2/1:

$$\frac{2}{1} \cdot \left(\frac{1}{2} \cdot \frac{3}{4}\right) = \frac{2}{1} \cdot \left(\frac{1}{2} \cdot \frac{x}{y}\right)$$

$$\left(\frac{2}{1} \cdot \frac{1}{2}\right) \cdot \frac{3}{4} = \left(\frac{2}{1} \cdot \frac{1}{2}\right) \cdot \frac{x}{y}$$ [by the associative law for nonnegative rationals]

$$\frac{3}{4} = \frac{x}{y}$$ [by the properties of the inverse and the identity of multiplication]

Try Exercise 25

25. Use the method of the example to solve for x:

$$\frac{1}{4} \cdot \frac{3}{5} = \frac{1}{4} \cdot \frac{x}{15}$$

DISTRIBUTIVE LAW. The distributive law for multiplication over addition is another that holds for both the whole numbers and the nonnegative rational numbers. The whole-number properties are inherited by the nonnegative

26. Verify the distributive law for the product of $1/5$ and the sum $1/3 + 4/3$.

rationals because the definitions for the numbers themselves and for the operations with them have been made to preserve, in the extended number system, the desirable properties of the original whole-number system.

Since fractions with the same denominator can always be found for performing the addition of two rational numbers, the *distributive law* can be stated without any loss in generality as follows:

THE DISTRIBUTIVE LAW. For any nonnegative rational numbers a/b, c/d, and e/d, the following is true:

$$\frac{a}{b}\cdot\left(\frac{c}{d}+\frac{e}{d}\right)=\left(\frac{a}{b}\cdot\frac{c}{d}\right)+\left(\frac{a}{b}\cdot\frac{e}{d}\right)=\frac{a\cdot c}{b\cdot d}+\frac{a\cdot e}{b\cdot d}.$$

As has been the case in many of the previous proofs, we find that a careful examination of a special case will provide the pattern or argument for a general proof (this is left for the exercises).

Example 7 Verify that the distributive law holds for the product of $3/11$ and the sum of $4/7$ and $2/7$.

We must prove that

$$\frac{3}{11}\cdot\left(\frac{4}{7}+\frac{2}{7}\right)=\left(\frac{3}{11}\cdot\frac{4}{7}\right)+\left(\frac{3}{11}\cdot\frac{2}{7}\right).$$

1. $\frac{3}{11}\cdot\left(\frac{4}{7}+\frac{2}{7}\right)=\frac{3}{11}\cdot\left(\frac{4+2}{7}\right)$ [by the definition of addition for nonnegative rational numbers]

$\quad=\frac{3\cdot(4+2)}{11\cdot 7}$ [by the definition of multiplication for nonnegative rational numbers]

2. $\left(\frac{3}{11}\cdot\frac{4}{7}\right)+\left(\frac{3}{11}\cdot\frac{2}{7}\right)=\left(\frac{3\cdot 4}{11\cdot 7}\right)+\left(\frac{3\cdot 2}{11\cdot 7}\right)$ [by the definition of multiplication for nonnegative rational numbers]

$\quad=\frac{3\cdot 4+3\cdot 2}{11\cdot 7}$ [by the rule for addition of nonnegative rational numbers]

$\quad=\frac{3\cdot(4+2)}{11\cdot 7}$ [by the distributive law for whole numbers]

Therefore, by the transitive property for equals for nonnegative rational numbers,

$$\frac{3}{11}\cdot\left(\frac{4}{7}+\frac{2}{7}\right)=\left(\frac{3}{11}\cdot\frac{4}{7}\right)+\left(\frac{3}{11}\cdot\frac{2}{7}\right)$$

Try Exercise 26

Division

DEFINITION. The division of two nonnegative rational numbers can be defined in a way that is analogous to the definition given for whole numbers, that is, in terms of multiplication.

> **DEFINITION. For any two nonnegative rational numbers a/b and c/d ($c/d \neq 0$), the *quotient* a/b divided by c/d is the number e/f if and only if the product of e/f and c/d is a/b. More briefly:**
>
> $$\frac{a}{b} \div \frac{c}{d} = \frac{e}{f} \leftrightarrow \frac{e}{f} \cdot \frac{c}{d} = \frac{a}{b}.$$

Just as for whole numbers, this definition does not permit division by 0.

Example 8 Use the definition of division to find the quotient $1/3 \div 5/8$.

Solution $\frac{1}{3} \div \frac{5}{8} = \frac{e}{f}$ if and only if $\frac{1}{3} = \frac{e}{f} \cdot \frac{5}{8}$. We multiply both terms by the reciprocal of $5/8$:

$$\frac{1}{3} \cdot \frac{8}{5} = \left(\frac{e}{f} \cdot \frac{5}{8}\right) \cdot \frac{8}{5}$$

$$\frac{8}{15} = \frac{e}{f} \cdot \left(\frac{5 \cdot 8}{8 \cdot 5}\right) = \frac{e}{f} \cdot \frac{40}{40}$$ [by the definition of multiplication and the associative property for nonnegative rationals]

$$\frac{8}{15} = \frac{e}{f}$$ [by the identity property]

Try Exercise 27

27. Use the definition of division to find $\frac{4}{3} \div \frac{5}{9}$.

An interpretation of division with an area model in a sixth-grade textbook

Use the pictures to help you solve each division problem.
Check by multiplying.

1.

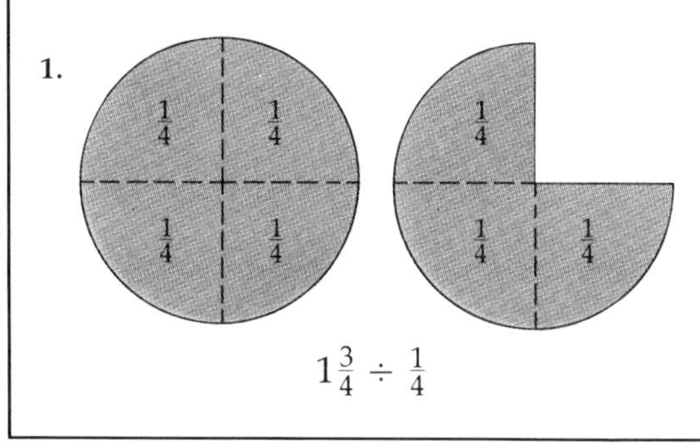

2.

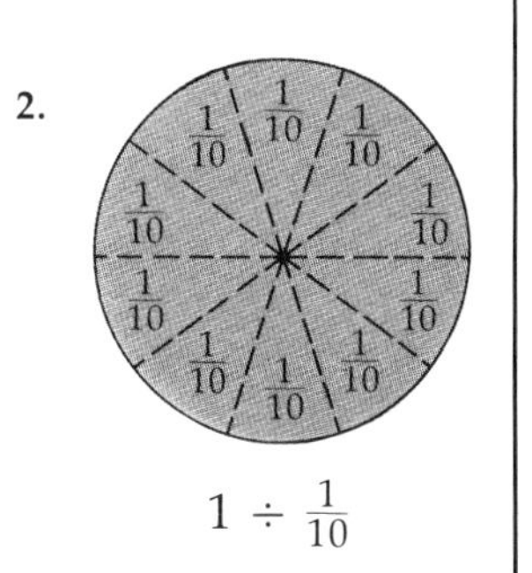

RULE FOR DIVISION. It is always possible to find the quotient directly from the definition, as was done in Example 8. That procedure is followed exactly to derive a general *rule for the division* of nonnegative rational numbers.

The quotient of any two nonnegative rational numbers $\frac{a}{b}$ and $\frac{c}{d}\left(\frac{c}{d} \neq 0\right)$ is the product $\frac{a}{b} \cdot \frac{d}{c}$. That is:

$$\frac{a}{b} \div \frac{c}{d} = \frac{a}{b} \cdot \frac{d}{c} = \frac{a \cdot d}{b \cdot c}.$$

We start with the definition of division:

$$\frac{a}{b} \div \frac{c}{d} = \frac{e}{f} \leftrightarrow \frac{e}{f} \cdot \frac{c}{d} = \frac{a}{b}.$$

Since any nonnegative rational number (except 0) has a multiplicative inverse, c/d has such an inverse; it is d/c. Both terms of the last equation above can be multiplied by d/c to give:

$$\left(\frac{e}{f} \cdot \frac{c}{d}\right) \cdot \frac{d}{c} = \frac{a}{b} \cdot \frac{d}{c}$$ [by the property of equals]

$$\frac{e}{f} \cdot \left(\frac{c \cdot d}{d \cdot c}\right) = \frac{e}{f} = \frac{a}{b} \cdot \frac{d}{c} = \frac{a \cdot d}{b \cdot c}$$ [by the associative property, the identity, and the equals properties]

This rule looks like the familiar "invert and multiply" rule you may have learned at one time.

The last rule indicates how important the reciprocals are. All divisions, except by 0, are possible in the set of nonnegative rational numbers and we have established closure for division on the set of nonzero nonnegative rational numbers.

COMPOUND FRACTIONS. By analogy, since we write $a \div b$ as $\frac{a}{b}$, we generalize the idea of a fraction and for $\frac{a}{b} \div \frac{c}{d}$ use $\frac{\frac{a}{b}}{\frac{c}{d}}$. If we can show that

$$\frac{\frac{a}{b}}{\frac{c}{d}} = \frac{\frac{a}{b} \cdot \frac{p}{q}}{\frac{c}{d} \cdot \frac{p}{q}},$$

this will be a useful way to think about the division of nonnegative rational numbers. We do this easily.

1. $\frac{a}{b} \div \frac{c}{d} = \frac{\frac{a}{b}}{\frac{c}{d}} = \frac{a}{b} \cdot \frac{d}{c}$ [by the rule for division]

2. $\frac{\frac{a}{b} \cdot \frac{p}{q}}{\frac{c}{d} \cdot \frac{p}{q}} = \frac{\frac{a \cdot p}{b \cdot q}}{\frac{c \cdot p}{d \cdot q}}$ [by the definition of multiplication for nonnegative rational numbers]

$= \frac{a \cdot p}{b \cdot q} \cdot \frac{d \cdot q}{c \cdot p}$ [by the rule for division]

$= \frac{a \cdot d \cdot p \cdot q}{b \cdot c \cdot p \cdot q}$ [by the definition of multiplication and the associative and commutative laws for nonnegative rational numbers]

$= \frac{a}{b} \cdot \frac{d}{c}$ [by the definitions for equality and multiplication]

This allows us to treat division as follows.

Example 9 Find the quotient $3/8 \div 5/6$.

Solution We write the compound fraction: $\frac{\frac{3}{8}}{\frac{5}{6}}$; multiply numerator and denominator by $\frac{6}{5}$: $\frac{\frac{3}{8} \cdot \frac{6}{5}}{\frac{5}{6} \cdot \frac{6}{5}}$; find the indicated products: $\frac{\frac{18}{40}}{\frac{1}{1}} = \frac{18}{40}$ (by the identity property).

Try Exercise 28

28. Find $3/4 \div 3/8$ by writing the quotient as a compound fraction and then simplifying.

EXERCISE SET 6.4

1. Use the area model to show:

a) $\frac{3}{5} \cdot \frac{3}{4}$ b) $\frac{3}{4} \cdot \frac{3}{5}$

2. Use the area model to show:

a) $\frac{5}{4} \cdot \frac{2}{3}$ b) $\frac{2}{3} \cdot \frac{5}{4}$

3. Show that the set of nonnegative rational numbers is closed under multiplication.

4. Verify the commutativity of multiplication by using

a) 4/7 and 8/3 b) 2/9 and 5/7

5. Show that the commutative law holds for multiplication of any two nonnegative rational numbers.

6. Verify the associativity of multiplication by using
a) 3/5, 4/9, 8/13 b) 4/7, 3/8, 5/11

7. Simplify:

a) $\dfrac{105}{90}$ b) $\dfrac{60}{126}$

8. Prove that the distributive law for multiplication over addition holds for nonnegative rational numbers.

9. Compare the following:

a) $\dfrac{5}{8} \cdot \left(\dfrac{3}{7} - \dfrac{1}{7}\right)$ and $\dfrac{5 \cdot 3}{8 \cdot 7} - \dfrac{5 \cdot 1}{8 \cdot 7}$

b) $\dfrac{4}{7} \cdot \left(\dfrac{7}{12} - \dfrac{5}{12}\right)$ and $\dfrac{4 \cdot 7}{7 \cdot 12} - \dfrac{4 \cdot 5}{7 \cdot 12}$

10. On the basis of Problem 9, what is your conjecture about the distribution of multiplication over subtraction? Prove your conjecture.

11. To find x, use the fact that every nonnegative rational (except 0) has a reciprocal:

a) $\dfrac{1}{3} \cdot \dfrac{5}{12} = \dfrac{1}{3} \cdot \dfrac{x}{24}$ b) $\dfrac{1}{5} \cdot \dfrac{7}{8} = \dfrac{1}{5} \cdot \dfrac{x}{32}$

12. Show that the cancellation law for multiplication holds for nonnegative rational numbers: If a/b, c/d, and e/f are any nonnegative rational numbers ($a/b \neq 0$) and if $\dfrac{a}{b} \cdot \dfrac{c}{d} = \dfrac{a}{b} \cdot \dfrac{e}{f}$, then $\dfrac{c}{d} = \dfrac{e}{f}$.

13. Use the definition of division to complete the following table:

$\dfrac{a}{b} \div \dfrac{c}{d} =$	$\dfrac{a \div c}{b \div d} =$
$\dfrac{3}{5} \div \dfrac{7}{5} = \dfrac{3}{7}$	$\dfrac{3 \div 7}{5 \div 5} = \dfrac{3}{7}$
$\dfrac{4}{5} \div \dfrac{11}{5} =$	$\dfrac{4 \div 11}{5 \div 5} =$
$\dfrac{25}{13} \div \dfrac{3}{2} =$	$\dfrac{25 \div 3}{13 \div 2} =$
$\dfrac{70}{15} \div \dfrac{5}{4} =$	$\dfrac{70 \div 5}{15 \div 4} =$

14. Make a conjecture on the basis of the pattern in the table of Problem 13. You may want to extend the table. Prove your conjecture.

Problems 15–18 are similar to those that might be found in an elementary textbook. Some of them involve ratio and proportion as well as multiplication and division. Solve them as if you were going to explain them to an elementary pupil. Make appropriate sketches or models that could be used for each.

15. We have $2\frac{3}{4}$ cups of fresh strawberries. One serving is 3/8 of a cup. How many servings do we have?

16. If $1\frac{1}{2}$ cans of water are needed for $\frac{1}{2}$ can of frozen juice concentrate, how many cans of water are needed for 4 cans of concentrate?

17. We have 2/3 of the date cake left over. If one serving is 1/9 of the cake, how many servings are left?

18. A recipe for punch calls for 2 liters of soft drink and 1/2 liter of cranberry juice. What is the ratio of soft drink to cranberry huice? How much soft drink will be needed to make punch from an entire bottle of juice that holds $1\frac{1}{2}$ liters?

6.5 ORDER

6.5 AIMS

You should be able to:

a) Determine which of two nonnegative rational numbers is the greater.
b) Find the number halfway between two nonnegative rational numbers.

The same definition for order that was used for whole numbers can be used with the appropriate changes for nonnegative rationals.

DEFINITION. For any nonnegative rational numbers a/b and c/d, a/b *is less than c/d if and only if there is a nonnegative rational e/f ($e/f \neq 0$) such that $a/b + e/f = c/d$. That is,**

$$\frac{a}{b} < \frac{c}{d} \leftrightarrow \frac{a}{b} + \frac{e}{f} = \frac{c}{d}, \quad \frac{e}{f} \neq 0.$$

As with the whole numbers, c/d is *greater than* a/b if and only if a/b is less than c/d.

It is not difficult to show that if the denominators of the two fractions are the same, then $a/c < b/c$ if and only if $a < b$. This reduces the problem to determining which of two whole numbers is greater. Actually, in practice it may be easier to find fractions with the same denominator before determining the order of two nonnegative rational numbers.

29. Determine which of 5/9 and 7/11 is greater by finding a common denominator.

Example 1 Which is greater, 5/7 or 2/3?

Solution Obtain a common denominator:

$$\frac{2}{3} = \frac{2 \cdot 7}{3 \cdot 7} = \frac{14}{21}; \qquad \frac{5}{7} = \frac{5 \cdot 3}{7 \cdot 3} = \frac{15}{21}.$$

$\frac{15}{21} > \frac{14}{21}$ because $15 > 14$, and thus $\frac{5}{7} > \frac{2}{3}$.

Try Exercise 29

DENSITY. The rational numbers have another important property that the whole numbers do not, because between any two rational numbers there is another such number. Indeed, it is possible to find the number halfway between two nonnegative rational numbers by simply averaging the two of them.

* The symbols < for "less than" and > for "greater than" were introduced about 1600 by Thomas Harriot, an Englishman who later moved to the colony of Virginia.

In a classroom

Teacher: (drawing a number line on the board and pointing to 1): What is the number halfway between 0 and 1?

Maria: 1/2.

Teacher: Right (draws and labels the point for 1/2). What number is halfway between 1/2 and 0?

Manuel: 1/4?

Teacher: That's right. How did you get it?

Manuel: I took half of a half.

Teacher: That's the way to do it. Now, what number is halfway from 1/4 to 0?

Cecile: 1/8.

Sam: I see a pattern. The denominators . . .

Teacher: Don't tell us. Let everybody look for the pattern. When you see one, raise your hand.

(Pause. Most pupils raise their hands.)

All right, Sam, what pattern do you see?

Sam: The numerators are all 1 and the denominators double every time.

Teacher: Is that what all of you see?

(Class indicates general agreement.)

Now, think about this. What is the smallest number greater than zero? Or, what number is right next to 0?

(Pause. Several pupils raise their hands.)

Teacher: Serena, you are so excited. Tell us what you think.

Serena: Well, there can't be any number right next to 0, because no matter how small a number is you can take half of it and get a smaller one.

Sam: It seems like it has to be, but that doesn't seem right.

Teacher: You all think about it some more. Tomorrow we'll see if we can find the biggest number which is less than 1.

Example 2 Find the number halfway between 1/4 and 1/3.

Solution First find a common denominator; 12 will do. Thus,

$$\frac{1}{4}+\frac{1}{3}=\frac{1\cdot 3}{4\cdot 3}+\frac{1\cdot 4}{3\cdot 4}=\frac{3}{12}+\frac{4}{12}.$$

The average is:

$$\frac{1}{2}\cdot\left(\frac{3}{12}+\frac{4}{12}\right)=\frac{1}{2}\cdot\frac{7}{12}=\frac{7}{24}.$$

Now check to see if 7/24 is between 3/12 and 4/12:

$$\frac{3}{12}=\frac{6}{24} \text{ and } \frac{4}{12}=\frac{8}{24}; \qquad \frac{6}{24}<\frac{7}{24}<\frac{8}{24}.$$

Try Exercise 30

By using a procedure similar to that of Example 2 it can be proved that the average of two numbers is between them. Between any two nonnegative rational numbers, however close together, there is actually an infinite number of such numbers. This reflects the fact that the nonnegative rational numbers have the property of being *dense* on the number line.

Try Exercise 31

30.
a) Find the nonnegative rational halfway between 1/3 and 1/2.
b) Find a whole number halfway between 6/1 and 10/1.
c) How many whole numbers are between 11/1 and 12/1?

31.
a) What sequence of numbers is generated in this class?
b) What is the tenth number of the sequence?
c) Will any term of this sequence ever be 0?

EXERCISE SET 6.5

1. Use the definition of "greater than" to determine which is greater:
a) 3/7 or 5/8 b) 21/5 or 27/7

2. Find the number halfway between
a) 4/3 and 5/4 b) 17/5 and 18/7

3. Show that the average of any two nonnegative rational numbers is between them.

4. Prove that for any nonnegative rational numbers a/b and c/b, $a/b < c/b$ if and only if $a < c$.

5. Find four numbers between 2/3 and 3/4.

6. Find five numbers between 9/5 and 11/5.

7. Find 20 numbers between 5/3 and 7/5. They need not be equally spaced in the interval.

■ **8.** Describe a procedure for finding some specified number of nonnegative rational numbers between two given ones.

9.
a) Find the number halfway between 0 and 1.
b) Find the number halfway between 1 and your answer to (a).
c) Proceed in this way to find a sequence of numbers and describe the sequence.

■ **10.** Prove that for any nonnegative rational numbers a/b and c/d the following is true:

$$\text{If } \frac{a}{b}<\frac{c}{d}, \text{ then } \frac{a}{b}<\frac{a+c}{b+d}<\frac{c}{d}.$$

Note that $\dfrac{a+c}{b+d}$ is not the average.

11. Here is one that might be in a sixth-grade textbook: Which of these numbers: 3/14, 1/5, 2/9, is smaller than one of the numbers but greater than the other?

■ **12.** A pupil says that to show $\frac{1}{3} < \frac{4}{5}$ you need only to look at the criss-cross products and observe that $1 \cdot 5 < 3 \cdot 4$. Is the pupil's observation correct for all nonnegative rational numbers?

■ **13.** Prove the cancellation law for addition for the "less than" relation: For any nonnegative rational numbers, if

$$\frac{a}{b} + \frac{c}{d} < \frac{a}{b} + \frac{e}{f}, \text{ then } \frac{c}{d} < \frac{e}{f}.$$

6.6 AIMS

You should be able to:
Use the method of "working backwards" as a technique in problem solving.

6.6 PROBLEM SOLVING

There are recognized algebraic techniques (along with other ways) that could be used to solve the next problem.* As the problem is stated, it seems that the nonnegative rational numbers are involved.

Example 1 Suzanne and Charles were going to sell candy bars as a fund-raising project for their high school Drum and Bugle Corps. They set up a table in a busy shopping area on a Saturday morning. The first customer was very generous and bought one-half of their entire stock and $\frac{1}{2}$ bar more. To the next customer they also sold one-half of the remaining stock and $\frac{1}{2}$ bar more. They did the same thing to the third customer. Strangely enough, in all three of these transactions not a single candy bar had to be cut! It was now getting close to lunch time, and Suzanne and Charles decided to share one of the bars they had left. After this bar was eaten they noticed there were only 8 bars left to sell. How many bars did they start out with in the morning?

What are the *relevant facts* for this problem?

Try Exercise 32

32. Write out the facts for the problem in Example 1.

What is the *problem* that *needs to be solved*?

1. How many bars did the students start with in the morning?
2. How many bars did they sell? (Then add 8)
3. How many bars did they have at each stage of the morning's transactions?

What are some *ideas* for getting the solution?

Try Exercise 33

33. List three ideas for the solution of the problem.

Here are some ideas that may have also occurred to you:

Idea 1 Choose a quantity to be represented by x and write some equations.

Idea 2 Get some counters and pretend to be the students with 8 bars left.

Idea 3 Work backwards from the 8 bars and reconstruct the events.

* As a reminder

THE FIVE-STEP PROCESS
1. What are the relevant *facts*?
2. What actually is the *problem*?
3. What are my *ideas*?
4. How do I get a *solution*?
5. *Check* and *review* the solution.

We use Idea 3. There are many ways to use this technique. Sometimes it is possible to work back from a desired answer. Here we work backwards in time and trace the events.

The Solution Since Suzanne and Charles ate 1 bar just before they counted and got 8, they must have had: $8 + 1 = 9$.

What they had (9) was what was left after they sold one-half of their remaining stock and $\frac{1}{2}$ bar more to the third customer. Let us "undo" the third transaction:

replace the $\frac{1}{2}$ candy bar: $9 + \frac{1}{2}$;
multiply by 2 to undo the one-half: $2 \cdot (9 + \frac{1}{2}) = 19$.

We now know that after the second transaction they had 19 bars. Let us "undo" the second transaction to see what that gives:

replace the $\frac{1}{2}$ candy bar: $19 + \frac{1}{2}$;
multiply by 2: $2 \cdot (19 + \frac{1}{2}) = 39$.

After the first transaction they had 39 bars. When the first transaction is undone, we should get the answer:

replace the $\frac{1}{2}$ candy bar: $39 + \frac{1}{2}$;
multiply by 2: $2 \cdot (39 + \frac{1}{2}) = 79$.

This looks like a solution, but it must be checked.

Try Exercise 34

34. Check the solution of the problem by using the answer and working forward through the conditions of the problem.

The technique of *working backward* is *especially useful* when the following circumstances hold:

1. There is one final goal and there are many statements and facts given.
2. The reversal steps are one-to-one: in each step going backward there is one possibility for the previous step.

This does not mean that the technique cannot be used with other types of problems. There are many variations of so-called water-jar problems which can be attacked successfully with the technique of *working backward* even though the reversal steps are not one-to-one. Working backward is also useful for trying to develop a proof of some statement because it is often helpful to think of working in reverse from the statement you are trying to prove to see how a direct proof might be developed.

Example 2 You have a jar that holds 7 liters of water and another that holds 3 liters of water. These jars are not marked in any way. There is an infinite supply of water. Find the steps you must take to measure out exactly 5 liters of water.

What are the *relevant facts*?

35. What are the relevant facts in Example 2?

Try Exercise 35

Exactly what is the *problem*?

Try Exercise 36

36. State the problem.

Ordinarily we would continue and have you think of some ideas, but in this case we will just choose the idea of *looking backward* to show how it can be successfully adapted even in a problem where it might not be possible to have a one-to-one reversal.

Idea Look back to the second-last step just before the goal is accomplished.

The Solution The goal here is 5 liters of water. What could have been the conditions just before the goal was accomplished?

In the 7-liter jar	*In the 3-liter jar*
2 ←	← 3
3 ←	← 2
4 ←	← 1
*7 → (−2)	→ 1
6 → (−1)	→ 2

The direction of the arrows indicates which way the water would have to be poured to reach the required 5.

In this problem there is not just one possibility for the previous stage; however, the backward look can still help to decide which route to take. This amounts to a restatement of the problem, since the problem now is to find how each of the situations just short of the goal can be set up so that the final solution is possible.

37. Think about how to achieve the condition marked by *7. Start there and solve the problem.

Try Exercise 37

EXERCISE SET 6.6

The *working backward* technique will not necessarily work for all of the following problems; use the five-step process and any ideas that occur to you.

Solve Problems 1–3 as if you were using the five-step process and were explaining them to elementary pupils. Use diagrams when you can.

1. Ann gave Bob half of her peanuts; Bob then gave Cathy half of the peanuts he got from Ann. Cathy gave Dave half of the peanuts she got from Bob. Poor Dave got only 8 peanuts. How many did Ann have before the peanut-sharing began? How many peanuts did each person have after the sharing was completed?

2. If you subtract 8 from a number and then add 12 you have 21. What was the original number?

3. Two walls measure 10 by 12 feet each. They can both be painted satisfactorily with 3/4 of a gallon of paint. How many gallons of paint are needed to paint the 9-foot-high walls of a room that measures 15 by 20 feet? Ignore the fact that there might be doors and windows in the room.

4. Solve Example 2 of the text using the first line of the table. Is it possible to get a solution from any of the other lines?

5. You wish to measure 4 liters of water but have only a 5-liter container and a 3-liter container. How can this be done if neither container is marked in any way for measuring?

6. Find the smallest number that each of these will divide evenly: 3/8, 5/16, 19/24.

7. The reciprocal of 2 less than a certain number is twice the reciprocal of the number itself. What is the number?

■ 8. Find an exact division in which one subtracts in succession the numbers 1296, 432, and 864. As the problem is set up the letters w, p, r, and s do not represent one-digit numbers. You will have to determine how many digits they each have. Is more than one answer possible?

```
          XYZ      XYZ is a 3-digit number
    w)  p
      1296
         r
        432
          s
         864
```

9. If you add the digits of a number and continue to do this until only one digit is left, then you have the *digital root* of that number. For example, the digital root of 16 is 7, and the digital root of 256 is 4. Is it possible for a square number to have any of the nine digits 1, 2, 3, 4, 5, 6, 7, 8, 9 for a digital root? Make a conjecture and then provide a proof for your conjecture.

REVIEW TEST

1. Use the ratio model to represent 2/3.

2. Use the number line to represent 7/8.

3. Represent 2/9 with a counting model that has 18 objects as the unit.

4. Use the definition of equality for nonnegative rational numbers to show that the following are equal: 2/3, 4/6, and 6/9.

5. Find the least common denominator and subtract: $\frac{15}{21} - \frac{30}{126}$.

6. Prove the following cancellation law. For any nonnegative rational numbers a/b, c/d, and e/f, if $\frac{a}{b} + \frac{c}{d} = \frac{a}{b} + \frac{e}{f}$, then $\frac{c}{d} = \frac{e}{f}$.

7. Illustrate the associative property of addition with 3/7, 5/8, and 1/4.

8. Use the area model to represent $\frac{3}{8}$ of $\frac{4}{5}$.

9. Use the definition of division to find 3/5 divided by 3/8.

10. Is the system of nonnegative rational numbers closed under the operation of addition? subtraction? multiplication? division?

11. Multiply $3\frac{1}{8}$ by $2\frac{1}{4}$.

12. Use the definition of "greater than" to determine which is greater: 13/15 or 14/20.

13. Put in correct order from smallest to largest:

19/5, 23/6, 27/7.

14. Prove the following cancellation law for multiplication and the "less than" relation. For any nonnegative rational numbers a/b, c/d, and e/f $(a/b \neq 0)$, if $\frac{a}{b} \cdot \frac{c}{d} < \frac{a}{b} \cdot \frac{e}{f}$, then $\frac{c}{d} < \frac{e}{f}$.

15. Use the definition of subtraction to show that the difference of two nonnegative rational numbers $\frac{a}{b} - \frac{c}{d}$ is $\frac{a \cdot d - b \cdot c}{b \cdot d}$.

16. Some of the following statements are always true; mark those with T. Provide a counterexample for the others.

a) If two numbers a/b and c/d are equal, then $a = c$ and $b = d$.

b) If $\frac{a}{b}$ is a nonnegative rational number, then $\frac{a \cdot p}{a \cdot p}$, where p is any whole number, is equal to it.

c) A fraction is in lowest terms if and only if both the numerator and the denominator are prime.

d) In finding the sum of two nonnegative rational numbers it does not matter which fraction is used to represent them, since the sums will always be equal.

e) The nonnegative rational numbers provide closure for subtraction, a property the whole-number system lacks.

17. Show that for any two nonnegative rational numbers a/b and c/d, $\frac{a}{b} > \frac{c}{d}$ if and only if $a \cdot d > b \cdot c$.

18. There is a right distributive law for division over addition. Prove that for any rational numbers a, b, and c,

$$(a + b) \div c = (a \div c) + (b \div c).$$

■ **19.** The lowest-terms fraction for $\frac{16}{64}$ can be found (incorrectly) by simply wiping out both numerals for 6: $\frac{1\not{6}}{\not{6}4}$ to get $\frac{1}{4}$. Find another rational number for which this incorrect "wipeout" rule gives the right result.

20. The reciprocal of 2 more than a number is $\frac{1}{2}$ the reciprocal of the number. What is the number?

21. The ratio of freshmen to seniors at the party was 6 to 7. There were 21 seniors there, how many freshmen?

22. In a group of 60 people, 2 out of 6 have blue eyes. How many people have blue eyes?

23. Five oranges will yield 3 small glasses of juice. How many oranges are needed to fill 8 small glasses?

FOR FURTHER READING

1. Hunter Ballew has a common-sense approach to the use of terminology in the elementary classroom in his article "Of fractions, fractional numerals, and fractional numbers" in the *Arithmetic Teacher*, 21:5, May 1974, pp. 442–444.

2. James K. Bidwell shows that the inverse-operation method is significantly better than the common-denominator method or the complex-fraction method. See "Some Consequences of Learning Theory Applied to the Division of Fractions" in *School Science and Mathematics, 71*, May 1971, pp. 426–434.

TEXTS

Swenson, Esther J., *Teaching Mathematics to Children*, Second Edition. The Macmillan Company, New York, New York (1973). (A methods text)

The National Council of Teachers of Mathematics, *Topics in Mathematics for Elementary School Teachers.* Twenty-ninth Yearbook of the NCTM, Washington, D.C. (1964).

chapter 7

nonnegative rational numbers—decimals

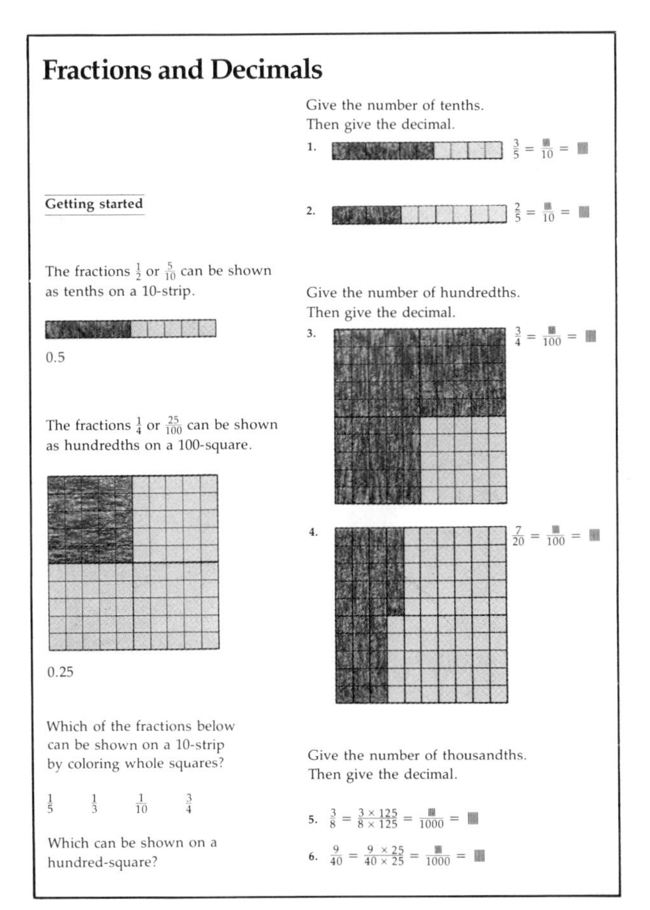

Fractions and Decimals

Getting started

The fractions $\frac{1}{2}$ or $\frac{5}{10}$ can be shown as tenths on a 10-strip.

0.5

The fractions $\frac{1}{4}$ or $\frac{25}{100}$ can be shown as hundredths on a 100-square.

0.25

Which of the fractions below can be shown on a 10-strip by coloring whole squares?

$\frac{1}{5}$ $\frac{1}{3}$ $\frac{1}{10}$ $\frac{3}{4}$

Which can be shown on a hundred-square?

Give the number of tenths.
Then give the decimal.

1. $\frac{3}{5} = \frac{■}{10} = ■$

2. $\frac{2}{5} = \frac{■}{10} = ■$

Give the number of hundredths.
Then give the decimal.

3. $\frac{3}{4} = \frac{■}{100} = ■$

4. $\frac{7}{20} = \frac{■}{100} = ■$

Give the number of thousandths.
Then give the decimal.

5. $\frac{3}{8} = \frac{3 \times 125}{8 \times 125} = \frac{■}{1000} = ■$

6. $\frac{9}{40} = \frac{9 \times 25}{40 \times 25} = \frac{■}{1000} = ■$

Many manipulative materials, such as base-ten blocks and colored rods, can be adapted to provide physical models for fractions. The lesson here uses an area model to develop the decimals for basic fractions.

7.1 AIMS

You should be able to:

a) Find the terminating decimal for any number a/b that has one.
b) Recognize which rational numbers a/b have a terminating decimal.
c) Prove that every rational number a/b represented by a lowest-terms fraction has a terminating decimal if and only if the denominator of that fraction has only 2 and 5 as prime factors.

7.1 DECIMAL NUMERALS FOR BASIC FRACTIONS

A point often missed by pupils in elementary school is that though only one number system—the nonnegative rational numbers—is used for ordinary arithmetic, it is possible to represent these numbers in several different ways. In order to make a problem easier, we are free to choose from a variety of notation as follows:

Notation type	*Examples*		
1. Fraction	1/2	3/4	1/400
2. Decimal	0.5	0.75	0.0025
3. Percent	50%	75%	0.25%
4. Scientific	5×10^{-1}	7.5×10^{-1}	2.5×10^{-3}

TERMINATING DECIMAL NUMERALS. Any natural number can be the denominator of a fraction but some (such as those in 1/3, 1/7, or 1/9) do not seem to fit nicely with base-ten numeration. However, as a beginning some ingenious thinker had the idea of considering just those fractions that have a power of ten (such as 10^1, 10^2, 10^3, and so on) for the denominator. Because this subset consists of all possible fractions which have denominators that are powers of the base of the decimal system, they are known as the *basic fractions*.

Simon Stevin (1584–1620), a Dutch army quartermaster, published one of the earliest arithmetic books suggesting the extension of the decimal place-value table for whole numbers to numbers that are less than 1. This extension makes it possible to find decimal numerals for all rational numbers that have a basic-fraction representation. The double line in the table below marks the end of the original chart and the start of the extension to it. The new entries follow the same pattern as before and are symmetric about the ones place.

The Extended Place-Value Chart for Base Ten

Ten thousands	*Thousands*	*Hundreds*	*Tens*	*Ones*	*Tenths*	*Hundredths*	*Thousandths*		
10^4 or 10 000	10^3 or 1 000	10^2 or 100	10^1 or 10	10^0 or 1	$\frac{1}{10^1}$ or $\frac{1}{10}$	$\frac{1}{10^2}$ or $\frac{1}{100}$	$\frac{1}{10^3}$ or $\frac{1}{1000}$		

With this table, any rational number

$$\frac{c}{10^j} = a_k 10^k + a_{k-1}10^{k-1} + a_{k-2}10^{k-2} + \cdots + a_0 10^0$$
$$+ b_1 10^{-1} + b_2 10^{-2} + \cdots + b_j 10^{-j},$$

where every a and b is one of the digits from 0 to 9, can be expressed as the *terminating decimal*

$$a_k a_{k-1} a_{k-2} \cdots a_0 \,.\, b_1 b_2 b_3 \cdots b_j,$$

which has j numerals to the right of the decimal point between a_0 and b_1.

Try Exercise 1

With the extended place-value table we are able to write the decimal numeral* for any nonnegative rational number represented by a basic fraction.

Example 1 Write the decimal numeral for

a) $\frac{3}{10}$ b) $\frac{32}{10}$ c) $\frac{197}{1000}$ d) $\frac{237}{10000}$

Solution From the definitions of addition and multiplication for rational numbers and the extension of the numeration system we can state:

a) $\frac{3}{10} = \frac{3}{1} \cdot \frac{1}{10} = 3 \cdot \left(\frac{1}{10}\right)$ and write the terminating decimal 0.3 for the nonnegative rational $\frac{3}{10}$. The 0 indicates there is no whole number; the decimal point between "0" and "3" indicates that the following digits refer to place values in the extended chart.

Similarly for the others:

b) $\frac{32}{10} = \frac{30}{10} + \frac{2}{10} = 3 + \frac{2}{10} = 3.2$

c) $$\frac{197}{1000} = \frac{100}{1000} + \frac{90}{1000} + \frac{7}{1000} = \frac{1}{10} + \frac{9}{100} + \frac{7}{1000}$$
$$= \frac{1}{10} + 9 \cdot \frac{1}{100} + 7 \cdot \frac{1}{1000} = 0.197$$

d) $$\frac{237}{10\,000} = \frac{200}{10\,000} + \frac{30}{10\,000} + \frac{7}{10\,000}$$
$$= \frac{2}{100} + \frac{3}{1000} + \frac{7}{10\,000}$$
$$= 0 \cdot \frac{1}{10} + 2 \cdot \frac{1}{100} + 3 \cdot \frac{1}{1000} + 7 \cdot \frac{1}{10\,000}$$
$$= 0.0237$$

1. Complete the place-value chart in the text.

* For convenience, the decimal numeral is often referred to as a *decimal*; the word numeral is understood. In the notation used here, 0.abc means $\frac{a}{10} + \frac{b}{100} + \frac{c}{1000}$. The use of the decimal point distinguishes this from the notation of algebra in which xy is used to mean x multiplied by y.

2 Write the decimal numeral for $\frac{627}{1000}$.

Try Exercise 2

This procedure can be used for any rational number a/b that is equal to $c/10^j$, where c and j are any whole numbers.

Example 2 Find the decimal for 7/20.

Solution We need to find an equivalent fraction with a power of ten as the denominator. Because $\frac{7}{20} = \frac{7 \cdot p}{20 \cdot p}$, we need a p such that $20 \cdot p$ is a power of ten:

$$20 \cdot p = 2 \cdot 2 \cdot 5 \cdot p = 10 \cdot 2 \cdot p$$

and if we replace p by 5 we have powers of ten:

$$\frac{7}{20} = \frac{7 \cdot 5}{10 \cdot 2 \cdot 5} = \frac{35}{100},$$

$$\frac{35}{100} = \frac{30}{100} + \frac{5}{100} = \frac{3}{10} + \frac{5}{100}$$

$$= 0.35 \quad \text{by place-value notation.}$$

3. Find the decimal for 9/40. *Hint*: Find p such that

$$40 \cdot p = 2 \cdot 2 \cdot 10 \cdot p = 10^k.$$

Try Exercise 3

Example 3 Find the terminating decimal numeral for 7/80.

Solution We find the factors of 80:

$$80 = 2 \cdot 2 \cdot 2 \cdot 2 \cdot 5 = 2 \cdot 2 \cdot 2 \cdot 10.$$

Because only 2 and 5 appear, we merely need three more factors of 5 to get all factors of 10:

$$\frac{7}{80} = \frac{7 \cdot 5 \cdot 5 \cdot 5}{80 \cdot 5 \cdot 5 \cdot 5} = \frac{875}{10\,000}$$

$$= \frac{800}{10\,000} + \frac{70}{10\,000} + \frac{5}{10\,000}$$

$$= \frac{8}{100} + \frac{7}{1000} + \frac{5}{10\,000} = 0.0875.$$

4. Find a basic fraction for each of the following:
a) 3/50
b) 7/125
c) 39/160

5. Write the decimal for each of the fractions in Exercise 5.

Try Exercises 4 and 5

These examples suggest that it is easy to determine if a nonnegative rational number has a basic fraction representation and thus has a terminating decimal.

Example 4 Prove the following theorem.

> **If a/b is a nonnegative rational number such that b has no prime factors other than 2 or 5, then a/b can be expressed as a terminating decimal.**

We know that:

$$\frac{a}{b} = \frac{a}{2^m \cdot 5^n} \quad \text{[by the hypothesis]}$$

$$= \frac{a \cdot 2^n \cdot 5^m}{2^m \cdot 5^n \cdot 2^n \cdot 5^m} \quad \text{[by the “equals” relation for nonnegative rational numbers]}$$

$$\left.\begin{aligned} &= \frac{a \cdot 2^n \cdot 5^m}{2^{n+m} \cdot 5^{n+m}} \\ &= \frac{a \cdot 2^n \cdot 5^m}{10^{m+n}} \end{aligned}\right\} \quad \text{[by the associative and commutative properties for multiplication of whole numbers and the meaning of exponents]}$$

Because $a \cdot 2^n \cdot 5^m$ is a whole number (by closure for multiplication) and $m + n$ is a whole number also, we have

$$\frac{a}{b} = \frac{c}{10^k},$$

where c and k are both whole numbers. Thus a/b can be represented by a basic fraction and has a terminating decimal.

Not all fractions are equivalent to some basic fraction and hence there are some rational numbers that cannot be expressed as $c/10^j$ or a terminating decimal. We show that at least one such number exists.

Example 5 Prove that $1/3$ cannot be expressed as $c/10^j$, where c and j are whole numbers.

First we assume that there is such a representation and then arrive at a contradiction.

Assume that $\frac{1}{3} = \frac{c}{10^j}$, where c and j are whole numbers. By the definition of equality for nonnegative rational numbers, $1.10^j = 3 \cdot c$. Now $3 | 3 \cdot c$ and hence $3 | 10^j$ also; but $10^j = 2^j \cdot 5^j$ and hence $3 \nmid 10^j$ (a contradiction). Thus we conclude that there is no basic fraction representation for $1/3$.

Does this mean that every number a/b, such that b has a factor other than 2 or 5, cannot be expressed as a terminating decimal? The following statement answers this question.

If a/b is a nonnegative rational number in lowest terms, such that the denominator b has a prime factor other than 2 or 5, then a/b cannot be expressed as a terminating decimal.

The argument for this general proof is similar to that of Example 5 and is left for the exercises. We can now recognize all those nonnegative rational numbers that have terminating decimals; those that do not will be considered in a subsequent section of the chapter.

6. Use what you know related to factors of 2 and 5 in the denominator to determine which of the following have terminating decimals:

a) $\frac{3}{6}$ b) $\frac{13}{39}$

c) $\frac{31}{70}$ d) $\frac{24}{192}$

Try Exercise 6

TERMINATING DECIMALS TO FRACTIONS. In order to read a decimal numeral, one must decide whether it is tenths, hundredths, or thousandths, and so on. Once this has been done, the matter of conversion to a fraction is obvious.

5.1 is read "five and one-tenth" (the word *and* indicates the location of the decimal point); we can write $5\frac{1}{10}$. 0.15 is read "fifteen-hundredths" and we write $\frac{15}{100}$. 0.236 is read "two hundred thirty-six thousandths" and we write $\frac{236}{1000}$.

7. Find a fraction for each of the following:
a) 37.5
b) 0.23
c) 0.0012
d) 65.315

Try Exercise 7

More formally, this conversion can be done by using expanded notation.

Example 6 Find the fraction for 0.713.

Solution By the place-value notation:

$$0.713 = 7 \cdot \frac{10}{10} + 1 \cdot \frac{1}{100} + 3 \cdot \frac{1}{1000}.$$

By the definition of equality for nonnegative rational numbers and the definition of addition and multiplication we get:

$$\frac{700}{1000} + \frac{10}{1000} + \frac{3}{1000} = \frac{713}{1000}.$$

Such expanded notation is helpful in establishing the algorithms for the operations using decimal notation.

8. Find the fraction for each of the following by using expanded notation and performing the addition:
a) 27.6
b) 0.37
c) 0.012
d) 0.0026

Try Exercise 8

EXERCISE SET 7.1

1. Write the decimal numeral for the following basic fractions:

a) $\frac{129}{10}$ b) $\frac{6}{10^2}$ c) $\frac{26}{100}$

d) $\frac{278}{10^3}$ e) $\frac{379}{10^4}$ f) $\frac{121}{10^5}$

2. Find decimal numerals for the following:
a) 9/20 b) 5/16 c) 13/200 d) 33/60
e) 11/125 f) 3/32 g) 11/80 h) 7/35

3. Prove that the rational number 5/6 cannot be expressed as a terminating decimal.

4. Repeat Problem 3 for 5/11.

■ **5.** Prove the theorem stated in the text:
If a/b is a nonnegative rational number in lowest terms, such that the denominator b has a prime factor other than 2 or 5, then a/b cannot be expressed as a terminating decimal.

6. Find all the nonnegative rational numbers $1/b$ such that b is less than or equal to 100 and $1/b$ can be expressed as a terminating decimal.

7. Show that the proof in Exercise 5 in combination with the theorem of Example 4 prove the following statement: Any nonnegative rational number expressed in the lowest-term fraction can be represented by a terminating decimal if and only if the denominator b has only the prime factors 2 and 5.

8. Is the set of rational numbers that have terminating decimal representation closed under addition? multiplication? subtraction? division? State a reason or give a counterexample for each.

9. Is there a difference in the numerals for the following:
a) five hundred seventeen-thousandths
b) five hundred and seventeen-thousandths

10. Write the basic fraction for each of the following terminating decimals:
a) 0.368 b) 0.15 c) 4.127
d) 13.023 e) 90.165 d) 37.302

11. Make a flowchart for finding the decimal numeral of any basic fraction.

12. Make a flowchart for finding the decimal numeral of any fraction equivalent to a basic fraction.

■ **13.** The basic fractions for base four have denominators that are powers of 4. Extend the place-value table for base four. Use the extension to convert each of the following to a numeral based upon that extension:

a) $\left(\frac{3}{4}\right)_{\text{ten}}$ b) $\left(\frac{2}{16}\right)_{\text{ten}}$ c) $\left(\frac{3}{64}\right)_{\text{ten}}$

■ **14.** Repeat Problem 13 for base five. This time convert the following:

a) $\left(\frac{4}{5}\right)_{\text{ten}}$ b) $\left(\frac{3}{25}\right)_{\text{ten}}$ c) $\left(\frac{2}{125}\right)_{\text{ten}}$

15. Write each of the following as a fraction (base ten) in lowest terms.
a) 0.023_{four} b) 0.333_{four}
c) 0.024_{five} d) 0.333_{five}

7.2 ALGORITHMS FOR THE OPERATIONS

Addition

The usual algorithms for whole numbers can be adapted for addition with terminating decimals. The value of each place of the extended chart is ten times that of the place to its right, so it is possible to use the same reasoning about regrouping as was done with the whole numbers.

In Chapter 6, the nonnegative rational numbers have been shown to have the same properties as whole numbers (associative and commutative laws for both addition and multiplication and the distributive law for multiplication over

7.2 AIMS

You should be able to:
a) Use and justify the addition and subtraction algorithms for decimals.
b) Use and justify the multiplication and division algorithms for decimals.

either addition or subtraction). The very same arguments used for the whole numbers to justify the rearranging of terms and the order of performing certain operations along with the repeated use of the distributive law could thus be carried out for the nonnegative rational numbers. In fact, the identical proofs used in Chapter 4 would suffice if the words "nonnegative rational number" were substituted for "whole number" each time it was appropriate. We shall therefore not repeat these arguments in detail.

The remaining details for the establishment of the algorithms depend upon the conventions of the numeration system. The properties of the nonnegative rational numbers are not altered with a change in notation.

Example 1 Add 7.23 and 22.92.

By applying the notation system we have

$$\left(7 + \frac{2}{10} + \frac{3}{100}\right) + \left(20 + 2 + \frac{9}{10} + \frac{2}{100}\right).$$

By the repeated use of the associative and commutative laws for addition of nonnegative rational numbers we have:

$$(7 + 22) + \left(\frac{2}{10} + \frac{9}{10}\right) + \left(\frac{3}{100} + \frac{2}{100}\right).$$

By the definition of addition:

$$(29) + \left(\frac{11}{10}\right) + \left(\frac{5}{100}\right).$$

Because $\frac{11}{10} = \frac{10}{10} + \frac{1}{10} = 1 + \frac{1}{10}$, we regroup to conform to the notation scheme:

$$(29 + 1) + \left(\frac{1}{10}\right) + \left(\frac{5}{100}\right) = (30) + \left(\frac{1}{10}\right) + \left(\frac{5}{100}\right).$$

By the numeration system the sum is then 30.15. This procedure is reflected in the steps of the intermediate algorithm.

The intermediate algorithm		*Explanation*
7.23		
22.92		
.05	1. Add the hundredths	2/100 + 3/100 = 5/100
1.10	2. Add the tenths	
9.00	3. Add the ones	$2/10 + 9/10 = 11/10 = 1\frac{1}{10}$
20.00	4. Add the tens	
30.15	5. Repeat steps (1) to (4)	

Steps (1) through (4) correspond to the rearranging that is possible due to the associative and commutative properties. The explanations correspond to the operations done; fractions are used to represent the numbers.

The final algorithm is of course:

$$\begin{array}{r} 7.23 \\ +22.92 \\ \hline 30.15 \end{array}$$

Try Exercise 9

9. Use the intermediate algorithm to add 18.91 and 314.78.

Except for the fact that there is a decimal point, such a calculation is exactly like the addition of whole numbers. Just as with the whole-number addition algorithms, the corresponding place values must be aligned. Since the decimal point clearly marks the boundary between the ones and the tenths, it is sufficient to see that these points are aligned; then all other places will correspond in each column.

Example 2 Add 4.105 and 17.2. Justify the procedure.

Solution We do not use expanded notation to justify the algorithm, since we need only convert each decimal to a basic fraction:

$$4.105 = \frac{4105}{1000}, \qquad 17.2 = \frac{172}{10}.$$

Hence

$$\begin{aligned} 4.105 + 17.2 &= \frac{4105}{1000} + \frac{172}{10} \\ &= \frac{4105}{1000} + \frac{17200}{1000} && \text{by the definition of equality for nonnegative rational numbers} \\ &= \frac{21305}{1000} && \text{by the definition of addition} \\ &= 21.305 && \text{by the numeration system} \end{aligned}$$

This addition problem can be treated as one with whole numbers, and it is not necessary to think about adding thousandths, hundredths, and tenths. However, attention must be paid to the location of the decimal points.

Write: $\begin{array}{r} 4.105 \\ +17.200 \\ \hline \end{array}$ Think: $\begin{array}{r} 4105 \\ +17200 \\ \hline 21305 \end{array}$ Write: $\begin{array}{r} 4.105 \\ +17.200 \\ \hline 21.305 \end{array}$

(add as whole numbers)

Try Exercise 10

10. Convert to fractions to justify the addition:

$$\begin{array}{r} 8.41 \\ 17.37 \\ 12.94 \\ \hline \end{array}$$

11. Subtract as in the example in the text (use expanded notation):

$$282.14 - 13.87.$$

Subtraction

Any subtraction algorithm that can be used for whole numbers can be extended for the case of terminating decimals. Parallel arguments for regrouping (borrowing) can be made, since $1 = 10/10$, $1/10 = 10/100$, and so on.

Example 3 Subtract and justify the procedure: $46.24 - 13.79$.

Solution By the place-value notation this means:

$$\left(40 + 6 + \frac{2}{10} + \frac{4}{100}\right) - \left(10 + 3 + \frac{7}{10} + \frac{9}{100}\right).$$

The arguments used for subtraction of whole numbers still hold, so we can rearrange and write:

$$(40 - 10) + (6 - 3) + \left(\frac{2}{10} - \frac{7}{10}\right) + \left(\frac{4}{100} - \frac{9}{100}\right).$$

Just as with whole numbers, regrouping is required to subtract the hundredths:

$$(40 - 10) + (6 - 3) + \left(\frac{1}{10} - \frac{7}{10}\right) + \left(\frac{10}{100} + \frac{4}{100} - \frac{9}{100}\right).$$

We need to regroup again to subtract the tenths:

$$(40 - 10) + (5 - 3) + \left(\frac{10}{10} + \frac{1}{10} - \frac{7}{10}\right) + \left(\frac{10}{100} + \frac{4}{100} - \frac{9}{100}\right).$$

It is now possible to perform all the indicated operations:

$$(30) + (2) + \left(\frac{4}{10}\right) + \left(\frac{5}{100}\right).$$

By the notation system we have: 32.45.

The usual algorithm is:

$$\begin{array}{r} 46.24 \\ -13.79 \\ \hline \end{array} \qquad \begin{array}{r} {\scriptstyle 1\,1} \\ 46.\not{2}4 \\ -13.79 \\ \hline \end{array} \qquad \begin{array}{r} {\scriptstyle 5\;1\,1\,1} \\ 4\not{6}.\not{2}4 \\ -13.79 \\ \hline \end{array} \qquad \begin{array}{r} 46.24 \\ -13.79 \\ \hline 32.45 \end{array}$$

(regrouping 1/10 to 10/100) (regrouping 1 to 10/10)

Try Exercise 11

A flowchart from a sixth-grade textbook makes the procedure simple and clear.

Subtracting with decimals

Teri jumped 4.97 meters. The world's record is 8.90 m. How many more meters is the world's record?

Finding the answer

Copy. Line up the decimal points → Subtract the hundredths → Subtract the tenths → Subtract the whole numbers

Copy. Line up the decimal points	Subtract the hundredths	Subtract the tenths	Subtract the whole numbers
8.90	8.9̸0̸ (8 10)	8̸.9̸0̸ (7 18 10)	8̸.9̸0̸ (7 18 10)
− 4.97	− 4.97	− 4.97	− 4.97
	3	.93	3.93

The world's record is 3.93 m more.

Multiplication

The usual whole-number algorithms can likewise be extended to the use of terminating decimal numerals for nonnegative rational number multiplication.

Example 4 Find the product of 51.2 and 3.6. Justify the procedure.

By the notation system $(51.2) \cdot (3.6)$ means

$$\left(50 + 1 + \frac{2}{10}\right) \cdot \left(3 + \frac{6}{10}\right).$$

By repeated use of the distributive law, just as for whole numbers, we have:

$$\left(50 + 1 + \frac{2}{10}\right) \cdot (3) + \left(50 + 1 + \frac{2}{10}\right) \cdot \left(\frac{6}{10}\right)$$

and with the commutative law for multiplication we get:

$$3 \cdot 50 + 3 \cdot 1 + 3 \cdot \frac{2}{10} + \frac{6}{10} \cdot 50 + \frac{6}{10} \cdot 1 + \frac{6}{10} \cdot \frac{2}{10}.$$

This procedure is reflected in the intermediate algorithm.

The intermediate algorithm		*Explanation*
51.2		
3.6	Multiply by 0.6 then by 3:	
.12	1. $(6/10) \cdot (2/10) = 12/100$	1. See procedure above for this term, also the notation system, and the definition of multiplication.
.60	2. $(6/10) \cdot 1 = 6/10$	2–4. Same as (1).
30.00	3. $(6/10) \cdot 50 = 300/10$	
.60	4. $3 \cdot (2/10) = 6/10$	
3.00	5. $3 \cdot 1 = 3$	5, 6. By the whole-number algorithm.
150.00	6. $3 \cdot 50 = 150$	
184.32	7. Use the addition algorithm for terminating decimals.	

12. Convert to fractions, multiply, and then convert back to a decimal:

$273.4 \times 2.13.$

The final algorithm is:

Write:	Think:	Think:	Write:
51.2	(1-place decimal)	512	51.2
3.6	(1-place decimal)	36	3.6
		3072	3072
		15360	15360
		18432	184.32
		(Treat as whole numbers)	(2-place decimal)

The algorithm can be justified by showing that if the numerals are converted directly to basic fractions, the problem is reduced to a problem in whole numbers.

Example 5 Find the product of 30.2 and 1.76.

Solution By using the notation system, we can convert immediately to fractions:

$$(30.2) \cdot (1.76) = \frac{302}{10} \cdot \frac{176}{100}.$$

$$= \frac{53152}{1000} \qquad \text{By the definition of multiplication and the whole-number algorithm}$$

$$= 53.152 \qquad \text{by the numeration system}$$

Try Exercise 12

LOCATION OF DECIMAL POINT IN MULTIPLICATION. There is no need to think through each time where the decimal point should be located in the product.

If one factor has *m* places after the decimal point and the other factor has *n* places after the decimal point (when both factors are expressed as terminating decimals), then their product will have $(m + n)$ places to the right of the decimal point.

By the notation system, the factor with m decimal places can be expressed as $k/10^m$, where k and m are both whole numbers. Similarly, the one with n places can be expressed as $p/10^n$ with p and n whole numbers. Their product is

$$\frac{k}{10^m} \cdot \frac{p}{10^n} = \frac{k \cdot p}{10^{n+m}}$$

and hence it can be expressed as a terminating decimal with $(m + n)$ places after the decimal point.

Example 6 Find the product of 1.275 and 2.25.

Solution We treat the problem as a whole-number problem and locate the decimal point by the rule above:

$$\begin{array}{rl} 1.275 & \text{(3-place decimal)} \\ 2.25 & \text{(2-place decimal)} \\ \hline 6375 & \\ 25500 & \\ 255000 & \\ \hline 286875 & \text{By the rule, the product is a 5-place decimal 2.86875.} \end{array}$$

Try Exercise 13

Estimation can be used as an alternative to the rule for locating the decimal point in a product. In any case, the estimate will serve as a check. For the product 30.2×1.76 we round 30.2 to 30 and 1.76 to 2. The product of 60 indicates that 53.152 must be the correct answer.

Try Exercise 14

Division

TWO WHOLE NUMBERS WITH A DECIMAL QUOTIENT. The division algorithm for whole numbers can be used to find other names for quotients. For example, $\frac{104}{8}$ can be found by:

$$\begin{array}{r} 13 \\ 8\overline{)104} \\ 8 \\ \hline 24 \\ 24 \\ \hline \end{array}$$

Another name for $\frac{104}{8}$ is 13.

Suppose that the same thing is done for $\frac{108}{8}$:

$$\begin{array}{r} 13 \\ 8\overline{)108} \\ 8 \\ \hline 28 \\ 24 \\ \hline 4 \end{array}$$

This can be written as

$13\frac{4}{8}$ or 13.50.

Imagine that we try the same thing for $\frac{1}{8}$; now we get:

$$\begin{array}{r} 0 \\ 8\overline{)1} \end{array}$$ or $0\frac{1}{8}$ (which does not help).

13. Use the rule to locate the decimal point in each of the following products:

a)
$$\begin{array}{r} 1.35 \\ \times\ 2.5 \\ \hline 3375 \end{array}$$

b)
$$\begin{array}{r} 17.53 \\ \times\ 3.65 \\ \hline 639845 \end{array}$$

c)
$$\begin{array}{r} 18.731 \\ \times\ 1.265 \\ \hline 23694715 \end{array}$$

14. Use the method of estimating to locate the decimal point in each of the following products:

a)
$$\begin{array}{r} 1.47 \\ \times\ 3.5 \\ \hline 5145 \end{array}$$

b)
$$\begin{array}{r} 37.5 \\ \times\ 1.75 \\ \hline 65625 \end{array}$$

c)
$$\begin{array}{r} 15.75 \\ \times\ 13.25 \\ \hline 2086875 \end{array}$$

Observe, however, that $\frac{1000}{8}$ is $1000 \times \frac{1}{8}$, so that if we could find another name for $\frac{1000}{8}$, it would be 1000 times the quotient $\frac{1}{8}$. We do this:

$$\begin{array}{r} 125 \\ 8\overline{)1000} \\ \underline{8} \\ 20 \\ \underline{16} \\ 40 \\ \underline{40} \end{array}$$

Since $\frac{1000}{8} = 125$, we multiply both terms by $\frac{1}{1000}$:

$$\frac{1}{8} = \frac{125}{1000},$$

which is a basic fraction we know how to write. Thus,

$$\frac{1}{8} = 0.125.$$

The corresponding algorithm for this statement is:

$$\begin{array}{r} 0.125 \\ 8\overline{)1.000} \end{array}$$

The algorithm procedure is applied to $8\overline{)1000}$, but the placement of the decimal point makes it apply to $8\overline{)1}$.

Example 7 Find the decimal numeral for 13/40, using division.

Solution

$$\begin{array}{r} 0\tfrac{13}{40} \\ 40\overline{)13} \end{array}$$

Try $\frac{1300}{40}$:

$$\begin{array}{r} 32 \\ 40\overline{)1300} \\ \underline{120} \\ 100 \\ \underline{80} \end{array} \quad \text{(still a remainder)}$$

Try $\frac{13000}{40}$:

$$\begin{array}{r} 325 \\ 40\overline{)13000} \\ \underline{120} \\ 100 \\ \underline{80} \\ 200 \\ \underline{200} \end{array} \qquad \frac{13000}{40} = 325.$$

Multiply by 1/1000. Therefore, $\frac{13}{40} = 0.325$.

The algorithm is applied to $40\overline{)13000}$, but the placement of the decimal point makes it apply to $\begin{array}{r} 0.325 \\ 40\overline{)13.000} \end{array}$.

It is not necessary to rewrite the algorithm each time we have to multiply by a larger power of ten; zeros can be annexed as needed.

Try Exercise 15

15. Find the decimal numeral for 11/80 by division.

DIVIDEND OR DIVISOR ARE NOT WHOLE NUMBERS. So far we have considered an extension of the division algorithm for two whole numbers in which a terminating decimal is obtained for the answer. If the dividend or divisor is not a whole number, the procedure above can still be used, as the following example shows.

Example 8 Find the quotient: $9.98\overline{)134.73}$.

By the notation system we can write the quotient $134.73 \div 9.98$ as:

$$\frac{13473}{100} \div \frac{998}{100}.$$

By the rule for division of nonnegative rational numbers and the identity property we have

$$\frac{13473}{100} \div \frac{998}{100} = \frac{13473}{100} \cdot \frac{100}{998} = \frac{13473}{998}.$$

Thus if 13473 is divided by 998, we have the quotient wanted originally.

We have, in effect, multiplied both the divisor and the dividend by 100 and then gone on to perform the division.

We can think of $134.73 \div 9.98$ as the compound fraction

$$\frac{134.73}{9.98} = \frac{(134.73) \cdot (100)}{(9.98) \cdot (100)} \quad \text{(by Section 6.4).}$$

16. Prove (by rewriting the quotient in terms of fractions) that $4.12\overline{)43.054}$ is 10.45.

17. Use estimation to locate the decimal point of the quotient Q for each of the following:
a) $8.36\overline{)4.2845}$; $Q = 512500$
b) $0.024\overline{).00876}$; $Q = 36500$
c) $152.51\overline{)1849.9463}$; $Q = 121300$

Therefore the quotient is really $\frac{13473}{998}$, as in Example 8.

The whole-number problem $998\overline{)13473}$ has already been taken care of:

$$\begin{array}{r} 13.5 \\ 998\overline{)13473.0} \\ 998 \\ \hline 3493 \\ 2994 \\ \hline 4990 \\ 4990 \\ \hline \end{array}$$

To show this in the original problem, a caret (‸) is used to indicate the new locations of the decimal points:

$$9.98_\wedge\overline{)134.73_\wedge}$$

The carets show that the equivalent problem is: $998\overline{)13473}$.

Try Exercise 16

DECIMAL POINT LOCATION IN DIVISION. The preceding discussion for the location of the decimal point in the quotient is the basis of the following rule:

Multiply both the divisor and the dividend by the power of ten to make the divisor a whole number. Divide as you would for whole numbers. Locate the decimal point in the quotient to correspond to the new position of the decimal point in the dividend.

The decimal point can also be located in division by finding an estimate of the quotient. In the example above, 134.73 can be approximated by 130, the divisor by 10, so the quotient should be about 13, which means that 13.5 is the correct answer.

Try Exercise 17

EXERCISE SET 7.2

1. Use expanded notation and fractions to prove that the indicated sums are correct:

a) $\begin{array}{r} 16.12 \\ 7.53 \\ \hline 23.65 \end{array}$ b) $\begin{array}{r} 3.141 \\ 2.06 \\ \hline 5.201 \end{array}$

2. Use the intermediate algorithm to add:

a) $\begin{array}{r} 37.45 \\ 6.37 \\ \hline \end{array}$ b) $\begin{array}{r} 48.76 \\ 53.45 \\ \hline \end{array}$

3. Write a flowchart for the usual algorithm for the addition of numbers represented by decimals.

4. Use the flowchart of Problem 3 to find the following sums:

a) $\begin{array}{r} 9.71 \\ 4.37 \\ \hline \end{array}$ b) $\begin{array}{r} 13.61 \\ 213.423 \\ 7.56 \\ \hline \end{array}$

5. Find the difference for each of the following by converting to fractions, subtracting, and converting back to decimals:

a) $\begin{array}{r} 7.32 \\ -1.98 \\ \hline \end{array}$ b) $\begin{array}{r} 8.11 \\ -2.34 \\ \hline \end{array}$ c) $\begin{array}{r} 19.21 \\ -1.23 \\ \hline \end{array}$ d) $\begin{array}{r} 21.31 \\ -1.54 \\ \hline \end{array}$

6. Explain how the conversion to fractions justifies the ordinary subtraction algorithm for decimal numerals.

7. Write a flowchart that could be used for the difference

$$121.736 - 15.149.$$

8. Use the flowchart found in Problem 7 to find the difference

$$167.325 - 16.789.$$

9. The digits of the following products are correct, but the decimal point is not located. Locate it correctly, using the usual rule:

a) $31.82 \times 7.15 = 22751300$
b) $0.612 \times 81.7 = 50000400$

10. The digits of the following products are correct, but the decimal point is not located. Locate it correctly by estimation:

a) $1800 \times 2.02 = 363600$
b) $17000 \times 0.0301 = 511700$

11. Find the terminating decimal for each of the following by using division.

a) 9/20 b) 5/16 c) 13/200 d) 11/125

12. Find the quotient for each of the following by converting to fractions and finding an equivalent whole-number division:

a) $2.53\overline{)12.9789}$ b) $4.15\overline{)17.43}$
c) $47.1\overline{)282.6}$ d) $0.014\overline{)0.4564}$

13. In each of the following, the correct digits are given for the quotient Q; but the decimal point is not located. Use estimation to locate the decimal point in each case:

a) $3.75\overline{)1923.75}$; $Q = 51300$
b) $2.56\overline{)772.352}$; $Q = 301700$
c) $82.036\overline{)345.37156}$; $Q = 42100$
d) $16.5\overline{)53.0475}$; $Q = 321500$

Problems 14, 15 are similar to problems given to elementary pupils.

14. Three friends who play tennis decided it might be a good idea if they split the cost of two tennis rackets three ways. Each of the rackets they decided to buy cost \$27.99. How much did each of them need to contribute to the purchase?

15. Bill and Sue were in charge of the Wiener Roast being held by their hiking club. They bought 5 pounds of wieners at \$1.55 per pound, $3\frac{1}{2}$ dozen rolls at \$0.88 per dozen, and 2 gallons of milk at \$1.65 per gallon. If there are a total of 20 people in the club, how much does each person owe Bill and Sue?

7.3 AIMS

You should be able to:

a) Use division to find the decimal numeral for any fraction.
b) Convert a repeating decimal to a fraction.
c) Show that every member of the set of nonnegative rational numbers can be represented by either a terminating decimal or a nonterminating repeating decimal.

7.3 NONTERMINATING REPEATING DECIMALS

Not all nonnegative rational numbers have lowest-terms fractions with denominators that consist of only the prime factors 2 and 5. For example, 1/6 and 1/9 have denominators with a factor 3, and we already know that such rational numbers can not be represented by a terminating decimal. The division algorithm can be used to find the decimal representation for such numbers also.

Example 1 Find the decimal numeral for $\frac{13}{7}$.

```
        1 .8 5 7 1 4 2
    7)1 3 .0 0 0 0 0 0 0 0
        7
      →6 0
       5 6
         4 0
         3 5
           5 0
           4 9
             1 0
               7
             3 0
             2 8
               2 0
               1 4
               →6
```

These remainders are the same. When the remainder is obtained for the second time, the repeating has begun: the same succession of digits will occur over and over.

To indicate that the digits will continue to repeat, the numeral is written with a bar* over the repeating part:

$$\frac{13}{7} = 1.\overline{857142}.$$

EVERY RATIONAL NUMBER HAS A DECIMAL NUMERAL. Think about the division above. Because the divisor is 7, at any stage in the division the remainder must be one of the set $\{0, 1, 2, 3, 4, 5, 6\}$. If 0 is obtained, the process ends and the numeral is a *terminating decimal.* If the remainder is never 0, then there are only six possible different remainders. In the example above the remainders in order were: 6, 4, 5, 1, 3, and 2, and then the repetition started. The same remainder will be obtained for the second time (in this case, by the seventh step or before) at which point the digits must repeat. Furthermore, the length of the repeating cycle will be no more than six digits.

* Sometimes the repetition is indicated by ellipses (···) after the last given digits. The repeating cycle of digits can be determined by visual inspection.

Example 2 The decimal for $\frac{58}{22}$ repeats. What is the maximum possible length of the repeating cycle?

Solution First, the simplest fraction must be found: $\frac{58}{22} = \frac{29}{11}$. To find the decimal we would divide by 11. There are just ten possible nonzero remainders, and the length of the repeating cycle is *at most ten* but of course can be shorter.

Try Exercises 18 and 19

Thus every nonnegative rational number does have a decimal numeral, and it can always be found by performing the division. If the denominator of the lowest-terms fraction is b, then the possible remainders for the division are the members of the set $\{0, 1, 2, 3, 4, \ldots, b-1\}$. If some remainder is 0, then the decimal terminates. If the remainder is never 0, then a given remainder will occur for the second time by the bth step or before. The numeral will then repeat and will not terminate. This argument establishes the result:

If the decimal numeral for a nonnegative rational number does not terminate, it must repeat. If a/b is the lowest-terms fraction for such a nonterminating decimal, then the repeating cycle will have a maximum length of $b - 1$.

CONVERSION OF NONTERMINATING REPEATING DECIMALS TO FRACTIONS. Every nonterminating repeating decimal numeral names a rational number because each one can be converted back to a fraction from which it was obtained.

Example 3 Find the fraction for $0.412\overline{412}$.

Solution Call the nonnegative rational number n. Then $n = 0.412\overline{412}$. The length of the repeating cycle is 3, so we multiply by 10^3 and obtain:

$$\begin{aligned} 1000n &= 412 \cdot 412\overline{412} \\ n &= 0 \cdot 412\overline{412} && \text{(Rewriting from above)} \\ \hline 999n &= 412 && \text{(Subtracting } n \text{ from } 1000n\text{)} \\ n &= \frac{412}{999} && \text{(Dividing by 999)} \end{aligned}$$

Try Exercise 20

Example 4 Find the fraction for $7.2334\overline{34}$.

Solution First multiply by 100 to make the repeating cycle begin at the decimal point. Let

$$n = 7.2334\overline{34}$$

$$100n = 723.34\overline{34}. \qquad \text{(This is now like Example 3.)}$$

18. Find the decimal for 29/11.

19. What is the maximum possible length of the repeating cycle of the decimal numeral for 720/39?

20. Find the fraction for:

a) $0.2341\overline{2341}$

b) $6.27\overline{27}$

21. Find the fraction for:

$5.41\overline{616}$

Multiply by 100 again (the length of the cycle is two):

$$10\,000n = 72334.34\overline{34}.$$

Subtract $100n$ from $10\,000n$:

$$\begin{aligned} 10\,000n &= 72334.34\overline{34} \\ 100n &= 723.34\overline{34} \\ \hline 9900n &= 71611 \\ n &= \frac{71611}{9900} \qquad \text{(Dividing by 9900)} \end{aligned}$$

Try Exercise 21

This can be done for a repeating decimal of any arbitrary length.

Example 5 Find the fraction for the decimal $0.\text{abcd}\overline{\text{abcd}}$,* where a, b, c, and d are one of the digits from 0 to 9.

22.

a) Find the fraction for $1.99\overline{9}$.

b) Find the fraction for $3.99\overline{9}$.

Let $n = 0.\text{abcd}\overline{\text{abcd}}$.

The cycle length is 4 digits We multiply by 10^4:	It could be k digits. We would multiply by 10^k:
$10^4 n = \text{abcd.abcd}\overline{\text{abcd}}$.	$10^k n = \text{abcd} \ldots \text{k}.\overline{\text{abcd} \ldots \text{k}}$

We subtract n from both members:

$$\begin{aligned} 10000n &= \text{abcd.abcd}\overline{\text{abcd}} \\ -n &= 0.\text{abcd}\overline{\text{abcd}} \\ \hline 999n &= \text{abcd} \end{aligned}$$

Thus $n = \dfrac{\text{abcd}}{9999}$, where abcd is a 4-digit whole number.

In the general case we would have:

$$n = \frac{\text{abcd} \ldots \text{k}}{10^k - 1},$$

where abcd . . . k is a k-digit whole number.

In this way every nonterminating repeating decimal can be converted to a fraction. Every decimal numeral that either terminates or repeats endlessly represents a rational number. If the endlessly repeating 9 (as in 0.9999 . . . or in 0.39999 . . .) is ruled out, then every nonnegative rational number has just one decimal numeral.

Try Exercise 22

* Remember what this notation means from the opening pages of this chapter.

EXERCISE SET 7.3

In Problems 1 and 2 convert to fractions and simplify each of the following:

1. a) $0.5555\overline{5}$ b) $0.7777\overline{7}$ c) $3.12\overline{12}$

2. a) $6.516\overline{16}$ b) $13.7651\overline{651}$ c) $27.2414\overline{414}$

3. In what sense could the terminating decimal 0.513 be regarded as nonterminating?

4. Make a flowchart for converting the repeating decimal $0.abc\overline{abc}$ to a fraction.

5. Make a flowchart for converting the repeating decimal $0.abcbc\overline{bc}$ to a fraction.

6. Show that the number 0.67 is not equal to $0.66\overline{6}$.

7. Show that the number 0.33 is not equal to $0.33\overline{3}$.

8. Study the patterns found in the following sums:

a) $0.3131\overline{31} + 0.2424\overline{24}$

b) $4.5267\overline{67} + 9.5214\overline{14}$

c) $6.123\overline{123} + 4.365\overline{365}$

9. Is the set of nonnegative rational numbers represented by nonterminating repeating decimals closed under addition? subtraction? multiplication? division? Give a proof or a counterexample for each one.

10. Find the decimal for each of the nonnegative rationals $n/7$, where $n = 1, 2, 3, 4, 5, 6$. Study the patterns of each of these for any unusual features.

11. Find the decimal for each nonnegative rational number $n/9$, where $n = 1, 2, 3, 4, 5, 6, 7, 8$. Study the patterns for any unusual features.

12. Find the decimal for each of the numbers $n/11$, where $n = 1, 2, 3, 4, 5, 6, 7, 8, 9, 10$.

13. Find the decimal for each of the numbers $n/13$, where $n = 1, 2, 3, 4, 5, 6, 7, 8, 9, 10, 11, 12$.

14.

a) Show that $0.667 \neq 2/3$, but the difference between them is less than 0.001.

b) Show that $0.6667 \neq 2/3$, but the difference between them is less than 0.0001.

15.

a) Show that $0.3333 \neq 1/3$, but the difference between them is less than 0.0001.

b) Show that $0.33333 \neq 1/3$, but the difference between them is less than 0.00001.

16. It is possible to indicate how decimals could extend indefinitely and yet still not repeat. For example:

0.1010010001000010000010000001 . . .
3.5566555666655556666655555566666 . . .

Construct five more such decimals. If such symbols represent numbers, what can be said about these numbers?

■ **17.** Prove that every nonterminating repeating decimal numeral represents a number that can be represented by a fraction.

7.4 AIMS

You should be able to:

a) Use a calculator to find the decimal numeral for those fractions that have a repeating cycle exceeding the length of the display.
b) Adapt other calculations to a fixed calculator display, even though the results exceed the capacity of the screen.

7.4 THE CALCULATOR AND DECIMALS

Any pupil who has access to a calculator will learn a great deal about conversion from fractions to decimals simply by observing the calculator display. Any child who sees $1 \div 3$ appear on the display as 0.33333333 is going to raise some questions, perhaps not about a nonterminating decimal, but just about the fact that all those 3's show up in contrast to $1 \div 4$, which just gives 0.25 on the display. For that matter, youngsters who are at the stage where all they have heard of is $\frac{1}{4}$ of a pizza and have not worked with decimals at all may find 0.25 just about as mysterious as the repeating patterns 0.12121212 or 0.12312312 they might have chanced to see.

All calculators use finite decimals and simply chop off, or truncate, the answer to $2 \div 3$ at some point or else round it off, so that the screen displays 0.6666666667. We know that 0.6666666667 is not exactly 2/3, but it is closer to it than 0.0000000001, which is more than good enough for most purposes.

Try Exercise 23

23.

a) Show that the difference between

0.6666666667 and 2/3

is less than 0.0000000001.

b) Experiment with your calculator. What is 0.667×3? 0.6667×3? 0.66667×3? and so on. What is $(2 \div 3) \times 6$? $\frac{(2 \times 6)}{3}$?

If there is a difference, can you account for it?

In a calculator more digits are carried internally than appear on the display, so that in the course of a long sequence of operations large errors due to truncation will not build up. You may find on your calculator that the answer will be different, depending upon the order in which the operations are performed. For example:

$$\left(\frac{1506}{7} - 215\right) \cdot 7 = 1506 - (215 \cdot 7).$$

You may find that on your calculator:

$$\left(\frac{1506}{7} - 215\right) \cdot 7 = 0.9999999994,$$

while

$$1506 - (215 \cdot 7) = 1.0000000000.$$

By the time pupils reach the seventh grade or even earlier, they will have studied repeating decimals formally. With the additional power provided by a calculator, all sorts of interesting questions can be raised that are no longer too time-consuming to investigate. These investigations can add interest and excitement to the study of mathematics and also promote deeper understanding of some of the basic concepts the pupils have learned.

In a classroom

Teacher: Do you have some problem, Emmy?

Emmy:* I did all those exercises where you find the decimals for fractions like 3/7 and 2/11.

* We like to think that a more famous Emmy was encouraged in her early mathematics classroom in this way. Emmy Noether (1882–1935) was born in Erlangen, Germany, but after receiving her doctorate moved to Göttingen and became a vital part of the group of mathematicians there. Because she was a woman, Noether never became a part of the regular faculty although she did give lectures and work with students just as if she were but received no pay. Noether was a gifted mathematician, and her work in algebra has turned out to be of fundamental importance in the research being done in the area of algebra at the present time.

Teacher: You did them *all*? Wasn't that quite a few?

Emmy: Well, I did some at school and the rest at home on the calculator. I checked the ones I did in school, too.

Teacher: Were your answers right?

Emmy: Yes, but I think there is something wrong with the calculator. When the denominator gets big, something goes wrong.

Teacher: Can you explain to us what happened?

Emmy: I thought of all the fractions with a numerator of 1, like this (writing on the board). I took the denominators up to 20. Then I divided to find the decimal numeral:

$$1/2 = 0.5,$$
$$1/3 = 0.33333333,$$
$$1/4 = 0.25,$$
$$1/5 = 0.2.$$

24. Find all the results that Emmy found. Look at $1/n$ for $n = 1, 2, 3, 4, \ldots, 19, 20$.

Try Exercise 24

Emmy: (continuing) When I got to 1/17, something didn't work.

Teacher: How do you know? Did you check your answer in some way?

Emmy: No, but my answer didn't repeat, and I know it should.

Teacher: How do you know that?

Emmy: Well, the denominator has a prime factor besides 2 and 5. The number 17 is prime.

Teacher: Class, do you agree that Emmy's reasoning is correct? Should her answer repeat?

Sophia: First we have to make sure that 1/17 is the simplest fraction. We know there can't be any common factors other than 1.

Carl: There aren't any, so Emmy is right.

Teacher: That seems correct. Emmy, what did you get on your calculator?

Emmy: I got 0.0588235294, and that doesn't repeat.

Maria: I think I know why. We know it has to start repeating before 17 digits, but the calculator doesn't go out that far.

Emmy: Sure! I guess you just can't find the repeating decimal on the calculator; it's too long.

❞

Furthermore, the use of calculators requires students to focus on the analysis of problems and the selection of the appropriate operations. The effective use of calculators can improve student attitudes toward, and increase interest in, mathematics.

A Position Statement on Calculators in the Classroom by The National Council of Teachers of Mathematics, September 1978.

Teacher: Don't give up too easily. Think for a minute: how would you find the answer with a pencil? I mean, how do you do long division? Yes, Michael?

Michael: First you estimate the quotient, then multiply and subtract, and then keep doing that.

Teacher: That will be close enough for a beginning. Let's try this.

$$17\overline{)1.000000000}$$

We have already found that 1/17 is about 0.0588235294. Let's leave off the last two digits, because we can't be sure how the calculator is rounding. So we multiply 0.05882352 by 17 and subtract the product.

$$\begin{array}{r} 0.05882352 \\ 17\overline{)1.00000000000} \\ .99999984 \\ \hline 16 \end{array}$$

What should be done next?

Mary: Divide 16 by 17. I just did that on my calculator and I get 0.9411764706. I'll drop the last two digits . . . .

Teacher: Can you show us up here what to write next?

Mary: (writing) I multiplied 94117647 × 17.

$$\begin{array}{r} 0.05882352|94117647 \\ 17\overline{)1.0000000000000000} \\ .99999984 \qquad\qquad \\ \hline 1600000000 \\ 1599999999 \\ \hline 1 \end{array}$$

Emmy: I got it! That 1 is going to be the same as in the beginning. It's repeating now.

In solving the problem the basic division algorithm was used and adapted to the calculator so that it could be done more quickly. A pupil who did not understand the algorithm would not be able to understand, much less figure out independently, what went on here. This is an example of how one needs to know more, not less, mathematics to get the most out of a calculator, and how a calculator can lead to greater understanding and insight in mathematics.

EXERCISE SET 7.4

In Problems 1–5 imagine that you have a calculator whose display shows only three digits. How could you use the calculator to help you in the following:

1. To add 56.71 + 89.52 + 72.89.

2. To subtract 156.39 − 27.65.

3. To multiply 124.5 × 75

4. To divide $16.5\overline{)957.8}$

5. To find the decimal for 3/88.

6. How would the methods used in Problems 1–5 change if you had a screen that displayed four digits?

In Problems 7–10 find the decimal numeral for each:

7. 1/23 **8.** 1/43 **9.** 13/53 **10.** 11/53

In Problems 11–13 convert each to decimal notation. Look for a pattern that relates the number of times 2 or 5 is used as a factor to the number of digits in the decimal.

11. a) $\frac{1}{2^1 \cdot 5^2}$ b) $\frac{3}{2^2 \cdot 5^1}$

12. a) $\frac{77}{2^2 \cdot 5^3}$ b) $\frac{99}{2^3 \cdot 5^2}$

13. a) $\frac{1111}{2^3 \cdot 5^4}$ b) $\frac{1365}{2^4 \cdot 5^3}$

■ **14.** If a and b are whole numbers such that GCF $(a, b) = 1$ and b is such that $b = 2^p \cdot 5^q$, where p and q are also whole numbers, then the length of the finite decimal representation of a/b is the larger of the two numbers p, q. Show that this statement is true (look at Problems 11–13 again).

15. Find the decimal for each of the fractions $n/11$, where $n = 1, 2, 3, 4, 5, 6, 7, 8, 9, 10$. Can you account for the unusual features of the patterns you find?

16. Solve Problem 15 for $n/111$. Choose n from 1 to as high as you care to go up to 110.

In Problems 17–20 find the decimal for each. Study the patterns with respect to the length of the nonrepeating parts and the length of the repeating parts. ■ Can you find any relationship?

17. a) $\frac{1}{5^1 \cdot 3}$ b) $\frac{1}{5^2 \cdot 3}$ c) $\frac{1}{5^3 \cdot 3}$

18. a) $\frac{1}{2^1 \cdot 3}$ b) $\frac{1}{2^2 \cdot 3}$ c) $\frac{1}{2^3 \cdot 3}$

19. a) $\frac{1}{11}$ b) $\frac{27}{33}$ c) $\frac{17}{111}$

20. a) $\frac{17}{5^1 \cdot 27}$ b) $\frac{11}{5^2 \cdot 27}$ c) $\frac{131}{5^3 \cdot 27}$

7.5 AIMS

You should be able to:

a) Convert from a percent to either a fraction or a decimal and the reverse.
b) Apply percent as a ratio.

7.5 PERCENT, RATIO, AND PROPORTION

The basic fractions have denominators that are powers of ten; in the case of percents, the denominators are restricted still more because they are a way of expressing ratios with denominators of 100. Percent means *per hundred*, which makes it a comparison to 100; therefore 50% means 50 per hundred or 50:100.

FROM PERCENT TO DECIMALS. Converting a percent to a decimal is easy if we remember that percent is a comparison to 100. In order to find the decimal we need to find the original fraction that made a comparison to 100.

Example 1 Convert 36.5% to a decimal.

Solution 36.5% means 36.5 per hundred or 36.5 compared to 100.

36.5:100 means $\frac{36.5}{100}$, which is 0.365.

Example 2 Convert 3/8% to a decimal.

Solution 3/8% means 3/8 per hundred or 3/8 compared to 100.

3/8:100 means $\frac{3/8}{100} = \frac{3}{800} = 0.00375.$

25. Convert each of the following to a decimal numeral:

a) 7.35%
b) 127%
c) $\frac{1}{2}\%$

Try Exercise 25

CONVERSION TO PERCENT. To convert to percent from decimals or fractions, we need to find an equivalent fraction that makes a comparison to 100, that is, it has 100 as a denominator.

Example 3 Convert 1.21 to percent.

Solution We first write a fraction that has 100 in the denominator:

$$1.21 = 1\frac{21}{100} = \frac{121}{100}.$$

Therefore 121 is being compared to 100 and 1.21 = 121%.

Example 4 Convert 3/8 to percent.

Solution Because the decimal automatically is in terms of powers of ten, we write 3/8 as the decimal 0.375. Then

$$0.375 = \frac{375}{1000} = \frac{37.5}{100}.$$

Therefore 37.5 is being compared to 100, which means 3/8 is 37.5%.

Try Exercise 26

Once this is understood, we can think of the percent symbol as a part of the numeral. From this point of view, $n\%$ means $n \cdot \frac{1}{100}$ or $n \cdot (0.01)$. This makes it easy to convert from one notation to the other because the symbol (%) can be replaced with the symbol ($\cdot\ \frac{1}{100}$), and vice versa. We now do this for the examples just above.

Example 5 Convert to decimals: a) 36.5% b) 3/8%

Convert to percent: c) 1.21 d) 3/8

Solution

a) $36.5\% = 36.5 \cdot \frac{1}{100} = 0.365$

b) $\frac{3}{8}\% = \frac{3}{8} \cdot \frac{1}{100} = \frac{3}{800} = 0.00375$

c) $1.21 = 1.21 \cdot \left(100 \cdot \frac{1}{100}\right)$ [multiplying by 1]

$= (1.21 \cdot 100) \cdot \frac{1}{100}$

$= 121 \cdot \frac{1}{100}$

$= 121\%$ $\left[\text{replacing } \cdot \frac{1}{100} \text{ by } \%\right]$

d) $\frac{3}{8} = \frac{3}{8} \cdot \left(\frac{100}{1} \cdot \frac{1}{100}\right)$ [multiplying by 1]

$= \frac{300}{8} \cdot \frac{1}{100} = 37.5\%.$

Try Exercise 27

PERCENT INTERPRETED AS A RATIO—PROPORTION. It seems easier for youngsters to understand percent if they think of it as a ratio, a special kind of ratio. For example, 13% means 13 per hundred, which can be written 13:100 or 13/100, the ratio of 13 to 100.

26. Convert each of the following to percent:
a) 2.43 b) 0.71
c) 0.032 d) 3/4
e) 5/8

27. Treat conversion as a symbol replacement and do the following:
a) Write 0.25% as a decimal.
b) Write $33\frac{1}{3}\%$ as a fraction.
c) Write 1/16 as a percent.
d) Write 3.35 as a percent.

Using physical models to show the ratio 3 to 1

In making conversions from percent to decimals or fractions or the reverse, it is possible to think of the two ratios that are involved and write a statement called a proportion using these two ratios.

DEFINITION. The statement that a ratio a/b is equal to a ratio c/d is called a *proportion*. Thus $\frac{a}{b} = \frac{c}{d}$ is a proportion; the terms of the proportion are a, b, c, and d.

Proportions can be used to solve a wide variety of problems, and because percents are a special type of ratio, any problem that involves percents can be interpreted by means of a proportion.

Example 6 Six students out of the 48-member student band were in bed with the flu and did not play in the concert. What percent of the students in the band missed the concert?

Solution We form a ratio that compares the number of students sick to the total number in the band. Then $n/100$ is another ratio that compares the number of sick students to 100. But these ratios are equal, so we form the proportion:

$$\frac{6}{48} = \frac{n}{100}.$$

$$48 \cdot n = 6 \cdot 100 \qquad \text{[by equality for nonnegative rationals]}$$

$$n = \frac{600}{48} = 12.5,$$

12.5 per hundred is 12.5%.

The discussion of ratio, proportion, and percent will be continued in the following section on problem solving.

Try Exercise 28

28.

a) Six students were absent from the concert. This is 12.5% of the band. How many members are in the band?

b) If 12.5% of the band of 48 members were absent, how many students was that?

EXERCISE SET 7.5

1. The following rules are sometimes stated for converting from decimals to percent and the reverse:

a) To convert a decimal to percent, move the decimal point of the numeral two places to the right and annex a % symbol.

b) To convert a percent to a decimal, drop the % symbol and move the decimal point of the numeral two places to the left.

Show that these rules are correct.

2. Complete the following table. These frequently-used conversions are worth memorizing.

	Fraction	*Decimal*	*Percent*
a)	1/2		
b)		$0.33\overline{3}$	
c)			$66\frac{2}{3}\%$
d)		0.25	
e)	3/4		
f)			20%
g)		0.4	
h)	3/5		
i)		0.8	
j)	1/8		
k)			37.5%
l)			62.5%
m)	7/8		

3. If your calculator has a % key, get acquainted with its use if you have not already done so. Make a flowchart or charts for each procedure that is possible with your calculator using the % key.

4. Complete the following table:

7% of 1.43 = 0.1001
14% of 1.43 =
21% of 1.43 =
28% of 1.43 =
35% of 1.43 =

Study the pattern formed in the table. Can you explain this?

■ **5.** Create a problem similar to Problem 4 in which a pattern is developed in a special sequence of operations.

The following problems are similar to those that elementary pupils solve. Work out the solutions as if you were going to explain them to such pupils.

6. In a class of 30 students, 4 out of 5 do not have blue eyes. What percent of the class has blue eyes?

7.
a) During a special sale, a leather coat which originally cost $180 was reduced 20%. Later the coat was further reduced by 10%. What was the final cost of the coat?
b) If you were buying the coat would you have preferred to purchase the coat at an initial reduction of 30% with no further reduction? Explain your answer.

8. Imagine that you have $20.00 with you and you happen to see a pair of shoes marked $21.75 on sale for 10% off. There is also a 4% sales tax on the item. Can you buy the shoes with the money you have with you?

9. The junior-high chess team won 27 and tied 3 out of 43 matches they played this year. What percent of the games did they win? What percent of the games did they lose?

7.6 AIMS

You should be able to:
Apply the five-step solving process to problems involving ratio, proportion, and percent.

7.6 PROBLEM SOLVING

Both this and the previous chapter have been concerned with the different types of symbols that are possible for nonnegative rational numbers; different types would not persist if they were not all useful in some way. Symbols do play an important role in problem solving;* one of the advantages of the mathematical approach comes from the use of convenient, effective, and concise notation.

Elementary pupils often find ratio, proportion, and percent concepts difficult to work with. They would find it less so if they used a general problem-solving approach instead of trying to memorize some standard way of doing certain types of problems. We examine in detail some typical problems.

Example 1 The girls' basketball team won 14 out of 19 games. What percent of their games did they win?

* As a reminder:

THE FIVE-STEP PROCESS
1. What are the relevant *facts*?
2. What actually is the *problem*?
3. What are my *ideas*?
4. How do I get a *solution*?
5. *Check* and *review* the solution.

What are the relevant *facts* in the problem?

Try Exercise 29

How can the *problem* be restated?

1. 14 is what percent of 19?
2. Convert 14/19 to percent.
3. What number x compared to 100 is the same ratio as 14 to 19?

Try Exercise 30

Some *ideas* for the solution arise out of these problem statements.

Idea 1. Write an equation to correspond to the problem statement (1); solve the equation.

Idea 2. Use any method to convert 14/19 to percent.

Idea 3. Write two ratios; form a proportion. If only one term is unknown, solve for that term.

Each of the ideas will be used in turn to find the *solution.*

Idea 1 Use the problem statement:

Solution 14 is what percent of 19?

$$14 = (x\%) \cdot 19 \quad \text{[direct translation]}$$

$$14 = x \cdot \frac{1}{100} \cdot 19$$

$$1400 = x \cdot 19 \quad \text{[multiply both members by 100]}$$

$$x = \frac{1400}{19} = 73.7 \quad \text{[to the nearest tenth]}$$

They won 73.7% of the games.

Idea 2 Because 14/19 is the ratio of the number of games won to the total number played, the problem is to convert 14/19 to percent.

Solution

$$\frac{14}{19} = \frac{14}{19} \cdot \frac{100}{1} \cdot \frac{1}{100}$$

$$= \frac{1400}{19} \cdot \frac{1}{100} \quad \left[\text{replace} \cdot \frac{1}{100} \text{ by } \%\right]$$

$$= 76.7\%.$$

Idea 3 What number x compared to 100 is the same as 14 to 19?

29. What are the relevant facts for Example 1?

30. Restate the following problem in three different ways: Out of 37 customers interviewed in a sales survey, 27 had never purchased Brand-X soft drink. What percent of this sample had never purchased Brand-X?

31. What are the relevant facts for Example 2?

Solution We write the proportion that follows from the problem statement:

$$\frac{14}{19} = \frac{x}{100},$$

$$14 \cdot 100 = 19x, \quad \text{[by equality for nonnegative rational numbers]}$$

$$x = \frac{1400}{19} = 76.7. \quad \text{[to the nearest tenth]}$$

Therefore, $\frac{14}{9} = \frac{76.7}{100}$; that is 76.7% of the games were won. All of these solutions are correct but they each require a slightly different approach to the problem.

Example 2 A nurse needs to give a dosage of 2500 units of anticoagulant Heparin. One cubic centimeter (cc) has 5000 units of the drug. How many cc must be used for the required dosage? What are the relevant *facts* in this problem?

Try Exercise 31

How can the *problem* be restated?

1. The number of cc used multiplied by the dosage per cc is the total dosage needed. Find the number of cc.
2. In a ratio table of number of cc to number of units of the drug, what is the entry corresponding to 2500?
3. Find what percent 2500 is of 5000. Find the same percent of 1 cc.

What are some *ideas* that might be used? We use ideas that are directly inspired by the ways we looked at the problem just above.

Idea 1. Write a sentence that translates to an equation that can be solved.

Idea 2. Write ratios and set up a table of ratios.

Idea 3. Solve two problems:
   a) 2500 is what percent of 5000?
   b) Find the same percent of 1 cc for the answer.

Each of these can be used to find the solution to the problem. We use Idea 2.

Idea 2. Make a table of ratios. Write a proportion and solve for the unknown term.

Solution Write the ratio table:

Number of cc	1	?
Number of dosage units	5000	2500

One proportion that can be written is:

$$\frac{1}{5000} = \frac{x}{2500}$$

$$\frac{1}{2} = \frac{2500}{5000} = x$$

Therefore $\frac{1}{2}$ cc of Heparin is needed.

Try Exercise 32

32. Use each of the ideas 1 and 3 to solve Example 2 in two more ways.

A glance toward the classroom

At one time pupils learned about (memorized?) three "cases" of percent. The problem they had then was to try to recognize which case the problem at hand fit. It would seem far better to have youngsters get practice at analyzing and thinking about the problem in a common-sense way and not depend upon imperfectly understood rote solutions.

EXERCISE SET 7.6

1. Three cans of baked beans sell for \$0.79. How many cans can be purchased with \$3.00? (It is not possible to buy a part of a can of beans!)

2. A teacher who earns \$13 000 per year receives a 4 raise for next year and an additional raise of 3% the following year. Would a 7% raise the first year and none the second have been more advantageous for this teacher?

3. The ratio of black marbles to red marbles in a bag is 1 to 5. There are 186 marbles in the bag. How many must be black? What percent of the total number of marbles are red?

4. At the end of the day a cash register in a department store showed an intake of \$1240.73. This amount includes a 6% sales tax on all sales.

a) What is the amount of tax that was collected that day?

b) What was the amount of the sales for the day?

5. This fall, 646 freshmen were registered in a certain school. They formed 38% of the entire student body. How many students were in the entire student body of the school?

6. A box contains 425 grams of a certain brand of cereal. One serving of the cereal is $28\frac{1}{3}$ grams. If each serving contains 5 grams of protein, what percent of the cereal is protein? Each serving contains 20 grams of carbohydrate; what percent of the cereal is carbohydrate?

7. There are 13 servings of rye crackers in a 234-gram package. Each serving contains 3 grams of protein and 13 grams of carbohydrates. What percent of the cracker is protein? carbohydrate?

8. One centimeter on the map represents 85 kilometers on the highway. What is the ratio of map distance to highway distance? If city A is 2.3 cm away from city B on the map, what is the actual distance between them?

REVIEW TEST

1. Find the decimal numeral for each of the following:
a) 3/16 b) 12/125 c) 17/32

2. Write T after the statement if it is always true; otherwise write F.
a) 17/68 has a terminating decimal. ______
b) Every nonnegative rational number can be represented by either a terminating decimal or a nonterminating repeating decimal. ______
c) The maximum possible number of digits in the repeating cycle of the decimal for 1/7 is 7. ______
d) If a cake recipe calls for $1\frac{1}{3}$ cups of sugar and 4 cups of flour, then the same recipe used with 7 cups of flour will require $2\frac{1}{3}$ cups of sugar. ______
e) The product of a 2-place decimal by a 3-place decimal is a 6-place decimal. ______

3. Convert to the lowest-terms fraction:
a) 0.3078 b) 90.16

4. Write 0.0105 in expanded notation.

5. Convert to fractions to justify the algorithm for the difference: 7.23 – 1.267.

6. In each of the following the correct digits for the computations are given. Locate the decimal point in each case as indicated.
a) 5.92 × 63.2 = 374144 (use the rule)
b) 49.723 × 64.36 = 320017228 (by estimation)
c) $0.0015\overline{)0.5127}$ = 34180 (by the rule)
d) $561\overline{)47}$ = 83778966 (by estimation)

7. Find the decimal for 4/99.

8. Find the decimal for the reciprocal of 0.36.

9. Convert each of the following to the lowest-terms fraction:
a) $5.27\overline{27}$ b) $90.32141\overline{141}$

10. Find a number that can be added to $0.33\overline{3}$ so that the sum is a whole number.

11. The screen of your calculator displays only 5 digits. How could you use the calculator to help you do the following calculations?
a) Multiply: 32768 × 5.67 b) Divide: $16.3\overline{)201.142}$

12. Use a calculator to find the decimal numeral for 2/17. Be sure you have found the entire repeating cycle.

13. Convert to decimal numerals:
a) 13.4% b) 0.54% c) $\frac{2}{3}$% d) 154%

14. What percent of 94 is 37?

15. 137 is 120 percent of what number?

16. Find 27% of 150.

17. In a certain classroom, there are 4 chairs with left-handed writing desks. The remaining 31 chairs have right-handed arms. What percent of the desks in the room are for left-handed people?

18. An auto travels 148 kilometers on 15.7 liters of gasoline. At this rate, how far will this auto travel on 76 liters of gasoline?

■ **19.**
a) Explain how division could be used to convert $\left(\frac{1}{4}\right)_{\text{ten}} = 0.25$ to $0.abcd\ldots_{\text{three}}$.
b) Does the answer in (a) terminate or is it nonterminating and repeating?
c) Write expanded notation for 24.13_{five}. Convert to a fraction in base ten.

FOR FURTHER READING

1. A less serious approach to the use of a pocket calculator is found in *Games with the Pocket Calculator* by S. Thiagarajan and H. Stolovitch, Dymax, Box

310, Menlo Park, California, 94025. This contains about 50 pages of games, some hard and some easy; they are a fun way to use the calculator to learn some mathematics.

2. For some implications for the classroom see the November 1976 issue of the *Arithmetic Teacher*, 23:7. It contains much information on minicalculators in the classroom as well as many imaginative ideas for classroom use.

3. For a reasonable perspective on the use of calculators try: "A modest proposal concerning the use of hand calculators in schools" by Edwin E. Hopkins in the *Arithmetic Teacher*, 23:8, December 1976.

TEXTS

Heimer, Ralph T., and Cecil R. Trueblood, *Strategies for Teaching Children Mathematics*. Addison-Wesley Publishing Co., Reading, Massachusetts (1977). (A methods text.)

Meserve, Bruce E., and Max A. Sobel, *Contemporary Mathematics*. Prentice-Hall, Inc., Englewood Cliffs, New Jersey (1972).

chapter 8 integers, rationals, and real numbers

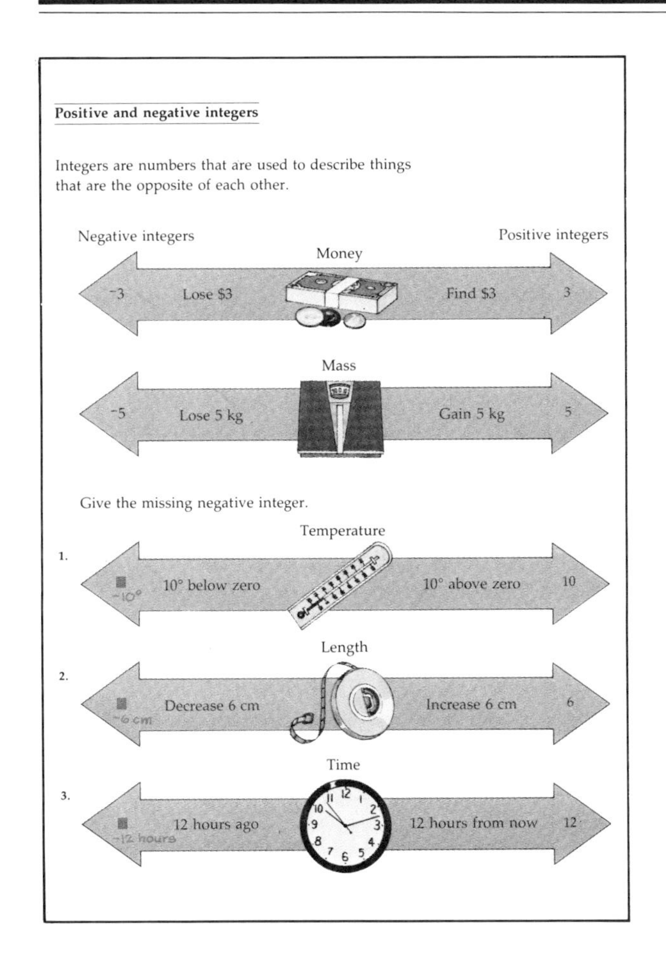

Positive and negative integers

Integers are numbers that are used to describe things that are the opposite of each other.

Negative integers

Positive integers

Money

$^-3$ Lose $3 | Find $3 3

Mass

$^-5$ Lose 5 kg | Gain 5 kg 5

Give the missing negative integer.

Temperature

1. ■ 10° below zero | 10° above zero 10

$^-10°$

Length

2. ■ Decrease 6 cm | Increase 6 cm 6

$^-6$ cm

Time

3. ■ 12 hours ago | 12 hours from now 12

$^-12$ hours

This lesson provides familiar and readily understood applications or illustrations for positive and negative integers. These examples could easily be expanded and developed into a variety of activities for the pupils.

8.1 AIMS

You should be able to:

a) State the definition of additive inverse and integers.
b) Interpret addition on a number line.
c) Define and use the definitions for addition and subtraction of integers.
d) Define the absolute value of a number.

8.1 THE INTEGERS

DEFINITION. We have already mentioned that the nonnegative rational numbers were treated before the integers because that is the sequence commonly used in elementary textbooks. If one considers the difficulty that children often have with fractions, it would probably make more sense to introduce integers first, and some textbook series are now doing just that.

In this chapter we make a fresh start and return to the system of whole numbers with the intention of extending it to a system that is closed under subtraction, because it is annoying not to have the difference $a - b$ exist for all pairs of whole numbers a and b. Providing closure for subtraction is equivalent to requiring that for any statement $x + b = a$, where a and b are whole numbers, there is an x that makes the statement true.

Try Exercise 1

1. Read page one of this chapter if you have not already done so.

First we need the concept of *additive inverse.*

DEFINITION. For any number a, the *additive inverse* of a is the unique number ^{-}a such that

$$a + {}^{-}a = {}^{-}a + a = 0.$$

Now the overall scheme is to create an additive inverse for each whole number. For example, the additive inverse of 3 is $^{-}3$, which is often described in the elementary school as "the opposite of 3."* Joining all such additive inverses with the whole numbers, we get the set of integers.

Try Exercises 2 and 3

2. What is the additive inverse of each of the following?

a) 5 b) $^{-}7$ c) $^{-}(^{-}18)$

3. Find a in each of the following:

a) $6 + a = 0$
b) $^{-}10 + a = 0$
c) $(x + 1) + {}^{-}a = 0$

DEFINITION. The set of integers I is the union of the set of all the whole numbers with the set of the additive inverses of the whole numbers:

$$I = \{\ldots, -3, -2, -1, 0, 1, 2, 3, \ldots\}.$$

The natural numbers are now called the *positive integers* and their additive inverses are called the *negative integers.*

The integers are pictured on a number line with the negative numbers on the left side of 0 and the positive numbers on the right.

$^{-}7$ $^{-}6$ $^{-}5$ $^{-}4$ $^{-}3$ $^{-}2$ $^{-}1$ 0 1 2 3 4 5 6 7

* Other notation such as -3 is also used for this element, but it is less confusing for youngsters, when they first learn about "opposites", if the upper dashes are used exclusively.

Addition of Integers

We wish to define all operations for integers, so that the whole-number operations remain the same and the desirable properties of the whole-number operations (such as the commutative, associative, and distributive laws) are retained in the extended system.

NUMBER-LINE INTERPRETATION. Addition that makes sense in terms of real-world applications of integers can be represented on a number line by using arrows that point to the right for positive numbers and to the left for negative numbers.

To perform the addition $a + b$, we draw an arrow for a and then one for b and find the sum indicated by the second arrowhead.

Example 1 Use a number line to add $5 + {}^-3$.

Draw an arrow for 5; at point 5 draw the one for ${}^-3$. The sum is 2.

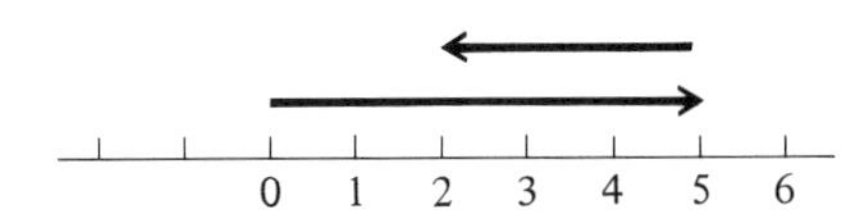

Example 2 Use a number line to add ${}^-9 + 4$.

Draw the arrow for ${}^-9$; at point ${}^-9$ draw the one for 4. The sum is ${}^-5$.

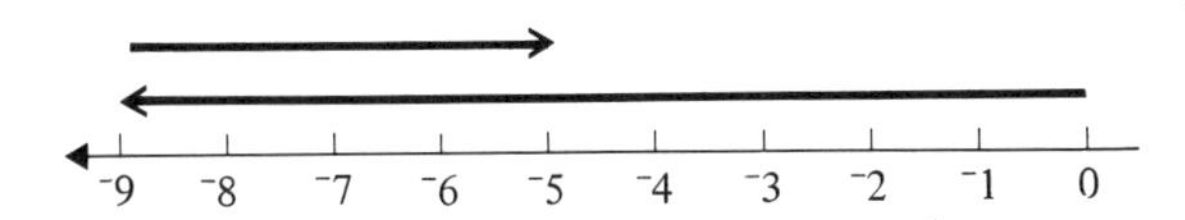

Example 3 Use a number line to add ${}^-2 + {}^-4$.

Draw the arrow for ${}^-2$; at point ${}^-2$ draw the one for ${}^-4$. The sum is ${}^-6$.

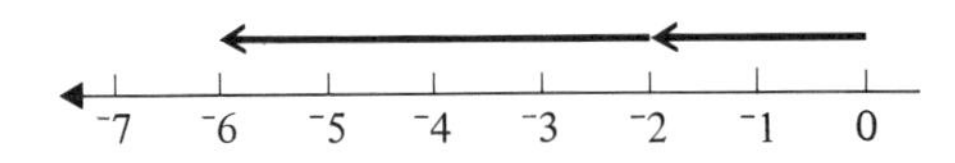

Try Exercise 4

4. Use a number line to add:

a) ${}^-3 + 5$
b) $4 + {}^-9$
c) ${}^-4 + {}^-2$

We find some sums for particular pairs of integers under the assumption that both the commutative property and the associative property hold for addition of integers.

Example 4 Find the following sums, assuming that properties of the whole numbers have been inherited by the integers:

a) ${}^-2 + {}^-4$ b) ${}^-3 + 8$ c) ${}^-6 + 4$

5. If the laws for whole numbers are to hold, then the following sums are really determined. Show this.

a) $^-4 + 7$
b) $^-8 + 5$

Solution

a) The number line would indicate that the sum is $^-6$; if so, then $(^-2 + {}^-4) + 6$ would be 0. We show that it is:

$(^-2 + {}^-4) + 6 = (^-2 + {}^-4) + (4 + 2)$ [by renaming 6]
$= {}^-2 + ((^-4 + 4) + 2)$ [by the associative property of addition]
$= {}^-2 + (0 + 2)$ [by the definition of additive inverse]
$= {}^-2 + 2$ [by the definition of 0]
$= 0$ [by the definition of additive inverse]

Hence $^-2 + {}^-4$ must be $^-6$.

b) $^-3 + 8 = {}^-3 + (3 + 5)$ [by renaming 8]
$= (^-3 + 3) + 5$ [by the associative law for addition]
$= 0 + 5$ [by the definition of additive inverse]
$= 5$

c) $^-6 + 4 = (^-2 + {}^-4) + 4$ [by part (a)]
$= {}^-2 + (^-4 + 4)$ [by the associative law for addition]
$= {}^-2 + 0$ [by the definition of additive inverse]
$= {}^-2$ [by the definition of 0]

A study of the sums in Example 4 shows that the requirement that the properties be retained in effect determines what the sums must be.

Try Exercise 5

DEFINITION. The number line and the considerations of the illustrations above are consistent with the following definition for addition.

ADDITION OF INTEGERS. For all whole numbers a and b:

1. $a + b$ **is the usual whole-number sum;**
2. $^-a + {}^-b = {}^-b + {}^-a = {}^-(a + b)$;
3. $^-a + b = b + (^-a) = {}^-(a - b)$ **for** $a > b$;
4. $^-a + b = b + (^-a) = b - a$ **for** $a \leqslant b$.

Example 5 Use the definition of integer addition in the following:

a) $^-16 + 7$ b) $13 + {}^-6$ c) $^-11 + {}^-13$

Solution

a) By part (3) of the definition the sum is $^{-}(16 - 7) = {}^{-}9$.
b) By part (4) of the definition the sum is $13 - 6 = 7$.
c) By part (2) of the definition the sum is $^{-}(11 + 13) = {}^{-}24$.

Try Exercise 6

6. Use the definition for integer addition to find the following:
a) $5 + 8$
b) $7 + {}^{-}4$
c) $5 + {}^{-}8$
d) $^{-}3 + {}^{-}6$

We need to show that the basic properties for addition of whole numbers have been passed along to the integers. Closure and commutativity follow immediately from the definition, since it is obvious that the defined sums are either a whole number or the additive inverse of one, and the commutative property is a part of the definition itself.

We need to show that associativity has been preserved as well. This means that for any integers a, b, and c, $(a + b) + c = a + (b + c)$.

Example 6 Verify that the associative property holds for $^{-}3$, 2, and $^{-}5$.

We must show that $^{-}3 + (2 + {}^{-}5) = ({}^{-}3 + 2) + {}^{-}5$:

$$^{-}3 + (2 + {}^{-}5) = {}^{-}3 + {}^{-}3 \quad \text{[by integer addition]}$$
$$= {}^{-}6;$$
$$({}^{-}3 + 2) + {}^{-}5 = {}^{-}1 + {}^{-}5 \quad \text{[by integer addition]}$$
$$= {}^{-}6.$$

The proof of this property in general is somewhat more difficult than the verification done in Example 6. Since many cases must be considered, the proof is rather long and is left for the exercises.

Try Exercise 7

7. Verify the associative property in the following:
a) $({}^{-}6 + {}^{-}4) + {}^{-}8 =$
b) $(7 + {}^{-}3) + 10 =$

ABSOLUTE VALUE. If we think in terms of distance along the number line, then each natural number is the same distance from the point 0 as its additive inverse, and 8 and $^{-}8$ are both at a distance 8 from the point 0. The *absolute value* can be thought of as a function that assigns to each integer its distance from the point 0. The definition for this functions is as follows:

DEFINITION. The *absolute value* of a number n is $|n|$:

1. $|n| = n$ if n is a positive integer or 0;
2. $|n| = {}^{-}n$ if n is a negative integer.

Example 7 Find the absolute value of each of the following:
a) 3 b) $^{-}5$ c) $({}^{-}8 + 4)$ d) $(a - a)$

8.
a) Find the absolute value of each of the following:

$8 \quad {}^-17 \quad 0$

b) Find each of the following:

$|6| \quad |{}^-3 + 2| \quad |{}^-({}^-3)|$

9. Use the rule in terms of absolute value to find the following:
a) ${}^-20 + 12$
b) $24 + {}^-13$

Solution

a) $|3| = 3$
b) $|{}^-5| = {}^-({}^-5) = 5$
c) $|{}^-8 + 4| = |{}^-4| = 4$
d) $|a - a| = |0| = 0$

Notice that although the domain of the absolute-value function is the set of integers, the range is the set of nonnegative integers.

Try Exercise 8

The absolute value provides a way to restate parts 3 and 4 of the definition for integer addition with a rule that is possibly more familiar:

To find the sum of a negative and a positive integer, take the absolute values of each of the two numbers and subtract the smaller whole number from the larger. This is the sum, unless the negative number has the greater absolute value, in which case the sum in the additive inverse of the result.

This rule can be shown to be equivalent to parts (3) and (4) of the original definition above, and its proof is left for an exercise.

Try Exercise 9

The definition for addition stated that for any whole numbers a and b, ${}^-a + {}^-b = {}^-(a + b)$, that is, the additive inverse of a whole number added to the additive inverse of another whole number is the same as the additive inverse of the sum of the two whole numbers. We might wonder if this statement applies to any pair of integers.

Example 8 Prove that for any integers a and b,

$$ {}^-a + {}^-b = {}^-(a + b). $$

If we show that $({}^-a + {}^-b) + (a + b) = 0$, we will have proved what is required.

1. $({}^-a + {}^-b) + (a + b)$
   $= ({}^-a + {}^-b) + (b + a)$ [by the commutative property of integer addition]
2. $= ({}^-a + ({}^-b + b) + a)$ [by the associative property for integer addition]
3. $= ({}^-a + (0 + a))$ [by the definition of additive inverse]
4. $= ({}^-a + a)$ [by the definition of 0]
5. $= 0$ [by the definition of additive inverse]
6. Therefore ${}^-a + {}^-b = {}^-(a + b)$ [by step (3)]

Subtraction

DEFINITION. Subtraction, just as for whole numbers, is defined in terms of addition.

DEFINITION. **For any integers a and b, the *difference* $a - b$ is the number c if and only if $a = c + b$. More briefly:**

$$a - b = c \leftrightarrow a = c + b.$$

Example 9 Use the definition of subtraction to find the following differences:
a) $5 - {}^{-}3$ b) ${}^{-}14 - {}^{-}11$ c) ${}^{-}7 - 3$

Solution

a) $5 - {}^{-}3 = c \leftrightarrow 5 = c + {}^{-}3$; c must be 8.
b) ${}^{-}14 - {}^{-}11 = c \leftrightarrow {}^{-}14 = c + {}^{-}11$; c must be ${}^{-}3$.
c) ${}^{-}7 - 3 = c \leftrightarrow {}^{-}7 = c + 3$; c must be ${}^{-}10$.

Try Exercise 10

10. Use the definition of subtraction to find:
a) $4 - {}^{-}2$
b) $5 - 7$
c) ${}^{-}6 - 3$

SUBTRACTION AND ADDITIVE INVERSES. There is a useful connection between subtraction and the additive inverses. Observe the patterns in the following:

$$5 - {}^{-}3 = 8 \qquad 5 + 3 = 8$$
$${}^{-}14 - {}^{-}11 = {}^{-}3 \qquad {}^{-}14 + 11 = {}^{-}3$$
$${}^{-}7 - 3 = {}^{-}10 \qquad {}^{-}7 + {}^{-}3 = {}^{-}10.$$

This holds for all differences of integers, and thus every such difference can be transformed to an addition problem.

For any integers a and b, $a - b = a + {}^{-}b$.

One can subtract by adding the inverse of the subtrahend.

As so often before, we depend upon a definition to do this.

1. a and b are any integers; their difference is c:
   $a - b = c \leftrightarrow a = c + b$ [by the definition of integer subtraction]
2. $a + {}^{-}b = (c + b) + {}^{-}b$ [since every integer has an additive inverse and closure for addition]
3. $\quad = c + (b + {}^{-}b)$ [by the associative property for integer addition]
4. $\quad = c + 0$ [by the definition of additive inverse and the definition of 0]
   $\quad = c$
5. Therefore $a + {}^{-}b = a - b$ [by renaming c as $a - b$]

Example 10 Subtract (use the theorem just proved):
a) $17 - ({}^{-}3)$ b) ${}^{-}5 - 4$ c) ${}^{-}3 - ({}^{-}8)$

11. Use the theorem relating subtraction to the additive inverse:
a) $15 - (^-7)$
b) $^-12 - 4$
c) $^-7 - (^-8)$

Solution

a) $17 - (^-3) = 17 + 3 = 20$
b) $^-5 - 4 = {^-5} + {^-4} = {^-9}$
c) $^-3 - (^-8) = {^-3} + 8 = 5$

Try Exercise 11

The fact that for any integers a and b, $a - b = a + {^-b}$, is important not only for convenience in computation. This statement establishes that the integers are closed under subtraction. Because every integer has an additive inverse, we can always find for any integer b an inverse ^-b. Since addition is defined for all integers, this means that every subtraction is possible in the system of integers.

A glance toward the classroom

Negative number and additive inverse are two distinct concepts, although they are related. Additive inverses are often denoted ^-n, but ^-n is a negative number if and only if $n > 0$. If n is 3, then ^-n is the negative number $^-3$. In algebra and higher mathematics the upper dash is hardly ever used and negative numbers are written -3, -5, and so on, with the dash on the same line. Elementary pupils may confuse this notation with the operation symbol for subtraction.

EXERCISE SET 8.1

1. What is the additive inverse of 0? Why?

2. What is the additive inverse of $^-(^-a)$? Why?

3. Prove that for any integer a, $a + 0 = 0 + a = a$.

4. If a is a positive integer, can ^-a be a positive integer? Why?

5. Show that ^-a is not always a negative integer.

6. Complete the following table:

a	b	c	$a + (b + c)$	$(a + b) + c$
3	5	7	$3 + 12 = 15$	$8 + 7 = 15$
$^-3$	$^-4$	$^-2$		
$^-5$	$^-6$	3		
$^-4$	3	2		
11	$^-15$	$^-20$		
2	$^-7$	0		

■ **7.** Show that for any integers a, b, and c, $a + (b + c) = (a + b) + c$.

8. Provide a counterexample to prove that subtraction of integers is
a) not associative
b) not commutative

9. Prove that for any integers a and b,

$$^-(a - b) = (b - a).$$

11. The temperature yesterday went from 10 degrees above 0 in the afternoon to 5 below 0 at night. What was the maximal difference between the temperatures?

13. Prove that if a and b are any integers, there is always an integer x that is a solution to the equation $x + a = b$.

15. In each of the following determine if the statement is true. For those that are true, state the property or definition that guarantees the truth of the statement in general.

a) ${}^-3 + {}^-4 = {}^-4 + {}^-3$ b) ${}^-9 - {}^-4 = 4 - {}^-9$

c) $(3 + 7) + {}^-2 = 3 + (7 + {}^-2)$

d) $(3 - 5) + 6 = 3 - (5 + 6)$

e) ${}^-3 + {}^-5 = {}^-(5 + 3)$ f) $6 - ({}^-3) = 6 + 3$

g) $(6 - 7) - 9 = 6 - (7 - 9)$ h) ${}^-14 + 21 = {}^-(14 - 21)$

17. Show that if a, b, and c are any integers, and $a + c = b + c$, then $a = b$.

19. On the basis of Problem 18 make a conjecture about the relation between $|a - b|$ and $|b - a|$. ■ Prove that your conjecture is true.

■ **20.** Prove that the rule for addition of a negative integer and a positive integer given in terms of the absolute value is equivalent to the corresponding statements in the definition of integer addition.

21. Addition and subtraction for integers can be modeled with red and white counters; each red chip is ${}^-1$, while each white one is 1. Thus (R)(R)(R) means ${}^-3$.

One red chip with one white chip is the same as 0 chips, so

also means ${}^-3$. Use red and white chips to represent the sum in each of the following:

a) $3 + 2$ b) $4 + 3$ c) $3 + {}^-6$ d) $5 + {}^-3$

10. Use the theorem of the text to express the following differences as sums:

a) ${}^-8 - 3$ b) ${}^-a - {}^-a$ c) $15 - {}^-3$
d) $x - y$ e) ${}^-12 - {}^-2$ f) ${}^-3 - 6$

12. For each of the following find the integer x that makes the statement true:

a) $x + 27 = {}^-3$ b) ${}^-36 + x = {}^-4$ c) $x - 2 = {}^-8$

14. Use arrows and a number line to illustrate the following:

a) $5 - 9$ b) $4 - {}^-5$ c) ${}^-4 - 3$ d) ${}^-4 - {}^-7$
e) $3 + 7$ f) $3 + {}^-7$ g) ${}^-6 + 2$ h) ${}^-3 + {}^-1$

16. Show that the set of integers is closed under subtraction.

18. Compare the answers in each of the following pairs:

a) $|3 - 5|$ and $|5 - 3|$ b) $|{}^-3 - {}^-4|$ and $|{}^-4 - {}^-3|$
c) $|{}^-4 - 2|$ and $|2 - {}^-4|$ ■ d) $|x - 3|$ and $|3 - x|$

Using a model for integers to demonstrate addition and subtraction

22. Subtraction can be demonstrated with the scheme of Problem 21 by the removal of appropriate chips.

a) (R)(R)(R)(R)(R) → (R)(R)(R) shows $^-5 - {}^-2 = {}^-3$

(R)(R)(R)(R)(R)(R) =

b) (R)(R)(R)(R)(R)(R) (R)(W)(R)(W)(R)(W)

→ (R)(R)(R)(R)(R)(R)(R)(R)(R) shows $^-6 - 3 = {}^-9$.

Explain why the model in b) is correct. Why must white chips be added?

Use red and white chips to model the following:

c) $^-20 - {}^-15$
e) $^-21 - 3$
d) $14 - {}^-6$
f) $12 - {}^-5$

■ **24.** Assume that a different definition had been made for the sum of two integers and that for any whole numbers a and b, $^-a + {}^-b = a + b$. Prove that with this definition at least one whole-number property would fail to be true for the integers.

■ **23.** It is possible to name integers by means of differences of whole numbers. For example: $11 - 8$, $12 - 9$, $13 - 10$, $14 - 11$, and so on, are all ways to name 3. Similarly, $8 - 11$, $9 - 12$, $10 - 13$, ... all name $^-3$. Two pairs (a, b) and (c, d) are equal if and only if $a + d = b + c$.

a) What integer is represented by each of the difference pairs in the set: $\{(0, 5)(1, 6)(2, 7)(3, 8)(4, 9)\ldots\}$?

b) Show that it is consistent with the usual definitions for integers to define the sum of two integers represented by (a, b) and (c, d) as the pair $(a + c, b + d)$. For example, $(7, 4) + (8, 4) = (15, 8)$, which corresponds to $3 + 4 = 7$.

c) Create a definition for subtraction that uses the ordered pairs for differences to represent integers.

8.2 AIMS

You should be able to:

a) State and apply the definition of multiplication and division of integers.

b) Provide proofs for the properties of integer multiplication and division.

8.2 MULTIPLICATION AND DIVISION OF INTEGERS

Multiplication

We wish to define multiplication for negative integers, so that the familiar properties of whole numbers are retained. Patterns such as those represented on a number line can be used to investigate some possibilities.

$^-15$	$^-10$	$^-5$	0	5	10	15	20
$5\cdot{}^-3$	$5\cdot{}^-2$	$5\cdot{}^-1$	$5\cdot 0$	$5\cdot 1$	$5\cdot 2$	$5\cdot 3$	$5\cdot 4$

Since the multiples of 5 are five units apart on the whole-number portion of the number line, we would like to have this pattern continue for the negative-integer portion. The same holds for multiples of $^-6$; they should all be 6 units aparts.

Try Exercise 12

12. Use the idea of the pattern for multiples to complete the sequence:

$$3\cdot({}^-6) = {}^-18$$
$$2\cdot({}^-6) = {}^-12$$
$$1\cdot({}^-6) = {}^-6$$
$$0\cdot({}^-6) = 0$$
$$^-1\cdot({}^-6) =$$
$$^-2\cdot({}^-6) =$$
$$^-3\cdot({}^-6) =$$

We also find some products for particular pairs of integers under the assumption that the desirable properties for the whole numbers are retained and see that the definition for integer multiplication cannot be made arbitrarily or whimsically.

Example 1 Find the product $8 \cdot (^-2)$.

We suspect that to be consistent with the number-line interpretation $8 \cdot (^-2) = {}^-16$. Does this also fit the requirement that the important laws are retained?

1. We assume that the distributive law holds:

$$8 \cdot (^-2 + 2) = 8 \cdot {}^-2 + 8 \cdot 2.$$

2. However, $^-2 + 2 = 0$ [by the definition of additive inverse].
3. Therefore, $8 \cdot (^-2 + 2) = 8.0 = 0$ [by renaming].
4. Thus, $8 \cdot {}^-2 + 8 \cdot 2 = 8 \cdot {}^-2 + 16 = 0$ and $8 \cdot {}^-2 = {}^-16$ [by the definition of additive inverse].

Example 2 Find the product $^-2 \cdot {}^-8$ under the assumption that the whole-number properties are retained.

1. We assume the distributive law is retained:

$$^-2 \cdot (^-8 + 8) = {}^-2 \cdot {}^-8 + {}^-2 \cdot 8 = {}^-2 \cdot {}^-8 + {}^-16 \text{ [by Example 1].}$$

2. However, $^-8 + 8 = 0$, hence $^-2 \cdot (^-8 + 8) = 0$.
3. Thus, $^-2 \cdot {}^-8 + {}^-16 = 0$ [by the transitive property of equals].
4. Therefore $^-2 \cdot {}^-8 = 16$ [by the definition of additive inverse].

Try Exercise 13

The following *definition* for *multiplication* of integers in terms of whole-number products is consistent with the number-line interpretation and the particular examples just described.

DEFINITION. For any two whole numbers a and b,

1. **$a \cdot b$ is the same whole-number product as before,**
2. **$^-a \cdot b = b \cdot {}^-a = {}^-(a \cdot b)$,**
3. **$^-a \cdot {}^-b = {}^-b \cdot {}^-a = a \cdot b$.**

Example 3 Use the definition to find the products:

a) $6 \cdot 7$ b) $^-2 \cdot 8$ c) $^-2 \cdot {}^-8$

Solution

a) $6 \cdot 7 = 42$ [by part (1) of the definition]
b) $^-2 \cdot 8 = {}^-(2 \cdot 8) = {}^-16$ [by part (2) of the definition]
c) $^-2 \cdot {}^-8 = 2 \cdot 8 = 16$ by part (3) of the definition]

Try Exercise 14

13. Find the products under the assumption that the whole-number properties are retained:

a) $5 \cdot (^-3)$
Hint: Start with $5 \cdot (^-3 + 3)$.
b) $^-3 \cdot {}^-2$

14. Use the definition to find the products:

a) $3 \cdot {}^-12$
b) $^-12 \cdot {}^-11$
c) $^-1 \cdot 13$

15. Compare the following pairs of products:

a) $3 \cdot (^{-}12 \cdot 2)$ with $(3 \cdot {}^{-}12) \cdot 2$

b) $(^{-}2 \cdot {}^{-}11) \cdot 3$ with $^{-}2 \cdot (^{-}11 \cdot 3)$

c) $^{-}5 \cdot (^{-}2 \cdot {}^{-}4)$ with $(^{-}5 \cdot {}^{-}2) \cdot {}^{-}4$

The definition for integer multiplication is given in terms of whole numbers, and the closure, commutative, and identity properties are taken care of immediately. The distributive and associative properties follow from the definition for multiplication of integers without much difficulty.

The arguments used in the general proofs are similar to those in the following case in that they depend upon the definitions made for the integers and the properties of whole numbers.

Example 4 Prove that for any whole numbers a, b, and c,

$$^{-}a \cdot (b + c) = {}^{-}(a \cdot b) + {}^{-}(b \cdot c).$$

Proof

1. $^{-}a \cdot (b + c) = {}^{-}(a \cdot (b + c))$ [by part (2) of the definition of integer multiplication]
2. $= {}^{-}(a \cdot b + a \cdot c)$ [by the distributive property for whole numbers]
3. $= {}^{-}(a \cdot b) + {}^{-}(a \cdot c)$ [by Example 8 of Section 8.1]

Try Exercise 15

Division

DEFINITION. The whole-number definition for division is extended to the set of integers and is again in terms of multiplication.

DEFINITION. For any integers a, b, and c ($b \neq 0$), a divided by b ($a \div b$) is the number c if and only if $a = c \cdot b$. More briefly:

$$a \div b = c \leftrightarrow a = c \cdot b.$$

There are integer pairs for which there is no integer quotient, since we already know that there are many positive integer pairs that do not have such a quotient. In this extension of the whole-number system we have not gained any improvement for division as we did when we obtained closure for subtraction.

Example 5 Use the definition to find the quotients.

a) $^{-}72/^{-}9$ b) $^{-}18/3$ c) $45/^{-}7$ d) $0/18$

Solution

a) The quotient is 8 because $8 \cdot {}^{-}9 = {}^{-}72$.
b) The quotient is $^{-}6$ because $^{-}6 \cdot 3 = {}^{-}18$.
c) There is no integer c such that $c \cdot {}^{-}7 = 45$, so there is no quotient in the set of integers.
d) The quotient is 0, since $18 \cdot 0 = 0$.

Division by 0 is still undefined, for if there were an integer c such that $17 \div 0 = c$, then $17 = 0 \cdot c$ would have to be true. This is true for any other nonzero integer divided by 0. Also $0 \div 0$ is not defined, just as it was not defined in the case of whole numbers, and for the same reason: the quotient would not be unique.

Try Exercise 16

16. Use the definition to find the quotients:

a) $\frac{^-49}{^-7}$ b) $\frac{^-36}{6}$

c) $\frac{24}{^-12}$ d) $\frac{0}{^-9}$

PROPERTIES OF INTEGERS—A SUMMARY. In the system of integers the familiar properties of the whole numbers are retained, but a gain has been made in that all subtractions are possible. We summarize some important properties:

1. *Closure* holds for addition, subtraction, and multiplication.
2. *Commutativity* and *associativity* hold for addition and multiplication.
3. The *distributive law* of multiplication over addition holds.
4. Every integer a has a unique *additive inverse* ^-a.
5. There are two *identities*:

0, the additive identity,
1, the multiplicative identity.

Many other properties can be deduced from these basic ones along with the definitions for the operations with integers. We include a list of the more important ones that have either been proved in the text or have appeared as an exercise in the first two sections of this chapter:

For any integers a, b, and c:

1. $^-a + {^-b} = {^-(a+b)}$
2. $a + c = b + c \leftrightarrow a = b$
3. $a - b = a + {^-b}$
4. $|a - b| = |b - a|$
5. $a \cdot (b - c) = a \cdot b - a \cdot c$
6. $^-(a - b) = b - a$
7. $^-1 \cdot a = {^-a}$
8. $a \cdot {^-b} = {^-(a \cdot b)}$
9. $^-a \cdot {^-b} = a \cdot b$
10. If $a \cdot b = 0$, then either $a = 0$ or $b = 0$
11. If $a \neq 0$, then $0 \div a = 0$.

EXERCISE SET 8.2

1. Use the definition for multiplication to prove that for any integer a, $a \cdot 0 = 0$.

2. Prove that for any integer a, $^-a = (^-1) \cdot a$.

3. Verify that the following statements are true.

a) $^-5 \cdot (3 + 2) = {^-5} \cdot 3 + {^-5} \cdot 2$

b) $6 \cdot (^-1 + 3) = 6 \cdot {^-1} + 6 \cdot 3$

c) $^-6 \cdot (3 + {^-6}) = {^-6} \cdot 3 + {^-6} \cdot {^-6}$

d) $7 \cdot (^-11 + {^-2}) = 7 \cdot {^-11} + 7 \cdot {^-2}$

e) $^-8 \cdot (^-2 + {^-3}) = {^-8} \cdot {^-2} + {^-8} \cdot {^-3}$

f) $^-1 \cdot (^-5 + {^-6}) = {^-1} \cdot {^-5} + {^-1} \cdot {^-6}$

4. Prove that for any integers a, b, and c, $a \cdot (b + c) = a \cdot b + a \cdot c$

In Problems 5–12 the operation symbols ★, □, △, ○ are to be replaced by +, −, ×, or ÷. If two symbols of the same problem are alike, they must be replaced by the same symbol; there are no other restrictions.

5. (4 ★ ⁻6) △ 10 = 1

6. ⁻7 ★ ⁻8 ○ 6 = 7

7. (2 △ ⁻7) □ 6 = ⁻20

8. (⁻8 ★ 4) ★ ⁻2 = 1

9. 6 ○ (3 ★ 2) □ [4 ○ (⁻3 □ ⁻2)] = ⁻14

10. (⁻3 △ ⁻2 △ 5) □ (⁻6 ○ 3) = 0

11. {[(⁻28 △ ⁻6) □ 17] □ ⁻2} △ ⁻1 = 0

12. [(⁻15 △ 2) □ 3] ★ [2 △ (⁻9 ★ 1)] = 10

13. Prove that for any integers a, b, and c, $a \cdot (b \cdot c) = (a \cdot b) \cdot c$.

14.

a) Complete the following table:

a	b	c	$a \cdot (b - c)$	$a \cdot b - a \cdot c$
3	4	5	$3 \cdot (4 - 5) = {}^-3$	$3 \cdot 4 - 3 \cdot 5 = {}^-3$
⁻2	3	6		
2	⁻3	4		
4	2	⁻5		
2	⁻3	⁻7		
⁻11	⁻2	0		

b) Make a conjecture based on the pattern in the table.

15. Prove that for any integers a and b, $a \cdot (b - c) = a \cdot b - a \cdot c$.

16. Complete the following table. How are $m \cdot {}^-n$ and ${}^-(m \cdot n)$ related?

m	n	$m \cdot {}^-n$	${}^-(m \cdot n)$
2	7	$2 \cdot {}^-7 = {}^-14$	${}^-(2 \cdot 7) = {}^-14$
1	⁻2		
⁻2	5		
⁻3	⁻4		

17. Prove that for any integers m and n, $m \cdot {}^-n = {}^-(m \cdot n)$.

18. Prove that for any integers m and n, ${}^-m \cdot {}^-n = m \cdot n$.

19. Show that the set of integers is closed under multiplication.

20. Suppose that p and q are both positive integers; explain why the following statement is true: ${}^-p \cdot {}^-q = |{}^-p| \cdot |{}^-q|$.

21. Is it true that for any integers a, b, c, and d:

$$(a - b) + (c - d) = (a + c) - (b + d)? \text{ Why?}$$

22. Do not perform the multiplication but state YES if the product is a positive integer and NO if it is not.

a) (⁻320) · (⁻650) · (⁻709) · (⁻1100)

b) (2) · (⁻5001) · (3201)

c) (⁻1762) · (⁻1980) · (0) · (705)

d) (⁻676) · (782) · (⁻124) · (562) · (⁻131)

23. Find an integer x that makes the statement true in each of the following:

a) $x \div 7 = 3$
b) $8 \div x = 4$
c) $4x \div {}^{-}4 = 2$
d) ${}^{-}30 \div x = {}^{-}10$
e) ${}^{-}16 \div {}^{-}x = {}^{-}4$
f) $33 \div {}^{-}3x = 1$
g) $x \div {}^{-}6 = {}^{-}3$
h) $({}^{-}12 \cdot x) \div {}^{-}3 = 4$

24. Show that for all nonzero integers a, $0 \div a = 0$.

25. Show that for all integers a and b, if $a \cdot b = 0$, then either $a = 0$ or $b = 0$.

26. Prove that for any integers a and b such that $a \div b$ is defined, $a \div ({}^{-}b) = {}^{-}(a \div b)$. That is, show that

$$\frac{a}{{}^{-}b} = {}^{-}\left(\frac{a}{b}\right).$$

27. Prove that if $a \div b$ is defined for integers a and b, then $(a \div b) \cdot b = a$.

■ **28.** Refer back to Problem 24 of Exercise Set 8.1. Create a definition for multiplication that uses ordered pairs to represent integers.

29. The following example is part of a lesson from a seventh-grade textbook. Three other flowcharts are also supplied in the text to illustrate the products $6 \cdot 3$, ${}^{-}6 \cdot 3$, and ${}^{-}6 \cdot {}^{-}3$.

a) Write the three flowcharts that correspond to these three products.
b) What would a student conclude about multiplication of integers from these flowcharts?

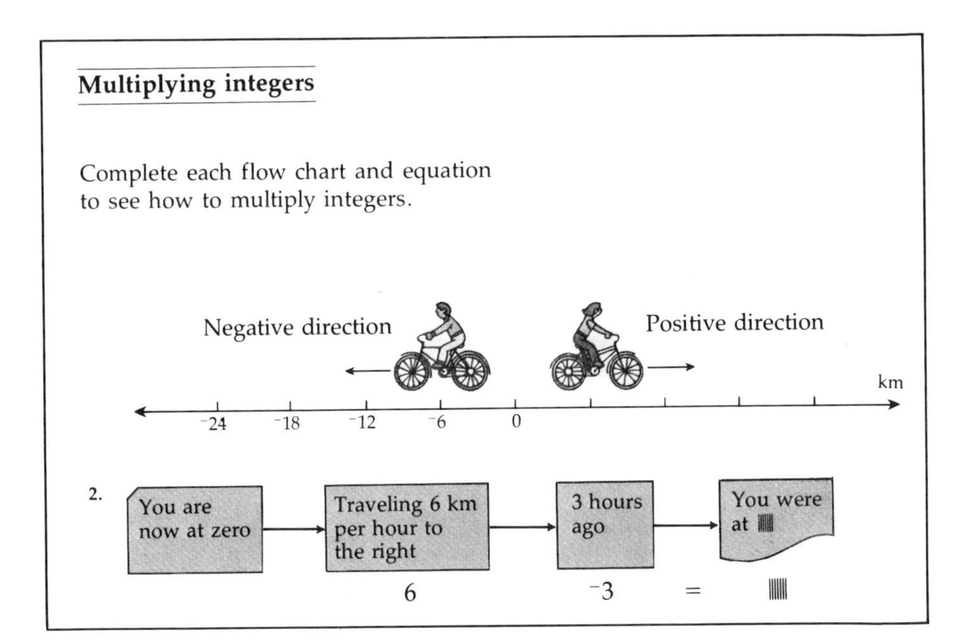

8.3 AIMS

You should be able to:

a) Define and apply the definitions of the "less than" and "greater than" relations for integers.
b) Verify or prove the following properties for inequalities: trichotomy, transitivity, addition, and multiplication.

8.3 ORDER OF INTEGERS

DEFINITION. The definitions for the order relations of whole numbers can be extended to include the integers.

DEFINITION. For any integers a and b, a is *less than* b ($a < b$) if and only if there is a positive integer c such that $a + c = b$. That is, $a < b \leftrightarrow a + c = b$, where c is a positive integer.

As before, b is *greater than* a ($b > a$) if and only if a is *less* than b ($a < b$).

It is not difficult to see that any positive integer is greater than its additive inverse.

Example 1 Prove that if a is a positive integer, then a is greater than ^-a.

Solution We must show that there is a positive integer c such that $^-a + c = a$.

1. For c we use $a + a$, which is a positive integer because of closure for whole-number addition.
2. Then $^-a + (a + a) = (^-a + a) + a$ [by the associative property for integers]
   $= 0 + a = a$ [by the definition of additive inverse and the identity 0]
3. Therefore $a > {}^-a$ [by the definition of "greater than" for integers]

Try Exercise 17

17.

a) Use the definition to show that $^-5 > {}^-16$.
b) Put the following in order from the smallest to the largest:

$0, 1, {}^-2, 17, {}^-30, 45, {}^-17, 2.$

PROPERTIES OF INTEGER RELATIONS. A number of properties for inequalities follow directly from the definitions of "less than" and "greater than."

Example 2 Show that the trichotomy property holds for integers:

For any integers a and b, one and only one of the following is *true*: $a = b$, $a < b$, $a > b$.

1. We look at the difference $a - b$ [by all differences are possible in the set of integers]
2. $a - b$ is either 0, a positive integer, or a negative integer [by closure for subtraction]
3. If $a - b = 0$, then $a = 0 + b$ [by definition of subtraction]
4. $a = b$ [by definition of 0]
5. If $a - b = k$ is a positive integer, then $a = b + k$ and $a > b$ [by definition of $<$ relation for integers and the definition of subtraction]
6. If $a - b = {}^-k$, where k is a whole number, then $a + k = b$ and $a < b$ ____________

Try Exercise 18

Some properties for the "less than" relation are almost identical to those for the "equals" relation. There are some important differences, however, as we see in the next example.

Example 3 Prove that the following multiplication properties hold for integers:

> **For any integers a, b, and c:**
>
> **a) if $a < b$ and $c > 0$, then $a \cdot c < b \cdot c$**
>
> **b) if $a < b$ and $c < 0$, then $a \cdot c > b \cdot c$.**

Proof of (a) is similar to the proof for whole numbers and is left for the exercises.

Proof of (b):

1. $a < b \leftrightarrow a + k = b$, where k is a positive integer [by the definition of "less than"];
2. $c \cdot (a + k) = c \cdot a + c \cdot k = c \cdot b$ [by the distributive law and the "equals" property];
3. $c \cdot a + (c \cdot k + {}^{-}(c \cdot k)) = c \cdot b + {}^{-}(c \cdot k)$ [by the "equals" property and the definition of integers];
4. $c \cdot a + 0 = c \cdot a = c \cdot b + {}^{-}(c \cdot k)$ [by the definition of additive inverse and the identity 0];
5. since c is a negative integer and k is a positive integer, the product $c \cdot k$ is a negative integer, and so ${}^{-}(c \cdot k)$ must be a positive integer;
6. therefore $c \cdot b < c \cdot a$ or $c \cdot a > c \cdot b$ [by the definition of "less than"].

This theorem shows that in an inequality there is an important difference between multiplying by a negative integer and multiplying by a positive one.

The following properties can be proved in exactly the same way as for whole numbers, and their proofs are left for exercises.

Transitive property: For any integers a, b, and c, if $a < b$ and $b < c$, then $a < c$.

Addition property: For any integers a, b, and c, if $a < b$, then $a + c < b + c$.

Try Exercise 19

APPLICATIONS. These properties for the order relations along with those already summarized in the previous two sections can, of course, be used to deduce still more properties of the integers. What is just as important, they form the basis for finding in a systematic way those integers that transform an open sentence in the variable x into a true sentence. This problem is often encountered in algebra.

18. Justify step 6 of the proof in Example 2.

19. Replace the blank in each of the following with one of the relations: less than, greater than, equals.

a) $6 < 11 \rightarrow 2 \cdot 6$ ____ $2 \cdot 11$
b) $6 < 11 \rightarrow 0 \cdot 6$ ____ $0 \cdot 11$
c) $6 < 11 \rightarrow {}^{-}5 \cdot 6$ ____ ${}^{-}5 \cdot 11$

From a seventh-grade textbook

Comparing integers

1.

A temperature of 20 degrees above zero is ___?___ (higher, lower) than a temperature of 25 degrees below zero.

20 ● ⁻25

20. Justify steps (1) through (4) of the solution of Example 4.

21. Find all the integers x that make the following statement true: $|x| \leqslant 3$.

By this time you should be able to supply the reasons that justify each of the steps in the following.

Example 4 Find all the integers x that make the following true: $3x + 7 = {}^-23$.

Solution

Since $3x + 7 = {}^-23$, then

$(3x + 7) + {}^-7 = {}^-23 + {}^-7$; 1. ____________

$3x + (7 + {}^-7) = {}^-30$; 2. ____________

$3x + 0 = 3x = 3 \cdot ({}^-10)$; 3. ____________

$x = {}^-10$. 4. ____________

Try Exercise 20

Example 5 Find all the integers x that make the following true: $7x + 3 = 4$.

Solution Since $7x + 3 = 4$, then

$$(7x + 3) + {}^-3 = 4 + {}^-3;$$
$$7x + (3 + {}^-3) = 1;$$
$$7x + 0 = 1.$$

There is no integer x such that $7x = 1$, and there is no solution in the set of integers.

Example 6 Find the integers x that make the following statement true: $|x| < 5$.

Solution The cases when x is nonnegative and when it is negative have to be considered separately.

If $x \geqslant 0$, then $|x| = x$ and $x < 5$. Now, $x \geqslant 0$ and also $x < 5$, so x is either 0, 1, 2, 3, or 4.

If $x < 0$, then $|x| = {}^-x$ and ${}^-x < 5$, so $x > {}^-5$. Now, x is less than 0 and also greater than ${}^-5$, so x is either ${}^-4$, ${}^-3$, ${}^-2$, or ${}^-1$.

The set of all the solutions is $\{{}^-4, {}^-3, {}^-2, {}^-1, 0, 1, 2, 3, 4\}$.

Try Exercise 21

EXERCISE SET 8.3

1. In each of the following verify the addition property for inequalities: If $a < b$, then $a + c < b + c$.

a) $5 < 13$, $c = 0$

b) ${}^-6 < {}^-3$, $c = {}^-2$

c) ${}^-10 < 32$, $c = 3$

d) $17 < 34$, $c = {}^-17$

2. Prove that for any integers a, b, and c, if $a < b$, then $a + c < b + c$.

3. Because of the "equals" property for any integers a, b, c, and d, if $a = b$ and $c = d$, then $a + c = b + d$. Does an analogous result hold for inequalities? If $a < b$ and $c < d$, is it true that $a + c < b + d$? Give either a proof or a counterexample to justify your answer.

4. Complete the following table; replace the relation R of the last column by one of the following: $<$, $>$, $=$, to make the statement correct.

a	b	$a < b$	c	$a \cdot c \text{ R } b \cdot c$
5	8	$5 < 8$	$^-3$	
$^-3$	4		$^-2$	
$^-7$	$^-1$		$^-4$	
$^-12$	$^-11$		0	
0	5		2	
$^-13$	2		$^-1$	

5. Prove that for any integers a, b, and c, if $a < b$ and $c > 0$, then $a \cdot c < b \cdot c$.

6. Prove the transitive law for the "less than" relation for integers: for any integers a, b, and c, if $a < b$ and $b < c$, then $a < c$.

7. In each of the following find all of the integers x that make the statement true:
a) $|x| \leqslant 6$ b) $|x - 2| < 5$

■ **8.** Find three integers x that satisfy each of the following inequalities:
a) $(x - 5) \cdot (x - 3) > 0$ b) $(x - 4) \cdot (x - 2) > 0$

■ **9.** Solve Problem 8 for:
a) $(x - {}^-8) \cdot (x - 7) > 0$ b) $(x - {}^-2) \cdot (x - 3) > 0$

■ **10.** How many integers x can you find that satisfy the following inequalities?
a) $(x + 3) \cdot (x + 6) < 0$ b) $(x + 7) \cdot (x + 4) < 0$

■ **11.** Solve Problem 10 for:
a) $(x - 1) \cdot (x - 2) < 0$ b) $(x - 4) \cdot (x - 5) < 0$

8.4 THE RATIONAL NUMBERS

With the extension of the whole numbers to the nonnegative rational numbers in Chapter 6 we gained closure for division, and in this chapter, with the extension to the integers, we gained closure for subtraction. We now want the best of all possible worlds—closure for both operations. The plan is similar to what we did for nonnegative rational numbers, except now we start with the integers and define the rational numbers to be all those that can be represented as $\frac{a}{b}$, where $b \neq 0$ and a and b are integers.

It is not necessary to work through all of the details of this extension because we have Chapter 6 to build on and are able to use the fact that the system of integers has all of the whole-number properties; in fact, it has more because of the closure of subtraction. We are now in a position to benefit from the analysis of each of the systems considered so far.

8.4 AIMS

You should be able to:

a) Define a rational number.
b) Define and apply the definitions for addition, subtraction, multiplication, division, and equality for rational numbers.
c) Demonstrate that there is closure for subtraction on the set of rationals and closure for division on the set of nonzero rationals.
d) Determine which of the order relations: less than, greater than, or equals holds for any two rational numbers.
e) Make suitable modifications to the proofs for the properties of nonnegative rationals and integers, so that corresponding properties can be shown to hold for rational numbers.

22.

a) Show that $\frac{{}^{-}6}{8} = \frac{{}^{-}9}{12}$.

b) Show that $\frac{6}{{}^{-}8} = \frac{12}{{}^{-}16}$.

If we are able to prove a particular theorem in the system of nonnegative rational numbers by applying certain properties of whole numbers, then we are able to prove the identical theorem for rational numbers because the integers have the same properties needed in the proof as did the whole numbers. We can use definitions that are analogous to those given for nonnegative rational numbers and then proceed by making minor changes such as replacing "nonnegative rational" with "rational" and justifying the arguments in a proof by means of properties of integers instead of whole numbers.

If a theorem can be proved by means of certain properties of a mathematical system, then it will automatically be true in any other mathematical system with identical properties. This is an important reason for analyzing and comparing the properties of various mathematical systems.

DEFINITION OF A RATIONAL NUMBER. The definition of a rational number is that used for a nonnegative rational number with the suitable modifications we have mentioned. The reason for the definition is again to supply the "missing quotients."

> **DEFINITION. A *rational number* is any number that can be represented in the form $\frac{a}{b}$, where a and b are any integers and $b \neq 0$.**

The integers are all included in this definition since, for example, ${}^{-}3 \div 1 = \frac{{}^{-}3}{1}$ is the integer ${}^{-}3$, and ${}^{-}5 \div 1 = \frac{{}^{-}5}{1}$ is the integer ${}^{-}5$. The nonnegative rationals, for example, $\frac{2}{3}$ and $\frac{4}{5}$ are included together with such numbers as $\frac{{}^{-}2}{3}$ and $\frac{{}^{-}4}{5}$, which are neither integers nor nonnegative rationals.

The definition for equality remains the same, except that this time it applies to rational numbers.

> **DEFINITION. The rational number $\frac{a}{b}$ is *equal to* the rational number $\frac{c}{d}$ if and only if $a \cdot d = b \cdot c$.**

Example 1 Show that $\frac{{}^{-}6}{8}$ is equal to $\frac{}{{}^{-}12}$.

Because ${}^{-}6 \cdot {}^{-}12 = 8 \cdot 9$, since $72 = 72$, they are equal by the definition of equality for rational numbers.

Try Exercise 22

Rational numbers can be named with fractions in an infinite number of ways, just as the nonnegative rationals can. However there is, as before, a simplest representation for a rational number.

DEFINITION. The fraction "$\frac{a}{b}$" for a rational number, where a and b are both integers, $b \neq 0$, is the *lowest-terms fraction* if and only if GCD$(a, b) = 1$, that is, a and b are relatively prime and $b > 0$.

The only change in this definition is that the denominator b must be a positive integer. With this definition, the lowest-terms fraction for $\frac{2}{3}$ is still $\frac{2}{3}$ and the lowest-terms fraction for the rational $\frac{^-9}{6}$ is $\frac{^-3}{2}$, while that for $\frac{7}{^-49}$ is $\frac{^-1}{7}$.

Try Exercise 23

These definitions pave the way for proving that if a/b is any rational number, then for any integer p ($p \neq 0$), $\frac{a \cdot p}{b \cdot p} = \frac{a}{b}$, and that if, in addition, $\frac{a}{b}$ is represented by the lowest-terms fraction, then all the fractions equivalent to "$\frac{a}{b}$" can be found in this way. The proofs for these statements follow directly from the definitions for equality and will not be repeated here since they appear in Chapter 6. The "equals" relation, as defined for the rational numbers, can likewise be shown to be an equivalence relation just as it was for the nonnegative rationals.

Definition of the Operations

ADDITION. The definition for addition is the same as for the nonnegative rationals with the exception of the word *rational*.

ADDITION. For any rational numbers $\frac{a}{b}$ and $\frac{c}{d}$, their sum is $\frac{a}{b} + \frac{c}{d} = \frac{a \cdot d + b \cdot c}{b \cdot d}$.

By the same arguments as were used for the nonnegative rational numbers, we can show that this definition guarantees closure for the addition operation, that equivalent fractions can be used to represent the individual rational numbers of a sum without changing the result, and that the operation of addition is both commutative and associative. All of the proofs now depend upon the corresponding properties of the integers and upon the definitions given so far in this development.

23. Find the lowest-terms fraction for each of the following:

a) $\frac{^-11}{22}$

b) $\frac{^-12}{^-48}$

c) $\frac{7}{^-14}$

24. Add:

a) $\frac{^-3}{11} + \frac{^-2}{11}$

b) $\frac{0}{1} + \frac{^-5}{12}$

c) $\frac{^-3}{8} + \frac{2}{5}$

Example 2 Add $\frac{3}{5}$ and $\frac{^-5}{8}$.

By the definition: $\frac{3}{5} + \frac{^-5}{8} = \frac{3 \cdot 8 + 5 \cdot {^-5}}{5 \cdot 8}$

$$= \frac{24 + {^-25}}{40} = \frac{1}{40}.$$

It is probably more efficient to find equivalent fractions with the same denominator.

Example 3 Add $\frac{3}{9}$ and $\frac{5}{^-27}$.

Solution We use the denominator 27:

$$\frac{3}{9} = \frac{3 \cdot 3}{9 \cdot 3} = \frac{9}{27}, \qquad \frac{5}{^-27} = \frac{5 \cdot {^-1}}{^-27 \cdot {^-1}} = \frac{^-5}{27}.$$

The sum is: $\frac{9}{27} + \frac{^-5}{27} = \frac{9 + {^-5}}{27} = \frac{4}{27}$. It can be obtained by using the rule for addition with like denominators for nonnegative rationals, that can also be established for the rational numbers.

Try Exercise 24

As before, $\frac{0}{1}$ is the additive identity, and its behavior can be easily established by means of the appropriate definitions.

SUBTRACTION AND ADDITIVE INVERSES. The same definition for subtraction has been used for whole numbers, nonnegative rational numbers, and integers and will be used again here. We repeat it merely for completeness.

DEFINITION. For any rational numbers $\frac{a}{b}$ and $\frac{c}{d}$, their *difference* $\frac{a}{b} - \frac{c}{d}$ equals $\frac{e}{f}$ if and only if

$$\frac{a}{b} = \frac{e}{f} + \frac{c}{d}.$$

Every rational number also has a unique additive inverse, which is a property not enjoyed by the nonnegative rational numbers.

Example 4 Show that every rational number $\frac{a}{b}$ has a unique additive inverse.

1. $\frac{^-a}{b}$ is a likely candidate for the required inverse. It is a rational number because $b \neq 0$ and a has an additive inverse ^-a.

2. We look at the sum of $\frac{a}{b}$ and $\frac{^-a}{b}$ to see if it is 0; if it is, we are done. Thus,

$$\frac{^-a}{b} + \frac{a}{b} = \frac{a}{b} + \frac{^-a}{b} = \frac{a + {}^-a}{b} = \frac{0}{b} = 0$$

[because of the definition of addition for rational numbers and integers]. Since the sum is 0, then $\frac{a}{b}$ and $\frac{^-a}{b}$ are additive inverses of each other.

Because $\frac{^-a}{b} = \frac{a}{^-b}$, we can say $^-\left(\frac{a}{b}\right) = \frac{^-a}{b} = \frac{a}{^-b}$.

Try Exercise 25

25. State the additive inverse for each of the following:

$$\frac{2}{3}, \frac{^-2}{55}, \frac{2}{^-9}, {}^-\left(\frac{2}{9}\right).$$

CLOSURE FOR SUBTRACTION. Since all rational numbers have an additive inverse, all subtractions are possible. The same proof that was developed for integers can be suitably modified for rational numbers to show that for any rational numbers a/b and c/d, their difference $a/b - c/d$ is equal to the sum $a/b + {}^-(c/d)$.

If we represent the rational number a/b by r and the rational number c/d by s, then we need to prove that $r - s = r + {}^-s$. The proof of the corresponding result for integers can be used with only minimal changes of the wording and justifications.

Just as with the nonnegative rational numbers, it is usually more convenient to find fractions with a common denominator before subtracting.

26. Subtract:

a) $\frac{4}{5} - \frac{^-2}{7}$

b) $\frac{2}{^-3} - \frac{1}{6}$

Example 5 Subtract $\frac{^-3}{5} - \frac{2}{7}$.

Solution

$$\frac{^-3}{5} = \frac{^-3 \cdot 7}{5 \cdot 7} = \frac{^-21}{35}, \qquad \frac{2}{7} = \frac{2 \cdot 5}{7 \cdot 5} = \frac{10}{35}.$$

We change the difference to the appropriate sum:

$$\frac{^-21}{35} - \frac{10}{35} = \frac{^-21}{35} + \frac{^-10}{35} = \frac{^-21 + (^-10)}{35} = \frac{^-31}{35}$$

Try Exercise 26

MULTIPLICATION. The definition of multiplication is an extension of the one used for nonnegative rational numbers in Chapter 6.

MULTIPLICATION. If $\frac{a}{b}$ and $\frac{c}{d}$ are any two rational numbers, their *product* is $\frac{a \cdot c}{b \cdot d}$.

27. Multiply:

a) $\frac{^-6}{8} \cdot \frac{3}{2}$

b) $\frac{6}{^-8} \cdot \frac{^-3}{^-2}$

c) $\frac{3}{^-5} \cdot \frac{5}{5}$

d) $\frac{^-7}{8} \cdot \frac{^-1}{^-1}$

Again we can use the same type of arguments as in the case of nonnegative rational numbers to prove that multiplication is closed on the set of rational numbers, that it is a commutative and associative operation, and that the products do not depend upon the particular representation (fraction) of the individual rational numbers. Similarly, the distributive laws can be proved. All proofs and arguments depend upon the corresponding properties for integers instead of whole numbers.

Try Exercise 27

MULTIPLICATIVE INVERSES. The definition for multiplicative inverse is the same as for nonnegative rationals, and we can easily show that every nonzero rational number has such an inverse.

Example 6 Prove that every nonzero rational number $\frac{a}{b}$ has a multiplicative inverse.

Proof

1. Since $a/b \neq 0$, then $a \neq 0$, and by definition $b \neq 0$, so there is a rational number b/a that is also not zero.
2. We look at the product: $\frac{a}{b} \cdot \frac{b}{a} = \frac{a \cdot b}{b \cdot a} = \frac{a \cdot b}{a \cdot b} = \frac{1}{1}$ [by the commutativity of integers and the definition of equality for rational numbers].
3. Therefore $\left(\frac{a}{b}\right)^{-1} = \frac{b}{a}$ and $\left(\frac{b}{a}\right)^{-1} = \frac{a}{b}$ [by the definition of multiplicative inverse].

These multiplicative inverses are crucial in the proof of closure for division on the set of nonzero rational numbers.

DIVISION. Division of two rational numbers can be defined in terms of multiplication in the same way it has been done for whole numbers, nonnegative rationals, and the integers, and it will not be repeated here.

The fact that every rational number (except zero) has a multiplicative inverse is the key to the proof of the following rule, just as it was for the corresponding rule for nonnegative rational numbers.

The quotient of two rational numbers a/b and c/d ($c/d \neq 0$) is given by the product $\frac{a}{b} \cdot \frac{d}{c}$.

That is, $\frac{a}{b} \div \frac{c}{d} = \frac{a}{b} \cdot \frac{d}{c} = \frac{a \cdot d}{b \cdot c}$.

With minor changes, the proof for nonnegative rational numbers applies to the case of rational numbers. It is this theorem that guarantees closure of the division operation for the nonzero rational numbers.

Example 7 Find $3/5 \div {}^{-}2/3$:

$$\frac{3}{5} \div \frac{^{-}2}{3} = \frac{3}{5} \cdot \frac{3}{^{-}2} = \frac{3 \cdot 3}{5 \cdot {}^{-}2} = \frac{9}{^{-}10}.$$

Try Exercise 28

ORDER. We use the same definition for order that was used for whole numbers, nonnegative rational numbers, and integers; there is no need to repeat it here. Just as with nonnegative rational numbers, it may be easier in determining the order relation that holds for two rational numbers if equivalent fractions with a common denominator that is a positive integer are compared.

Suitable adaptations of the proofs concerning the order relations for non-negative rationals and integers are used to prove the following properties:

Trichotomy Property: For any rational numbers a/b and c/d, one and only one of the following relations is true:

$$a/b = c/d, \qquad a/b < c/d, \qquad a/b > c/d.$$

Transitive Law: For any rational numbers a/b, c/d, and e/f, if $a/b < c/d$ and $c/d < e/f$, then $a/b < e/f$.

Addition Property: For any rational numbers a/b, c/d, and e/f, if $a/b < c/d$, then $a/b + e/f < c/d + e/f$.

Multiplication Property: For any rational numbers $a/b < c/d$,

a) if any rational $e/f > 0$, then $\frac{e}{f} \cdot \frac{a}{b} < \frac{e}{f} \cdot \frac{c}{d}$

b) if any rational number $e/f < 0$, then $\frac{e}{f} \cdot \frac{a}{b} > \frac{e}{f} \cdot \frac{c}{d}$.

Proofs of these properties are rather routine, since they essentially repeat the arguments of those done earlier for either the nonnegative rational numbers or the integers. Details for some of them are left for the exercises.

DENSITY. The rational numbers are dense in the entire number line, as were the nonnegative rationals in the half line to the right of 0. Between any two rational numbers there is always an infinite set of other rational numbers.

Try Exercise 29

SUMMARY. In the rational-number system all subtractions and all divisions (except by 0) are possible, so we have taken care of two major deficiencies of

28. Divide:

a) $\frac{^{-}2}{3} \div \frac{^{-}2}{3}$

b) $\frac{^{-}2}{8} \div \frac{1}{^{-}4}$

c) $\frac{27}{^{-}3} \div \frac{^{-}1}{9}$

29.

a) Which is greater: $\frac{^{-}7}{12}$ or $\frac{9}{^{-}14}$?

b) Find a rational number between $\frac{^{-}1}{3}$ and $\frac{2}{16}$.

the whole-number system. We shall see that even this system is not without flaws. The basic properties of the rational-number system are:

For the addition operation:	**For the multiplication operation:**
1. There is closure.	**6. There is closure.**
2. Commutativity holds.	**7. Commutativity holds.**
3. Associativity holds.	**8. Associativity holds.**
4. There is an additive identity (0).	**9. There is a multiplicative identity (1).**
5. Each element has an additive inverse.	**10. Each nonzero element has a multiplicative inverse.**

11. The distributive law of multiplication over addition holds.

Because subtraction is defined in terms of addition, while division is defined in terms of multiplication, there is no need to list these operations separately. Other properties can be deduced from these basic properties that characterize the rational numbers.

EXERCISE SET 8.4

1. For each of the following, find the lowest-terms fraction and three other fractions that represent the same rational number:
a) $^-3/^-16$ b) $^-5/^-7$ c) $^-6/9$ d) $^-5/11$

2. Prove that if a/b is any rational number, then for any integer p $(p \neq 0)$, $\frac{a}{b} = \frac{a \cdot p}{b \cdot p}$.

3. Prove that the equals relation for rational numbers is an equivalence relation.

4. Find the additive inverse and the multiplicative inverse of each of the following:
a) $3/16$ b) $^-5/^-11$ c) $^-3/8$
d) $2/^-9$ e) 17 f) $^-3$

5. For each of the following, answer TRUE if the statement is always true and FALSE otherwise.
a) In the set of rational numbers, division is always possible. ______
b) For any rational numbers a and b $(b \neq 0)$, $^-a \div b$ is the same as $a \div {}^-b$. ______
c) $9/8 > 3/4$ ______
d) $^-5/1 > 2/8$ ______
e) If p and q are two rational numbers such that $p < q$, then $^-p > {}^-q$. ______
f) It is impossible to find a fraction with a denominator of 14 that is equivalent to $^-5/35$. ______
g) The simplest fraction for negative two-thirds is $2/^-3$. ______

6. Absolute value has been defined for integers. Modify the definition so that it applies to rational numbers.

7. Show that for any two rational numbers a/b and c/d, the difference $a/b - c/d$ is also a rational number.

8. Prove that for any rational numbers a/b and c/d,

$$\frac{a}{b} - \frac{c}{d} = \frac{a}{b} + {}^{-}\left(\frac{c}{d}\right).$$

9. Show that the set of rational numbers is closed under multiplication.

10. Prove that the distributive law of multiplication over addition holds for rational numbers, that is, for any rational numbers a, b, and c, $a \cdot (b + c) = a \cdot b + a \cdot c$.

11. Observe that $5/3 < 15/3$ and (for the numerators) $5 < 15$;

$$^{-}3/8 < 6/8 \text{ and } ^{-}3 < 6.$$

Show that for any two rational numbers a/d and b/d (with $d > 0$), $a/d < b/d$ if and only if $a < b$. What can be said about the case when $d < 0$?

12. Find two rational numbers between $^{-}7/2$ and $13/4$.

13. Provide a counterexample to prove that subtraction for rational numbers is not associative.

14. Show that division is not associative for rational numbers.

15. Prove that the quotient of any two rational numbers a/b and c/d ($c/d \neq 0$) is

$$\frac{a}{b} \div \frac{c}{d} = \frac{a \cdot d}{b \cdot c}.$$

16. Find a rational number x that makes the statement true for each of the following:

a) $\frac{3}{4}x - 5 = \frac{1}{4}x + 4$ b) $\frac{^{-}6}{5}x + \frac{12}{5} = \frac{^{-}18}{5}$

17. Prove that if a, b, and c are rational numbers ($a \neq 0$), then there is always a rational number x that will make the following statement true: $a \cdot x + c = b$.

18. Use set-builder notation to describe the set of rational numbers that satisfy the inequality in each of the following:

a) $x + {}^{-}5/8 < 3/4$ b) $x + 1/3 > 1/2$

c) $3x + 4 < 5/8$ d) $5x - 2 > 4$

8.5 THE REAL NUMBERS

IRRATIONAL NUMBERS. Although the rational numbers are dense in the number line, they do not "fill" it; there are still many points for which there is no corresponding rational number. We can easily construct such a number by placing a right triangle along a number line with one end of the hypotenuse at 0, so that the other end will fall at point B of the line. We choose a triangle that has a length of 1 unit for each leg.

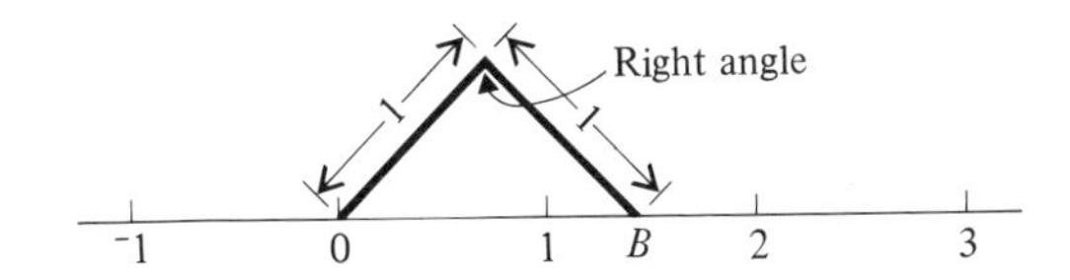

By the Pythagorean property of right triangles, we know that if a and b are each the length of one leg and c is the length of the hypotenuse, then $a^2 + b^2 = c^2$.

8.5 AIMS

You should be able to:

a) Describe the set of real numbers and be able to distinguish between rational and irrational numbers.

b) Use a proof by contradiction to prove that there is at least one irrational number.

c) Approximate an irrational square root by an averaging method.

d) Convert from standard decimal numerals to scientific notation.

Here each length is 1, and we can find what the distance from the point corresponding to 0 to the point B is

$$1^2 + 1^2 = c^2 \qquad \text{or} \qquad 2 = c^2.$$

We use proof by contradiction to show that there is no rational number c such that $c^2 = 2$. Since we write $\sqrt{2}$ to mean the nonnegative number whose square is 2, we will show that $\sqrt{2}$ is not a rational number.

PROOF THAT $\sqrt{2}$ IS NOT A RATIONAL NUMBER

1. We assume that $\sqrt{2}$ is a rational number. That is, there is a rational number a/b such that $\left(\frac{a}{b}\right)^2 = 2$ (where a and b are integers, $b \neq 0$).

2. This means that

$$\frac{a}{b} \cdot \frac{a}{b} = \frac{a \cdot a}{b \cdot b} = 2 \qquad \text{and} \qquad a \cdot a = 2 \cdot b \cdot b$$

(by the definitions of multiplication and equality for rational numbers).

3. We show that the factorization on the left cannot be the same as the one on the right: If the number a has some number x of prime factors, then $a \cdot a$ has $2x$ (an even number) of prime factors. If the number b has some number y of prime factors, then $b \cdot b$ has $2y$ such factors and $2 \cdot b \cdot b$ has $2y + 1$ (an odd number) of prime factors. However, the prime factorization of a number is unique, so this is impossible.

4. We have a contradiction in step 3, so the assumption in step 1 must be false; therefore $\sqrt{2}$ is not a rational number.

A similar proof can be used to show that $\sqrt{3}$, $\sqrt{7}$, $\sqrt{11}$, and so on, are also not rational numbers. Thus, although the statement $x^2 = 2$ is expressed in terms of rational numbers, there is no rational number that makes the statement true.

DEFINITION OF REAL NUMBERS. Since there are points on a number line with no corresponding rational numbers, we extend the rational-number system to include the numbers corresponding to the remaining points.

DEFINITION. The set of *real numbers* is the set of numbers such that there is a one-to-one correspondence between the elements of the set and the points of the number line. This set includes all of the *rational* numbers, and the remaining numbers are called *irrational*.

From Chapter 7 we know that any terminating decimal or a nonterminating repeating decimal is a numeral for a rational number. We assume that each of the remaining decimals represents an irrational number and that each irrational number has such a decimal. The real numbers are thus the set of all numbers represented by either terminating or nonterminating decimals. Nonrepeating decimals have already been constructed in an earlier exercise set.

Example 1 Construct a nonterminating nonrepeating decimal using the digits 3 and 4.

There are many ways to do this; here is one:

$$0.343343334333343333334\ldots$$

Each time one more numeral for 3 has been placed between the numerals for 4.

Try Exercise 30

30. Construct a nonterminating nonrepeating decimal using only the digits 2 and 3.

Some irrational numbers have special short names that arise in computation, and we have seen that one such is $\sqrt{2}$, which comes from the statement $x^2 = 2$. Another is $\sqrt[3]{17}$ (the principal cube root) and it comes from $x^3 = 17$. For some others we use a letter as, for example, the Greek letter π which represents an important irrational number we shall use in Chapter 10.

It is obviously impossible to do calculations with nonterminating nonrepeating decimal numerals; instead we approximate an irrational number with a terminating decimal or fraction. For example,

$$\pi \text{ is approximated by } 3.14 \text{ and } \frac{22}{7};$$

$$\sqrt{2} \text{ is approximated by } 1.41,\ 1.414, \text{ and } 1.4142.$$

The decimal used depends upon how close an approximation of the irrational number is needed.

31. How do we know that $\sqrt{2}$ is between 1 and 2?

APPROXIMATING SQUARE ROOTS. We can find such an approximation for $\sqrt{2}$ by using a method for approximating the square root of any number known as *Newton's method*:

1. Begin by making a guess. Any positive rational number will do, but since we know that $1 < \sqrt{2} < 2$ (that is, $\sqrt{2}$ is between 1 and 2), we choose the guess between 1 and 2; we take the average, which is 1.5 (the first estimate).

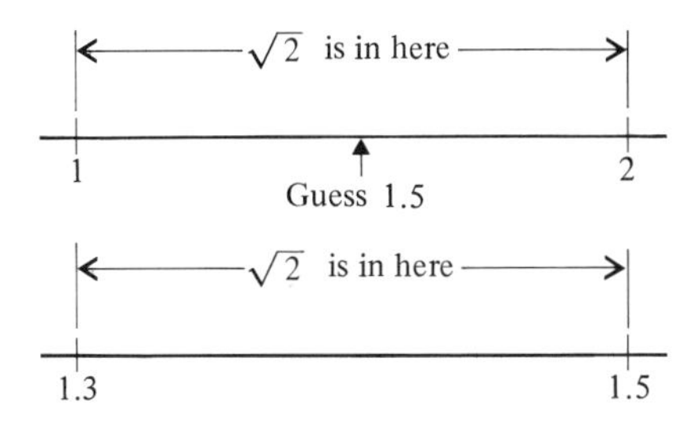

Try Exercise 31

32.
a) Obtain the next two approximations by repeating step 4.
b) If you have a calculator with a $\sqrt{x}$ key, find $\sqrt{2}$. Otherwise, square the number you have to find how close you are.

2. Divide 2 by the guess 1.5:

$$\frac{2}{1.5} \approx 1.3.$$

If the guess had been exactly $\sqrt{2}$, the quotient and the divisor would have been the same. Since the quotient is less than the divisor, the guess 1.5 is too large, and we know that $\sqrt{2}$ is between 1.3 and 1.5.

3. For the next guess we use the average of 1.3 and 1.5 (the midpoint of the interval):

$$\frac{1.3 + 1.5}{2} = 1.4 \text{ (2nd estimate)}.$$

We divide 2 by the second estimate:

$$\frac{2}{1.4} \approx 1.43.$$

$\sqrt{2}$ is in here

1.40 1.43

4. We find the midpoint of this smaller interval and divide 2 by it:

$$\frac{1.4 + 1.43}{2} = 1.415 \text{ (3rd estimate)}.$$

$$\frac{2}{1.415} \approx 1.413.$$

$\sqrt{2}$ is in here

1.413 1.415

5. This last step is repeated until the approximation is as close as desired. Each time, the interval containing $\sqrt{2}$ becomes smaller; the last interval in step 4 above is only 0.002 in width.

Try Exercise 32

Various methods, such as the use of infinite series, have been developed for approximating irrational numbers. The number π* can be approximated by the following infinite series:

$$\pi = 3 + \frac{4}{2 \cdot 3 \cdot 4} - \frac{4}{4 \cdot 5 \cdot 6} + \frac{4}{6 \cdot 7 \cdot 8} - \frac{4}{8 \cdot 9 \cdot 10} + \cdots.$$

* π cannot be named $\sqrt[n]{a}$ for any index n or rational a. This means that π is not the solution of any equation of the form $x^n = a$. Because it is not the solution of any polynomial equation, it is said to be nonalgebraic.

The terms of this series continue in this pattern, and enough of them are used to get the desired degree of approximation. This series was used to find the following approximations:

Number of terms	1	2	3	4
Approximation to π	3	3.1667	3.1334	3.1453

Try Exercise 33

33.

a) Obtain the next two approximations for the number π.

b) If your calculator has a $\boxed{\pi}$ key, compare your result with that given by the calculator.

PROPERTIES OF THE REAL-NUMBER SYSTEM. We state without proof that all of the discussed properties of the rational-number system are retained in the real-number system. That is, addition is commutative and associative, the distributive law holds, and so on.

The real numbers have an additional property called *completeness* (which we shall not define precisely); it guarantees that there is a number for every point of a number line (we assumed this earlier). It thus follows that every positive number has a square root, a cube root, and so on. The system of rational numbers does not have this important property.

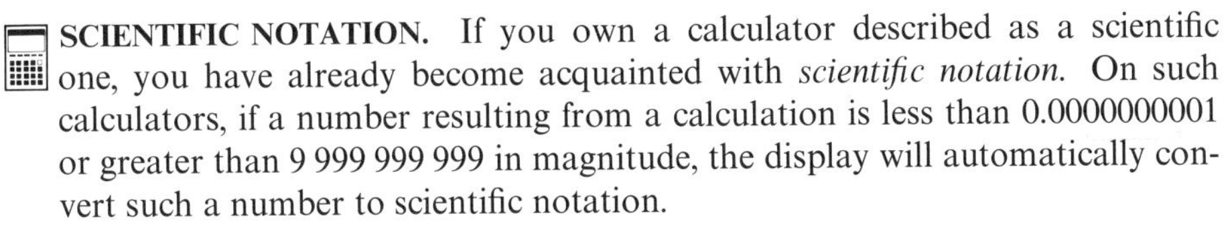

SCIENTIFIC NOTATION. If you own a calculator described as a scientific one, you have already become acquainted with *scientific notation.* On such calculators, if a number resulting from a calculation is less than 0.0000000001 or greater than 9 999 999 999 in magnitude, the display will automatically convert such a number to scientific notation.

Try Exercise 34

34.

a) Find $(125\,678)^2$ with your calculator.

b) Do you get an error signal? Why?

c) If you do not get an error signal, what was on the screen?

DEFINITION. ***Scientific notation*** **consists of a decimal numeral for a number between 1 and 10 (the mantissa), followed by a multiplication sign and** 10^n**, where** n **can be any integer.**

On a calculator with the scientific-notation feature the mantissa will be displayed on the screen followed by n, which is understood to mean 10^n.

Example 2 Divide $\dfrac{0.03785}{999\,999\,999}$.

On a 10 digit-screen the answer is:

$$\underset{\text{mantissa}}{\underset{\uparrow}{3.785000004}} \qquad \underset{n}{\underset{\uparrow}{-11}}$$

This means $3.785000004 \times 10^{-11}$ (a very small number).

35. Try the quotients of Example 2 and Example 3 on your calculator. Can you account for the results?

Example 3 Divide $\dfrac{127\,756}{0.0000009}$.

On a 10-digit screen the answer is:

$$\underset{\text{mantissa}}{\underset{\uparrow}{1.419511111}} \qquad \underset{n}{\underset{\uparrow}{11}}$$

This means $1.419511111 \times 10^{11}$ (a very large number).

Try Exercise 35

Scientific notation is most convenient for writing very large and very small numbers. To convert from a standard numeral to scientific notation we can multiply by the identity for multiplication.

Example 4 Find scientific notation for 382 400.

Solution We need a mantissa of 3.824; we divide and multiply by 10^5 (which is the same as multiplying by 1):

$$\frac{382\,400}{1} \cdot \frac{10^5}{10^5} = \frac{382\,400}{10^5} \cdot \frac{10^5}{1} = 3.824 \times 10^5.$$

Convert to scientific notation. Show how you multiplied by 1:

36. 841 500

37. 0.008952

Example 5 Find scientific notation for 0.000000412.

Solution We need a mantissa of 4.12; we multiply and divide by 10^7 (which is the same as multiplying by 1):

$$\frac{0.000000412}{1} \cdot \frac{10^7}{10^7} = \frac{(0.000000412 \times 10^7)}{1} \cdot \frac{1}{10^7} = 4.12 \times 10^{-7}.$$

The factor $\dfrac{1}{10^7}$ can be written as 10^{-7} because it is agreed that the multiplicative inverse of 10 can be written as either $\dfrac{1}{10}$ or 10^{-1}. Similarly the multiplicative inverse of 100 is $\dfrac{1}{100}$ or $\dfrac{1}{10^2}$ or 10^{-2}, and that for 1000 is $\dfrac{1}{1000}$ or $\dfrac{1}{10^3}$ or 10^{-3}. This is consistent with the usual positive integer exponents.

Try Exercises 36 and 37

To multiply or divide numbers expressed in scientific notation, we can compute with the numbers less than 10 separately and then calculate with the powers of ten, as in the following examples:

Example 6 Multiply $(3.1 \times 10^5) \times (4.5 \times 10^{-3})$.

Solution This is rearranged because the commutative and associative laws hold:

$$3.1 \times 4.5 \times 10^5 \times 10^{-3} = 13.95 \times 10^2$$

$\left(\text{since } 10^5 \times 10^{-3} = \dfrac{10^5}{1} \times \dfrac{1}{10^3} = 10^2\right)$. But this is not in scientific notation, so we write

$$\frac{13.95}{1} \times \frac{10}{10} \times 10^2 = 1.395 \times 10^3.$$

On a scientific calculator the answer would be given immediately in scientific notation:

$$(3.1 \quad 05) \times (4.5 \quad -03) = 1.395 \quad 03.$$

Example 7 Divide $\dfrac{3.41 \times 10^5}{1.1 \times 10^{-3}}$.

$$\frac{3.41 \times 10^5}{1.1 \times 10^{-3}} = \frac{3.41}{1.1} \times \frac{10^5}{10^{-3}}$$
$$= 3.1 \times 10^8,$$

since $10^5 \div \dfrac{1}{10^3} = 10^5 \times 10^3 = 10^8$.

Try Exercises 38 and 39

Scientific notation is very helpful in estimating, since the numbers in a product or quotient can be rounded and expressed in scientific notation and a quick calculation yields an approximate answer.

Dinosaurs first appeared on earth about 220 000 000 years ago. A number as large as this is often written in scientific notation.

Standard numeral		Scientific notation		
220 000 000	=	2.2	×	10^8
		a number between 1 and 10		a power of ten

What is the standard numeral for 3.6×10^7?

$3.6 \times 10^7 = 3.6 \times 10\ 000\ 000$
$= 36\ 000\ 000$

38. Multiply $(6.2 \times 10^8) \cdot (3.8 \times 10^{-3})$

39. Divide $\dfrac{8.24 \times 10^9}{2.5 \times 10^{-2}}$

EXERCISE SET 8.5

1. Prove that $\sqrt{3}$ is not a rational number.

2. Prove that $\sqrt{5}$ is not a rational number.

3. Explain why the method of proof used in the text will not work if you try to prove that $\sqrt{4}$ is irrational.

4. Which of the following have rational square roots? Why?

a) $\frac{^-239}{625}$ b) $\frac{14}{11}$ c) $\frac{25}{36}$ d) $\frac{22}{17}$

5. Approximate to four decimal places; use the method of averaging and dividing:
a) $\sqrt{3}$ b) $\sqrt{5}$ c) $\sqrt{23}$ d) $\sqrt{41}$

6. Make a sketch of the intervals found for the successive steps in the approximation of $\sqrt{3}$.

7. Repeat Exercise 6 for $\sqrt{5}$.

8. Is it possible for a negative number to have a square root? A fourth root? Why?

9. Make a flowchart for Newton's method of finding square roots.

10. Suppose that you did not know how to find square roots by Newton's method but wished to find $\sqrt{7}$. You know that $2 < \sqrt{7} < 3$. Devise a method that does not involve averaging.

11. Use your method of Problem 10 to find an approximation to $\sqrt{11}$.

12. Is it possible to subtract a rational number from a rational number and get an irrational number for the difference? Why?

13. Is the sum of two irrational numbers always an irrational? What about the product of two irrational numbers?

14. In the real number system it is possible to solve many equations such as $x^2 = 7$, $x^3 = 31$, and so on. Show that there is no real number x that will make the equation $x^2 + 1 = 0$ true.

15. Write standard decimal numerals for:
a) 6.3×10^6 b) 8.375×10^5
c) 7.25×10^{-5} d) 9.3×10^{-7}

16. Convert to scientific notation:
a) 61 400 000 b) 751 000
c) 0.00895 d) 0.00000214

Simplify and write scientific notation for your answer. A calculator will be helpful even if it does not have the scientific-notation feature.

17. $(6.3 \times 10^7) \cdot (2.1 \times 10^{-5})$

18. $(9.13 \times 10^{-4}) \cdot (8.2 \times 10^3)$

19. $\frac{3.26 \times 10^8}{2.13 \times 10^5}$

20. $\frac{9.8 \times 10^4}{3.1 \times 10^{-2}}$

21. Determine which of the following numbers is smaller without using either a square-root or a cube-root key of the calculator:

$$\sqrt[2]{8.7025}, \quad \sqrt[3]{22.02}$$

22. Without the use of the calculator put the following numbers in order from smallest to largest. Check with your calculator.

$$\frac{1}{1 + \sqrt{6.7}}, \quad \frac{1}{\sqrt{4.3} + \sqrt{7.9}}, \quad \frac{1 - \sqrt{3}}{\sqrt{6} - 1}$$

8.6 REAL-VALUED FUNCTIONS

Relations are defined as sets of ordered pairs, although we often emphasize the rule used to determine which pairs are members of a relation instead of the pairs themselves. Relations on the real numbers are subsets of $R \times R$, where R stands for the set of real numbers. To picture $R \times R$, we draw a pair of perpendicular axes to locate the origin (0, 0), and every point of the plane is thus associated with one ordered pair of real numbers.

Example 1 Graph the relation "greater than" on the set of real numbers.

The ordered pairs are:

$$\{(x, y) \mid x \text{ and } y \text{ are real numbers and } x > y\}.$$

We draw and label an x-axis and a y-axis. It helps to find some points such that $x = y$, which lie on the dashed line shown. Any point to the right of this line, for example $(5, {}^-3)$, satisfies the condition $x > y$, and we shade the half-plane as indicated.

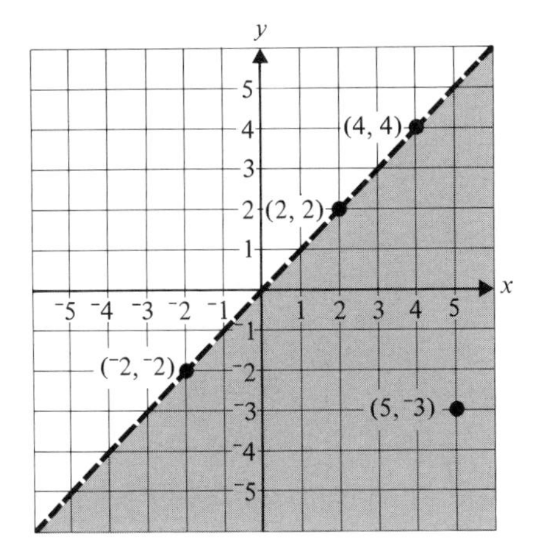

Try Exercise 40

FUNCTIONS AND THE CALCULATOR. Because a function is a relation in which no two ordered pairs have the same first coordinate and different second coordinates, we can think in terms of a hypothetical function machine that works on an element of the domain and gives out an element of the range.

A hand calculator is actually such a machine. Both the domain and the range for this machine are necessarily a subset of the rational numbers that can be expressed by a terminating decimal of some specified number of decimal places, which varies with the calculator. In addition to the keys that correspond to the four arithmetic operations, a calculator may also have such keys as $\boxed{\sqrt{x}}$, $\boxed{x^2}$, $\boxed{1/x}$, and $\boxed{\%}$. More elaborate calculators have many such function keys.*

8.6 AIMS

You should be able to:

a) Distinguish between relations and functions on the set of real numbers.
b) Given a rule for $f(x)$, plot a real-valued function.

40. Graph the relation "less than" on the set of real numbers, that is, the relation described by:

$$\{(x, y) \mid x \text{ and } y \text{ are real, } x < y\}.$$

* Some calculators can be *programmed*, that is, instructed to produce the output for an arbitrary rule such as $f(x) = 3x^2 + 3x + 32$. With such a program in use, the computation is done internally, and the display shows the final output.

41.
a) If your calculator has a $\boxed{\sqrt{x}}$ key and gives a different number of digits on the display, repeat Example 2.
b) Which of the outputs in Example 2 are exact square roots and which are approximations? Why?

42. A function on the real numbers is defined as follows: $g(x) = 3x - 1$, that is,

$$g = \{(x, y) \mid x \text{ is real and } y = 3x - 1\}.$$

Find the following outputs:
a) $g(0)$ b) $g(3)$ c) $g(^-2)$

If a calculator has a $\boxed{\sqrt{x}}$ key, then, for any nonnegative number that can be entered, the use of this key will cause $\sqrt{x}$ to appear on the screen.

Example 2 Use the $\boxed{\sqrt{x}}$ key to find the square root of

3, 4, 5, 6, 7, 8, 9, 10.

Enter	*Press*	*Display on a 10-digit screen*
3	$\boxed{\sqrt{x}}$	1.732050808
4	$\boxed{\sqrt{x}}$	2.
5	$\boxed{\sqrt{x}}$	2.236067977
6	$\boxed{\sqrt{x}}$	2.449489743
7	$\boxed{\sqrt{x}}$	2.645751311
8	$\boxed{\sqrt{x}}$	2.828427125
9	$\boxed{\sqrt{x}}$	3.
10	$\boxed{\sqrt{x}}$	3.16227766

Try Exercise 41

A function of the real numbers is most frequently defined by a rule such as the rule in Example 2 which was "Find the square root of the number used as input." This rule is condensed to $f(x) = \sqrt{x}$.

GRAPHS OF FUNCTIONS ON THE REAL NUMBERS

Example 3 A function f on the real numbers is defined as follows: $f(x) = 2x + 1$. Given the following inputs, find the outputs:

a) 0 b) 2 c) -2

Solution Thus: a) $f(0) = 1$ since $(2 \cdot 0 + 1)$
b) $f(2) = 5$ since $(2 \cdot 2 + 1)$
c) $f(^-2) = {}^-3$ since $(2 \cdot {}^-2 + 1)$

This function could also be described in the usual way as a set of ordered pairs:

$$f = \{(x, y) \mid x \text{ is a real number and } y = 2x + 1\}.$$

Since the domain is an infinite set, it is impossible to list all of the ordered pairs for this function.

Try Exercise 42

Example 4 Graph the function of Example 3.

We draw a pair of axes and label them. Then we plot the ordered pairs we have found: $(0, 1)(2, 5)(^-2, {}^-3)$. We plot more points, if necessary, to see a pattern. The graph in this case is a straight line.

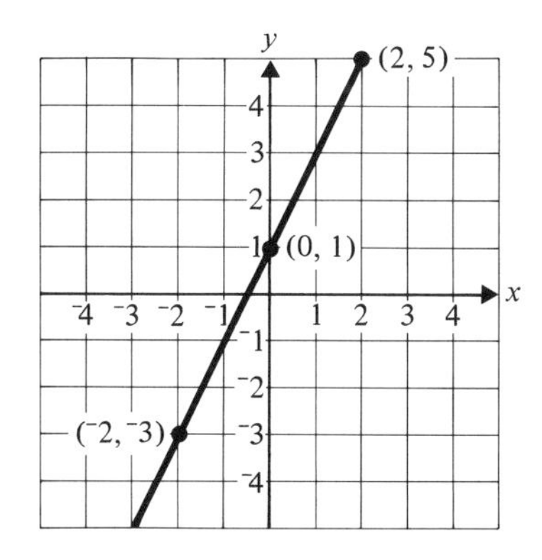

It can be shown that any function $f(x) = ax + b$, where a and b are real numbers, is a linear function, which means that the ordered pairs of the function lie on a straight line when graphed on the usual coordinate scheme for $R \times R$.

Try Exercise 43

Example 5 A function on the real numbers is defined as follows: $f(x) = x^2 + x - 2$. Find $f(^{-}2)$, $f(^{-}1)$, $f(^{-}\frac{1}{2})$, $f(0)$, $f(1)$, and $f(2)$.

Solution We make a table that shows the computation.

x	$x^2 + x - 2$	$f(x)$
$^{-}2$	$(^{-}2)^2 + {}^{-}2 - 2$	0
$^{-}1$	$(^{-}1)^2 + {}^{-}1 - 2$	$^{-}2$
$^{-}\frac{1}{2}$	$(^{-}\frac{1}{2})^2 + {}^{-}\frac{1}{2} - 2$	$^{-}(2\frac{1}{4})$
0	$0^2 + 0 - 2$	$^{-}2$
1	$1^2 + 1 - 2$	0
2	$2^2 + 2 - 2$	4

This function, described in the usual ordered-pair fashion, is:

$$f = \{(x, y) \mid x \text{ is a real number and } y = x^2 + x - 2\}.$$

Since the domain is an infinite set, it is impossible to list all of these ordered pairs.

Example 6 Graph the function of Example 5.

As before, we draw and label a pair of axes; the ordered pairs found above are plotted. The pattern for a smooth curve seems well enough established, and we sketch the remaining parts of the curve to approximate the graph.

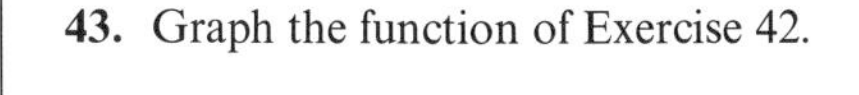
43. Graph the function of Exercise 42.

44.

a) If $x = {}^-2$, it is possible to find two values for y. Find them.
b) If $x = {}^-1$, it is possible to find two values for y. Find them.
c) What is the largest value possible for x? for y? The smallest value for x? for y?
d) What is the actual domain of this relation? the actual range?

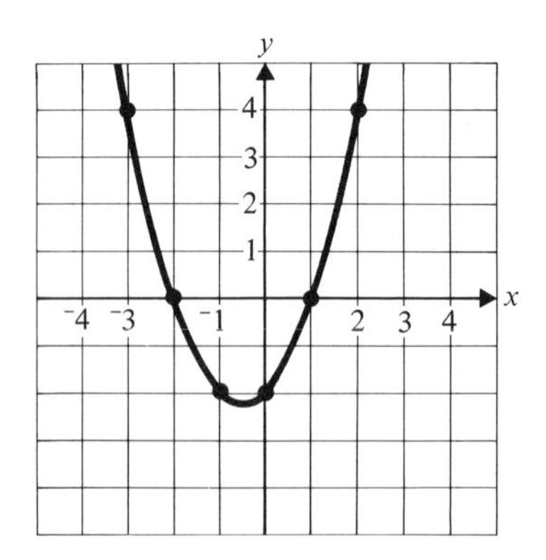

It can be shown that, when graphed, any function of x given by the rule

$$f(x) = ax^2 + bx + c, \text{ where } a, b, \text{ and } c \text{ are real numbers, } a \neq 0,$$

represents a curve called a parabola.

We easily recognize the graph of a real-valued function, since any vertical line will meet the graph of a function in no more than one point. The following are graphs of functions:

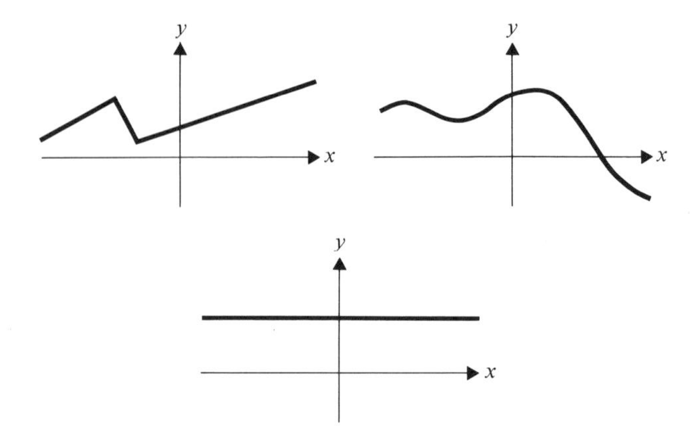

Example 7 Show that the set of ordered pairs is not a function: $\{(x, y) \mid x$ and y are real numbers and $x^2 + y^2 = 9\}$. The following are some pairs that satisfy the rule:

x	0	0	1	1	2	2	3	$^-3$
y	3	$^-3$	$\sqrt{8}$	$^-\sqrt{8}$	$\sqrt{5}$	$^-\sqrt{5}$	0	0

Try Exercise 44

All of these ordered pairs lie on a circle with a center at (0, 0) and a radius of 3. It is not a function, since for some values of x there are two values of y.

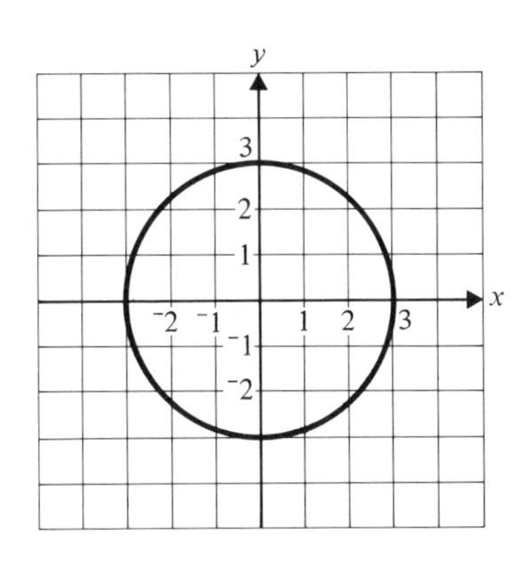

Try Exercise 45

45. Which of the following are graphs of functions?

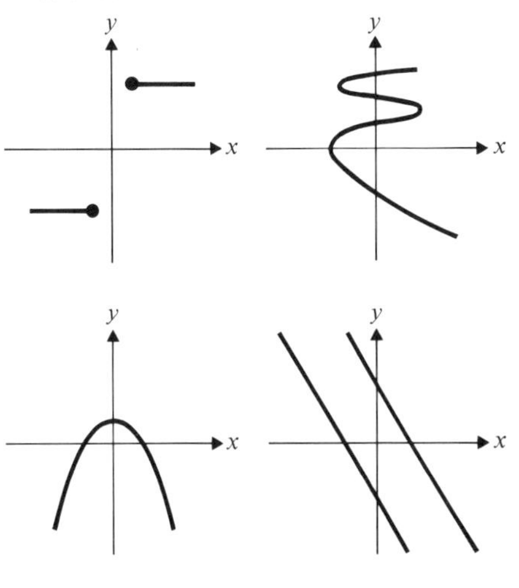

EXERCISE SET 8.6

1. If the following points are plotted on a graph and then connected in order, each figure is the graph of a set of ordered pairs for which it is not easy to write a rule. These are not functions, but you will recognize them. What are they?

a) $(2, 2)(^-4, 2)(^-4, ^-1)(2, ^-1)(2, 2)$
b) $(9, 3)(^-2, 3)(^-2, ^-3)(9, 3)$
c) $(^-3, ^-3)(2, ^-1)(2, 2)(^-3, 0)(^-3, ^-3)$
d) $(^-2, 4)(^-5, 4)(^-7, 1)(^-5, ^-2)(^-2, ^-2)(0, 1)(^-2, 4)$

2. Describe the relation $x > y + 1$ on the set of real numbers as a set of ordered pairs. Graph this set.

3. Describe the relation $x < y + 1$ on the set of real numbers as a set of ordered pairs. Graph the relation.

4. Which of the following are functions?

a) b)

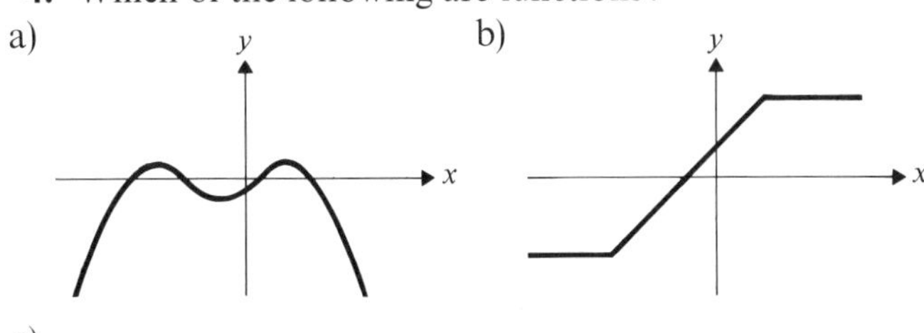

c)

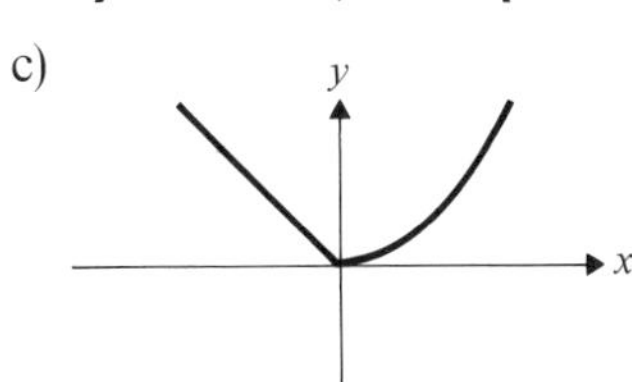

5.
a) Graph the relation $\{(x, y) \mid x$ is a real number and $y = x - 1\}$.
b) Is this relation a function?
c) What is the domain and range of this relation?

6.
a) Describe the function $f: f(x) = x + 2$ as a set of ordered pairs.
b) Graph the function f.
c) What are its domain and range?

7. For the function f with the rule $f(x) = \frac{x}{2} + 2$, find the following function values:

a) $f(2)$ b) $f(10)$
c) $f(3)$ d) $f(^{-}11)$
e) $f(\sqrt{2})$ to the nearest tenth
f) $f(\sqrt{7})$ to the nearest hundredth

8. For the function g defined by $g(x) = 3x - 10$, find the following function values:

a) $g(0)$ b) $g(^{-}5)$
c) $g(1/2)$ d) $g(^{-}3/4)$
e) $g(\sqrt{3})$ to the nearest tenth
f) $g(\sqrt{17})$ to the nearest hundredth

9. A function g has for its domain the set of real numbers and the range is $\{2\}$. From this information you can write the rule for the function. Graph the function.

10. If x is a real number, what is the domain of the function f given by the rule $f: f(x) = x + 1$? What is the range?

11. Graph the relation $g = \{(x, y) \mid y = |x|\}$. Is this relation a function? What is the range of this relation?

12. Here is a flowchart for finding function values with a calculator.

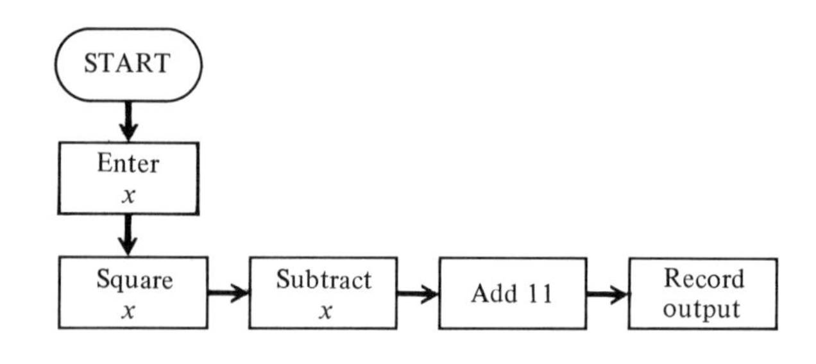

a) Describe f as a set of ordered pairs.

b) Find $f(x)$ for $x = 0, 1, 2, 3, 4, 5, 6, 7, 8, 9, 10$.

13. Investigate your own calculator as a function machine. How many functions does it have? (Even simple calculators have √x, %, +/−.) Learn how each of these keys is used on your calculator.

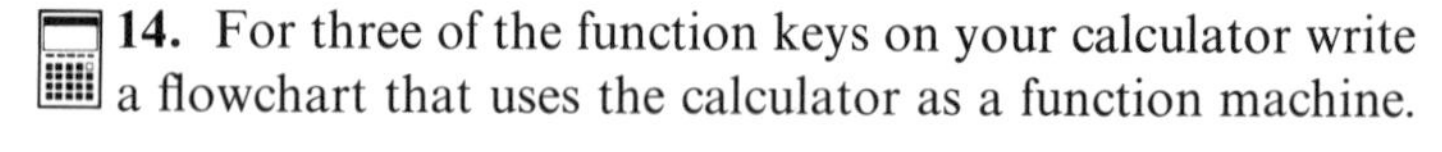

14. For three of the function keys on your calculator write a flowchart that uses the calculator as a function machine.

15. Write a flowchart for your calculator that will give the function value for the following function:

$$f(x) = \tfrac{1}{2}x^2 + 2x - 7.$$

16. Use your chart to find $f(-8)$, $f(-6.24)$, $f(-4)$, $f(-3)$, $f(-2)$, $f(-1)$, $f(0)$, $f(2.24)$ and $f(4)$ for the f of Exercise 15. Make a sketch of the function.

8.7 PROBLEM SOLVING

By now, you have had enough experience with the different ways of writing and thinking about the rational and the real numbers to realize that the type of symbols we choose for them are likely to affect both how we think about a problem and what steps we take to solve it.* Arithmetic, as it is now taught in the elementary and junior high school, is presented in such a manner that the transition to high-school algebra is not an abrupt change. Ordinary high-school algebra was far from ordinary 400 years ago. Algebraic notation was developed in three stages over a long period of time.

STAGE I—RHETORICAL ALGEBRA. In this type of algebra the solution of a problem is written in ordinary prose; no symbols are used. The following problem† is an adaptation of a Greek one that dates from about AD 500. We shall solve it by using ordinary prose rather than an algebraic equation, as we might ordinarily do.

Example 1 How many pieces of fruit are needed if two out of four people receive one-third and one-fifth of the total, respectively, while the third person is to get six pieces and the remaining fourth person gets one piece?

Try Exercise 46

If you had to go on and find the solution, it might occur to you to write some sort of equation that fits the conditions of the problem; however we intend to work out a prose solution.

A SOLUTION IN RHETORICAL ALGEBRA. In Western Europe problems such as this were solved in prose until about the 15th century. The Greeks, however, had moved on to the next stage much earlier due to the work of the mathematician Diophantus (ca 50?). All of the Greek algebra prior to Diophantus‡ was rhetorical.

Here is a possible *solution in prose*: Since we are going to assume that no person got a fractional piece of fruit, whatever the required number is, it must be divisible by three and five. The smallest number that has this property is fifteen. This amount of fruit can be checked to see if it satisfies all the conditions of the problem. Under the tentative solution, the first person will receive five pieces, the second will receive three pieces, the third will receive six, and the fourth will receive one. This adds up as follows: five and three are eight, added to six and one is fifteen. The solution is correct.

Unless one wished to make trial distributions of the actual fruit or models of the fruit, this is about the way the problem had to be solved. No one had yet thought of calling an unknown "x" or writing an equation in some sort of shorthand notation.

8.7 AIMS

You should be able to translate appropriate problems to mathematical language to facilitate problem solving, still using the five-step process.

46.

a) What are the facts of Example 1?
b) What is the problem?

* A reminder:

> THE FIVE-STEP PROCESS
> 1. What are the relevant *facts*?
> 2. What actually is the *problem*?
> 3. What are my *ideas*?
> 4. How do I get a *solution*?
> 5. *Check* and *review* the solution.

† This problem is an adaptation of one which appears in a book of problems collected about AD 500 and known as the *Greek Anthology*. The problems were assembled by the grammarian Metrodorus; it is believed that many of them are much more ancient and not all the work of one man. For more about this see *An Introduction to the History of Mathematics* by Howard Eves, 4th Edition. Holt, Rinehart, and Winston, New York (1976).

‡ Hypatia (AD 370–415) added to this work in her treatise *On the Astronomical Canon of Diophantus*, which contained alternative solutions as well as new problems that she posed. A number of women of her time were respected and superior mathematicians, but Hypatia is the only one for whom we have some precise information. Her father was a professor of mathematics at the University of Alexandria, the city that was the intellectual center of the world at the time.

Women in Mathematics by Lynn M. Osen. The MIT Press, Cambridge, Massachusetts (1974).

STAGE II—SYNCOPATED ALGEBRA. In this intermediate stage of development some symbols and abbreviations were adopted for operations or quantities that were used frequently. This would resemble the statements written for youngsters in which mathematical symbols are interspersed with English words. For example:

5 is greater than 3, instead of $5 > 3$.
The absolute value of $^{-}5$ is 5, instead of $|^{-}5| = 5$.
The additive inverse of $^{-}\frac{1}{2}$ is $\frac{1}{2}$, instead of $^{-}(^{-}\frac{1}{2}) = \frac{1}{2}$.

Interspersing words with mathematical symbols in this way is often helpful for pupils who have a great many new symbols to learn.

STAGE III—SYMBOLIC ALGEBRA. In this stage the solutions and often the problems are presented in mathematical "shorthand" and no English words appear in the solution.

Example 2 Consider only one *idea:* Use symbolic algebra to solve the problem stated in Example 1.

We need not repeat the stating of the facts and determining what the problem is, nor will we attempt to find more than one idea for the solution but use the given idea.

Let x be the total number of pieces of fruit that are needed. The facts can be presented in a table:

Person	1st	2nd	3rd	4th
Amount of fruit	$\frac{1}{3}x$	$\frac{1}{5}x$	6	1

From this chart we see that the total amount of fruit is also given by the sum

$$\frac{1}{3}x + \frac{1}{5}x + 6 + 1.$$

The following is then a true statement:

$$x = \frac{1}{3}x + \frac{1}{5}x + 6 + 1.$$

Once we have done this, there is still another decision to make. Should the fractions be retained or would it be better to convert to decimals? We choose to retain the fractions and solve for x by first finding a common denominator, which for 1/3 and 1/5 is 15. Multiplying both members by 15, we get

$$15 \cdot x = 15 \cdot \left[\frac{1}{3}x + \frac{1}{5}x + 6 + 1\right].$$

Thus,

$$15x = 5x + 3x + (15) \cdot (7)$$
$$15x = 8x + (7) \cdot (15)$$
$$7x = (7) \cdot (15) \quad \text{[by subtracting } 8x\text{]}$$
$$x = 15 \quad \text{[by multiplying both members by } 1/7\text{]}$$

If one is familiar with the arithmetic procedures, the entire solution is somewhat mechanical in contrast to the sharp attention that must be paid in the rhetorical solution.

Try Exercise 47

Symbolic algebra (such as this) did not reach Western Europe until the 16th century and was not widespread until the middle of the 17th century; thus the "modern math" of the 1690's was an equation written in x. Mathematicians took a long time to understand how helpful well-chosen symbols can be in problem solving, so it is no wonder that pupils sometimes have difficulty understanding and applying the various types of notation they are expected to use.

47. $x = \frac{1}{3}x + \frac{1}{5}x + 7$ can also be solved by first adding $^{-}\frac{1}{3}x + {}^{-}\frac{1}{5}x$ to both sides. Why? Finish the solution by using this method.

EXERCISE SET 8.7

Problems 1–9 are similar to those that elementary students are expected to solve. Use the five-step process and imagine that you will be explaining these to such students.

1. Pam has some quarters; denote the number of quarters she has by □.

a) Joe has twice as many quarters; how many does Joe have in terms of □?
b) Pat has 4/5 as many dimes as Pam has quarters; how many dimes does she have in terms of □?
c) Jim has half as many quarters as Pam; how many does he have in terms of □?
d) What is the value in cents of Jim's quarters? Pat's dimes (in terms of □)?

2. Let △ represent the age of a friend.

a) The mother of your friend is 21 years older than your friend. Represent the mother's age in terms of △.
b) Your friend's grandmother is 7 years more than 3 times as old. Represent the grandmother's age in terms of △.
c) Your friend has a brother and a sister who are two years apart. The sister is older than the brother. Their combined ages is $1\frac{1}{2}$ times your friend's age. Represent the brother's age in terms of △.

3. In a magic square the sums for all the rows (vertical, horizontal, and diagonal) are the same. Complete the magic square.

$112\frac{1}{8}$	?	$93\frac{3}{8}$
?	$87\frac{1}{4}$	?
?	?	$62\frac{3}{8}$

4. Find the number for x, y, and z in the following equations:

$$x \cdot x = 6\,241,$$
$$y \cdot y \cdot y = 39\,304,$$
$$z \cdot z \cdot z \cdot z = 28\,561,$$

5. It takes 12 minutes to saw a log into 5 pieces. How long would it take to saw the same log into 7 pieces?

6. Suppose you earned \$4.75 per hour, 8 hours per day, 5 days per week for 49 weeks per year, from now until you are 30. How much would you earn? Estimate and then calculate the answer.

7. Assume that you like each of the following cereals equally well. Cereal A costs 89 cents for a 425-gram box; cereal B costs 63 cents for a 198-gram box, and cereal C costs \$1.09 for a 680-gram box. Which of these cereals is the cheapest per gram? If the box of cereal A has 15 servings, that of cereal B has 7 servings, and cereal C has 24 servings per box, which one is the cheapest per serving?

8. One printing press can print 1/3 of the Sunday edition in 1 hour. A second press can do 2/7 of the job in 1 hour. How much of the printing job is done if both presses work for 1 hour? How long will both presses have to work in order to complete the entire edition?

9. The Luciano family has just purchased a new car that gets at least 13 kilometers per liter of fuel. They are planning an extensive motor trip of about 17 000 km. If fuel costs \$0.24 per liter, what will the gasoline cost for their trip?

■ **10.** Five people were on a tour of a grain mill. The scale in the mill would weigh a minimum of 200 pounds. One person in the group said it would still be possible for each person to find out how much they weighed although individually they all weighed under 200 pounds. They got weighed in pairs and a triplet: P and Q together weighed 255 pounds, Q and R together weighed 285, R and S together weighed 303, S and T together weighed 308, and P, R, and T together weighed 440. You can now find the weight of each person.

11. It makes a great deal of difference how a problem is represented:

a) Find the product $(1.5) \cdot (1.3\overline{3}) \cdot (1.25) \cdot (1.2) \cdot (1.1\overline{66})$. Would you be inclined to look for your calculator?

b) Find the product:

$$\left(1+\frac{1}{2}\right)\cdot\left(1+\frac{1}{3}\right)\cdot\left(1+\frac{1}{4}\right)\cdot\left(1+\frac{1}{5}\right)\cdot\left(1+\frac{1}{6}\right).$$

Would you get a different idea?

Try replacing $\left(1+\frac{1}{2}\right)$ by $\frac{2}{2}+\frac{1}{2}=\frac{3}{2}$.

■ c) Solve a harder problem:

$$\left(1+\frac{1}{2}\right)\times\left(1+\frac{1}{3}\right)\times\left(1+\frac{1}{4}\right)\times\left(1+\frac{1}{5}\right)\times\cdots\times\left(1+\frac{1}{n}\right).$$

■ **12.** A "trick or treat" problem. At Hallowe'en 3 youngsters come to the door for candy. If each child gets at least 1 piece of candy and no pieces are broken, in how many ways can the person treating "divvy up" n pieces of identical candy? If the person handed out 3 pieces of candy, each child would get 1 piece and there would be only one way to do this. If the person gave 4 pieces of candy, one of the children would get 2 pieces, so there would be three ways to do this. What happens if 5, 6, or 7 pieces are distributed? Look for a pattern and find a formula that yields the answer no matter how many candies are used.

REVIEW TEST

1. Write T for those of the following statements that are always true; for the others write F:

a) There are exactly five different pairs of whole numbers whose sum is 5. ______

b) There is an infinite set of pairs of integers whose sum is $^-5$. ______

c) There is no smallest integer. ______

d) The smallest possible absolute value of an integer is 1. ______

e) ^-a is always a negative number. ______

f) For any two rational numbers, the difference $a - b$ can always be found from the sum $a + {^-b}$. ______

g) The additive inverse of $3 \cdot x$ is either $^-3 \cdot x$ or $3 \cdot {^-x}$ for any rational number x. ______

h) The additive inverse of $^-(^-36)$ is $^-36$. ______

i) The additive inverse of $^-(^-b + c)$ is $b + {^-c}$. ______

j) If a/b is the multiplicative inverse of b/a, then $\left(\frac{a}{b}\right)^{-1}$ is $\frac{b}{a}$. ______

k) There is no integer that has a multiplicative inverse. ______

2. Use $<$, $>$, or $=$ in the blanks to make the statements true:

a) Any positive integer is ____ 0.

b) Any negative integer is ____ any positive integer.

c) Any integer is ____ another integer to the right of it on a number line.

d) If a and b are rational numbers and $a > b$, then $^-2 \cdot a$ ____ $^-2 \cdot b$.

e) If a and b are rational numbers and $a < b$, then $a + 3$ ____ $b + 3$.

f) For any real numbers a, b, and c, if $a = b$, then $a - c$ ____ $b - c$ and $a \cdot c$ ____ $b \cdot c$.

3. Show that the trichotomy property holds for rational numbers; that is, for any rational numbers a and b, one and only one of the following holds: $a = b$, $a < b$, or $a > b$.

4. Prove that for any rational numbers a, b, and c, if $a < b$, and $b < c$, then $a < c$.

5. Show that the associative property holds for multiplication of rational numbers.

6. Arrange in order from the smallest to the largest:

$$^-66/4,\ ^-17,\ 121/121,\ 19/21,\ ^-167/189,\ 561/121.$$

7. Show that for any rational numbers a, b, and c, if $a < b$, then $a + c < b + c$.

8. Show that for any rational numbers a, b, and c, with $c < 0$, if $a < b$, then $a \cdot c > b \cdot c$.

9. Prove that $\sqrt{7}$ is irrational.

10. Show that $\sqrt{2}/2$ is the multiplicative inverse of $\sqrt{2}$.

11. Show that the product $(1 - \sqrt{3})(1 + \sqrt{3})$ is a rational number.

12. Since $\sqrt{x^2}$ is always a nonnegative number, show that it is correct to define $\sqrt{x^2}$ as $|x|$.

13. Graph the function given by the rule $g(x) = \frac{1}{2}x + 3\frac{1}{2}$. Describe g as a set of ordered pairs.

14. Make a flowchart for your calculator to provide the output of the function rule $f(x) = x^2 + 3x + 24$. Using the chart, find

a) $f(0.25)$ to the nearest hundredth,

b) $f(175.67)$ to the nearest hundredth.

15. Find two real numbers whose product is 10^7, yet neither of the numbers has a final digit of 0.

16. Show that the product of a nonzero rational number and an irrational number must be irrational.

17. You are given the following two irrational numbers:

1.01001000100001 . . . ,
1.020020002000020 . . . .

a) Give the decimal for a rational number that lies between these two numbers.

b) Give the decimal for an irrational number that lies between these two numbers.

18. Each of the following is a restatement of some problems given by Diophantus in his book *Arithmetica* (of which only six of the original books are known). For each write the conditions of the problem in terms of equations for which a different variable is used for the different numbers referred to in the problem. Do not solve the problems (for more of these see Eves, op cit.).

a) From Book III, Problem 7: Find three positive rational numbers such that their sum is a square number and the sum of any two of them at a time is also a square number.

b) From Book III, Problem 15: Find three positive rational numbers such that the product of any two of them with the third subtracted is a square number.

FOR FURTHER READING

1. "A truck driver looks at square roots" by John W. Risoen and Jane G. Stenzel, in the *Arithmetic Teacher*, 26:3, November 1978, gives an interesting description of an actual incident and a nice result about square roots.

2. In "Fractions, decimals, and their futures," in the *Arithmetic Teacher*, 24:3, March 1977, the author, Donald H. Firl, questions that "division with fractions should have importance prior to a course in algebra . . ." and also proposes some changes in the instruction in decimals and fractions in the elementary school.

TEXTS

Begle, Edward G. *Mathematics of the Elementary School.* McGraw-Hill Book Company, New York, New York (1975).

Rising, Gerald R. and Joseph B. Harkin, *The Third "R": Mathematics Teaching for Grades K–8.* Wadsworth Publishing, Inc., Belmont, California (1978). (A methods text.)

chapter 9 finite mathematical systems

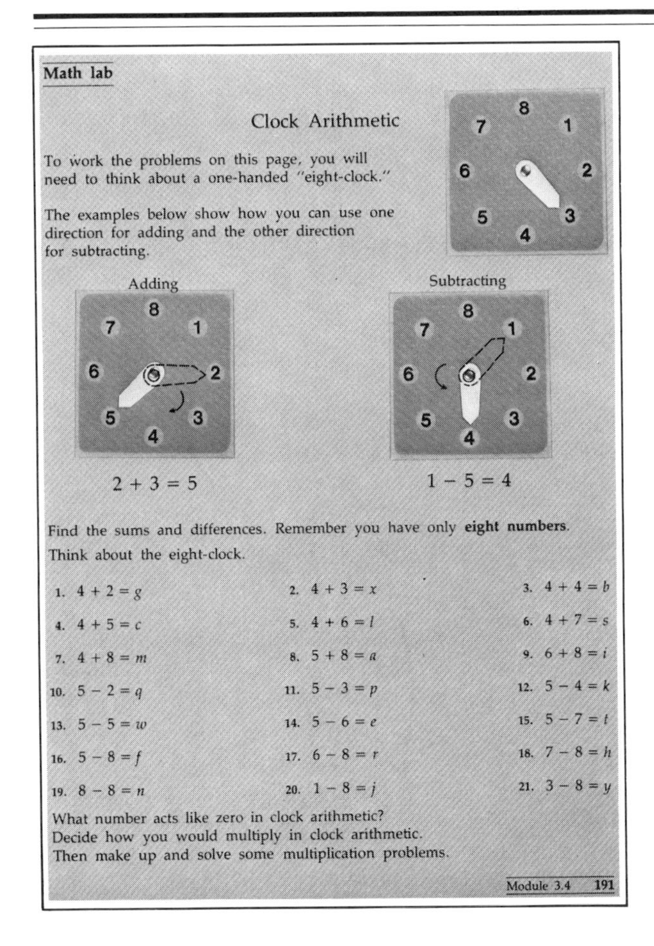
Math lab

Clock Arithmetic

To work the problems on this page, you will need to think about a one-handed "eight-clock."

The examples below show how you can use one direction for adding and the other direction for subtracting.

Adding

2 + 3 = 5

Subtracting

1 − 5 = 4

Find the sums and differences. Remember you have only **eight numbers**. Think about the eight-clock.

1. $4 + 2 = g$ 2. $4 + 3 = x$ 3. $4 + 4 = b$
4. $4 + 5 = c$ 5. $4 + 6 = l$ 6. $4 + 7 = s$
7. $4 + 8 = m$ 8. $5 + 8 = a$ 9. $6 + 8 = i$
10. $5 - 2 = q$ 11. $5 - 3 = p$ 12. $5 - 4 = k$
13. $5 - 5 = w$ 14. $5 - 6 = e$ 15. $5 - 7 = t$
16. $5 - 8 = f$ 17. $6 - 8 = r$ 18. $7 - 8 = h$
19. $8 - 8 = n$ 20. $1 - 8 = j$ 21. $3 - 8 = y$

What number acts like zero in clock arithmetic?
Decide how you would multiply in clock arithmetic.
Then make up and solve some multiplication problems.

Module 3.4 191

Clock arithmetic can be used, as it is done here with seventh graders, for individual investigations of a mathematical system. It is not difficult but gives pupils a chance to explore and invent on their own.

9.1 ADDITION AND MULTIPLICATION FOR CLOCK ARITHMETIC

So far we have considered a variety of number systems, each of which, as an extension of a previous one, remedied some deficiency of the whole-number system we started with.

This chapter is an about-face, since we are going to "think small" and discuss some mathematical systems that have only a finite number of elements. Surprisingly, there are many finite systems that have all the basic properties of the rational numbers. Some of these systems have a basis in a physical model, which is the case for the clock systems.

9.1 AIMS

You should be able to:

a) Use a clock model to construct addition and multiplication tables for a clock arithmetic.
b) Use the tables to determine the properties of the clock arithmetic.
c) Illustrate with examples the properties of a clock arithmetic and prove that certain properties hold.

A CLOCK SYSTEM. We shall start with a 4-hour clock which measures elapsed time and needs only 4 numbers, the set $\{0, 1, 2, 3\}$. If the hand starts at 0, after 4 hours it is back at 0.

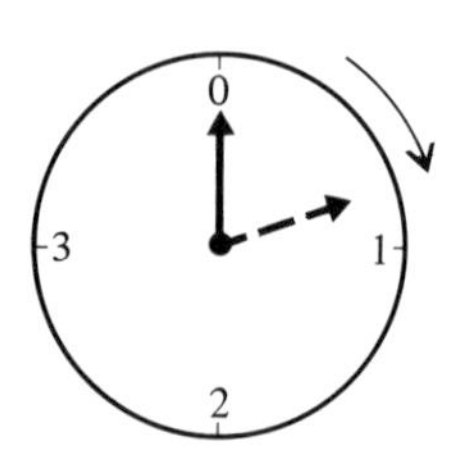

ADDITION ⊕. Addition for this clock can be defined from a study of the model in an obvious way. For example, to find the sum $1 \oplus 2$ we imagine that the hand starts at 0, goes 1 space clockwise, then two more spaces to stop at 3. We use the symbol $\oplus$ to indicate that this is not the usual whole-number addition. This system has so few elements that the complete definition of the addition operation can be given in this table of only 16 entries.

⊕	0	1	2	3
0	0	1	2	3
1	1	2	3	0
2	2	3	0	
3	3		1	2

$1 \oplus 2 = 3$ (tinted)

1. Use the 4-hour clock model to find the remaining entries in the $\oplus$ table. That is, find
a) $3 \oplus 1$ b) $2 \oplus 3$

Try Exercise 1

MULTIPLICATION ⊗. To define multiplication, we can either use the clock directly or else use repeated addition and refer to the addition table.

Example 1 Find $2 \otimes 3$ in the 4-hour clock arithmetic.

$2 \otimes 3$ means the hand starts at 0, goes 3 places in a clockwise direction, then three more and lands at 2. Thus $2 \otimes 3 = 2$.

Or else we look at the addition table and since $2 \otimes 3$, when interpreted as addition, means $3 \oplus 3$, this sum can be found in the table to be 2. Other entries are found in a similar manner, either by counting off on the clock or by referring to the addition table.

⊗	0	1	2	3
0	0	0	0	0
1	0	1	2	3
2	0	2		2
3	0		2	

$2 \otimes 3 = 2$ (tinted)

Try Exercise 2

2. Complete the ⊗ table. Find:
a) $2 \otimes 2$
b) $3 \otimes 1$
c) $3 \otimes 3$

Properties of 4-Hour Clock Arithmetic

Since the addition and multiplication tables completely define this mathematical system generally called J_4, we can use them to investigate some of the properties of the system.

CLOSURE. The set $\{0, 1, 2, 3\}$ used in J_4 is closed under addition, since the sum of any two of the numbers exists and is one of the numbers of the set. There are no blank spaces in the table and all the entries come from the set.

Try Exercise 3

3. Is the set $\{0, 1, 2, 3\}$ closed under the multiplication ⊗?

COMMUTATIVITY. Addition is commutative, since for all a and b in J_4 we have

$$a \oplus b = b \oplus a.$$

This can be proved by simple inspection of the addition table. Since the elements are listed by row and by column in the order 0, 1, 2, 3, the commutativity is evident from the symmetry with respect to the diagonal passing from the upper left to the lower right of the table.

⊕	0	1	2	3
0	0	1	2	3
1	1	2	3	0
2	2	3	0	1
3	3	0	1	2

4. Is the multiplication $\otimes$ commutative? Why?

Try Exercise 4

THE IDENTITIES. The number 0 is the additive identity, which can be seen by inspection of the addition table. For any element a of the set.

$$0 \oplus a = a \oplus 0 = a.$$

A similar inspection of the multiplication table shows that for any element a,

$$1 \otimes a = a \otimes 1 = a.$$

Thus 1 is the identity for multiplication.

Try Exercise 5

5. What properties must a multiplicative identity have? Is there such an identity?

ASSOCIATIVITY FOR ADDITION. It is not easy to tell from the table of an operation whether or not the operation is associative, but some instances give a clue.

Example 2 Is it true that $(2 \oplus 1) \oplus 3 = 2 \oplus (1 \oplus 3)$?

We calculate the two ways and compare:

$(2 \oplus 1) \oplus 3$	$2 \oplus (1 \oplus 3)$
$3 \oplus 3$ 2	$2 \oplus 0$ 2

The results are the same, so associativity holds at least in this one instance.

Try Exercise 6

6. Check two more cases for associativity of addition:

a) Compare $2 \oplus (3 \oplus 1)$ with $(2 \oplus 3) \oplus 1$

b) Compare $1 \oplus (2 \oplus 2)$ with $(1 \oplus 2) \oplus 2$

Unfortunately, if we chose to prove that addition is associative by examining all possible cases, we would need to check 64 of them. However, we could reason instead that the way the addition is performed on the clock insures associativity. Suppose we wish to show $(1 \oplus 2) \oplus 3$ on the clock. We need to count off $(1 + 2) + 3$ spaces or 6 spaces clockwise and land at 2. In order to show $1 \oplus (2 \oplus 3)$, we need to count spaces and move clockwise $1 + (2 + 3)$, or 6 spaces on the clock and land at 2. This argument, which depends upon the counting of the necessary spaces, hence upon the counting numbers, is valid for any of the elements of the system.

To show that $a \oplus (b \oplus c) = (a \oplus b) \oplus c$, we do as before. To show that $a \oplus (b \oplus c)$, we need to count off in the clockwise direction $a + (b + c)$ spaces. Similarly for $(a \oplus b) \oplus c$, we need to count off $(a + b) + c$ spaces in the clockwise direction. But the counting is done with the counting (natural) numbers, and we know that they are associative for addition.

ASSOCIATIVITY FOR MULTIPLICATION. Multiplication can be shown to be associative by working through all 64 possible cases that can occur in this system, but the definition of multiplication based on the clock makes this unnecessary. To show that $2 \otimes (3 \otimes 3) = (2 \otimes 3) \otimes 3$, we reason that $2 \otimes (3 \otimes 3)$ means $2 \otimes (3 \oplus 3 \oplus 3)$, which, by the addition on the clock, means we count and move clockwise $2 \cdot 9$ spaces to land on 2. Similarly, the product $(2 \otimes 3) \otimes 3$ means $(3 \oplus 3) \otimes 3$, which, when translated to the clockwise motion on the clock, means we count and move $6 \cdot 3 = 18$ spaces, in which case the hand will again be at 2. The counting of the spaces is done with the counting numbers, and they do have the associative property. The reasoning holds for any elements of the system, so the multiplication is associative in J_4.

Try Exercise 7

7. Check another instance of the associativity of multiplication. Is

$$3 \otimes (3 \otimes 2) = (3 \otimes 3) \otimes 2?$$

THE DISTRIBUTIVE LAW. The distributive law of multiplication over division could be proved by looking at every possible case (literally, a proof by exhaustion). Or else we can reason that the product $2 \otimes (3 \oplus 2)$ depends upon counting 5 spaces and doing it 2 times, so the final amount depends upon $2 \cdot (3 + 2) = 10$ in the counting numbers. The sum $(2 \otimes 3) \oplus (2 \otimes 2)$ depends upon counting 6 spaces, then 4 more for a total of 10 to land on 2. Since we are counting the spaces, we know that $2 \cdot (3 + 2) = 2 \cdot 3 + 2 \cdot 2$ is true for the counting numbers. The argument holds no matter which elements of J_4 are used.

Try Exercise 8

8. Use the table for the system J_4 and compare
a) $3 \otimes (1 \oplus 2)$ with $3 \otimes 1 \oplus 3 \otimes 2$
b) $2 \otimes (3 \oplus 1)$ with $2 \otimes 3 + 2 \otimes 1$

EXERCISE SET 9.1

1. Construct the addition table for the 5-hour clock ($\oplus$ for J_5).

2. Construct the multiplication table for the 5-hour clock ($\otimes$ for J_5).

3. For the system J_5 determine which of the following properties hold:
a) Closure for $\oplus$, for $\otimes$.
b) Commutativity for $\oplus$, for $\otimes$.
c) There is an identity element for addition, another for multiplication.

4. Compare the following pairs in J_5:
a) $1 \oplus (2 \oplus 3)$ with $(1 \oplus 2) \oplus 3$
b) $2 \oplus (4 \oplus 3)$ with $(2 \oplus 4) \oplus 3$
c) What property is illustrated in (a) and (b)?

5. Prove that addition is associative in J_5.

6. In J_5 show that the following statements are true:
a) $(1 \otimes 2) \otimes 3 = 1 \otimes (2 \otimes 3)$
b) $(2 \otimes 4) \otimes 3 = 2 \otimes (4 \otimes 3)$

7. Prove that multiplication is associative in J_4:

For any elements a, b, and c in J_4,
$a \otimes (b \otimes c) = (a \otimes b) \otimes c$.

8. What changes have to be made in the proof in Exercise 7 to make it apply to the system J_5?

9. Compare J_5 with the system of rational numbers. What are some points of difference? What are some similarities?

■ **10.** In the J_4 system, since $2 \oplus 1 = 3$, it would seem $1 < 3$, but also there is a c such that $3 \oplus c = 1$ and thus $3 < 1$. What is the explanation of this?

11. Prove that for any a, b, and c of J_4, the distributive law of multiplication over addition holds.

12. What changes in the proof of Problem 11 must be made to have a proof of the same property for J_5?

9.2 AIMS

You should be able to:

a) In a clock arithmetic find the additive inverse of each element and perform subtraction; state, use, and prove the relationship between subtraction and additive inverses.
b) Identify the multiplicative inverse of each element that has one and perform division; state, use, and prove the relationship between division and multiplicative inverses.

9.2 SUBTRACTION AND DIVISION FOR CLOCK ARITHMETIC

SUBTRACTION ⊖. For a 4-hour clock, subtraction can be defined so that it is consistent with the way addition is defined in terms of movement on the clock. We interpret counterclockwise motion as subtraction.

For $1 \ominus 2$ we imagine that the hand starts at 1 and then moves counterclockwise 2 spaces and comes to rest at 3. Thus $1 \ominus 2 = 3$.

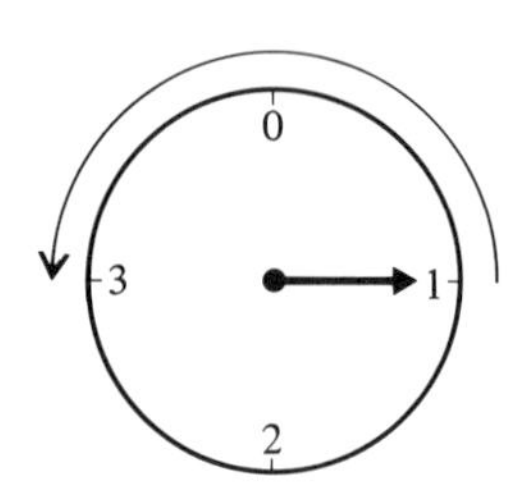

Try Exercise 9

9. Use the clock model to find the following differences:

a) $3 \ominus 2$ b) $0 \ominus 2$
c) $1 \ominus 2$ d) $1 \ominus 3$

Subtraction in a clock arithmetic could have been defined exactly as for integers: $a \ominus b$ is the number c which, when added to b, gives a. That is,

$$a \ominus b = c \leftrightarrow a = c \oplus b.$$

Example 1 In a 4-hour clock arithmetic, find $1 \ominus 2$.

Use the definition that depends upon $\oplus$. By this definition, $1 \ominus 2$ is the number c such that $1 = c + 2$.

From the $\oplus$ table for the 4-hour clock, that number is 3, as we indicate on the appropriate portion of the addition table:

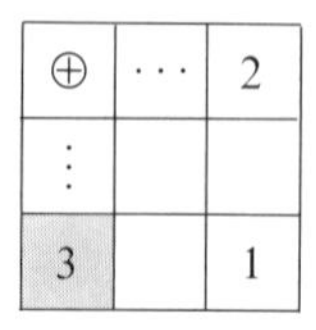

⊕	···	2
⋮		
3		1

$1 \ominus 2 = 3$ (tinted)

$$1 \ominus 2 = c \leftrightarrow 1 = c \oplus 2,$$

therefore, $c = 3$.

Try Exercise 10

SUBTRACTION AND ADDITIVE INVERSES. Just as in the set of integers, there are pairs of numbers whose sum is the additive identity 0:

$$3 \oplus 1 = 1 \oplus 3 = 0, \qquad 2 \oplus 2 = 0, \qquad 0 \oplus 0 = 0.$$

Just as in the system of integers, the pairs of numbers of J_4, whose sum is the additive identity, are called *additive inverses* of each other. The additive inverse of a number is again indicated by a raised dash sign; the additive inverse of 3 is $^{-}3$. This should not be called "negative three" in this system. There are no negative numbers (numbers less than 0) in clock arithmetics.

We suspect that just as before, there is a relation between subtraction and additive inverses, and the pattern in the following table confirms this:

$a \ominus b$	$a \oplus {}^{-}b$
$3 \ominus 2 = 1$	$3 \oplus {}^{-}2 = 3 \oplus 2 = 1$
$2 \ominus 3 =$	$2 \oplus {}^{-}3 = 2 \oplus 1 = 3$
$0 \ominus 1 =$	$0 \oplus {}^{-}1 =$
$1 \ominus 3 =$	$1 \oplus {}^{-}3 =$
$1 \ominus 2 =$	$1 \oplus {}^{-}2 =$

Try Exercise 11

In J_4 we can prove the following theorem:

Example 2 Prove that for any a and b of J_4,

$$a - b = a + {}^{-}b.$$

1. a and b are any elements of J_4; [by the definition of subtraction in J_4] their difference is c:

$$a \ominus b = c \leftrightarrow a = c \oplus b$$

2. $a \oplus {}^{-}b = (c \oplus b) \oplus {}^{-}b$ [since every element of J_4 has an additive inverse; closure for addition]

3. $= c \oplus (b \oplus {}^{-}b)$ [by the associative property for J_4 addition]

4. $= c \oplus 0$ [by the definition of additive inverse and the definition of 0]
$= c$

5. Hence $a \oplus {}^{-}b = a \ominus b$ [by renaming c as $a - b$, see step 1]

If you compare the proof of Example 2 with that of the corresponding theorem for integers, you will see that only very minor changes have been made. The

10. Use the definition of subtraction

$$a \ominus b = c \leftrightarrow a = c \oplus b$$

to find the following:

a) $3 \ominus 2$ b) $1 \ominus 3$

11. Complete the table in the text:

$0 \ominus 1 =$	$0 \oplus {}^{-}1 =$
$1 \ominus 3 =$	$1 \oplus {}^{-}3 =$
$1 \ominus 2 =$	$1 \oplus {}^{-}2 =$

only properties of the integers that were used are the associative property of addition, the existence of an additive inverse for each element, and a 0 element. All of these properties also hold in J_4, so the proof is identical.

DIVISION $\oplus$**.** In a clock arithmetic division is defined exactly as in the systems we have studied already; $a \oplus b$ is the number c such that $a = c \otimes b$. That is,

$$a \oplus b = c \leftrightarrow a = c \otimes b.$$

Example 3 Find $2 \oplus 3$ for the 5-hour clock arithmetic.

By the definition of division:

$$2 \oplus 3 = c \leftrightarrow 2 = c \otimes 3.$$

From the appropriate portion of the multiplication table for J_5 (see Problem 2 of Exercise Set 9.1) we see that c must be 4:

$\otimes$	$\cdots$	3
4		2

$2 \oplus 3 = 4$ (tinted)

$$2 \oplus 3 = 4 \leftrightarrow 2 = 4 \otimes 3.$$

Try Exercise 12

12. Use the definition of division to find the following:
a) $3 \oplus 2$
b) $3 \oplus 4$

DIVISION AND MULTIPLICATIVE INVERSES. The multiplication table for J_5 differs in a rather important way from that of J_4 because there are several pairs of numbers whose product is 1:

$$2 \otimes 3 = 3 \otimes 2 = 1, \qquad 4 \otimes 4 = 1, \qquad 1 \otimes 1 = 1.$$

As we have already defined for rational numbers, these pairs are *multiplicative inverses* of each other. The multiplicative inverse of 3 is 3^{-1} or 2.

Try Exercise 13

13. Find the following in the 5-hour clock system:
a) 2^{-1}
b) 4^{-1}
c) 1^{-1}

Just as for rational numbers, there is a relation between division and multiplicative inverses, which can be observed in the following table for elements of J_5:

$a \oplus b$	$a \otimes b^{-1}$
$2 \oplus 3 = 4$	$2 \otimes 3^{-1} = 2 \otimes 2 = 4$
$3 \oplus 2 = 4$	$3 \otimes 2^{-1} = 3 \otimes 3 = 4$
$1 \oplus 4 =$	$1 \otimes 4^{-1} =$
$4 \oplus 3 =$	$4 \otimes 3^{-1} =$

Try Exercise 14

14. Complete the table in the text:

$1 \oplus\!\!\!\!\div 4 =$ $\quad 1 \otimes 4^{-1} =$

$4 \oplus\!\!\!\!\div 3 =$ $\quad 4 \otimes 3^{-1} =$

In the system of rational numbers, division, except by 0, can be performed by multiplying by the appropriate multiplicative inverse. This theorem is true in J_5 and many other finite systems as well.

For any elements a and b ($b \neq 0$) of J_5, $a \oplus\!\!\!\!\div\, b = a \otimes b^{-1}$.

Example 4 Prove the theorem stated just above.

Proof

1. $a \div b = c \leftrightarrow a = c \otimes b$ [by the definition of J_5 division]
2. If $b \neq 0$, then there is an element b^{-1}, [since all nonzero elements of J_5 have a multiplicative inverse]
3. $a \otimes b^{-1} = (c \otimes b) \otimes b^{-1}$ [by the "equals" property]
4. $a \otimes b^{-1} = c \otimes (b \otimes b^{-1})$ [by the associative property for J_5 multiplication]
5. $a \otimes b^{-1} = c \otimes 1 = c$ [by the definition of multiplicative inverse and the definition of 1]
6. Thus $a \cdot b^{-1} = a \div b$ [by renaming c]

The proof of Example 4 is identical to that for the rational numbers except that the reasons refer to the properties of J_5.

EXERCISE SET 9.2

1. Compare the following pairs of numbers for J_4:
a) $3 \ominus 1$ with $3 \oplus {}^{-}1$ b) $1 \ominus 3$ with $1 \oplus {}^{-}3$

2. Which elements of J_4 have an additive inverse? a multiplicative inverse?

3. Is subtraction a binary operation in J_4?

4. Is subtraction in J_4 commutative? associative?

5. Construct the addition table for a 7-hour clock.

6. Construct the multiplication table for a 7-hour clock.

7. Which elements of J_7 have an additive inverse? a multiplicative inverse?

8. In the system J_5 compare the following pairs:
a) $2 \ominus 3$ with $2 \oplus {}^{-}3$ b) $4 \ominus 2$ with $4 \oplus {}^{-}2$
c) $1 \ominus 4$ with $1 \oplus {}^{-}4$ d) $3 \ominus 4$ with $3 \oplus {}^{-}4$

9. Prove that for any a and b elements of J_5, $a \ominus b = a \oplus {}^{-}b$.

■ **10.** Prove that for any elements a and b of J_n such that the element b has an additive inverse ${}^{-}b$, $a \ominus b = a \oplus {}^{-}b$.

In each of Problems 11–14 find an element x of J_5 to make the statement true:

11. a) $3 \ominus x = 4$ b) $2 \ominus x = 3$

12. a) $x \ominus 2 = 3$ b) $x \ominus 3 = 4$

13. a) ${}^{-}x \oplus 2 = 3$ b) $2 \ominus {}^{-}x = 1$

14. a) ${}^{-}x \ominus 4 = 2$ b) $3 \oplus {}^{-}x = 2$

15. Are all divisions defined in J_7?

16. Compare the system J_4 with the system J_7 with respect to:
a) the patterns in the $\otimes$ table,
b) the existence of multiplicative inverses.

17. Are all divisions possible in J_4?

18. In each of the following, find an element x in J_5 that makes the statement true:
a) $2 \otimes x = 3$ b) $4 \otimes x = 1$ c) $(2 \otimes x) \oplus 1 = 2$

19. In each of the following, find an element x in J_5 that makes the statement true:
a) $(4 \otimes x) \ominus 3 = 2$ b) $(^{-}2 \otimes x) \ominus 1 = 3$
c) $(^{-}3 \otimes x) \ominus 4 = 1$

20. Which of the equations in Problems 18 and 19 have solutions in J_4?

21. Which of the equations in Problems 18 and 19 have solutions in J_7?

■ **22.** Do all of the equations of the form $(a \otimes x) \oplus b = c$ for $a \neq 0$ have a solution in J_5? Prove your answer.

23. Compare the treatment of the "eight-clock" done for elementary pupils on the opening page of this chapter with the development in the first two sections. Solve some of the Problems 1–21.

9.3 AIMS

You should be able to:
a) Carry out the operations of addition and multiplication for a modular arithmetic J_n, $n \neq 1$.
b) Find numbers that are congruent mod m to a given whole number a.
c) Check computations with congruence.
d) Apply the concept of congruence to a simple problem.

9.3 MODULAR ARITHMETICS AND CONGRUENCE

MODULAR ARITHMETIC. The mathematical system J_n that corresponds to the n-hour clock has n elements and is usually referred to as *arithmetic modulo n*, or simply *arithmetic mod n*. Thus J_5 is arithmetic mod 5; the *modulus* of J_5 is 5.

Arithmetic mod 12 has the set of elements $\{0, 1, 2, 3, 4, 5, 6, 7, 8, 9, 10, 11\}$ and is a model of the 12-hour clock, although we are used to seeing 12 rather than 0 on an ordinary clockface. It does not matter which we use, 0 or 12, since the element is the additive identity in any case.

Arithmetic modulo 7 is a model of the days of the week, in which 1 corresponds to Sunday, 2 corresponds to Monday, and so on. If today is Thursday (5), then four days from now it will be Monday (2).

In all J_n systems each element has an additive inverse; the number 0 never has a multiplicative inverse, and in some systems there are other numbers as well that do not have a multiplicative inverse.

Try Exercise 15

15.
a) Find $7 + 8$ (mod 12)
b) Today is Monday (2); six days from now what day will it be? $2 + 6 = ?$

CONGRUENCE MOD 5. All modular arithmetics are based upon a relation called congruence. Think of an appropriately spaced whole-number line being wrapped about a 5-hour clock so that the line marks for 0, 1, 2, 3, and 4 fall on the corresponding marks on the clock. Then 5 would fall on 0, 6 on 1,

7 on 2, and so on. Geometrically, the congruent numbers are those that land on the same mark on the 5-hour clock, and these numbers are said to be congruent modulo 5; they have the same remainder after division by 5.

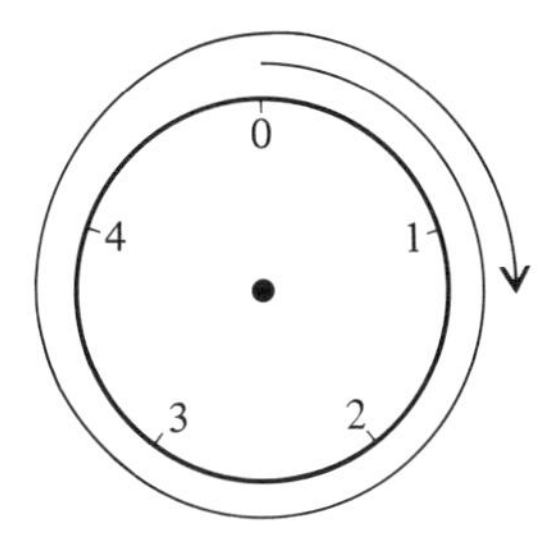

Example 1 Show that 17 is congruent to 22 modulo 5.

17 has a remainder of 2 after division by 5;
22 has a remainder of 2 after division by 5;
therefore 17 is congruent to 22 modulo 5.

Working with a clock

16. Where would 38 fall if the whole-number line were wrapped around the 5-hour clock? What is the remainder if you divide 38 by 5?

On a clock it would require 3 complete wraps of the whole number line and 2 more spaces to get 17. It would take 4 complete turns and 2 spaces more to get to 22.

Try Exercise 16

DEFINITION OF CONGRUENCE MOD m. The discussion and example above is the basis for the definition of the *congruence relation.**

DEFINITION. Any integer a is *congruent* to any integer b, modulo m (m a fixed integer) if and only if a and b have the same remainder after division by m. In symbolic form:

$$a \equiv b \pmod{m}.$$

This means that $a \equiv b \pmod{m}$ if and only if m divides the difference $a - b$. This congruence relation partitions the set of integers into classes of integers that are congruent to each other.

17.

a) Show that $14 \equiv 42 \pmod{4}$.

b) Find two numbers that are congruent to $34 \pmod{6}$.

Example 2 Show that $33 \equiv 47 \pmod{7}$ is true.

33 divided by 7 has a remainder of 5;
47 divided by 7 has a remainder of 5;
therefore $33 \equiv 47 \pmod{7}$.

Also the difference $47 - 33 = 14$ and $7 \mid 14$.

Try Exercise 17

The set $\{\ldots, {}^{-}16, {}^{-}9, {}^{-}2, 5, 12, 19, 26, 33, 40, 47, 54, \ldots\}$ consists of all those integers that have a remainder of 5 when divided by 7, and 33, 47, and $^{-}16$ are all members of this set. Also 7 divides the difference of any pair of numbers in this set.

The fact that the congruence relation has many of the familiar properties we have seen over and over in the systems studied so far makes it a useful concept for the solution of problems that are related to the concept of divisibility.

SOME PROPERTIES OF THE CONGRUENCE RELATION. The congruence relation satisfies the properties of an equivalence relation. For any integers a, b, c, and m the following hold:

1. *Reflexive property*: $a \equiv a \pmod{m}$.
2. *Symmetric property*: If $a \equiv b \pmod{m}$, then $b \equiv a \pmod{m}$.
3. *Transitive property*: If $a \equiv b \pmod{m}$ and $b \equiv c \pmod{m}$, then $a \equiv c \pmod{m}$.

* Karl Friedrich Gauss (1777–1855) invented this very efficient symbolism. With this concept of congruence and emphasis upon the remainder, new concepts concerning integers and divisibility were possible. In his book *Recreations in the Theory of Numbers—The Queen of Mathematics Entertains*, Dover Publications, 1964, Beiler refers to this as "the elegant invention of that 'Prince of Mathematicians,' Karl Friedrich Gauss."

The proof of these properties, since it depends only upon equality of the remainders in each case, is brief and follows immediately from the definition for congruence and the property of the equals relation:

1. If the remainder for a is r when divided by m, then $r = r$ and $a \equiv a$.
2. If the remainder for a is r_1 and that for b is r_2, then, if $r_1 = r_2$, it follows that $r_2 = r_1$ by the equals relation for numbers and thus the symmetric property holds for congruence.
3. If the remainder for a is r_1, that for b is r_2, that for c is r_3, and $r_1 = r_2$ and $r_2 = r_3$, then, by the equals property, $r_1 = r_3$, which establishes the transitive property for congruence.

Because of the fact that congruence is an equivalence relation it is possible in many cases to treat it as if it were the equals relation. The following theorem holds for the congruence relation.

ARITHMETIC OF CONGRUENCE. **If $a \equiv b \pmod{m}$ and $c \equiv d \pmod{m}$, then**

a) $a + c \equiv b + d \pmod{m}$,

b) $a - c \equiv b - d \pmod{m}$,

c) $a \cdot c \equiv b \cdot d \pmod{m}$.

Example 3 Use the theorem and add to obtain a new congruence:

$3 \equiv 12 \pmod{9}$ [the remainder is 3]
$5 \equiv 14 \pmod{9}$ [the remainder is 5]
$8 \equiv 25 \pmod{9}$ [by addition; the remainder is 8]

Before providing a proof, we use this example to illustrate why the theorem holds in this instance:

$3 \equiv 12 \pmod{9}$ means that $12 = 9 \cdot p + 3$ or $9 \mid (12 - 3)$,

$5 \equiv 14 \pmod{9}$ means that $14 = 9 \cdot q + 5$ or $9 \mid (14 - 5)$.

We add these equations: $26 = 9 \cdot (p + q) + 8$

This equation means $26 \equiv 8 \pmod{9}$ and $9 \mid (26 - 8)$. This illustration can be generalized to show that addition is possible for congruences.

Example 4 Show that if $a \equiv b \pmod{m}$ and $c \equiv d \pmod{m}$, then

$$(a + c) \equiv (b + d) \pmod{m}.$$

We start with the definition of the congruence relation and write:

1. $a \equiv b \pmod{m} \leftrightarrow m \mid (a - b)$
   $c \equiv d \pmod{m} \leftrightarrow m \mid (c - d)$

18. Use part (a) of the theorem to obtain another congruence:

$$12 \equiv 32 \pmod 4$$
$$8 \equiv 0 \pmod 4$$

19. Use part (b) of the theorem to obtain another congruence:

$$11 \equiv 26 \pmod 5$$
$$7 \equiv 12 \pmod 5$$

20. Use part (c) of the theorem to obtain another congruence:

$$19 \equiv 10 \pmod 3$$
$$14 \equiv 23 \pmod 3$$

2. Now $m \mid (a - b) + (c - d)$ [by a property of the "divides" relation]
3. But $(a - b) + (c - d) = (a + c) - (b + d)$ [by a property of the integers; see Exercise Set 8.2]
4. Therefore $m \mid (a + c) - (b + d)$ [by renaming]
5. Finally, $a + c \equiv b + d \pmod m$ [by the definition of congruence]

Proofs for parts (b) and (c) of the theorem on the arithmetic of the congruence relation can be done in a similar manner and are left for the exercises. We apply the theorem in the following examples.

Example 5 Obtain a new congruence from the following by subtracting $24 \equiv 13 \pmod{11}$ from $71 \equiv 16 \pmod{11}$.

We apply the theorem; subtract:

$71 \equiv 16 \pmod{11}$	$11 \mid 71 - 16$
$24 \equiv 13 \pmod{11}$	$11 \mid 24 - 13$
$47 \equiv 3 \pmod{11}$	$11 \mid 47 - 3$

Example 6 Obtain a new congruence from the following by multiplying $11 \equiv 43 \pmod 8$ and $2 \equiv 18 \pmod 8$.

We apply the theorem; multiply:

$$\begin{array}{r} 11 \equiv 43 \pmod 8 \\ 2 \equiv 18 \pmod 8 \\ \hline 22 \equiv 774 \pmod 8 \end{array}$$

Try Exercises 18, 19, and 20

Applications of the Congruence Relation

CHECKING COMPUTATIONS. Computations can be checked by using a congruence of any modulus, since, if the remainder of a sum is found to be the same as the sum of the remainders under some specified modulus m, then the addition has been checked. This is a generalization of casting out 9's.

Example 7 Check the following addition by using modulus 7.

Addition	Check	
125	R = 6 (mod 7)	
38	R = 3 (mod 7)	Add
15	R = 1 (mod 7)	
178	10	
$178 \equiv 3 \pmod 7$	Sum $10 \equiv 3 \pmod 7$	

Subtraction, multiplication, and division can be checked in a similar manner.

Try Exercise 21

PROBLEM SOLVING. Not only are congruences interesting in themselves, but they are also helpful in studying divisibility and prime numbers and can be used in solving certain problems related to these concepts.

Example 8 Rummaging in a desk drawer, Suzanne found some 11-cent stamps and some 10-cent stamps. She noticed that the total value of the stamps was $5.71. How many of each kind of stamp were there?

If you were searching for ideas to solve this problem, it might occur to you to translate the conditions of the problem to an equation. We do this: Let x be the number of 11-cent stamps and y be the number of 10-cent stamps; then

$$11x + 10y = 571.$$

The solutions (x, y) of this equation must be whole numbers, since stamps do not come any other way. One idea for finding the solutions might be by systematic trial and error, but we find that the concept of congruence can be applied.

We write the equation $11x + 10y = 571$ in arithmetic modulo 10. The 10 is chosen so that y will not appear in the final congruence.

Since $11 \equiv 1 \pmod{10}$, $10 \equiv 0 \pmod{10}$, and $571 \equiv 1 \pmod{10}$, then any x that satisfies $11x + 10y = 571$ will satisfy $x + 0 \equiv 1 \pmod{10}$.

The congruence has the following set of whole-number solutions:

$$\{1, 11, 21, 31, \ldots\},$$

any of which can be used in the original equation to find the corresponding y-values that are whole numbers:

x	1	11	21	31	41	51
y	56	45	34	23	12	1

All of these pairs turn out to be solutions of the problem. For example, $x = 21$ and $y = 34$. If the number of 11-cent stamps is 21, then their value is $2.31; the value of the 34 10-cent stamps is $3.40. The total value is $5.71.

Try Exercise 22

Equations that are of the same type as in Example 8 can be solved in the same way, provided that the coefficients of x and y are relatively prime and that the

21. Check the following addition, using congruence modulo 3:

$$\begin{array}{r} 40 \\ 16 \\ \hline 56 \end{array}$$

22.

a) Show that while 61 and 71 are possible solutions for x, the corresponding values for y are not.

b) Find five different solutions to the following:

$$x \equiv 2 \pmod{4}.$$

constant is not less than the product of these coefficients. It can be proved that an equation $ax + by = c$ for integers a, b, and c has one or more integer solutions if and only if the greatest common factor of a and b also divides c.

For example, $4x + 8y = 3$ does not have a solution in the set of integers because the greatest common divisor of 4 and 8 is 4, which does not divide 3.

EXERCISE SET 9.3

1. Construct addition and multiplication tables for J_6, arithmetic modulo 6.

2.
a) Find those elements that have multiplicative inverses in J_6.
b) Compare the elements that have inverses in J_6 with those that have them in J_4. Do you detect a pattern?
■ c) Make a conjecture that would enable you to predict what property an element must have in order to have a multiplicative inverse. Check your conjecture with J_7 and J_8.

3. This is part of a row from a multiplication table of a modular arithmetic. Which row and which arithmetic?

0	3	6	9	1	4

4. This is part of a row from a multiplication table of a modular arithmetic. Which row and which arithmetic?

0	5	10	0

5. What is the set of possible remainders for congruence mod 9?

6. What is the set of possible remainders for congruence mod 8?

7. Use the roster method to describe the seven sets of integers into which the set of integers is partitioned by the relation of congruence mod 7.

8. Repeat Problem 7 for congruence mod 5.

■ **9.** Show that the definition for the congruence relation modulo m is equivalent to the statement $a \equiv b \pmod{m}$ if and only if $m \mid (a - b)$.

10. If q is a whole number and $q \equiv 3 \pmod{9}$,
a) what is $q + 4 \pmod{9}$?
b) what is $q + q \pmod{9}$?

11. If n is a whole number and $n \equiv 5 \pmod{8}$,
a) what is $n + 7 \pmod{8}$?
b) what is $n + n + n \pmod{8}$?

12. If $z \equiv 2 \pmod{5}$, what is $z \cdot z \pmod{5}$?

13. If $w \equiv 1 \pmod{9}$, what is $w^3 \pmod{9}$?

■ **14.** Show that the addition check that uses the remainders for some modulus m is correct for the sum of three whole numbers.

15. Check the following computations by using congruence mod 7.

a) $\begin{array}{r} 372 \\ +513 \\ \hline 886 \end{array}$

b) $\begin{array}{r} 327 \\ -126 \\ \hline 202 \end{array}$

c) $\begin{array}{r} 316 \\ \times\ 21 \\ \hline 6636 \end{array}$

d) $17\overline{)128}$ with quotient 7, $R = 9$

■ **16.** The concept of congruence can be used to prove that the "casting out of 9's" serves as a check on whole-number computation. Prove that for whole numbers a and k,

$$10^k \equiv 1 \pmod 9 \qquad \text{and} \qquad a \cdot 10^k \equiv a \pmod 9.$$

Show how the proof follows from these two statements.

17. Which of the following have positive integer solutions? Find them.

a) $4x + 7y = 45$
b) $9x + 12y = 125$
c) $5x + 7y = 38$
d) $4x + 8y = 33$

■ **18.** A student who had taken a 100-point test on congruence noted a curious fact about the grade on the test. The grade had a remainder of 0 when divided by 3, a remainder of 1 when divided by 2, a remainder of 2 when divided by 5, and a remainder of 3 when divided by 4. What was the student's grade? Did you need all of the information? more information?

■ **19.** Is 999 999 divisible by 7? The first and probably easiest idea would be to perform the division. Do not do this. Look for a way to use congruences. *Hint*: $999\,999 = 10^6 - 1$. The solution to this problem will give you a clue about the application of congruences to divisibility problems.

■ **20.** One of the students in your class has been trying out some things on a calculator and has come up with a "magic number." The student asks you to choose a number from 1 to 9 and multiply it by 109, then asks you to find the sum of the digits. Knowing the sum of the digits, the student is able to tell you the number with which you started. Another student maintains there are lots of numbers that work in this trick. What is the magic about 109? Why does it work? Find some other such numbers that will work just as well.

21. Show that for any integers a, b, c, and d, if $a \equiv b \pmod m$ and $c \equiv d \pmod m$, then $a - c \equiv b - d \pmod m$.

22. Show that for any integers a, b, c, and d, if $a \equiv b \pmod m$ and $c \equiv d \pmod m$, then $a \cdot c \equiv b \cdot d \pmod m$.

■ **23.** Prove that if p is a prime number and a is a whole number, where $a < p$, then there is a unique whole number b such that $b < p$ and $a \cdot b \equiv 1 \pmod p$.

■ **24.** If a and m are any integers and $\text{GCD}(a, m) = 1$, then there is an integer b such that $a \cdot b \equiv 1 \pmod m$. Prove this statement.

■ **25.** Prove that for any integers a, m, and c ($m \neq 0$ or 1) such that $\text{GCD}(a, m)$ is 1, there is an infinite number of solutions to the equation $ax + my = c$.

9.4 PROBLEM SOLVING

AN OLD CHINESE PUZZLE. Number puzzles have fascinated professional mathematicians as well as others for a long time, and some of the concepts developed in this chapter are useful in solving many of them. Such is an old puzzle that dates at least as far back as the first century AD and is credited to a Chinese mathematician Sun-Tsu. The Chinese acquired considerable knowledge about such problems long before this type of mathematics was developed in Europe. Here is the puzzle.

Example 1 What are the two smallest positive integers that satisfy the following three conditions: they both have the same remainders of 2, 3, and 2 when divided by 3, 5, and 7 respectively?

Try Exercise 23

The facts and the problem statement are comparatively easy in this case; what we need is some ideas for finding the solution.*

Idea 1. Use systematic trial and error. Start with 7 and try each successive whole number.

Idea 2. Use a calculator to help in idea 1.

Idea 3. Find a set of positive integers that satisfies one condition, then another for the second condition, then another for the third. The intersection of these sets is the set of solutions.

Idea 4. Use a congruence relation, since there is a direct connection to remainders.

The Solution The first three ideas will certainly lead to a solution, but we choose instead to apply the concept of congruence (idea 4) hoping that it will be less work.

The three conditions that the two numbers must satisfy are:

a) $R = 2$ when the number is divided by 3.
b) $R = 3$ when the number is divided by 5.
c) $R = 2$ when the number is divided by 7.

We can use each of these conditions in turn. From (a) we know that if $R = 2$ when the number is divided by 3, then the number can be written as $3 \cdot n + 2$, where n is a whole number.

Now this same number must meet condition (b), so $3 \cdot n + 2 \equiv 3 \pmod 5$ or $3 \cdot n \equiv 1 \pmod 5$.

Try Exercise 24

9.4 AIMS

You should be able to:

a) Apply the problem-solving methods of this section to an appropriate problem.
b) Demonstrate increased problem-solving skill in examples that may not be directly related to this section.

23.
a) What are the relevant facts in Example 1?
b) State the problem.

24. Prove that if $3 \cdot n + 2 \equiv 3 \pmod 5$, then $3 \cdot n \equiv 1 \pmod 5$.

* A reminder:

THE FIVE-STEP PROCESS

1. What are the *relevant facts*?
2. What actually is the *problem*?
3. What are my *ideas*?
4. How do I get a *solution*?
5. *Check* and *review* the solution.

By the arithmetic of congruence we can write:

$$\begin{array}{l} 3 \cdot n \equiv 1 \pmod 5 \\ 3 \cdot n \equiv 1 \pmod 5 \\ \hline 6 \cdot n \equiv 2 \pmod 5 \end{array}$$

The 6 can be replaced by 1 in the congruence since $6 \equiv 1 \pmod 5$. Therefore, $n \equiv 2 \pmod 5$.

From this congruence we can conclude that $n = 2 + 5 \cdot y$, where y is any whole number. This expression for n can be used in condition (a), where we found that the number must be $3 \cdot n + 2$:

$$3 \cdot n + 2 = 3 \cdot (2 + 5 \cdot y) + 2 = 6 + 15 \cdot y + 2 = 8 + 15 \cdot y.$$

Hence a number that can be expressed as $8 + 15 \cdot y$ must have a remainder of 2 when divided by 3 and a remainder of 3 when divided by 5.

Try Exercise 25

25. Show that $8 + 15y \equiv 2 \pmod 3$ and also that $8 + 15y \equiv 3 \pmod 5$.

We have now established that $n = 8 + 15y$ satisfies the first two conditions of the problem, so we impose the third condition. For what values of y will n have a remainder of 2 when divided by 7?

This means $n = 8 + 15y$ must be congruent to 2 (mod 7). Since

$$8 \equiv 1 \pmod 7 \qquad \text{and } 15 \equiv 1 \pmod 7,$$

we get

$$8 + 15y \equiv 1 + y \equiv 2 \pmod 7.$$

Therefore $y \equiv 1 \pmod 7$. Thus y must be an element of the set

$$\{\ldots, {}^{-}6, 1, 8, 15, 22, 29, \ldots\}.$$

We can find $n = 8 + 15y$ by using each value for y in turn. If y is less than 0, then n is not a positive integer, so we will start with $y = 1$:

$$n = 8 + 15y = 8 + 15 \cdot 1 = 23.$$

Thus 23 is the smallest positive integer that satisfies the conditions of the Chinese puzzle.

Try Exercise 26

26. Since $n = 8 + 15 \cdot y$ and y must be an element of the set $\{1, 8, 15, 22, \ldots\}$, the values for n can be found easily. Find the first four positive integers n that satisfy the conditions of the Chinese puzzle.

Perhaps one of the other ideas would have taken less time in this problem; however, with idea 1 we would have had to do a fair amount of calculation before arriving at the second n that satisfied the problem conditions.

The solution just above depends upon a theorem we state without proof: If m and n are relatively prime, then it is always possible to find a number x such that $x \equiv r \pmod m$ and $x \equiv s \pmod n$, no matter what r and s are.

A variety of ideas could have been used in this Chinese puzzle; as a rule, it is not wise to rush into finding a solution before you have considered more than one approach. The solution based on the concept of congruence has possibilities for application in other situations, such as questions about divisibility.

A glance toward the classroom

It is important that prospective teachers have some personal experience in solving nontrivial problems, so that they will be more confident about guiding elementary pupils. Such experience also contributes to a teacher's ability to cope with unexpected and nonroutine questions in arithmetic.

EXERCISE SET 9.4

Not all the problems in this set depend directly upon the example developed in this section. Use the five-step process and any ideas from earlier chapters as well.

1. Solve the Chinese puzzle of Example 1 by applying ideas 1 and 2. Compare the time and effort required by this method with that used for the solution in the text.

2. Use idea 3 to solve the Chinese puzzle. Compare this solution with that done in the text. Is it longer or shorter? Is it easier?

3. A physical-education instructor was organizing a class into teams. First, groups of 9 were counted off, but the last team had only 7 pupils in it. The instructor then tried groups of 7, but the last team had only 4 pupils. This instructor knew that there were no more than 45 in the class and was then able to divide the class into teams with the same number of pupils. How many teams and how many per team were there?

4. Find an x that satisfies both congruences:

$$x \equiv 3 \pmod{7}, \qquad x \equiv 4 \pmod{9}.$$

5. Does the solution in Problem 4 also satisfy the congruence $x \equiv 1 \pmod{3}$?

■ **6.** The natives of Sunburst Island rate 2 spears worth 3 fishhooks and a knife. They will give 25 coconuts for 3 spears, 2 knives, and a fishhook together. How many coconuts will they give for each article separately?

7. Because $3^3 = 27$ and $27 \equiv 4 \pmod{23}$, we know that the number 3^3 when divided by 23 has the remainder 4. What is the remainder when $3^3 \cdot 3^3$ is divided by 23? when $3^3 \cdot 3^3 \cdot 3^3$ is divided by 23?

8. Is $3^{30} - 1$ divisible by 7? This is the same as asking if $3^{30} - 1 \equiv 0 \pmod{7}$.

9. Here is one for a bright fifth-grader: In your pocket you have \$1.08 in change, which consists of only cents, dimes, and nickels. You have twice as many coins as you have nickels. Find how many of each coin. Is there more than one solution to this problem?

10. Elementary pupils are often given patterns to examine in order to discover some general relationship or principle. Here is one for you:

n	n^2	$n^2 \pmod 4$
1	1	1
2	4	0
3	9	1
4	16	
5	25	
6	36	

a) What pattern do you see in this chart?
b) Extend the chart for a few numbers to see if the pattern holds.
■ c) Prove that this pattern holds for all square numbers.

11. Here is one that can be used to baffle elementary pupils: Tell the pupil to write down any 3-digit number with the first and third digits different. Have the pupil reverse the digits and subtract the smaller from the larger. Reverse the digits of the answer. Add the last two numbers. Here is where the pupils will be surprised: everyone will have the same answer, 1089. Investigate a few cases for yourself. ■ Show that it will work in every case.

Example

$$\begin{array}{r} 372 \\ -\ 273 \\ \hline 099 \\ +\ 990 \\ \hline 1089 \end{array}$$

12. This problem concerns an anecdote supposedly about Pythagoras and a slave he wished to punish. He told the slave to walk up and down past the 7 columns of the local Temple of Diana and count each one until he reached 1000. He was to start at one end, count one-by-one to the end, which was then 7, then turn. The column he had counted as 6 was now 8. In this way, by the time he reached the first column, it was counted as 13, whereupon he turned and the second column was counted as 14. He was to report to Pythagoras which column it was that was counted as 1000. Pythagoras knew a short way to find this out. What column should the slave have reported as the one that was counted 1000?

■ **13.** A primary child inadvertently spilled a box of 1 cm plastic cubes all over the floor. The teacher was trying to see that all of them had been picked up and asked how many were in the container. The pupil replied that if the cubes were counted by 2's, there was 1 left over, if counted by 3's there was 1 left over, if counted by 4's there was 1 left over, if counted by 5's there was 1 left over, if counted by 6's there was 1 left over, but that if the blocks were counted by 7's there was none left over. The teacher then said: "I can find out what the least number of cubes there must be." Find out the least number of cubes there can be under these circumstances.

REVIEW TEST

1. Describe and illustrate the relation between multiplicative inverses and division in the 5-hour clock arithmetic J_5.

2. Are all divisions possible in arithmetic mod 6? Explain and illustrate your answer with examples.

3. Show that $5 \cdot (6 - 4) \equiv (5 \cdot 6) - (5 \cdot 4) \pmod 7$.

4. Find the multiplicative inverse for each element of arithmetic mod 7 that has one.

5. Repeat Problem 4 for arithmetic mod 9.

6. Name a modular arithmetic in which $6 \otimes 4 = 1$. Is there more than one such arithmetic?

7. Name two modular arithmetics in which it is possible to have $3 \otimes x = 0$, although $x \neq 0$. In each case state a number that makes the equation true.

8. Does the following equation have a solution in arithmetic mod 5? in arithmetic mod 4? $(3 \otimes x) \ominus 2 = 0$.

9. Find five whole numbers that are congruent to 36 modulo 16.

10. What is the set of possible remainders for arithmetic mod 10?

11. If $p \equiv 3 \pmod 9$, what is $p^2 \pmod 9$, $(p + p) \pmod 9$?

12. Use congruences to check the following computation in mod 4:

a) $\begin{array}{r} 635 \\ +127 \\ \hline 752 \end{array}$ b) $\begin{array}{r} 516 \\ -322 \\ \hline 184 \end{array}$ c) $\begin{array}{r} 425 \\ \times \quad 24 \\ \hline 10190 \end{array}$

13. Does the following have a whole-number solution? If yes, find one such solution for $5x + 9y = 126$.

■ **14.** The equation $5x + 7y = 35k$ has positive solutions for some values of k. There is a value of k for which there are exactly 6 positive integer solutions for both x and y. Find them.

FOR FURTHER READING

1. "Let's play mod 7" by Christopher E. Niemann appears in *The Arithmetic Teacher*, 23:5, May 1976, pp. 348–350. This article relates how a game of mod 7 led to a simple method that allows you to "tell what day of the week a given date is." The author's small children learned to do it quickly.

2. If you learn this Chinese "trick" using modular arithmetic, you will be able to amaze your friends by guessing their ages. Read "An indirect method for calculating ages" by Carl Wang and Norman Woo in *The Mathematics Teacher*, 68:8, December 1975.

TEXTS

Koshy, Thomas, *An Elementary Approach to Mathematics*. Goodyear Publishing Co., Pacific Palisades, California (1976).

Myers, Nancy, *The Math Book*. Hafner Press, a division of Macmillan Publishing Co., New York, New York (1975).

chapter 10 geometry

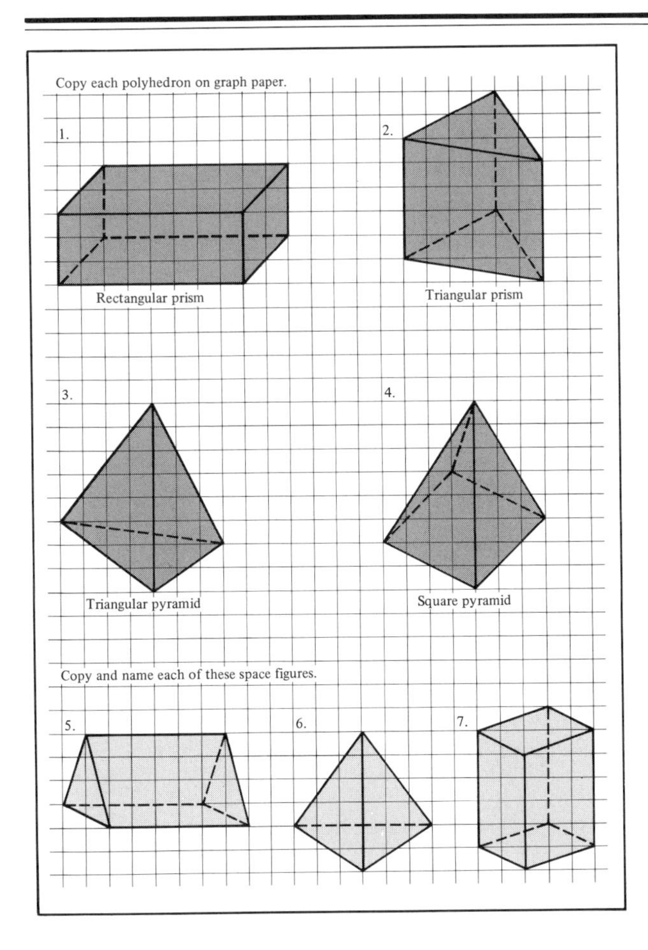

Graph paper can be used to good advantage, as in this sixth-grade text, to improve the pupils' observational, perceptual, and drawing skills.

10.1 INTRODUCTION

Ancient mathematicians spent a great deal of time on practical geometry problems, and we find that close to a quarter of the problems in the Rhind Papyrus* were of this type. The problems of the Papyrus were concerned with the formulas (based on trial and experiment) that were needed for finding land areas and the storage capacities of granaries.

Something closer to the formal geometry you are familiar with from high school began early in the 6th century BC. This geometry is credited to a merchant, Thales of Miletus (a town on the west coast of Asia Minor). Thales made a significant contribution to the mathematics of his time, since his work represented a basic change from inductive to deductive methods in geometry in that he established a number of theorems whose proofs did not depend upon illustrations but rather on logical reasoning from some sound basis to the conclusion.

Pythagoras, the famous Greek mathematician, also contributed to the development of deductive geometry and he is generally given credit for a theorem bearing his name. The development of Greek geometry gradually advanced for 300 years and finally, about 300 BC Euclid† produced *The Elements*, a book especially remarkable for its dominating influence on all the subsequent teaching of geometry. *The Elements* contains over 450 theorems, and although editorial changes have taken place the axioms (general truths) and postulates (geometric truths) Euclid used were probably the following:

Axioms

A_1 Quantities equal to the same quantity are equal to each other.
A_2 If equals are added to equals, the sums are equal.
A_3 If equals are subtracted from equals, the differences are equal.
A_4 Things which coincide are equal.
A_5 The whole is greater than any of its parts.

Postulates

P_1 A straight line can be drawn from one given point to another given point.
P_2 A finite line can be extended continuously in a straight line.
P_3 A circle can be drawn given any point as center and a finite straight line as radius.
P_4 All right angles are equal to one another.
P_5 If two lines are intersected by a third line so that the interior angles on one side are less than a straight angle, then the two lines, if infinitely extended, will intersect on that side.‡

Euclid claimed to derive all the 465 propositions (theorems) by logical deduction starting from these basic ten axioms and postulates. It is in a formal treatment of geometry that most American school children meet a logical, deductive, mathematical system for the first time. For many years it was also the first time these students had ever considered the ideas of geometry at all.

10.1 AIMS

You should be able to:
State the 10 axioms and postulates given by Euclid in the geometry he developed in *The Elements*.

* This Egyptian papyrus of about 1650 BC contains 85 problems and is one of the sources of our present-day knowledge of ancient Egyptian mathematics. It is named after A. Henry Rhind (1833–1863), the English Egyptologist who originally purchased it.

† Euclid was professor of mathematics at the University of Alexandria and probably received his training in Athens in the Platonic school. Apparently he was also the founder of the famous school of mathematics located at the University at Alexandria. Not much is known about the details of his life.

‡ Euclid's fifth postulate can be shown to be logically equivalent to the more familiar statement—*The parallel (or Playfair) postulate*: Through a point P not on a given line L there is exactly one line which can be drawn parallel to the line L. From the Playfair axiom and the other nine postulates and axioms given by Euclid, it is possible to derive Euclid's fifth postulate.

Fortunately, more geometry is now being taught in elementary school to children, who need to become familiar with geometric figures and ideas, to get experience in visualizing figures, and to interpret drawings of such figures although they are not ready for a formal axiomatic presentation of geometry.

EXERCISE SET 10.1

1. What property of equals does Euclid's axiom A_1 refer to?

2. State algebraically the assumption made in A_2.

3. Repeat Problem 2 for A_3.

4. If A•——•B can be made to coincide exactly with C•——•D, what do we know about segment AB compared to segment CD? Why?

5. A square piece of paper is cut into four pieces. Which axiom states that the original square is larger than any piece?

6. Which axiom or postulate guarantees that two points determine a straight line?

7. In Euclidean geometry A•——•B can be extended infinitely far in either direction. By what axiom (or postulate)?

8. Which axiom or postulate assures that circles can be drawn if the center and the radius are known?

9. If the angle a and the angle b together are less than a straight angle on this sketch, what does P_5 say will happen if line L and M are extended?

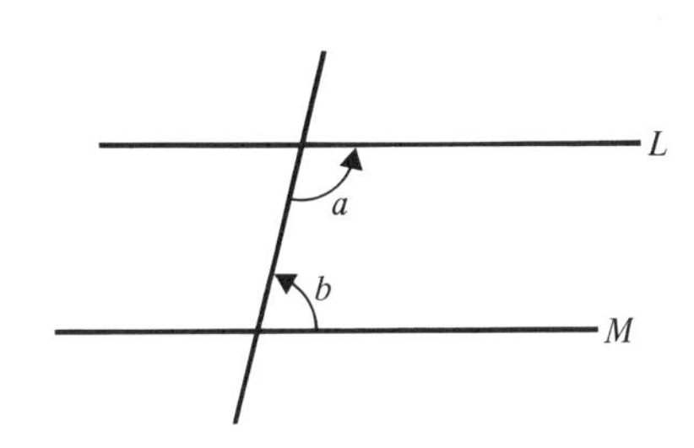

10. How many lines can be drawn through Q parallel to line L? By what axiom?

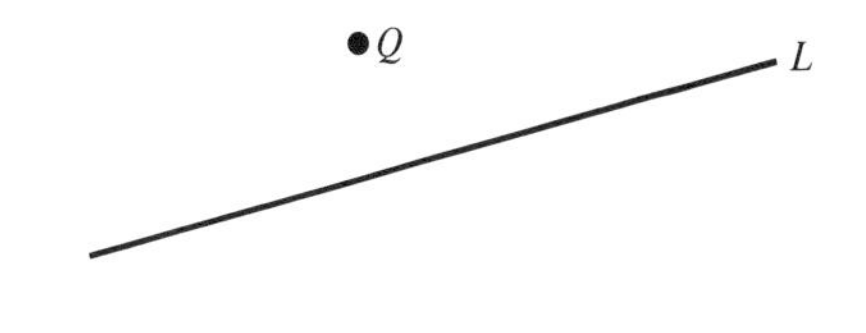

■ **11.** Show that the two adjacent squares with the cuts indicated by broken lines yield three pieces that can be rearranged to form a square with side c. What theorem could be proved with this dissection?

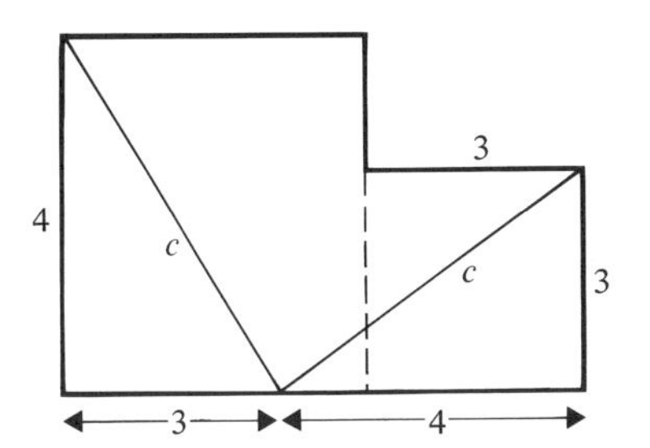

10.2 AIMS

You should be able to:

a) Identify, name, and draw the following kinds of geometric figures: line segments, rays, lines, and half-lines.
b) Identify, name, describe, and draw the following kinds of geometric figures: angles, curves, both closed and simple, polygons according to the number of sides, vertices, regular, convex, and polygonal regions.

10.2 GEOMETRY OF THE PLANE

It is not the purpose of this chapter to develop geometry formally as Euclid did and prove a long succession of theorems from a foundation of axioms and postulates. The informal geometry taught in the elementary school, which has been shown to help children make remarkable progress in their geometric understanding, depends upon experiment and intuition.

Just as we did not define the concept of set earlier, we shall not define the words *point, line, plane, space,* and *between,* since any attempt to do so would get us involved in logical difficulties.

POINTS, LINE. When we use a straight edge and join the points A and B, then we have a *line segment* $\overline{AB}$ with endpoints A and B:

If we imagine the line segment AB being extended infinitely far beyond B, then we have visualized the *ray* $\overrightarrow{AB}$:

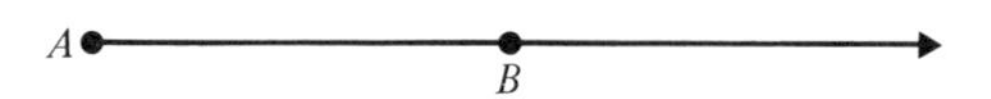

Of course, if we imagine it extended in the other direction as well, it is the *line* $\overleftrightarrow{AB}$:

OPEN HALF-LINES. A point A can be regarded as separating a line into three disjoint subsets.

The line L then consists of the point A, the half-line $\overset{\circ\!\!\longrightarrow}{AB}$, and the half-line $\overset{\circ\!\!\longrightarrow}{AD}$. The symbol $\circ\!\!\longrightarrow$ indicates that the point A is not included in the half-line.

THE PLANE. We imagine a plane as a flat smooth surface that extends infinitely far in all directions. A tabletop, a floor, or a sheet of paper models a portion of such a plane. One assumption about the plane is that any three points which do not lie on one line* determine one and only one plane. For example, points A, B, and C are not on the same line; they all lie on plane P.

* Points that do not lie on a single line are said to be *noncollinear*.

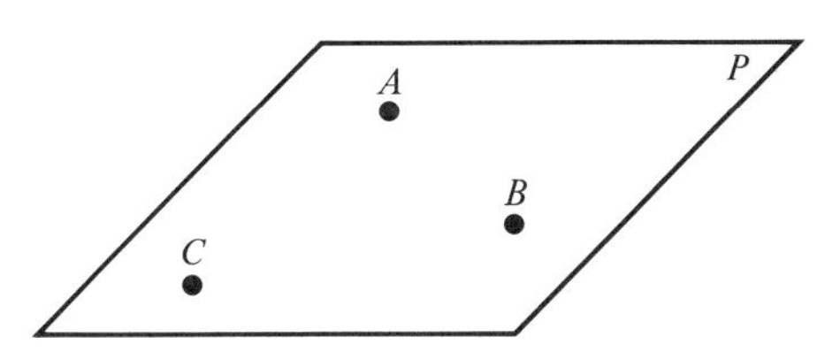

Another assumption is that if two points of a line lie in a plane, then the entire line is on the plane. For example, if points A and B are both on line L and in plane P, then the line L joining A and B is in P.

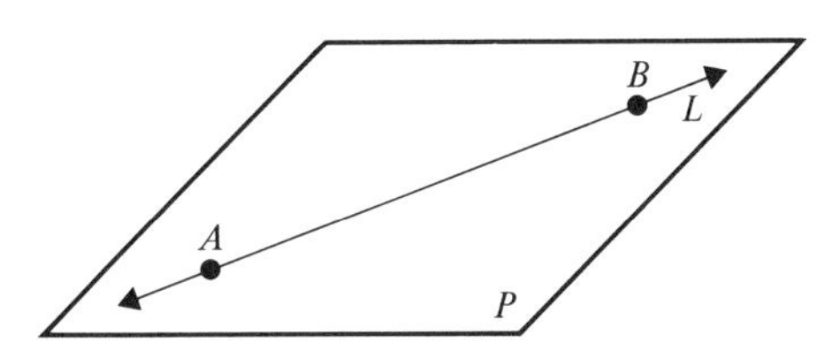

Try Exercise 1

Two lines in a plane need not intersect, since by the parallel postulate parallel lines in the plane do exist. Thus any two distinct lines in a plane are either parallel or they intersect.

HALF-PLANES. Just as a point of a line can be thought of as separating a line into three disjoint subsets, a given line of a plane can be thought of as separating the plane into three disjoint subsets: the line itself, the set of points on one side of the line, and the set of points on the other side of the line. In the figure, line L is the separating or boundary line, the shaded area represents one half-plane and the nonshaded area the other half-plane.

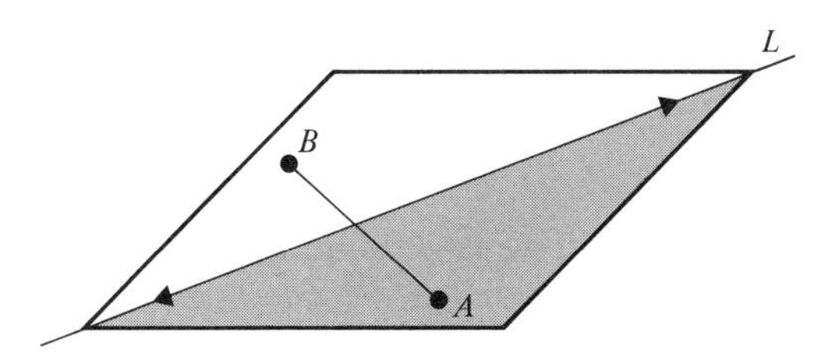

Two points are in different half-planes if the line segment joining them intersects the separating line L. Therefore the points A and B are not in the same half-plane.

Try Exercise 2

1. In the following sketch the three points A, B, and C all lie on the plane. Where are the points of $\overleftrightarrow{AB}$ and $\overrightarrow{AC}$ with respect to the plane?

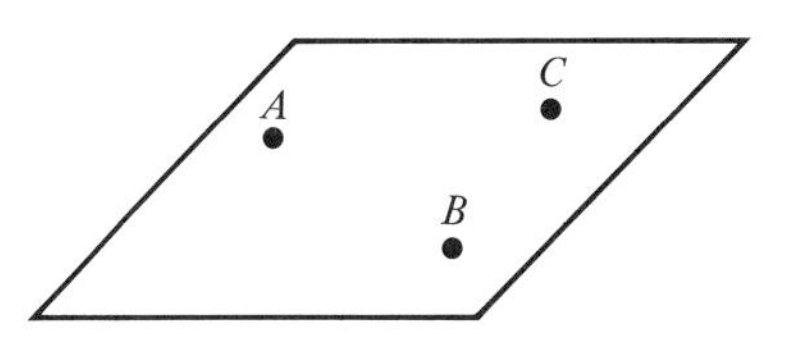

2. In the following sketch the line joining points C and D will intersect the line L that separates the plane into two half-planes, yet C and D seem to be in the same half-plane. Explain this.

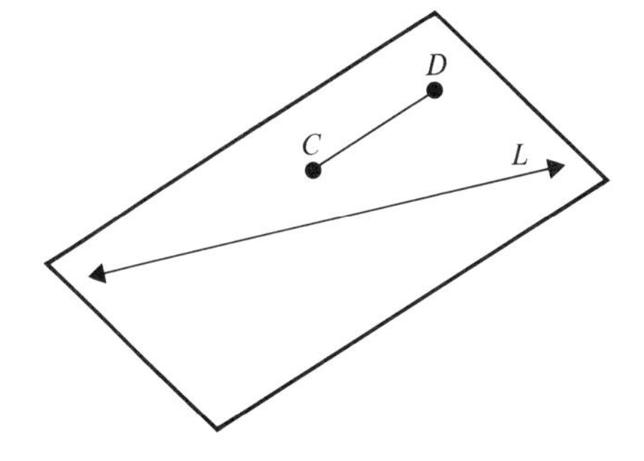

3. Complete the following statement for the obtuse angle ABC sketched below:

$$\measuredangle ABC = \_\_\_ \cup \_\_\_.$$

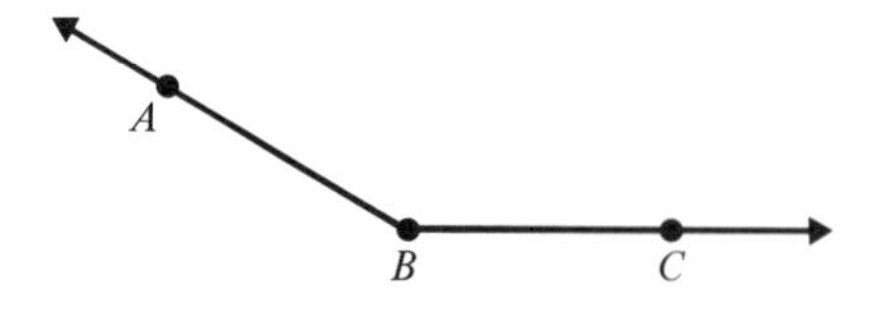

Subsets of the Plane

ANGLE. A line segment, a line, and a ray can be regarded as sets of points. An *angle* is a figure (set of points) that consists of the union of two rays with a common endpoint.

For example, $\measuredangle STR$ is the union of the rays $\overrightarrow{TS}$ and $\overrightarrow{TR}$. Angle STR can also be called $\measuredangle RTS$ or simply $\measuredangle T$, if that label is not ambiguous.

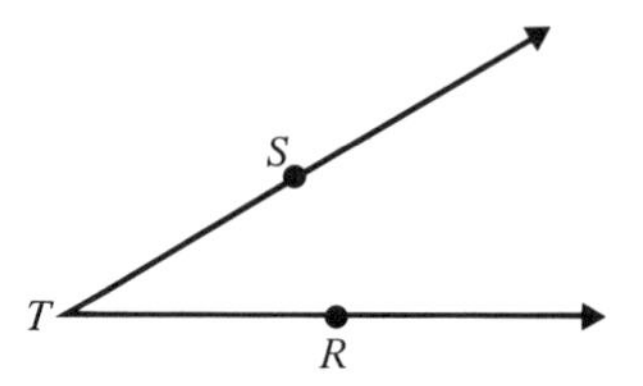

Try Exercise 3

CURVES. Roughly speaking, a *curve** is a figure that can be drawn starting at some point S without lifting the pencil and thus is said to be *connected.* The following are curves:

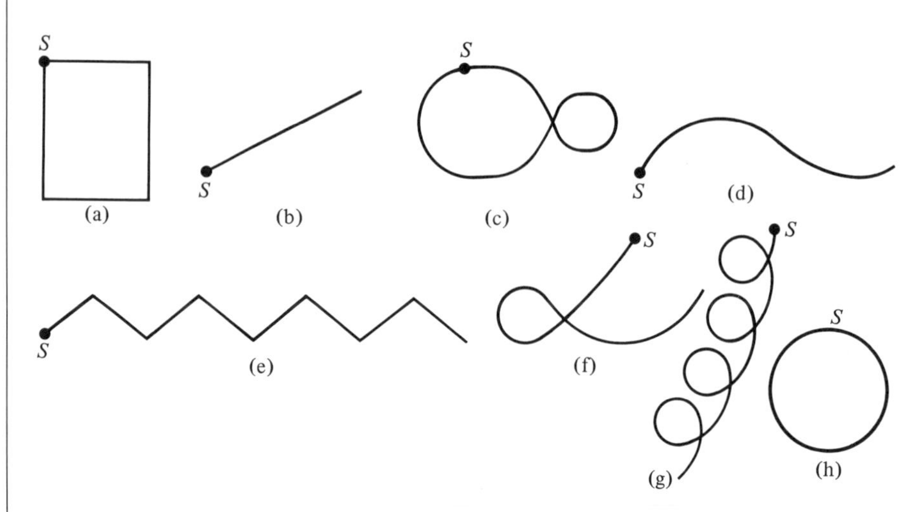

SIMPLE CURVES. A *simple curve* is one that never crosses itself, that is, the pencil never passes through a point of the curve a second time. In the curves above, (a), (b), (d), (e), and (h) are all simple curves, while the rest have at least one retraced point that is not the starting point.

CLOSED CURVES. A curve that goes back to the starting point and hence can be traced more than once in the same direction without lifting the pencil is called *closed.* All the others are called *open* curves. In the curves above, (a), (c) and (h) are closed curves.

* Some curves are important enough to have been given a special name. A famous one is the *versiera* studied by Maria Gaetena Agnesi (1718–1799), a distinguished Italian mathematician. (Due to an error in translation, the versiera has come to be known as the "witch" of Agnesi.) Far more important is her book *Analytical Institutions* (1748), a two-volume work on calculus that caused a sensation in the academic world at the time.

A curve can have more than one characteristic since, for example, a closed curve that does not cross itself is called a *simple closed curve.*

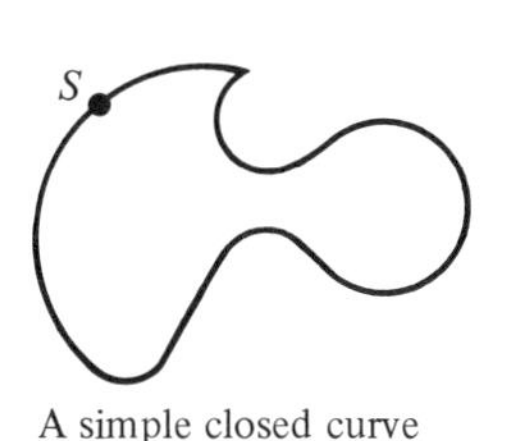

A simple closed curve

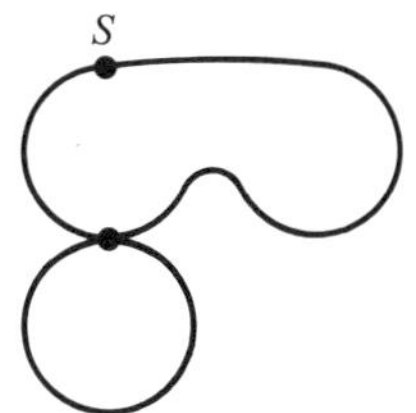

A closed curve that is not simple

Try Exercise 4

CONVEX REGIONS. A simple closed curve separates the plane into three disjoint sets of points: the curve itself, the interior of the curve, and the exterior of the curve.

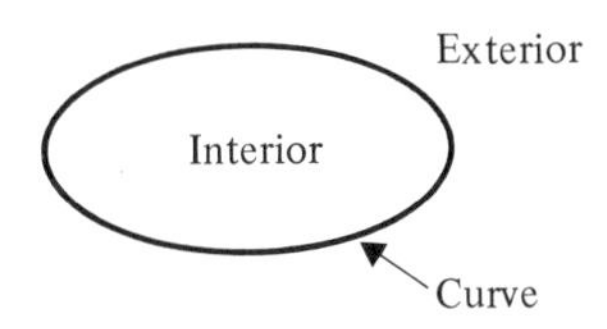

The set consisting of the interior and the boundary is often referred to as a *region.* The following are regions:

A triangular region

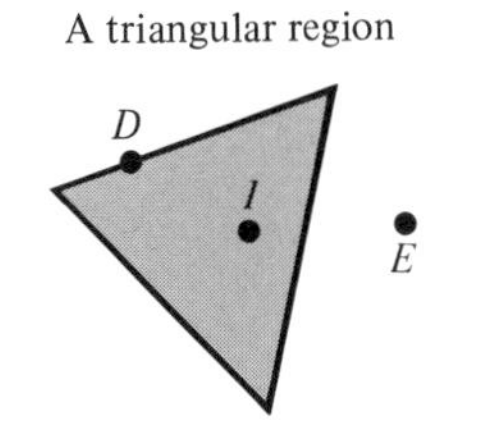

A circular region

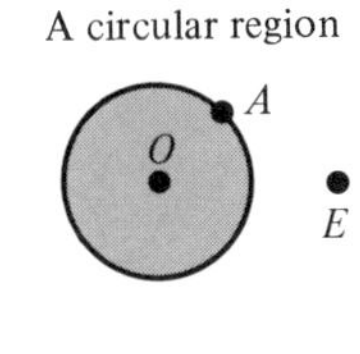

D is a point of the triangle, *I* is an interior point, *E* is an exterior point not in the region. The center *O* is an interior point, *A* is on the circle, and *E* is not in the region.

Some examples of other regions are:

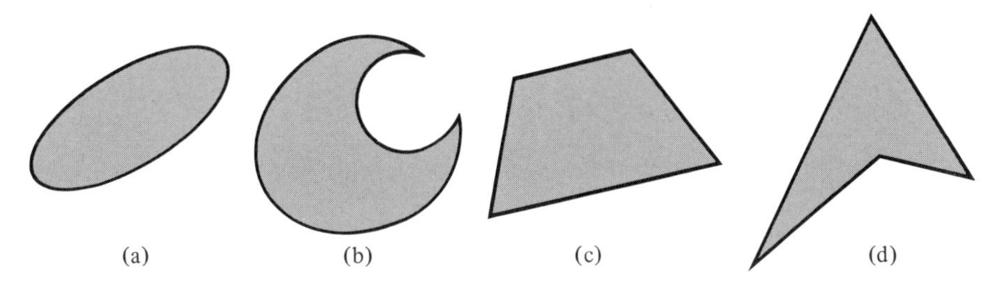

(a) (b) (c) (d)

4. Which of the following are:

a) closed curves?
b) simple curves?
c) simple closed curves?

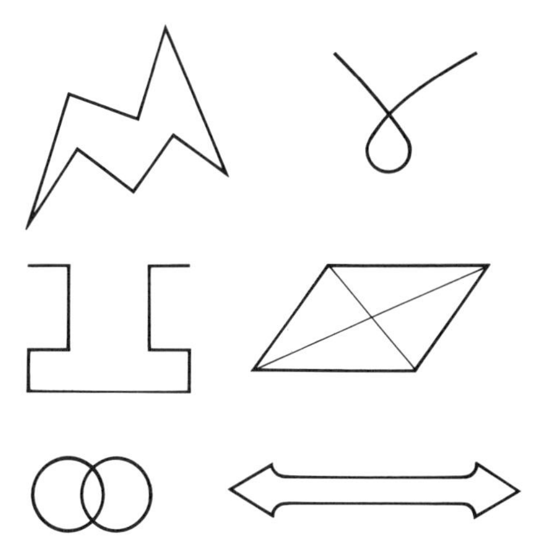

5. Study the figures (a), (b), (c), and (d) in the text. For which of these figures is it possible to choose two points A and B on the boundary of the region so that some part of the segment $\overline{AB}$ lies outside the region?

Try Exercise 5

Whenever it is impossible to find any segment $\overline{AB}$ that has points in the exterior (such as is asked for in Exercise 5), then the region is said to be *convex*.

DEFINITION. A *convex set* is a set of points S such that if points A and B are in set S, then $\overline{AB}$ is also in S.

Thus the above figures (a) and (c) are convex regions, as are any triangular or circular regions.

6. The boundary curve of a convex region is classified as a *convex curve*. For each of the following curves choose a point B to establish that the curve is not convex (it is thus *concave*).

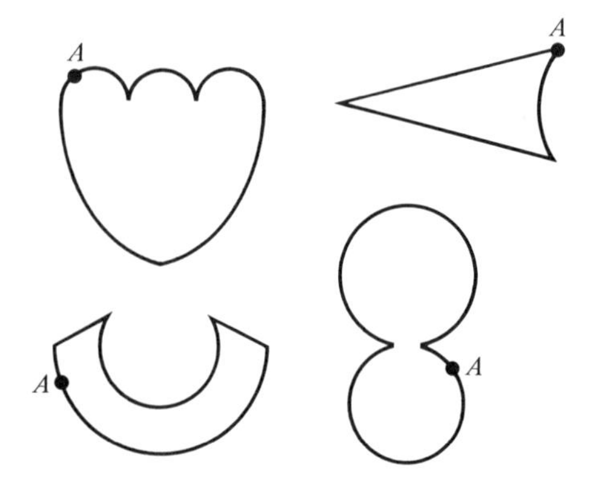

Try Exercise 6

POLYGONS. A simple closed curve which is a union of line segments is called a *polygon*. The common endpoints of a pair of segments is called a *vertex*, and the segments are called *sides*.

Example 1 Which of the following are polygons?

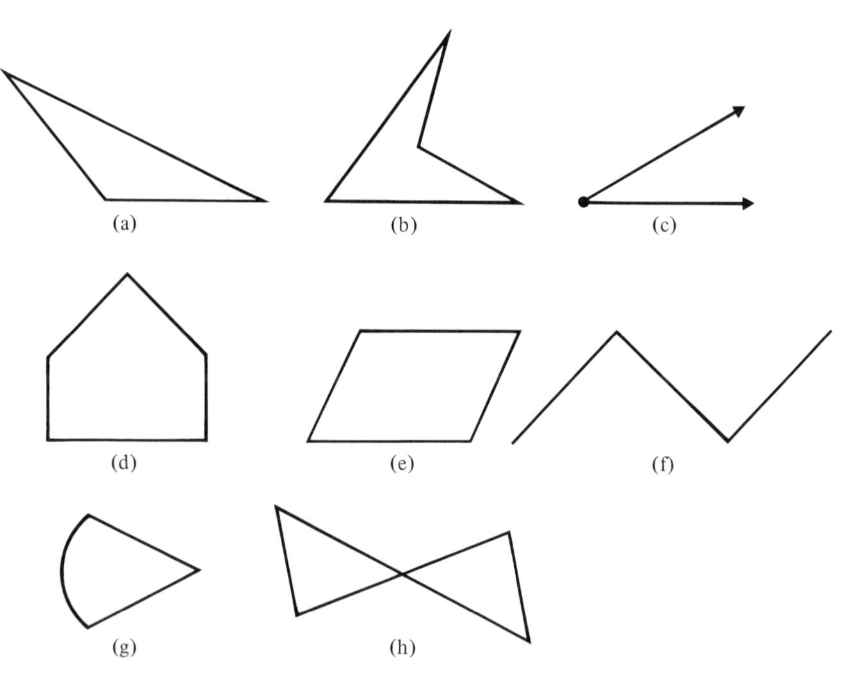

Solution (a), (b), (d), and (e) are all polygons; (c) and (f) are not because they are not closed; (g) is not because it does not consist only of line segments; (h) is not because it is not simple.

7. Answer TRUE for those statements that are always true and FALSE for the rest:

a) Every polygon is a convex figure.
b) Every polygon is a closed curve.
c) Every polygon has at least three sides and three vertices.
d) Every square is a polygon.

Try Exercise 7

A polygon is named according to the number of sides it has, although there are further classifications possible. A particular polygon can be named by listing its vertices in order. The rhombus here would be called the "rhombus $ABCD$" and the trapezoid would be called the "trapezoid $DEFG$."

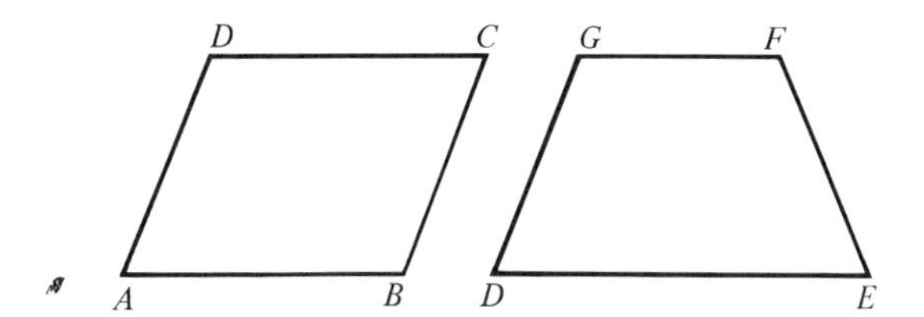

Classification by number of sides

Number of sides	*Example*	*Name*
3	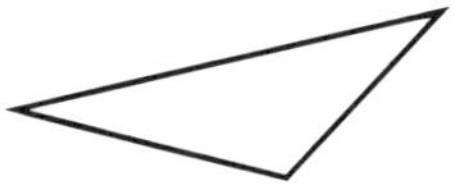	Triangle
4	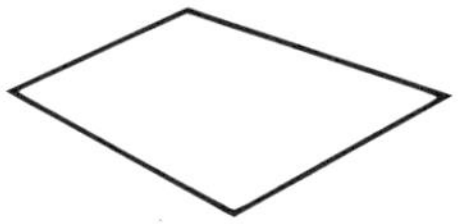	Quadrilateral
5	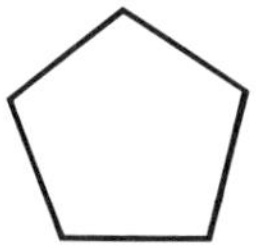	Pentagon
6	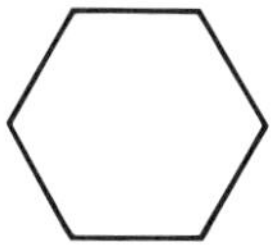	Hexagon
7	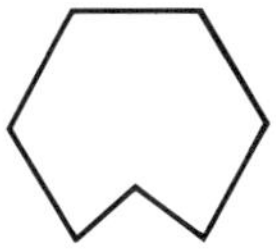	Heptagon
8	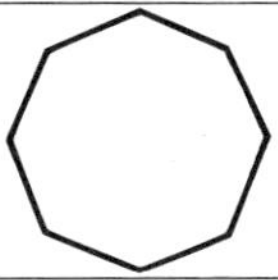	Octagon

If all the sides of a polygon are the same length and all of its angles are the same size, then it is called a *regular polygon.*

A glance toward the classroom

An important part of geometry is vocabulary; however, that should not be the focus of the geometry taught to elementary pupils. Before they are ready for formal definitions for quadrilateral, polygon, equilateral triangle, and the like, they should have a great deal of experience and many activities involving such figures.

EXERCISE SET 10.2

1. The line L has the points indicated. Give six other possibilities for naming the line L.

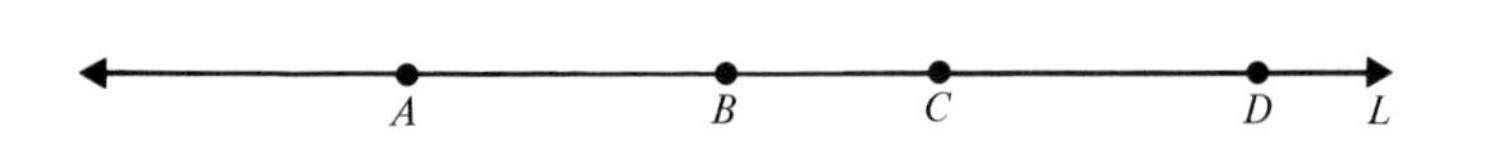

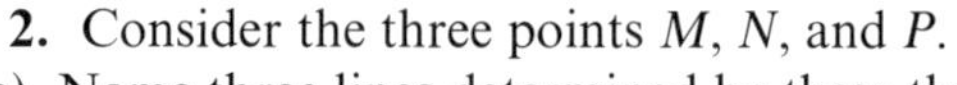

2. Consider the three points M, N, and P.
a) Name three lines determined by these three points.
b) How can three points be arranged so that only one line is determined by them? Make a sketch.
c) Is it possible to have three points arranged so that only two lines are determined? Explain.

•M

•N

•P

3. On the line below, M is the separation point.
a) Does $\overset{\circ}{\overrightarrow{MN}}$ contain point M? point N?
b) Does $\overrightarrow{NM}$ contain point M? point N?

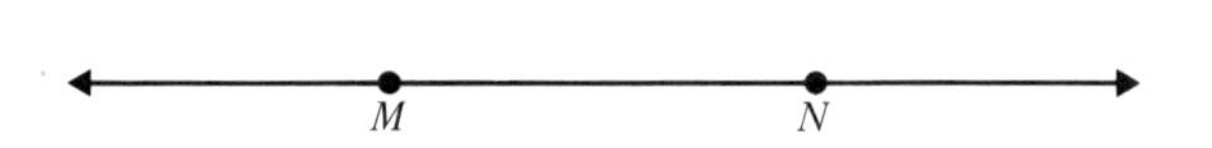

4. Consider the points M, N, and P on a line as follows:
a) How many line segments are determined by M, N, and P? Name them.
b) Is $N \in \overline{MN}$?
c) Is $\overline{MN}$ a subset of $\overrightarrow{MP}$? of $\overrightarrow{NM}$?

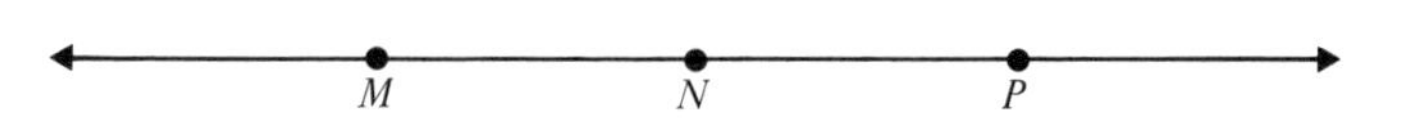

5. The line L contains two points M and N as shown:
a) Name the set of points $\overrightarrow{MN} \cap \overrightarrow{NM}$.
b) Name the set of points $\overrightarrow{MN} \cup \overrightarrow{NM}$.

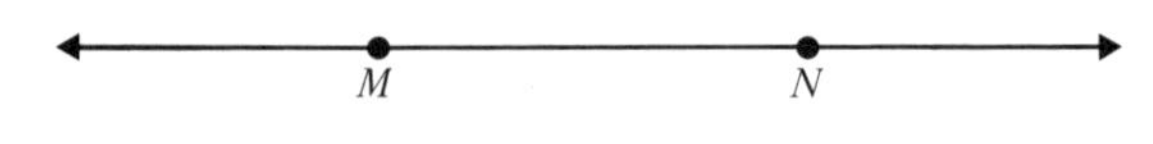

6. Is the intersection of two rays determined by two or more points on a line always a line segment? Is the union always the line?

7. For the following figure, find:
a) $\overrightarrow{BC} \cap \overline{CD}$ b) $\overrightarrow{BC} \cap \overrightarrow{DA}$ c) $\overline{AC} \cap \overline{BD}$

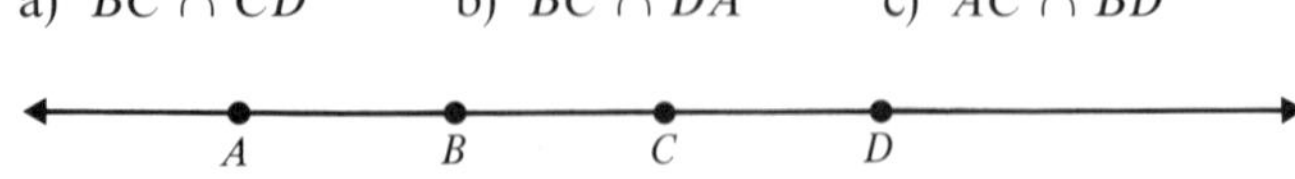

8. Think of any line. Is it possible to place one or more points on the line so that there are two rays determined and these two rays:
a) intersect in a point?
b) intersect in the empty set?
c) intersect in a ray?
d) intersect in a segment?
e) intersect in the whole line?

9. Repeat Problem 8 for the union of two rays.

10. The line $\overleftrightarrow{AB}$ and the line $\overleftrightarrow{CD}$ intersect in the point P. How do we know that all the points of both lines lie in the same plane?

11. Imagine that the plane you are considering is represented by a sheet of paper on your desk. Use a pencil as a model of a line. How many possibilities are there for the intersection of a line and a plane? Make a sketch of each one.

12. If two lines that lie in the same plane have no point of intersection in common, they are called parallel lines. Write an analogous definition for a pair of parallel planes.

13. Use folding to find a parallel line through a point P. On a sheet of paper draw a line L and a point P not on L. Put a fold through P such that the line L is folded back on itself. Draw the line PO of this first fold. Put another fold through P so that PO is folded back on itself. This second fold is parallel to the original line. Does it appear to be?

14.
a) Name all the triangles in the given figure.
b) Point I is in the interior of which triangles?

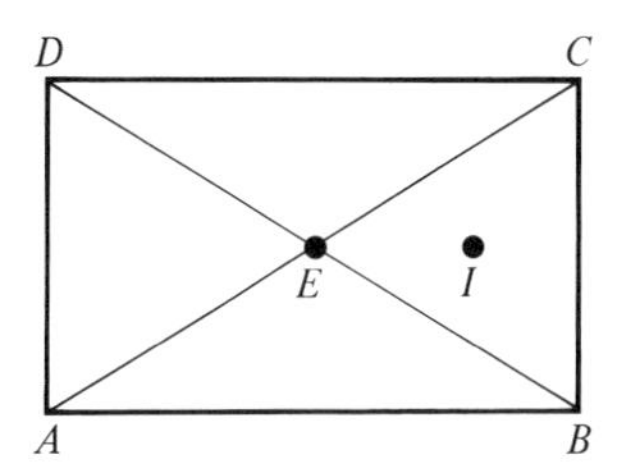

15. Explain why any triangle lies in exactly one plane.

16. Make a sketch to represent each of the following:
a) a concave pentagon
b) a closed convex hexagonal region
c) a regular triangle
d) a simple closed curve containing exactly one line segment
e) a simple closed curve containing exactly two line segments
f) a simple closed curve that has at least one interior point at the same distance from every point on the curve.

17. In the following figure many other closed figures can be isolated. Redraw the figures that you see. For example, ABC is a triangle that can be seen. There are many others. Find at least eight more.

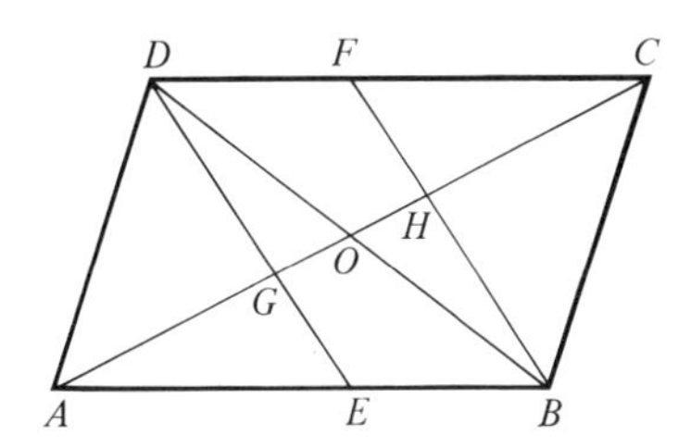

18. Repeat Problem 17 for the figure at the right.

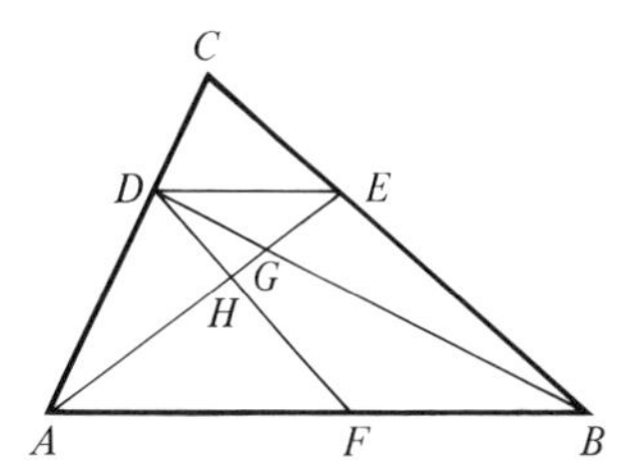

10.3 AIMS

You should be able to:

a) Describe and identify congruent figures.
b) Define a bisector of a segment, perpendicular, angle bisector, and circle in terms of congruence.
c) Find the lines of symmetry of figures that have them.
d) Determine if a figure has rotational symmetry.
e) Describe and identify similar figures.
f) State and apply the definition for similar polygons.

10.3 CONGRUENCE, SYMMETRY, AND SIMILARITY

An important part of the elementary-school geometry program is an informal study of the concepts of congruence, symmetry, and similarity. Pupils get a chance to visualize shapes being moved about in a plane and start to understand some relations that hold among various shapes.

Congruence

For the elementary-school child a technical definition of congruence is not necessary. It is sufficient to explain this concept as follows:

Two figures are *congruent* if they have the same size and shape. That is, two figures that fit exactly on each other are congruent.

In order to check congruence of figures, pupils are asked either to cut out shapes and superimpose them or to make tracings and do the same thing. The following excerpt from a sixth-grade book shows a lesson that could be carried out with tracing paper.

Segments can be congruent to each other.

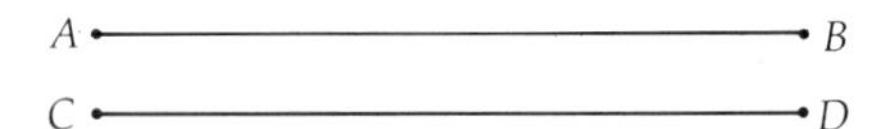

We write: $\overline{AB} \cong \overline{CD}$
We read: Segment AB is congruent to segment CD.

Angles can be congruent to each other.

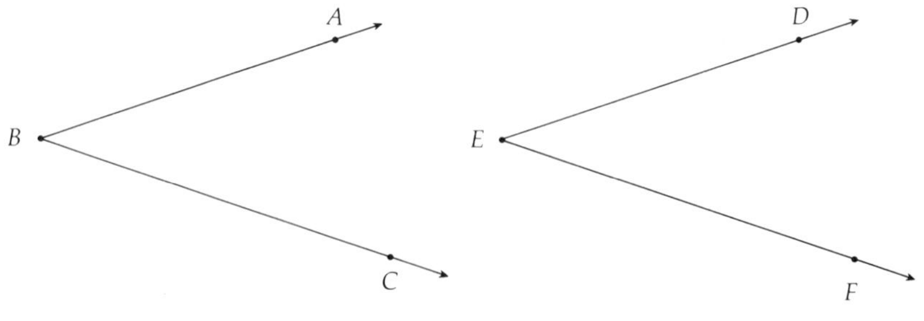

We write: $\angle ABC \cong \angle DEF$
We read: Angle ABC is congruent to angle DEF.

From a sixth-grade textbook

Notice that it is not correct to write for the above segments that $\overline{AB} = \overline{CD}$; these are different segments although they may be the same length. However, we can write $\overline{AB} = \overline{BA}$, since these do refer to the same segment. We can also write $AB = CD$, since AB and CD are the lengths of $\overline{AB}$ and $\overline{CD}$, respectively.

The congruency of two line segments or two angles, as we shall see later, can be checked by means of a compass. It is not necessary to use either a ruler or a protractor.

Try Exercise 8

BISECTORS AND PERPENDICULARS. The concept of congruence makes it easy to state the definition of the midpoint of a line segment.

DEFINITION. A point P of a line segment $\overline{AB}$ is the midpoint of the segment if and only if $\overline{AP} \cong \overline{PB}$.

Try Exercise 9

If two lines meet so that the angles of any adjacent pair of the four angles formed are congruent, then the lines are *perpendicular* and each of the angles is called a *right angle*. For example,

$$\measuredangle COB \cong \measuredangle BOD \cong \measuredangle DOA \cong \measuredangle AOC.$$

$\overleftrightarrow{CD}$ is perpendicular to $\overleftrightarrow{AB}$ ($\overleftrightarrow{CD} \perp \overleftrightarrow{AB}$) and all four angles are right angles.

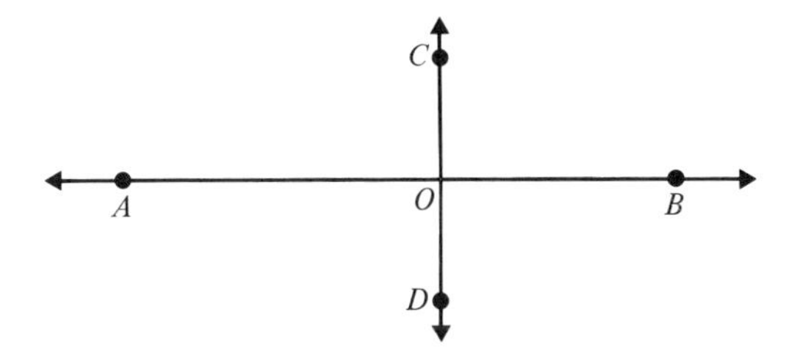

Try Exercise 10

CIRCLES. Although you are quite familiar with the figure called a circle, it can be thought of in terms of congruence, since a circle is a figure that has the following property: for any two points A and B on the curve and a fixed interior point O, $\overline{AO} \cong \overline{BO}$.

Symmetry

SYMMETRY ABOUT A LINE. The concept of symmetry is closely related to (but is not the same as) that of congruence. Roughly, a figure is symmetric about a line if a fold along that line would make the two halves of the figure

8. Use tracings to answer the questions asked in the sixth-grade textbook.

From a sixth-grade textbook.

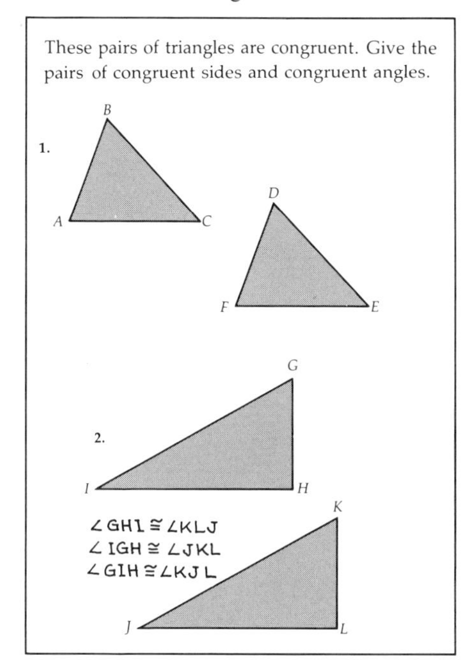
These pairs of triangles are congruent. Give the pairs of congruent sides and congruent angles.

9. Give a similar definition for the bisector of an angle. Complete the statement: $\overrightarrow{AD}$ is the bisector of the angle BAC if and only if ____.

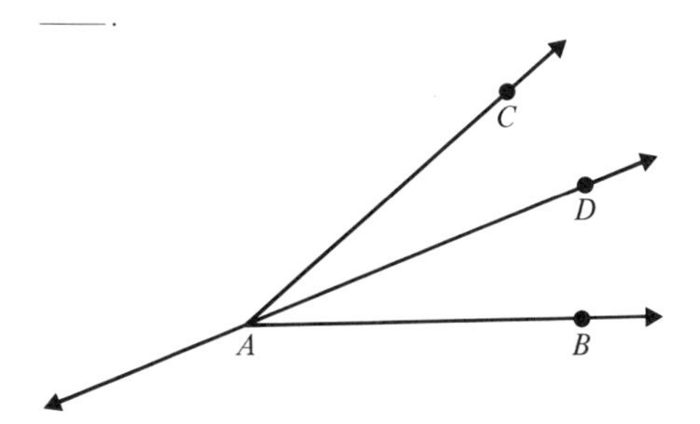

10. Why is it not correct to define perpendicular by requiring that two lines meet and a pair of angles is congruent?

match exactly. Thus the two halves are not only congruent, they are positioned so that the congruent pairs can be superimposed by folding.

Example 1 Which of the following figures or regions have one or more lines of symmetry?

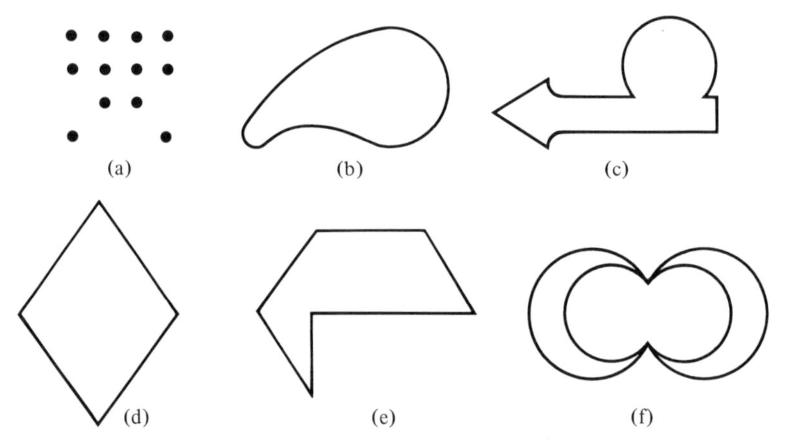

Solution (b), (c), and (e) have no lines of symmetry; (a) has one line, and (d) and (f) have two such lines each.

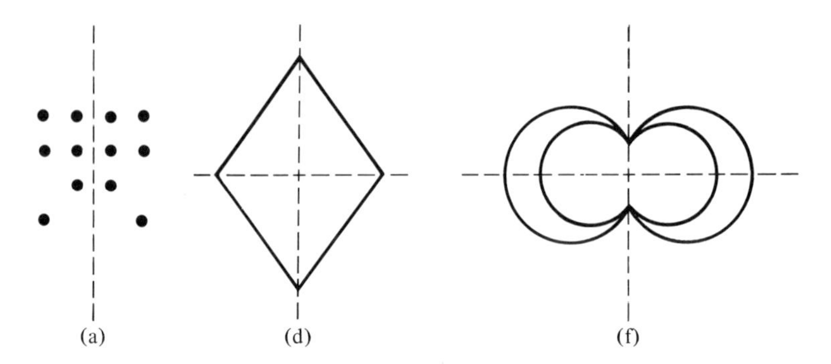

Although every symmetric figure contains one or more pairs of congruent figures, it is not true that if a figure consists of two congruent triangles, for example, then there is necessarily a line of symmetry. The rectangle *ABCD* can be used to show this:

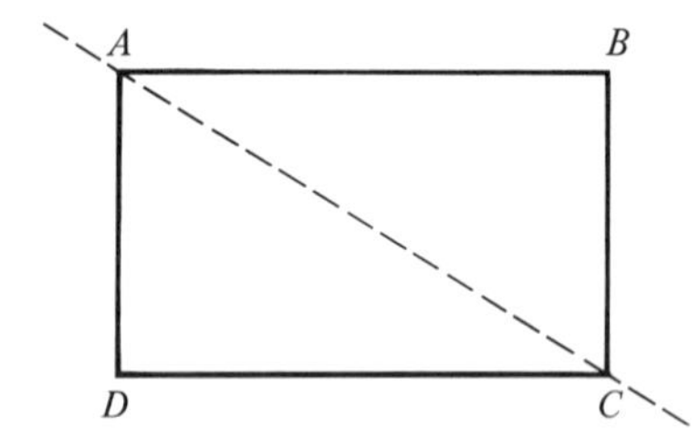

Since the opposite sides are equal and all the angles are right angles in a rectangle, then $\triangle ABC \cong \triangle CDA$. Yet if the paper were folded on the line $\overleftrightarrow{AC}$, then B would not fall on the point D.

Try Exercise 11

An investigation of the lines of symmetry of a figure can lead students to think about other properties of geometric figures as well. The following portion of a fifth-grade textbook is such a lesson.

Give the number of lines of symmetry.

	Polygon	Description	Picture	Number of lines of symmetry
1.	Rectangle	Opposite sides have the same length. All angles are right angles.		
2.	Square	All sides have the same length. All angles are right angles.		
3.	Parallelogram	Opposite sides are congruent. Opposite angles are congruent.		
4.	Rhombus	All 4 sides have the same length.		
5.	Regular pentagon	All 5 sides have the same length. All angles are congruent.		
6.	Regular hexagon	All 6 sides have the same length. All angles are congruent.		

Try Exercises 12 and 13

ROTATIONAL SYMMETRY. There is another type of symmetry that a plane figure may have. If it is possible to rotate a figure about a center point through less than a complete turn of 360° so that it looks to be in exactly the same position as at the start, then the figure is said to have *rotational symmetry*.

Example 2 Show that a square has rotational symmetry. The point O is the intersection of the four lines of symmetry for the square, and this is the center

11. Draw any rectangle $ABCD$ as in the text. Fold along the segment AC. Find the point on which D rests after the fold. Is it point B?

12. Do the lesson on symmetry taken from the fifth-grade textbook.

13.

a) Why is the rectangle not a regular figure?
b) Why is the parallelogram not a regular figure?
c) Why is the rhombus not a regular figure?

14. To test if a figure has rotational symmetry, first trace the figure on thin paper. Put a pin at the center points and hold the tracing firmly in place on the original figure; start to turn slowly in either direction. If the tracing fits the original figure at any time *before* you have completed a full turn of the tracing, the figure has rotational symmetry. Do this for the square in Example 2.

15. Multiply each pair of coordinates for points A, B, and C by 2 and draw a new triangle using these points. What is true about triangle ABC and the triangle you just drew?

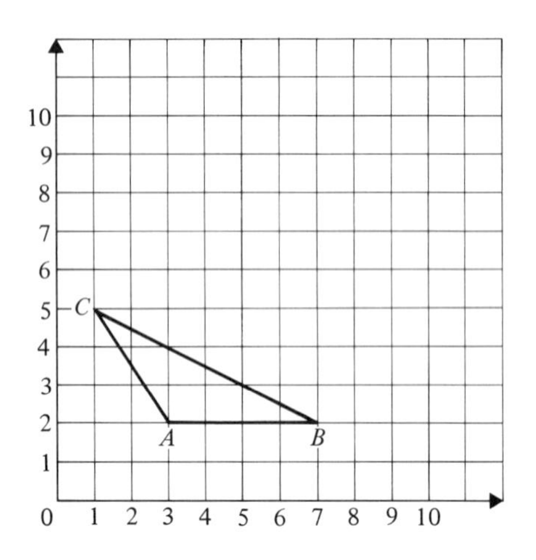

of rotation we will use. If the square is turned through a right angle counterclockwise, it will still look as it did originally; thus a square has rotational symmetry. Actually the square can be rotated three more turns of a right angle each time. After the fourth turn, a complete turn will be made about O and the square will be back in the original position, so the square is said to have rotational symmetry of *order* 4.

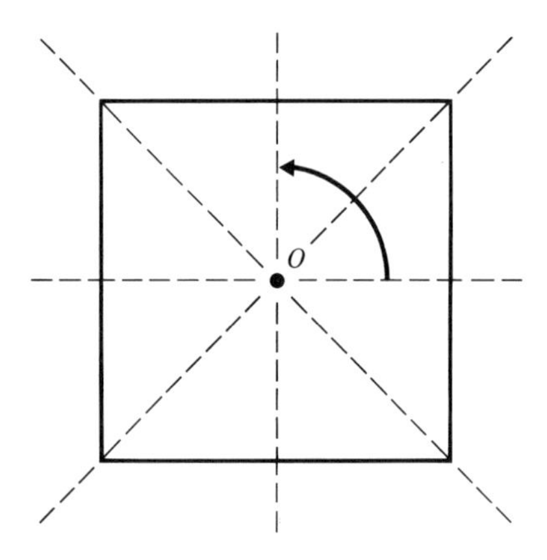

Try Exercise 14

SIMILARITY. In ordinary language the word "similar" means that two things are alike in some way. This is roughly the meaning in geometry although it must be made more precise. Two figures in geometry are *similar* if they have the same shape; they need not have the same size, so they are not necessarily congruent. The concept of similar figures is often developed with the help of figures drawn on the usual coordinate graph. The part of a sixth-grade lesson shown on p. 337 does this.

Try Exercise 15

If one shape can be obtained from another by magnification, then the shapes are similar.

SIMILAR TRIANGLES.

DEFINITION. Two triangles are *similar* if they can be labeled ABC and $A'B'C'$ in such a way that

1. $\angle A \cong \angle A'$, $\angle B \cong \angle B'$, and $\angle C \cong \angle C'$;
2. **the following proportion holds:** $\dfrac{AB}{A'B'} = \dfrac{BC}{B'C'} = \dfrac{AC}{A'C'}$.

In this case $\triangle ABC$ is similar to $\triangle A'B'C'$. In symbolic form: $\triangle ABC \sim \triangle A'B'C'$.

From a sixth-grade textbook.

Graphing similar figures

Each pair of coordinates for points *A, B,* and *C* was multiplied by 3. Then the new coordinates were used to draw the larger triangle.

Points	Coordinates	× 3
A	(1, 1)	(3, 3)
B	(2, 1)	(6, 3)
C	(2, 3)	(6, 9)

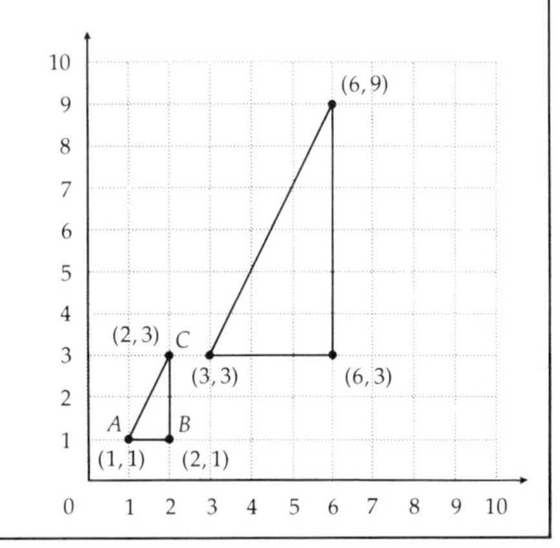

The two triangles have the same shape, but not the same size.

They are **similar triangles**.

Actually, for triangles if one of the conditions in the above definition holds, so does the other.

Example 3 Are any two equilateral triangles similar? Let $\triangle ABC$ be any equilateral triangle; hence the sides $\overline{AB}$, $\overline{BC}$, and $\overline{AC}$ are all congruent to each other. Let $\triangle DEF$ be any other equilateral triangle not identical to $\triangle ABC$; the sides $\overline{DE}$, $\overline{DF}$, and $\overline{EF}$ are all congruent to each other. If AB is the length of $\overline{AB}$, and so on, then the following must be true:

$$\frac{AB}{DE} = \frac{CB}{EF} = \frac{AC}{DF}.$$

Since the two triangles were chosen arbitrarily, all equilateral triangles are similar to each other.

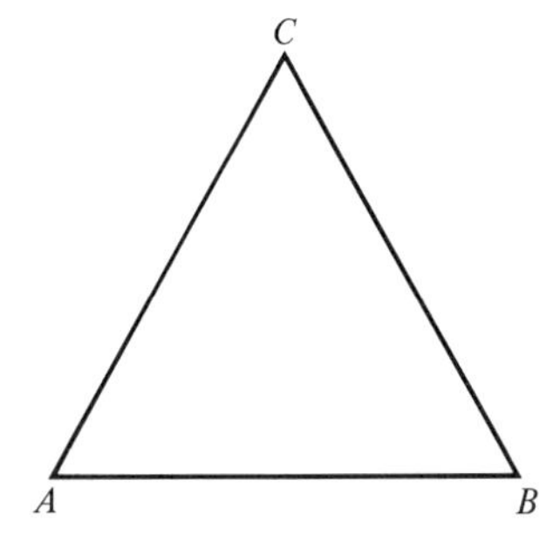

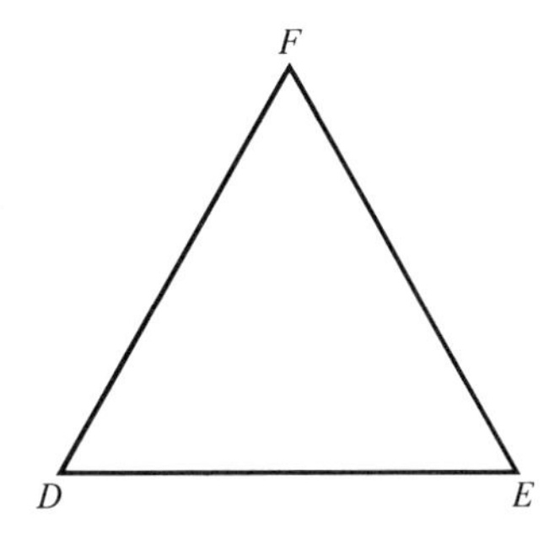

16. Show that the rectangle $ABCD$ is not similar to the rectangle $A'B'C'D'$.

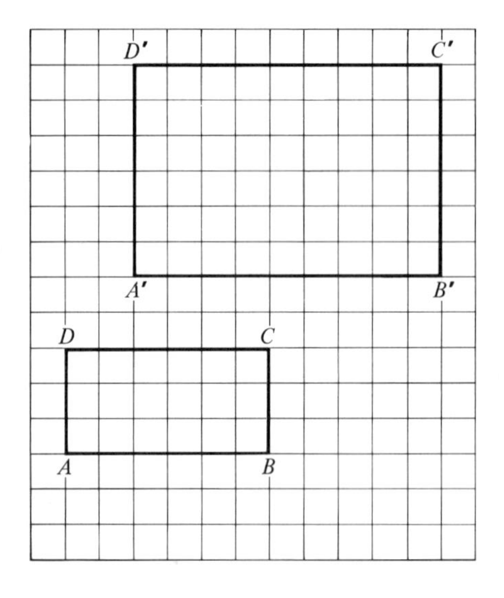

SIMILAR POLYGONS. Two polygons are similar if and only if for some pairing of vertices in order around the figure it can be shown that:

1. corresponding angles are congruent and
2. corresponding sides are proportional.

The two rectangles $ABCD$ and $A'B'C'D'$ sketched below are similar because $\measuredangle A \cong \measuredangle A'$, $\measuredangle B \cong \measuredangle B'$, $\measuredangle C \cong \measuredangle C'$, and $\measuredangle D \cong \measuredangle D'$ (since all the angles of a rectangle are right angles) and because

$$\frac{AB}{A'B'} = \frac{2}{3} = \frac{BC}{B'C'} = \frac{CD}{C'D'} = \frac{AD}{A'D'}.$$

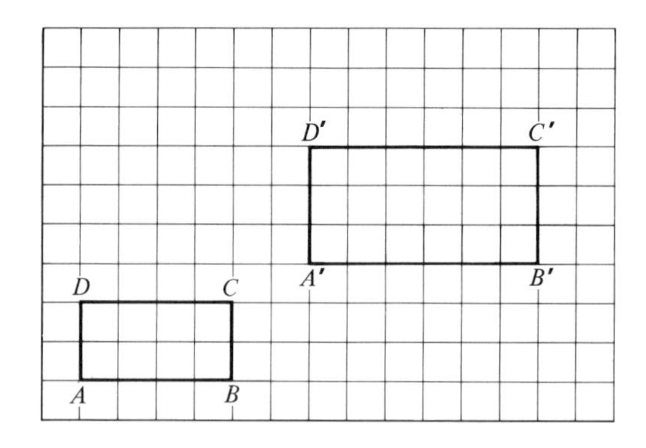

Try Exercise 16

EXERCISE SET 10.3

1. Use tracings to show that $\triangle PQR$ and $\triangle MON$ are congruent. Is it possible to superimpose one of these triangles on the other by sliding the tracing along the page?

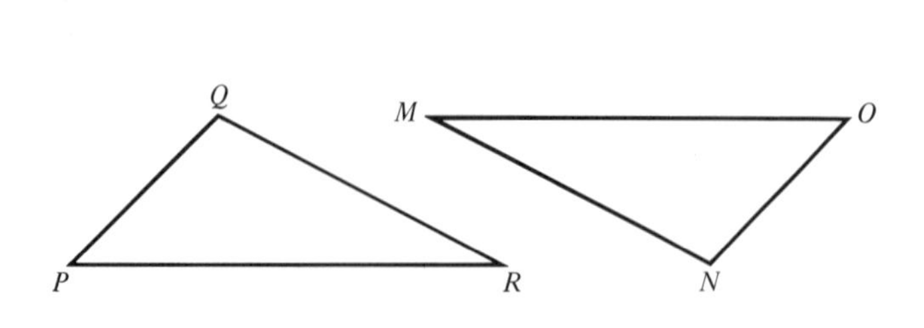

2. The following is a regular hexagon $ABCDEF$ with the diagonals $\overline{CF}$, $\overline{BD}$, $\overline{BE}$, and $\overline{AE}$ drawn. Find four pairs of congruent triangles in this figure. For one pair give the corresponding congruent angles and the congruent sides.

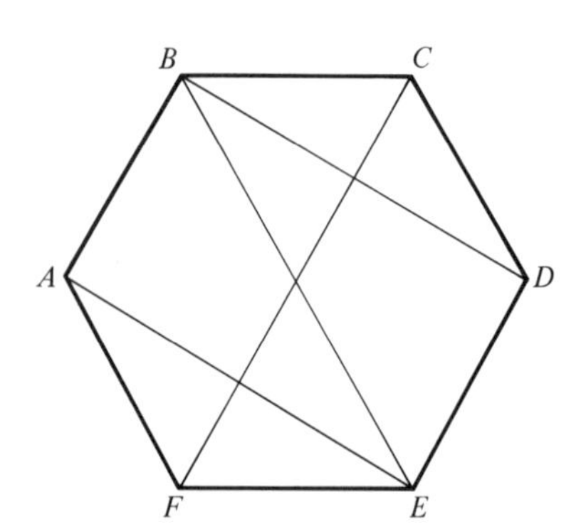

3. The following rhombus $ABCD$ has the two diagonals AC and BD drawn. (A rhombus has opposite angles congruent and all sides congruent to each other.) Find four pairs of congruent triangles in this figure. For one of these pairs give the corresponding angles that are congruent and the congruent sides.

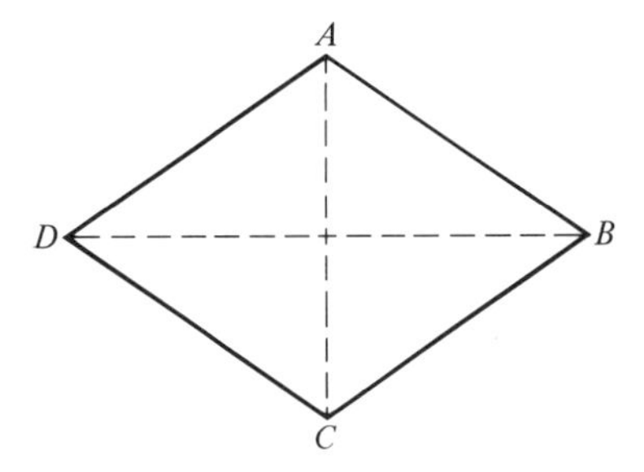

4. Complete the sketches to make a figure that is symmetric about the given line of symmetry:

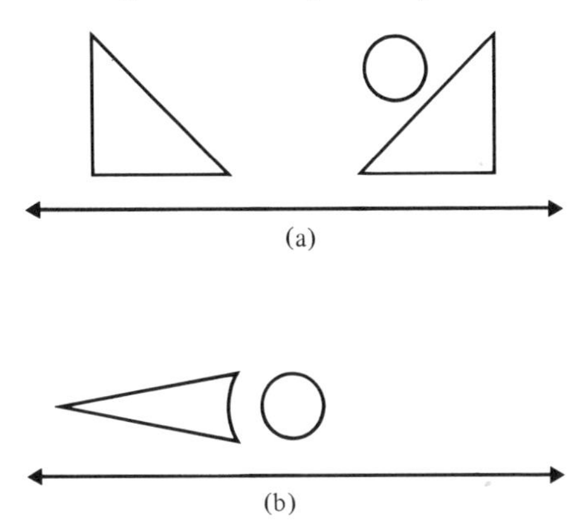

5. Complete the sketches to make a figure that is symmetric about both lines of symmetry:

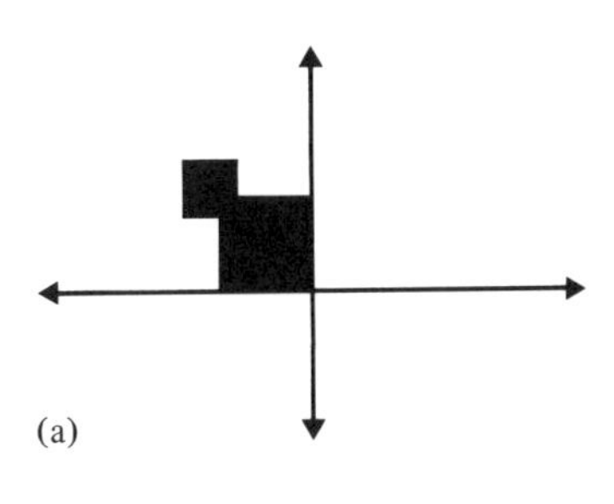

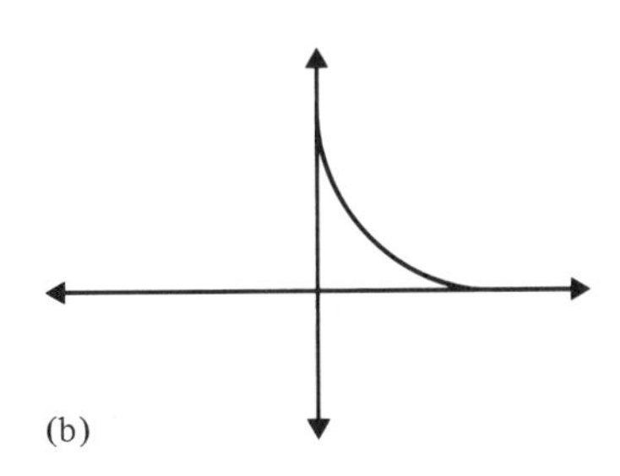

6. The following are sketches of a square sheet of paper that has been folded twice (into quarters). If the shaded region were cut out, what would it look like unfolded? Will the cutout be in a single piece? (Fold lines are $\overleftrightarrow{OA}$ and $\overleftrightarrow{OB}$ in each case.)

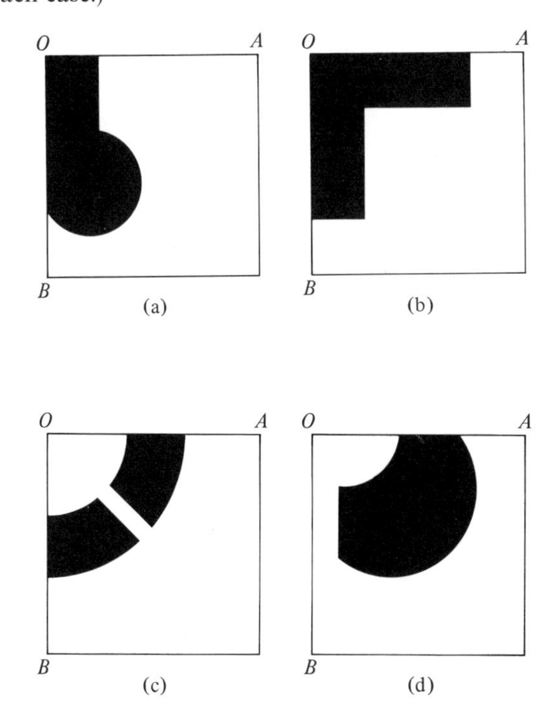

7. In the figures of Problem 6, how are the fold lines OA and OB related in each case? Why?

8. The following is a portion of a lesson in a fifth-grade textbook. Answer the questions.

Give the number of lines of symmetry of each triangle.

	Geometric figure	Description	Picture	Number of lines of symmetry
1.	Equilateral triangle	All sides have the same length.		
2.	Isosceles triangle	Two sides have the same length.		
3.	Scalene triangle	No two sides have the same length.		

9. Make a chart of all the regular polygons up to the regular heptagon showing the number of sides and the number of lines of symmetry. What patterns do you see? What conjecture can you make about the relation between the number of sides and the number of lines of symmetry? ■ Can you prove your conjecture?

10. Show that not all isosceles triangles are similar to each other.

11. If the two similar right triangles of the following sketch were rearranged so that $\triangle ABC$ was superimposed upon $\triangle A_1B_1C_1$ with point A on A_1 and point B on segment $\overline{A_1B_1}$, what could you say about the lines $\overleftrightarrow{BC}$ and $\overleftrightarrow{B_1C_1}$? Why?

12. If we move $\triangle ABC$ so that point B coincides with point B_1 and segment $\overline{BC}$ is contained in segment $\overline{B_1C_1}$, what observations can be made about this figure?

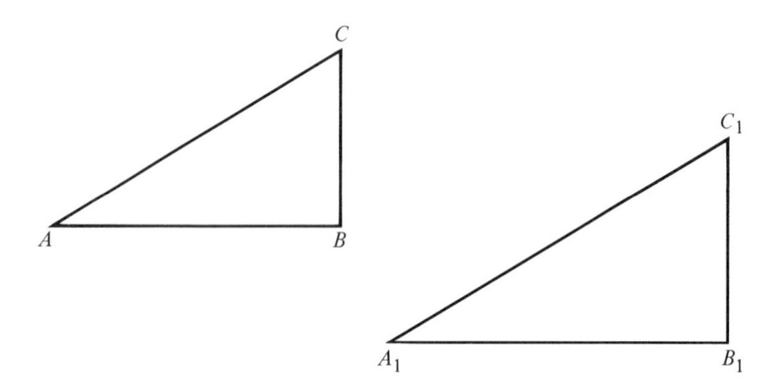

13. Does the situation in Problems 11 and 12 hold for any pair of similar triangles?

14. Prove that any two squares are similar.

15. Prove that any two rhombuses are not necessarily similar.

16. If a tracing of each of the following figures is rotated about the point O, how many times will the tracing fit on the original figure in one complete turn?

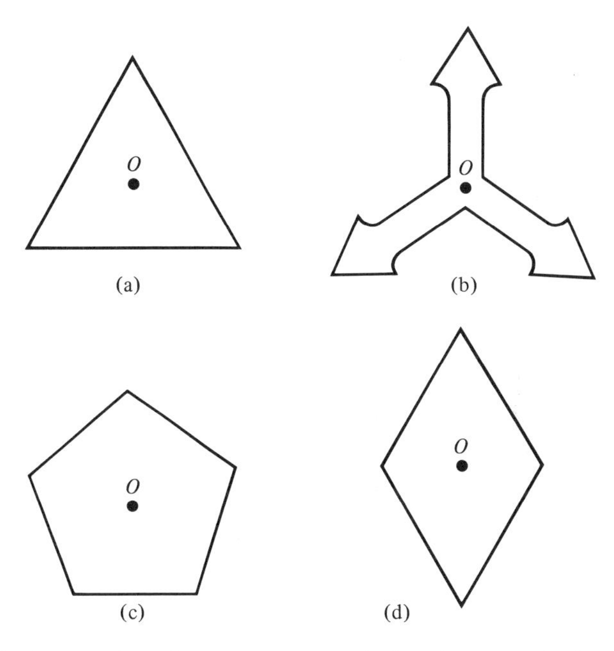

17. What is the order of the rotational symmetry of the figures in Problem 16?

18. Prove that the congruence relation is an equivalence relation, that is, it satisfies the reflexive, symmetric, and transitive properties.

10.4 TRANSFORMATION GEOMETRY

The concepts of congruence and symmetry are useful in thinking about other, more dynamic, aspects of geometry. *Transformation*, or *motion*, geometry, as it is frequently called in elementary school, builds upon the idea of rigid figures that "move" or are transformed to other positions. The transformations we are to discuss here are those in which a given geometric figure is transformed to its image that is congruent to it. They are called *rigid motions* because the figures do not change size or shape under the transformation. This is in contrast to transformations that magnify or minify the geometric figure or drastically change the appearance in some other way.

Translation

Intuitively, a translation moves a figure in a plane to another position in the plane, so that each point in the new position is the same distance in the same direction as every other point. For such a transformation it is necessary to be given both a distance and a direction that describes the translation.

10.4 AIMS

You should be able to:
Describe and perform each of the following transformations:
1. Translations
2. Reflections
3. Rotations

Example 1 Find the translation image of the segment $\overline{AB}$ in the direction and through the distance of the arrow $\overrightarrow{DE}$.*

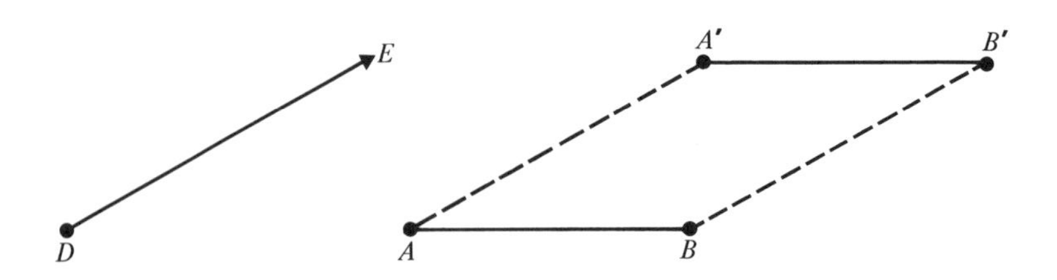

Solution We draw a line through A that is parallel to $\overrightarrow{DE}$ and imagine that point A is moved along this line to a point A' such that $\overline{AA'} \cong \overline{DE}$. At the same time B has of course landed at B'. Thus the segment $\overline{A'B'}$ is the image of $\overline{AB}$ under translation and $\overline{A'B'} \cong \overline{AB}$.

Try Exercise 17

17.
a) Since the point B is also moving along a line parallel to $\overrightarrow{DE}$, what can be said about the segments $\overline{AA'}$ and $\overline{BB'}$?
b) Why is $\overline{AA'} \cong \overline{BB'}$?

Segments are translated to congruent, parallel segments, and all the line segments connecting the image points to those of the original figure are congruent.

Example 2 Find the translation of $\triangle ABC$ in the direction of and through the distance of the arrow $\overrightarrow{DE}$.

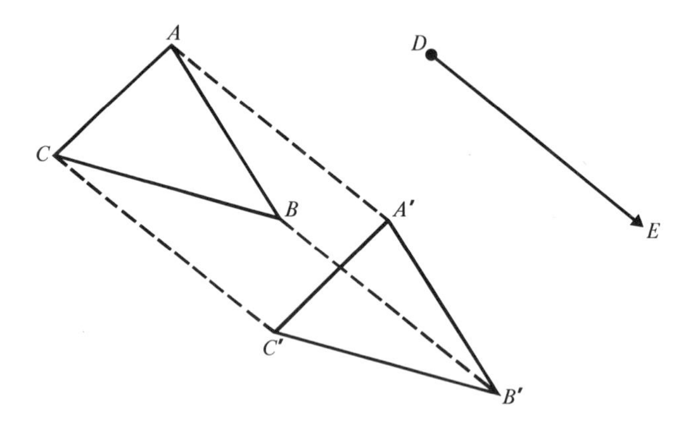

Solution We draw a line segment through A that is parallel to $\overline{DE}$ and has the length of $\overline{DE}$. Imagine that the triangle is then slid along this segment so that A lands at A' (A is translated to A'). At the same time the points B and C will have moved to B' and C'.

Try Exercise 18

18.
a) $\overline{AA'} \cong \overline{DE}$; is it also true that $\overline{CC'} \cong \overline{DE}$ and $\overline{BB'} \cong \overline{DE}$?
b) $\triangle ABC \cong \triangle A'B'C'$; what can be concluded about $\overline{AC}$ and $\overline{A'C'}$? $\overline{AB}$ and $\overline{A'B'}$? $\overline{BC}$ and $\overline{B'C'}$?
c) Imagine a point P on segment $\overline{AB}$ different from A or C; where is the image of P under the translation?

* Arrow DE means that travel starts at D and goes in the direction of E. The distance traveled is the length of segment $\overline{DE}$.

In the sixth grade the translations are often done on coordinate graph paper, so that it is easier to describe the direction and the distance. In the following portion of a sixth-grade lesson the arrow $\overrightarrow{DD'}$ is the direction of the slide, or translation. The slide can also be described as "over 4, up 3," since the new location of each point of the image $A'B'C'D'$ can be found in this way. From the graph we see that $\overline{D'C'} \cong \overline{DC}$, $\overline{C'B'} \cong \overline{CB}$, $B'A' \cong BA$, and $\overline{A'D'} \cong \overline{AD}$.

The slide is also called a **translation**. Rectangle $A'B'C'D'$ is the **slide image** or **translation image** of rectangle $ABCD$.

On your graph paper, draw rectangle $ABCD$ in the position shown here.

Then show the translation image of $ABCD$ after each of these translations.

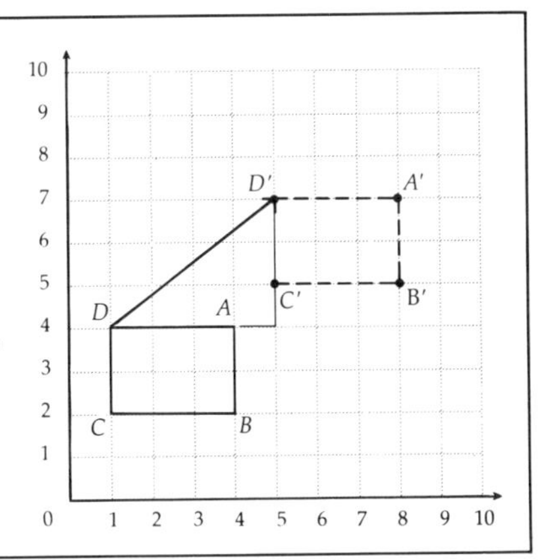

Try Exercise 19

From these examples it can be seen that each point in the figure moves to exactly one point of the image under the translation and the translated figure is congruent to the original (preimage) figure.

Reflection

The result of the *reflection transformation* resembles what we see in a mirror. A given line that acts as the "mirror" is called the *line of reflection* and is used as the axis of symmetry. The original figure along with its transformation image forms a symmetric figure about this line of symmetry.

Example 3 Find the reflection of the right triangle ABC about the line L.

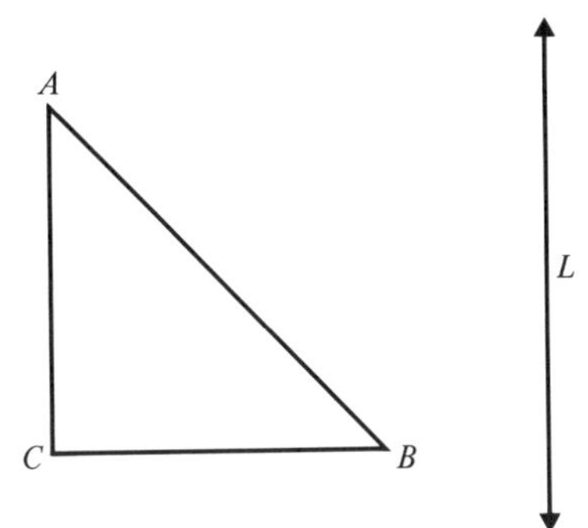

19.

a) Use the arrow $\overrightarrow{DD'}$ and show the image of $ABCD$ after the translation that corresponds to the arrow.

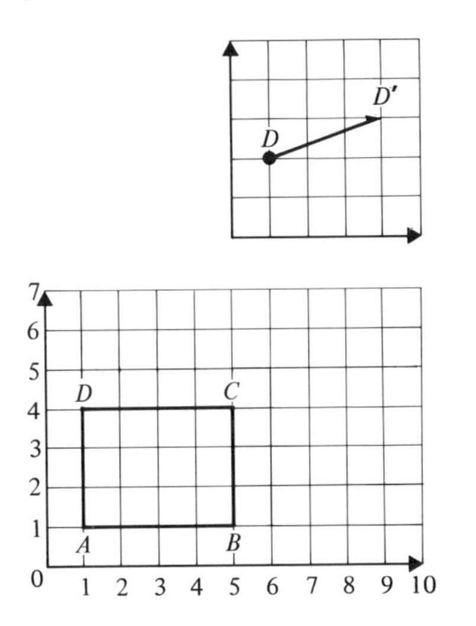

b) Use the arrow $\overrightarrow{DD'}$ and show the image of $ABCD$ after the translation.

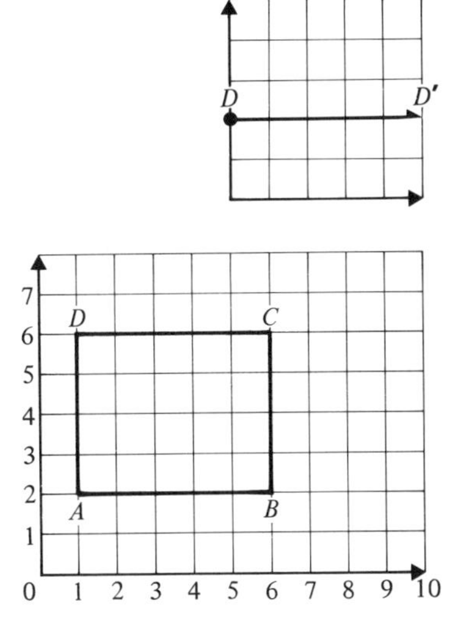

c) Is the transformed image in each case in (a) and (b) congruent to the original figure $ABCD$?

20. Draw any triangle ABC and a line L. Carry out the transformation of reflection for your triangle and line. If your paper is folded along the line L, what should happen to the triangle ABC? Put a small mirror along line L; what do you see?

We fold a sheet of thin paper and lay the fold along L. The triangle ABC can be traced on this sheet with the fold still in place along L. The paper can be unfolded and the diagram looks like this:

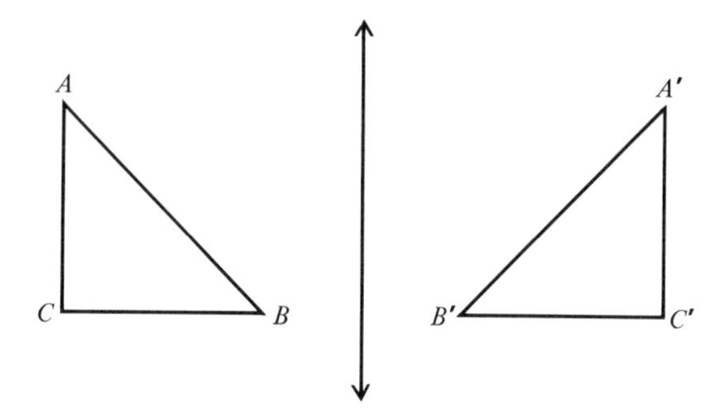

Try Exercise 20

Coordinate graph paper makes the job of finding the reflected image much easier. The following lesson from a sixth-grade textbook shows this usage:

21.

a) Show the reflection image in the following:

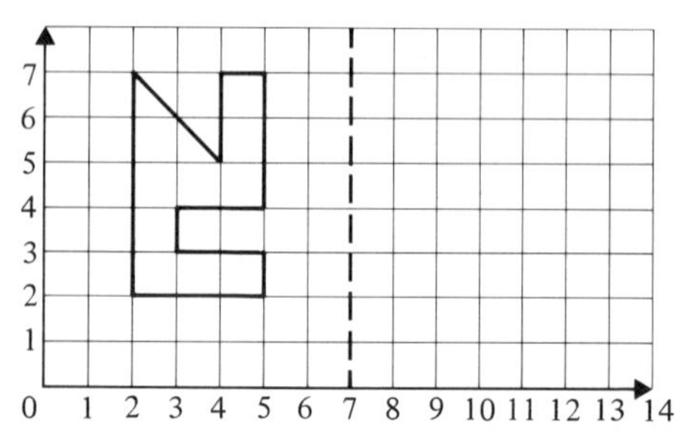

b) What are the coordinates of the reflection image in the exercise from the sixth-grade textbook? Is there any pattern to the new coordinates?

Suppose the dotted line on the grid is a mirror.
Copy each figure. Then show its reflection image.

Example:

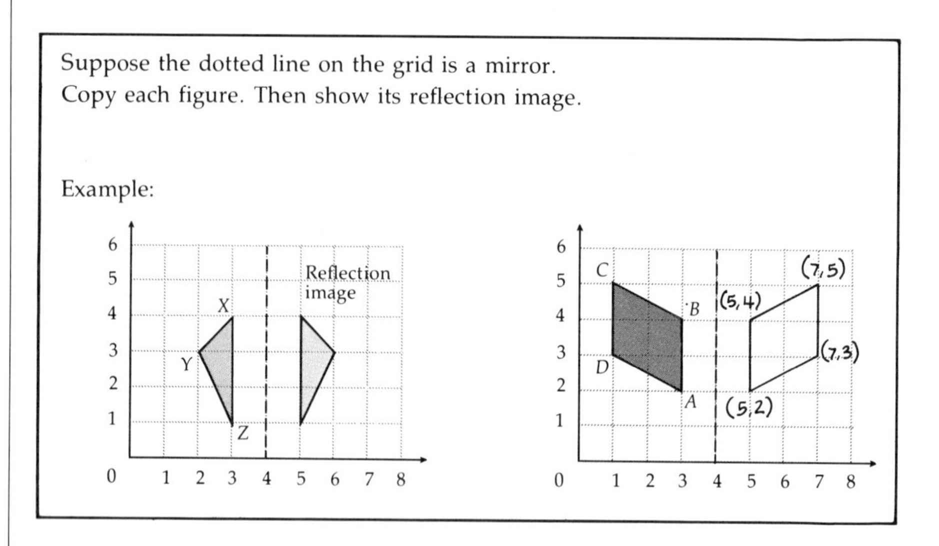

Again, just as for translations, each point A has assigned to it one image point A' under the reflection. If A' is the image of A and C' is the image of C, then the segment $\overline{AC}$ is congruent to its image $\overline{A'C'}$.

Try Exercise 21

Rotation

In the third transformation called a *rotation* the points of a given figure move in the plane along circles that have the same center called the *center of rotation.* For young children these are called *turning motions* and *turning points*, repectively.

For example, the following shows a segment and a triangle rotated through $\measuredangle AOA'$ about the point O. (It is conventional to have counterclockwise rotation called positive.) The ray $\overrightarrow{OA'}$ is drawn since the $\measuredangle AOA'$ is given. The radius OA is used to locate point A'. $\measuredangle BOB' \cong \measuredangle AOA'$ and $\overline{OB} \cong \overline{O'B'}$. If we imagine the points A and B actually rotating, the entire segment $\overline{AB}$ will be transformed to the position $\overline{A'B'}$.

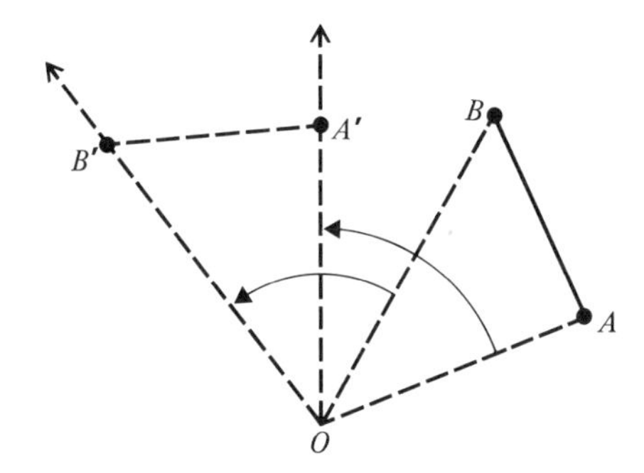

For a triangle, the three points A, B, and C are used. The three image points A', B', and C' are found; when these are joined, we have the entire image of the original triangle:

$$\measuredangle AOA' \cong \measuredangle BOB' \cong \measuredangle COC'$$

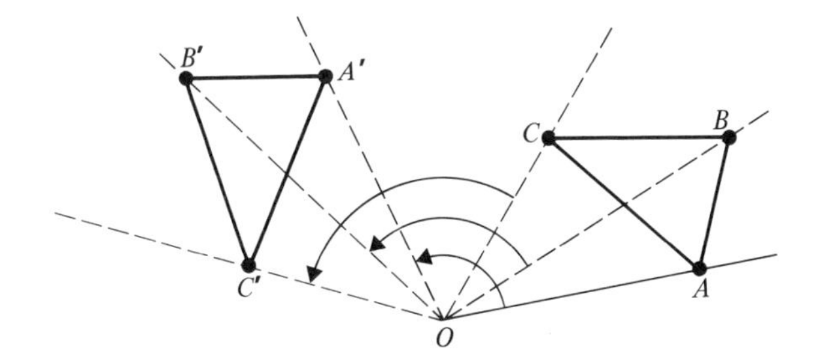

Example 4 Find the rotation image of $\triangle ABC$ after a right-angle rotation clockwise about point O ($\measuredangle AOF$ is a right angle).

22. Find the rotation image of the rectangle $ABCD$ after a rotation of a right angle in the clockwise direction about the point O.

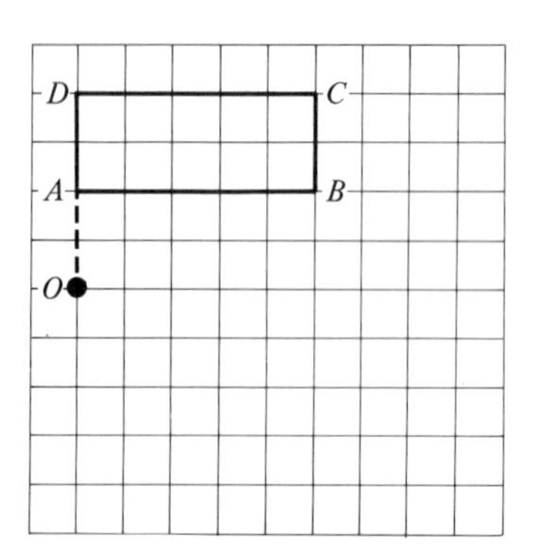

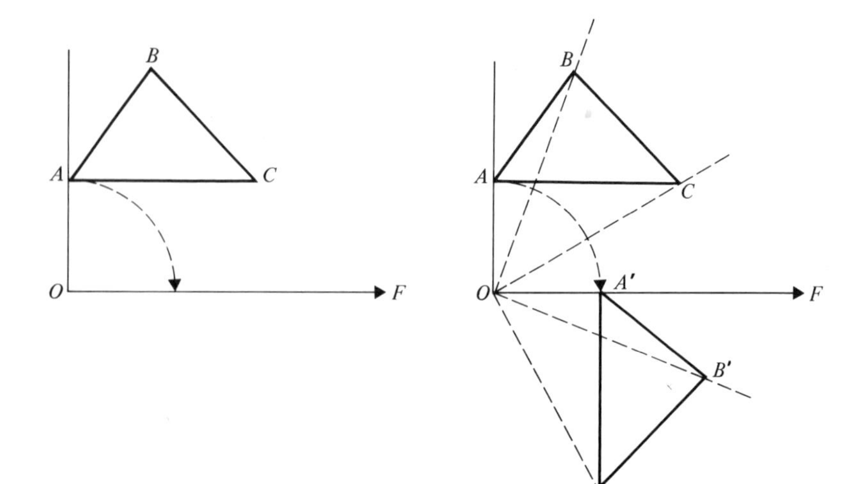

Solution This can be done with tracing paper. Place the paper over the drawing and trace $\triangle ABC$. Put a pin through the point O and gently rotate the triangle until point A lands on line $\overleftrightarrow{OF}$. Every point of the triangle is now in the image position.

Try Exercise 22

EXERCISE SET 10.4

1. Carry out the reflection indicated. Find the image of $\triangle ABC$ using line L as the line of reflection. Call the image $A'B'C'$.

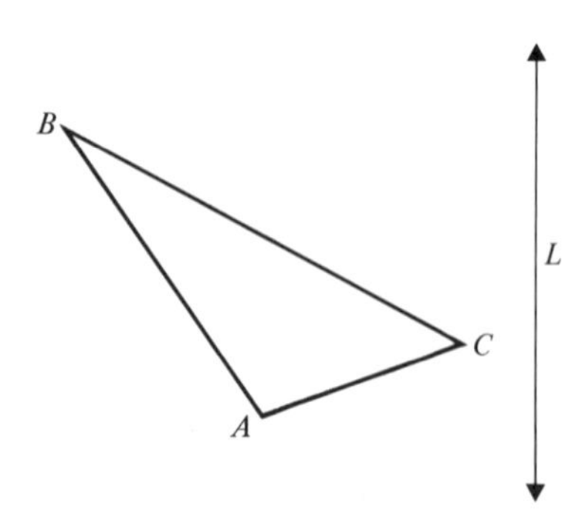

2.

a) In Problem 1, if you trace around $\triangle ABC$ starting at A and proceeding to B then to C, are you moving in a clockwise or counterclockwise direction? Trace $\triangle A'B'C'$ in a similar manner and answer the same question.

b) Is it possible to find the image $A'B'C'$ of Problem 1 starting with $\triangle ABC$ by using one or more translations instead of the reflection?

■ c) How is the answer to part (a) related to part (b)?

3. Find the line L such that P' is the image of P through the line L.

■ 4. In the following diagram, point A' is the reflection of point A through the line L. Use only a straightedge and find the reflected image of C through the same line. *Hint*: When a set of collinear points is reflected through a line, its image is also a set of collinear points.

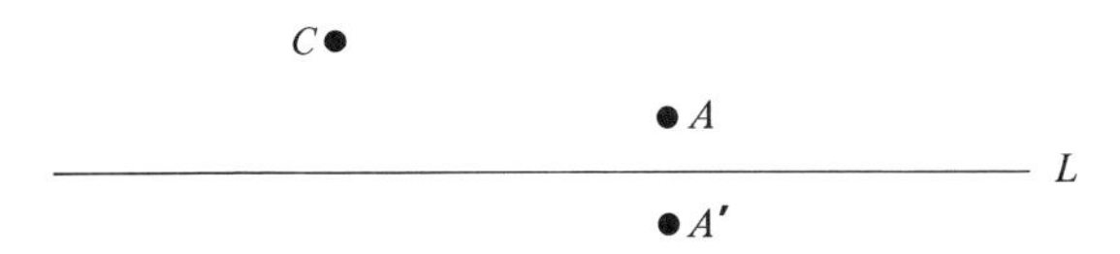

5. In the diagram, $\triangle A_1B_1C_1$ is transformed to $\triangle A_2B_2C_2$ under reflection through line L. Explain why the following are true:

a) $\triangle A_1B_1C_1 \cong \triangle A_2B_2C_2$
b) $\measuredangle OC_2P \cong \measuredangle OC_1P$
c) $\overline{A_1B_1} \cong \overline{A_2B_2}$

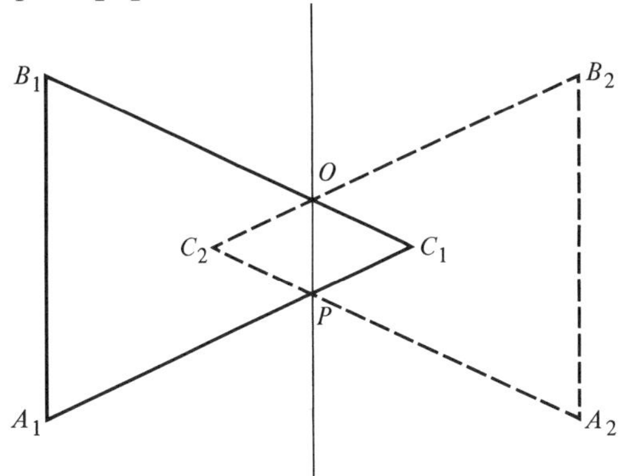

6. What transformation was used to transform the figure $ABCD$ to $A'B'C'D'$?

a)
b)

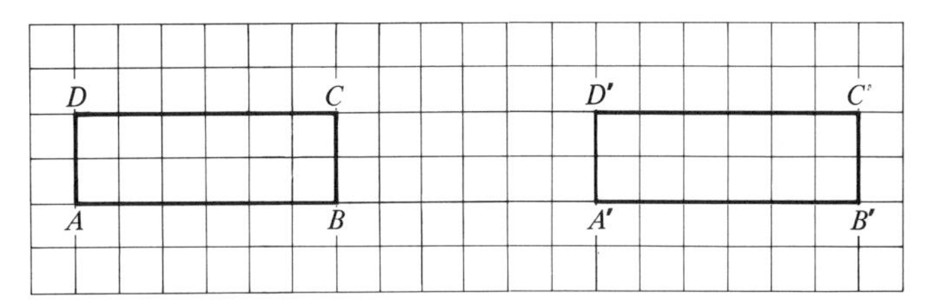

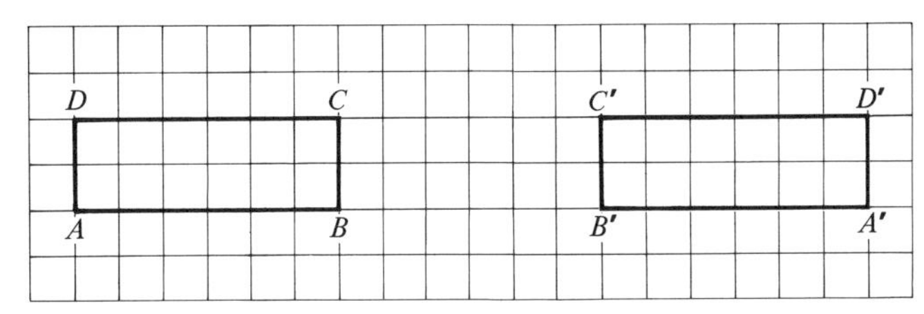

7.

a) Find the reflection image of $ABCD$ through line L_1. Call the image $A_1B_1C_1D_1$.
b) Find the reflection image of $A_1B_1C_1D_1$ through line L_2. Call the image $A_2B_2C_2D_2$.
c) Is there a single transformation that would have transformed $ABCD$ to $A_2B_2C_2D_2$?

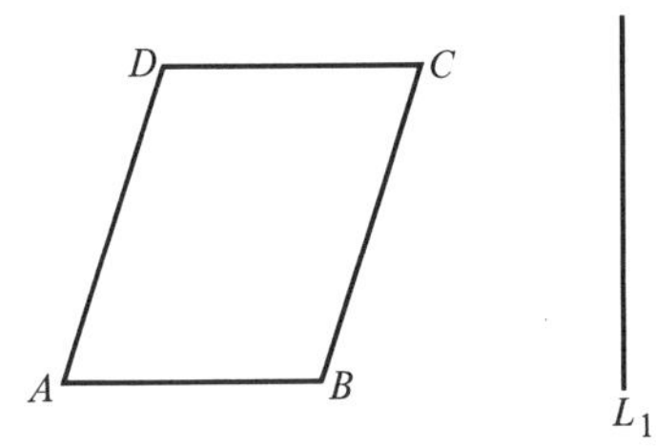

8. Repeat Problem 7 for the $\triangle ABC$ and the lines L_1 and L_2 given below.

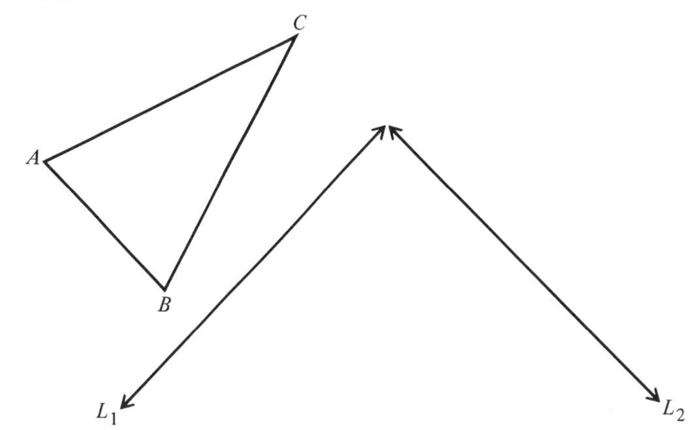

9. If two rigid motions are carried out in succession, it is called a *composition* of the two transformations. Find the composition of transformations (two or more) for each of the following, so that the second figure is the image of the first.

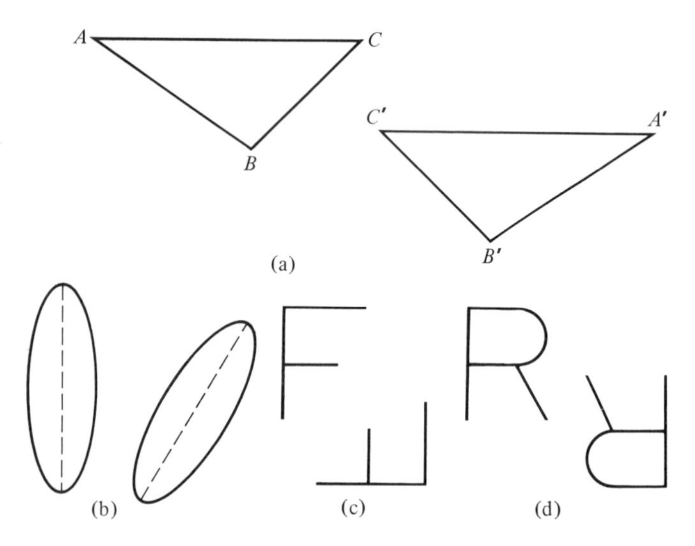

■ **10.** Is the composition of two translations always another translation? Why?

■ **11.** Is every rotation a composition of some other transformations that are not rotations? Why?

10.5 AIMS

You should be able to:

a) Apply the five-step problem-solving process to problems in geometry.
b) Demonstrate increased problem-solving skill in problems that may not be directly related to this section.

10.5 PROBLEM SOLVING

In the usual high-school geometry there are two important categories of problems: those referred to as constructions and those in which a proof of a theorem is demanded. These are not the problems* elementary pupils will work with, since such pupils are in the process of acquiring the basic concepts, learning to visualize geometric figures, and making inductive discoveries about some of the properties of Euclidean geometry.

The following is a problem in visualization.

Example 1 Show how the following figure can be cut into two congruent pieces. One is allowed to cut only along either a horizontal or vertical line, not across on a diagonal.

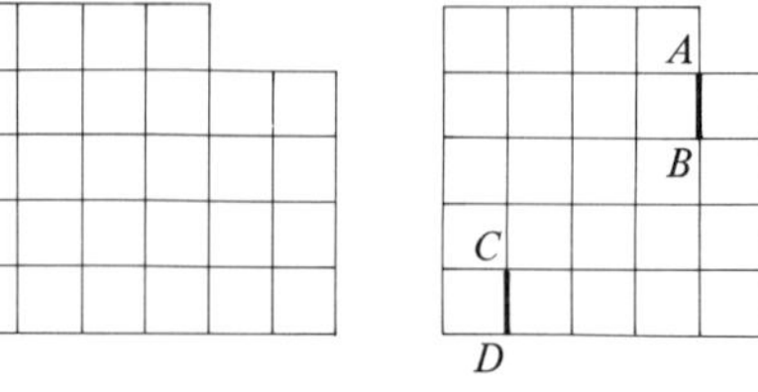

* A reminder:

THE FIVE-STEP PROCESS

1. What are the *relevant facts*?
2. What actually is the *problem*?
3. What are my *ideas*?
4. How do I get a *solution*?
5. *Check* and *review* the solution.

Try Exercise 23

What *ideas* might be useful here?

1. Count the squares; there are 28 of the unit squares, so each figure must have 14.
2. Make a tracing of the figure and try a few cuts to see what happens.
3. Try to find a line of symmetry.
4. Put the missing squares back into the figure; how could it be divided then? Will this help in the problem?
5. The diagonal of the rectangle formed in idea 4 would divide the figure. The closest thing to a diagonal is a zig-zag cut.
6. Use idea 5 and idea 4 combined.
7. The cut from point C to point D in the figure gives five unit squares in each of the horizontal and vertical rows.
8. The cut from A to B puts two unit squares in each figure.

Try Exercise 24

Even in such a problem it is possible to think and analyze and get various ideas, so that one does not have to "see a solution" immediately or not at all.

Pupils may learn some of the Euclidean constructions, and these should be justified to them and not presented as something to be memorized. For the following construction problem we assume the postulate* that if corresponding sides of two triangles are congruent, then the triangles themselves are congruent.

Example 2 Given a point P on a line L, construct a line perpendicular to the line L at P.

What are the *facts* here?

1. A point P on a line is given.

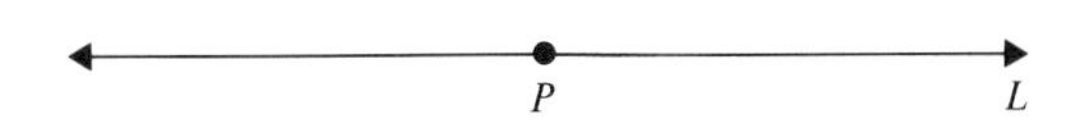

2. We are allowed to use a compass and straightedge (the Euclidean tools) to do the construction.

What is the *problem*?

1. To construct a line that is perpendicular to L and *passes through* P or
2. To find a line that intersects line L at P such that four congruent angles are formed.

23.

a) What are the facts in Example 1?

b) What is the problem? Can you state it in more than one way?

24.

a) Do you have any more ideas?

b) Solve the problem.

c) How do you check the solution in this problem?

* Since there are three sides, this postulate is referred to as the SSS-postulate for congruent triangles.

25.
a) Do you have any more ideas?
b) Can you solve the problem at this point? Do so.

26.
a) Draw $\overline{CA}$, $\overline{CP}$, and $\overline{CB}$.
b) Why is $\triangle APC \cong \triangle BPC$?
c) What can you say about the angles APB and BPC?

What are some *ideas*?

1. Since congruent angles are needed, find some congruent triangles.
2. We can fold the paper at P with line L folded back on itself. Can that be done with a ruler and compass?
3. We can think of the required perpendicular as a line of symmetry for some figure.
4. We can introduce some symmetric triangles into the figure so it looks like:

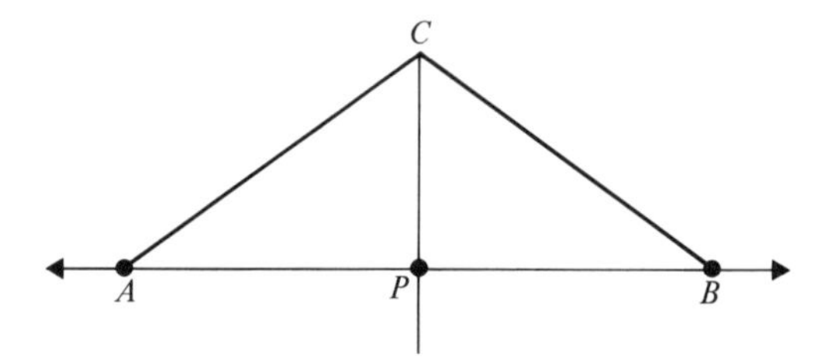

Try Exercise 25

Now for the *solution.* We use idea 4 and try to construct a symmetric figure that will resemble the sketch. We need points A and B such that $\overline{AP} \cong \overline{PB}$. We choose a radius of arbitrary length for the compass, put the center at P, and swing two arcs to get:

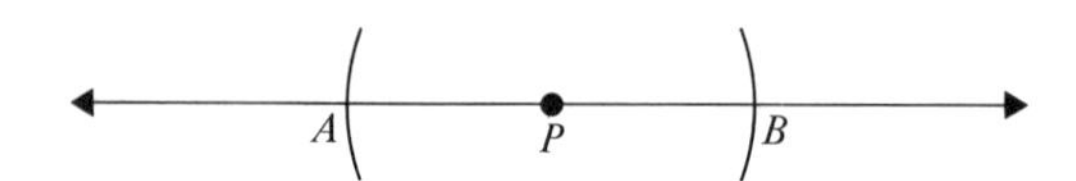

We now need to find a point C not on line AB so that $\overline{AC} \cong \overline{CB}$. We choose a larger radius for the compass, use A and B in turn as center, and swing two arcs to cross at point C.

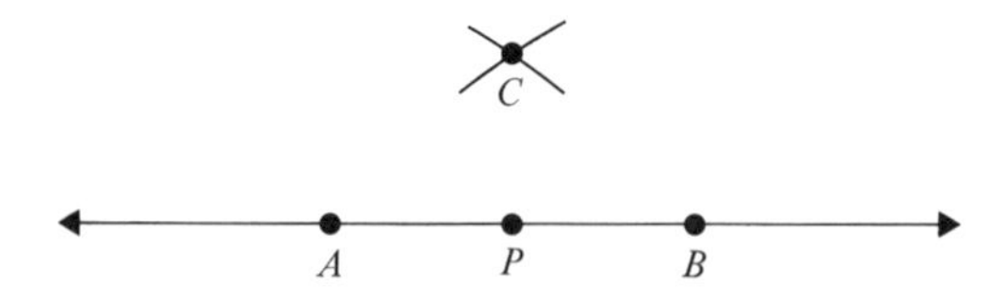

Try Exercise 26

Since $\measuredangle APC$ and $\measuredangle CPB$ are congruent adjacent angles formed by two intersecting lines, they are both right angles, and we are done. The construction has been shown to be correct.

Thinking about constructions in this way will help you work through and understand the next section.

EXERCISE SET 10.5

1. Construct a perpendicular to a line from a point P off the line. ■ Show that the construction is correct.

2. Use the postulate that two triangles are congruent if their corresponding sides are congruent to construct a triangle $A'B'C'$ that is congruent to the given triangle ABC.

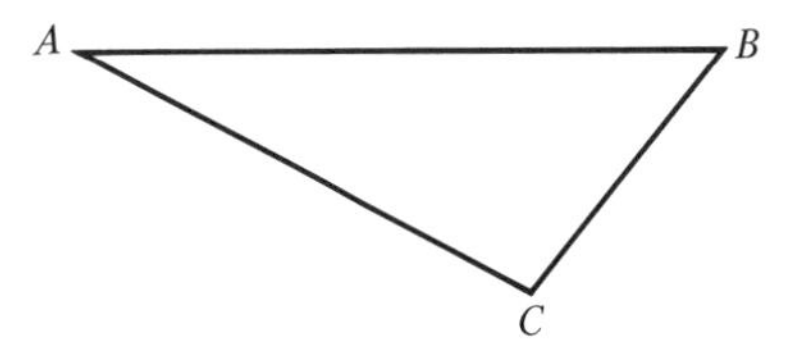

3. Suppose some pupils have discovered, by tearing off the corners of a triangular shape, the following:

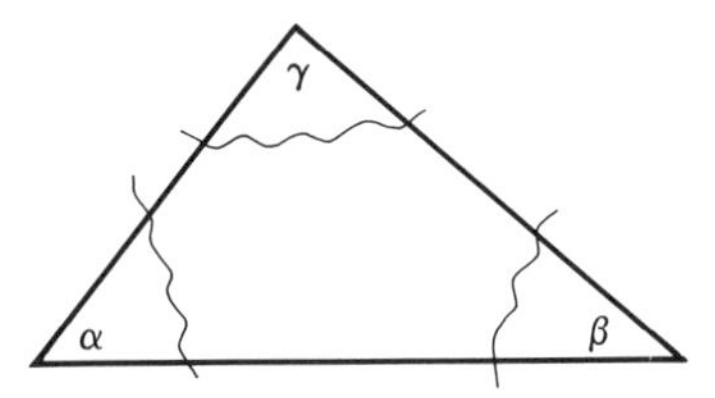

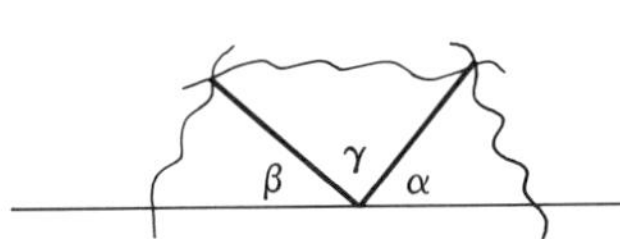

This looks as if the angle formed by the three angles placed to have a common vertex is a straight angle; however, it is difficult to tell if the two rays do form part of a straight line. In any case such a result needs a proof. ■ Use transformations to show that the three angles do form a straight angle.

4. Use the ideas of symmetry to help show that a line that is perpendicular to and bisects chord AB in the diagram also passes through the center of the circle.

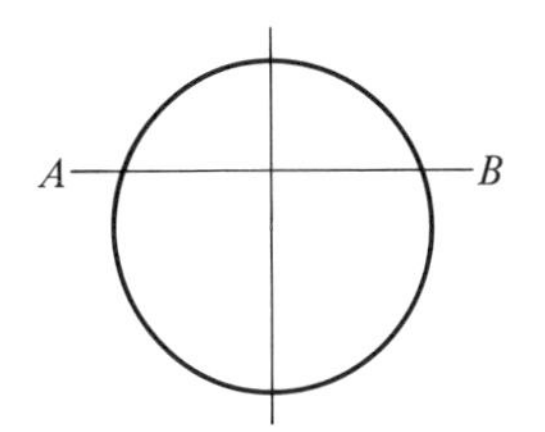

5. Use symmetry and reflection to help prove that the diagonals of a rhombus intersect at right angles.

6. Prove that the similarity relation is an equivalence relation on the set of all triangles.

■ 7. Use transformations and the SSS postulate to prove that if two triangles have two corresponding sides congruent and the corresponding included angles congruent, then the triangles are congruent (the SAS theorem for congruence).

■ 8. Use transformations to prove that if two triangles have two corresponding angles and the corresponding included side congruent, then the triangles are congruent (the ASA theorem for congruent triangles).

9. Are two triangles congruent if the three angles of one are congruent to the corresponding angles of the other? To put it in different words, is there an AAA theorem?

■ 10. Show that it is not correct to say that two triangles are congruent if two sides and an angle of one triangle are congruent to the corresponding parts of the other.

■ 11. How many different triangles do you see in the figure below? *Hint*: Be systematic in the way you count them.

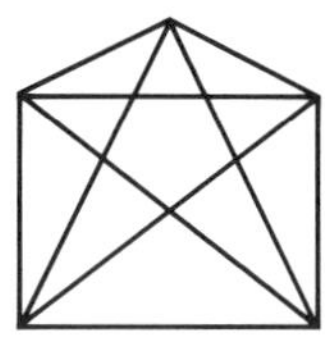

10.6 AIMS

You should be able to:

a) Perform the standard compass and straightedge constructions.
b) Make drawings by using other than the Euclidean tools in order to find patterns and discover properties of geometric figures.

10.6 CONSTRUCTION, DRAWING, AND DISCOVERY

As we have seen, Euclid's postulates allow only certain constructions to be done in the development of geometry.

1. With a straightedge one is allowed to draw a straight line of indefinite length through any two distinct points (the straightedge is not marked)
2. With a compass one is allowed to draw a circle with a given point as center and a given radius.

This restriction has provided an intellectual challenge for students ever since it became apparent that really difficult constructions can be accomplished with these simple tools.* These *constructions* of geometry must be distinguished from drawings that can be made under far less restrictive rules for other purposes. For the beginner, the making of drawings can be just as worthwhile as the Euclidean constructions in developing intuition and making discoveries.

Constructions

The following is a list of some basic compass and straightedge constructions, many of which are introduced to pupils as early as the sixth grade.

TO BISECT A LINE SEGMENT. It is required to find the midpoint of segment AB. Place the compass point at A and draw arcs of the same radius a and b. Without changing the compass radius, place the point at B and draw arcs c and d. Connect the two points of intersection C and D. Point P is the required point.

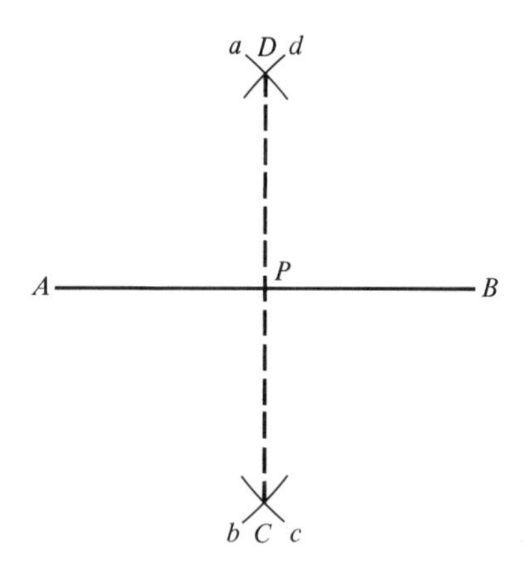

27.

a) If the paper were folded on line CD, where would point A fall?
b) Is $\overline{AD} \cong \overline{BD}$? Why?
c) Is $\triangle ADP \cong \triangle DPB$?

Try Exercise 27

Since $\triangle ADC \cong \triangle DBC$, it can be shown, by the SAS theorem for congruence, that $\triangle ADP \cong \triangle DPB$ (see Problem 7 of Exercise Set 10.5). It follows that $AP \cong PB$.

* Although many difficult constructions can be done, there are three classical problems that cannot be solved by Euclidean tools, as was finally shown in the 19th century. These problems are:

1. Constructing the edge of a cube that has twice the volume of a given cube.
2. Trisecting an arbitrary angle into congruent angles.
3. Constructing a square with the same area as that of a given circle.

TO CONSTRUCT AN ANGLE CONGRUENT TO A GIVEN ANGLE. If angle A is the given angle, place the compass point at A and draw an arc to determine B and C. Draw a line at A' and with the same compass setting draw an arc to determine P. Place the compass point at B and open so the pencil is at C. With this same setting, place point at P and draw the arc p to find Q. Connect Q to A'. Angle $QA'P$ is the required angle.

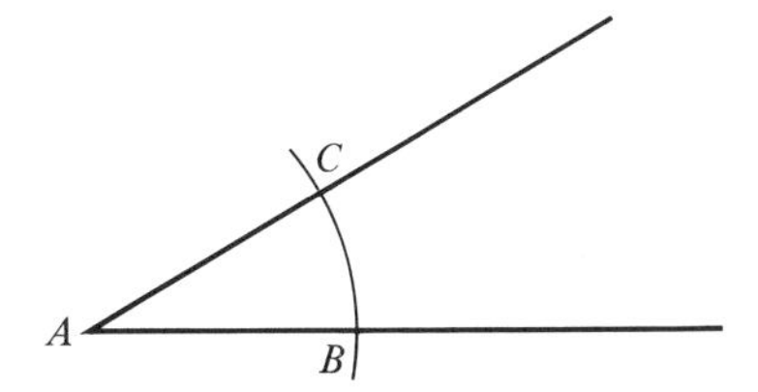

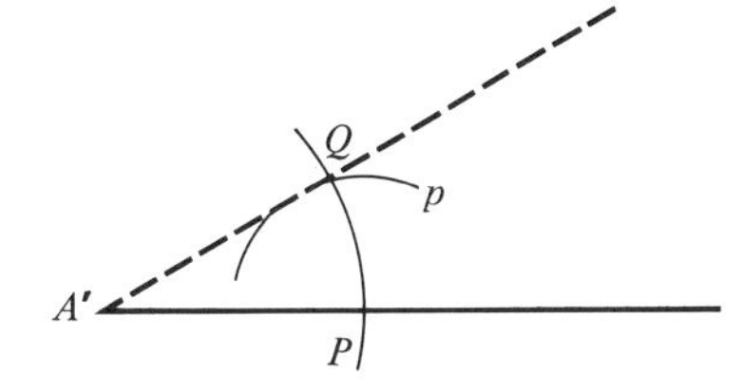

Try Exercise 28

28.

a) Show that $\triangle ACB \cong \triangle QA'P$.

b) Why is $\measuredangle CAB \cong \measuredangle QA'P$?

TO BISECT AN ANGLE. Place the compass point at A and draw an arc locating points B and C. Place point at B and draw arc b. Without changing the radius, place the point at C and draw arc c. Connect the intersection D with the vertex A; this line is the required bisector.

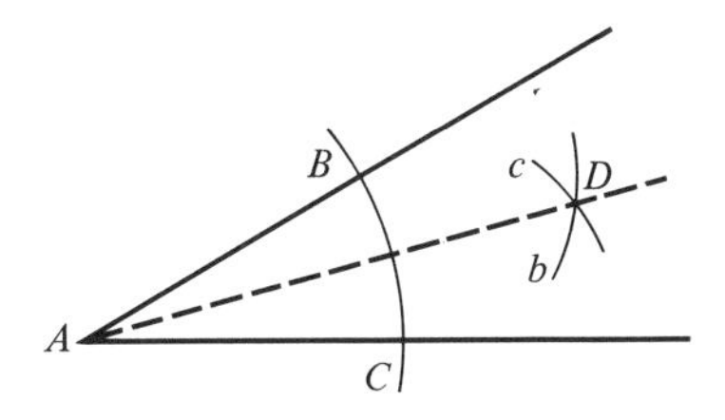

Try Exercise 29

29.

a) Imagine that the segments $\overline{BD}$ and $\overline{CD}$ are drawn. Is $\overline{BD} \cong \overline{CD}$?

b) What axiom or theorem is used to give $\triangle ABD \cong \triangle ACD$?

c) Why is $\measuredangle BAD \cong \measuredangle DAC$?

TO CONSTRUCT A LINE PARALLEL TO A LINE L THROUGH POINT P NOT ON L. Draw any line PQ through P that intersects the given line in Q. Construct $\measuredangle QPR$ so that it is congruent to $\measuredangle MQP$. The line PR is parallel to the given line.

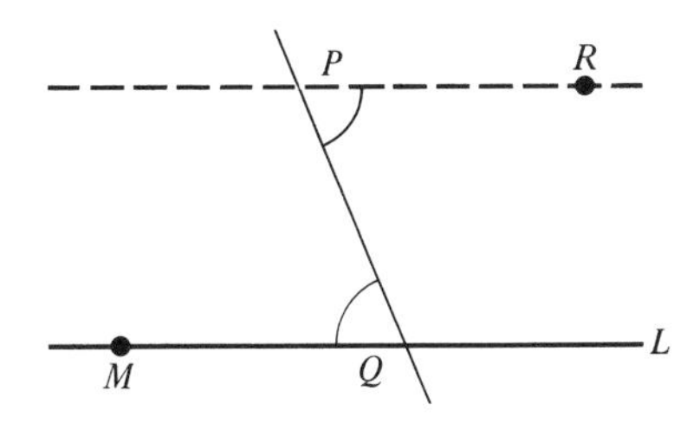

Try Exercise 30

30. What postulate, axiom, or theorem assures that line $\overleftrightarrow{PR}$ is parallel to the given line?

TO CONSTRUCT A TRIANGLE CONGRUENT TO A GIVEN TRIANGLE—SAS METHOD. Draw a line at A' and find C' so that $\overline{AC} = \overline{A'C'}$. At A' construct an angle that is congruent to $\measuredangle A$. Open the compass to fit AB, place the point at A', and draw arc a to find B'. Connect B' and C'.

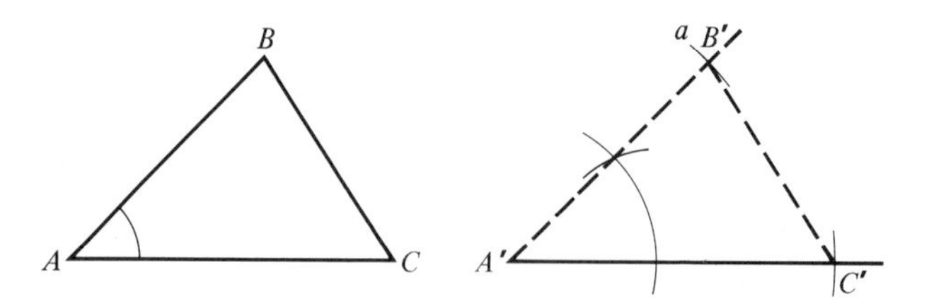

Young children can gain experience leading up to the Euclidean constructions by making drawings using rulers, protractors, coordinate paper, and so on. Carefully made drawings are useful for anyone in that they can lead to some interesting conjectures. For children such drawings form a way to introduce them to some important geometric concepts as they make "discoveries."

Learning by model building

EXERCISE SET 10.6

1. Construct a triangle congruent to a given triangle, using the ASA method. In the drawing use $\measuredangle A$, $\measuredangle C$, and side AC to construct the triangle $A'B'C'$.

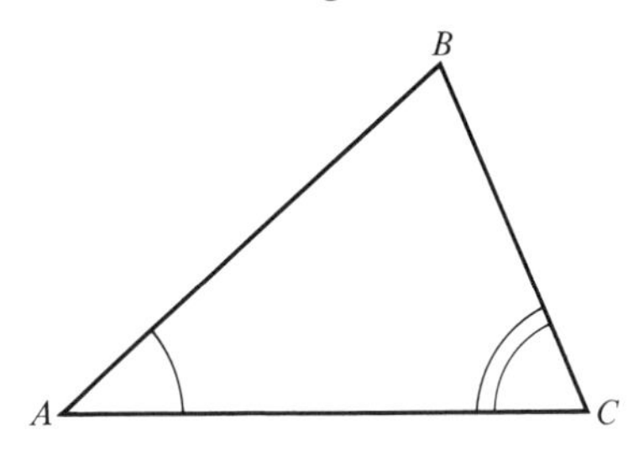

2. Show that the line bisecting a line segment (as in the construction in the text) is also perpendicular to that segment.

3.
a) Construct two triangles with their corresponding sides parallel. Use tracings or a compass to compare the angles.
b) Repeat (a) and generalize your result.

4.
a) Construct two angles such that their corresponding sides are perpendicular.
b) Repeat (a) and generalize.
■ c) Prove your generalization.

5.
a) Construct a right triangle.
b) Construct the altitude to the side opposite the right angle. This forms two triangles within the larger one.
c) Repeat for some other triangles; make a conjecture.
■ d) Prove it.

6.
a) Construct a parallelogram (opposite sides are parallel and congruent). Label it $ABCD$ and draw the diagonals intersecting at P.
b) Compare the lengths of $\overline{BP}$ and $\overline{PD}$, also of $\overline{AP}$ and $\overline{PC}$.
c) Repeat for several other parallelograms; what seems to hold in each case?
■ d) Prove your conjecture.

In the following problems it is not essential that you restrict yourself to the Euclidean constructions. These problems are intended for investigation and discovery. Make careful drawings in any way you choose.

7.
a) Draw a circle. Without changing the compass setting, mark off the radius length around the circle.
b) How many marks did you get? Join them to form a polygon.
c) Is the polygon a regular polygon?

8.
a) Draw a triangle and bisect one side; let the midpoint be P.
b) Through P construct a line parallel to one side of the triangle. Where does it meet the third side?
c) Repeat for some other triangles; what do you surmise is true?
■ d) Prove it.

9.
a) Draw a triangle and construct the perpendicular bisectors of each of its sides.
b) Repeat for some other triangles; what would you guess is true?
■ c) Prove that your guess always holds.

10.
a) Draw a triangle and construct its altitudes; extend them if necessary.
b) Repeat for some other triangles and make a conjecture if you can.

11.

a) Draw any triangle and find the bisectors of each angle.
b) Repeat for some other triangles and make a conjecture.
■ c) Prove your conjecture.

12.

a) Draw a quadrilateral and bisect all four sides.
b) Connect the midpoints in order to form another quadrilateral.
c) Repeat for some other quadrilaterals; is there a consistent pattern in your drawings?

10.7 AIMS

You should be able to:

a) Identify, draw, and describe the five regular polyhedra.
b) Make models of, describe, and draw the polyhedra commonly encountered in the elementary-school curriculum.
c) Draw edge, face, and vertex views of at least the tetrahedron and the cube.
d) Identify, draw, and describe a sphere, great circle, and small circle.

10.7 THE GEOMETRY OF THREE-DIMENSIONAL SPACE

Children are more likely to have developed an intuitive feeling for the geometry of three dimensions than for that of two dimensions (the plane), since they live in a three-dimensional world. Elementary pupils need experience with actual solids as well as practice in learning how to interpret drawings made of the various figures. There are some important three-dimensional geometric figures that are introduced during the elementary-school years.

Geometric solids

Polyhedra

A polyhedron can be thought of as a closed surface formed by planes that intersect in space. Such a surface thus consists entirely of polygonal figures. Here are some examples.

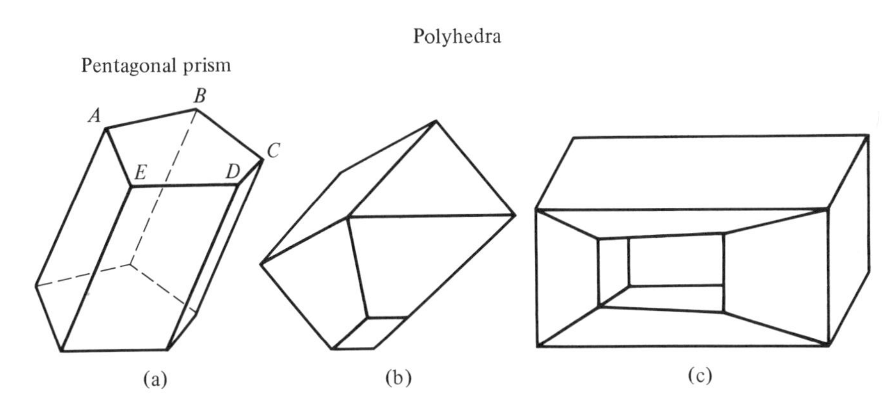

Here are some surfaces that are not polyhedra because not all of the faces are polygons.

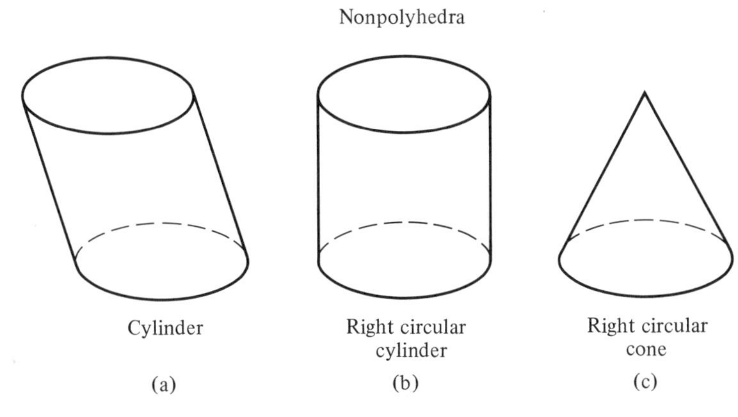

Each of the polygonal regions is called a *face*. The sides of the regions are called *edges* of the polyhedron, and the vertices of the regions are called *vertices*. In the pentagonal prism (figure (a) above), $\overline{AB}$, $\overline{BC}$, $\overline{CD}$, and so on are edges, A is a vertex, as are B, C, and so on. One face is the polygon $ABCDE$. If a polyhedron has no "holes" in it, it is called *simple*, so the polyhedron in figure (c) is not simple. For a simple polyhedron we can speak of the *interior* and the *exterior* in a manner that is similar to the situation for a polygon. A cube, for example, separates three-dimensional space into three disjoint sets: the set of points of the surface itself, the set of points in the interior of the cube, and the set of points in the exterior of the cube.*

Regular (Platonic) Polyhedra

A polyhedron is called *regular* if all of its faces are congruent regular polygons and all of its polyhedral angles are congruent. In such solids, the same number of polygons meet at each vertex. It turns out that there are only five different regular polyhedra.

* The polyhedra are sometimes thought of as the union of the surface and the interior points and thus are often referred to as *solids*.

31. Make a model of a tetrahedron. Cut four identical equilateral triangles out of something like a file card and tape the edges together. Save it for a future exercise.

The Five Regular Polyhedra

Number of faces	*Name*	*Sketch*	*Pattern for a model*
4	Tetrahedron	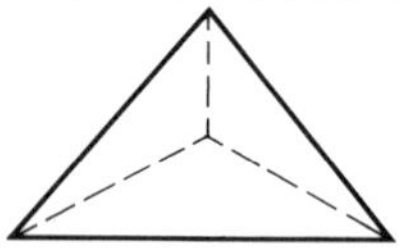	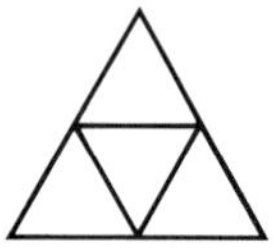
6	Cube	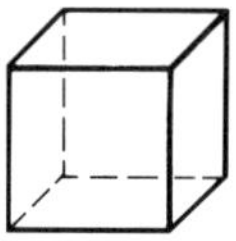	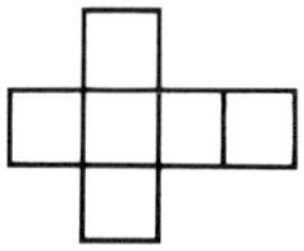
8	Octahedron		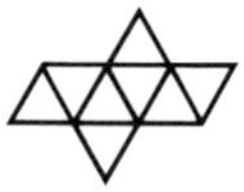
12	Dodecahedron	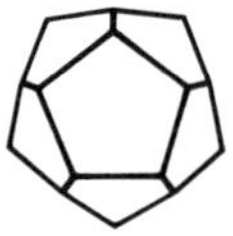	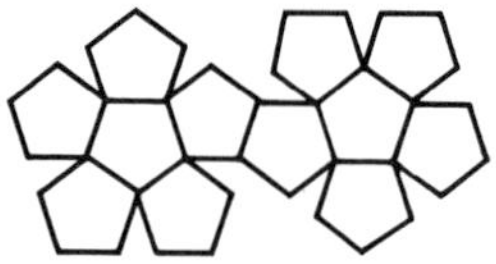
20	Icosahedron		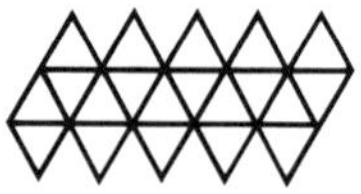

In making models of these polyhedra it is sometimes easier to cut out a sufficient supply of the regular polygons and then simply tape them together in the correct way rather than bother with making a pattern* and folding it properly.

Try Exercise 31

Drawing Polyhedra

PROJECTION DRAWINGS. Instead of trying to make perspective drawings (with vanishing points) of the Platonic solids, it is quicker to create a projection in which all the parts remain the same size regardless of the distance from the viewer. Such sketches can then be measured and used to make more complex representations.

* The patterns for the solids are properly called *net diagrams.*

The following figure shows a cube as seen from three different points of view:

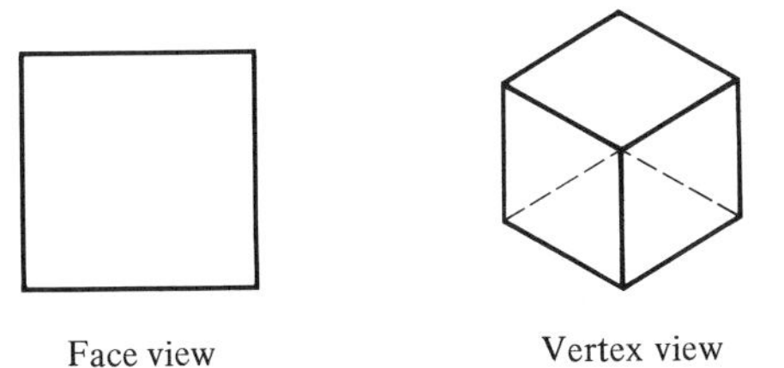

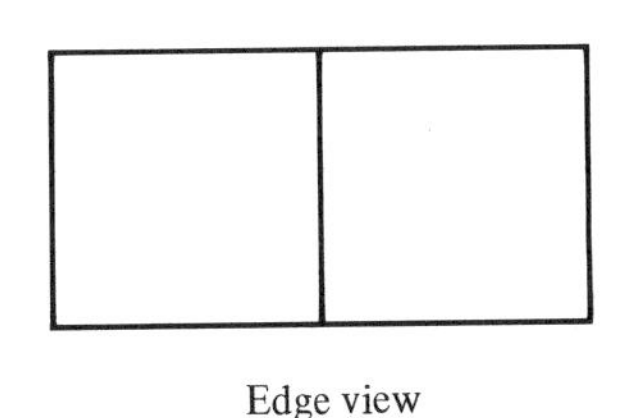

Face view Vertex view Edge view

Try Exercise 32

DRAWING A TETRAHEDRON. All polyhedra can be drawn by use of polygons. For example, to show the face view of a regular tetrahedron:

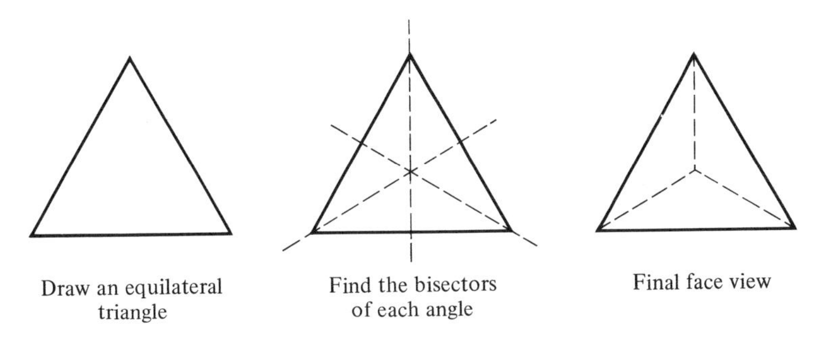

Draw an equilateral triangle Find the bisectors of each angle Final face view

To draw an *edge view* of the same figure:

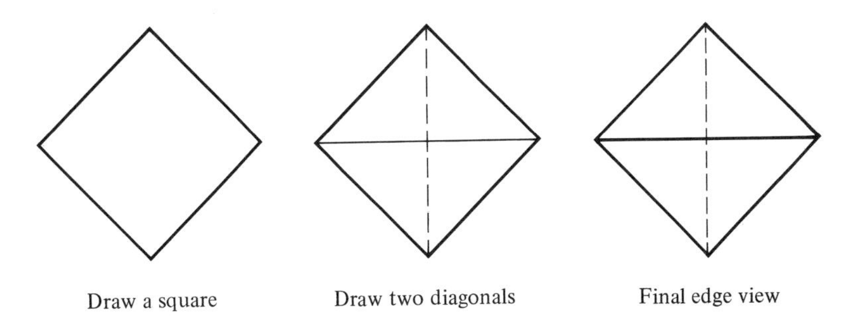

Draw a square Draw two diagonals Final edge view

Try Exercise 33

32. Get a model of a cube to look at. Hold it so the face is directly in front of you; you should see the face view of the text. Look down on a vertex so that it is the point of the cube closest to you. What do you see? Look at an edge so that all the points of the edge are closest to you. What do you see?

33. Use your model of a tetrahedron from margin Exercise 31 and look at the face view and the edge view. Make a sketch as to what the vertex view should be.

34.

a) Study the opening page of the chapter if you have not already done so.

b) Use coordinate graph paper and make a drawing of a triangular prism different from the one in the text.

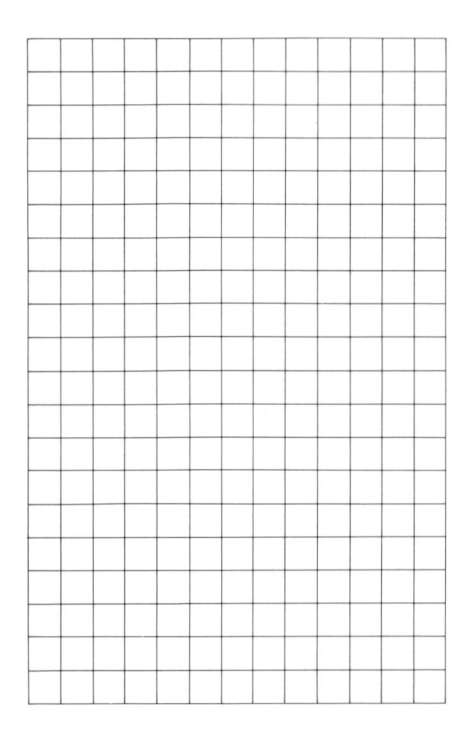

RECTANGULAR POLYHEDRON. A polyhedron with faces that are all rectangular regions is called a *rectangular polyhedron*; it can be drawn by starting with two rectangles.

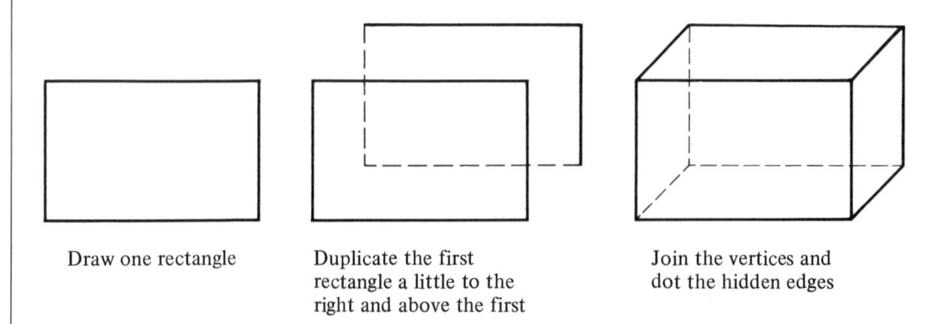

Coordinate graph paper can be very helpful to youngsters for making drawings of the polyhedra. Some examples of drawings that have been done this way appear on the opening page of this chapter.

Try Exercise 34

Spheres

A sphere is a closed surface such that all of the surface points are the same distance from a point inside called the *center*. A segment from the center to the sphere is called a *radius*, and a segment containing the center with both endpoints on the sphere is called a *diameter*. There is a direct analogy here to the corresponding plane figure, the circle.

Cross sections of a sphere (formed when a sphere is cut by a plane) are all circles. If the plane contains the center of the sphere, the cross section is called a *great circle*. All other cross sections are called *small circles*. On the Earth, the equator and the meridians are examples of great circles. The parallels of latitude are examples of small circles.

EXERCISE SET 10.7

1. Construct a model of a square pyramid. (The faces are isosceles triangles.) See the opening page of this chapter. Use your model to draw a face view, an edge view, and a vertex view.

2. Use a can for a model of a right circular cylinder. Make a drawing of the top view and the front view. Is there a side view?

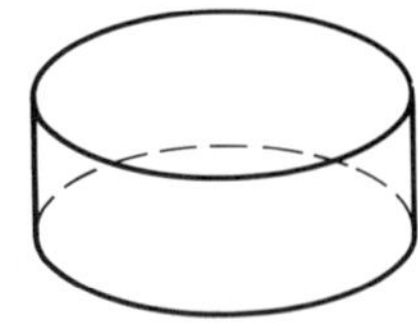

3. Make a model of a right circular cone by drawing a circle and removing a sector BOA shown unshaded in the sketch. The segments AO and OB are brought together and taped. A circular base of the correct size can be traced. Use your model to draw a vertex view, a side view, a face view (bottom).

4. Use the patterns in the text or cut out enough congruent shapes of each type needed and make models of the five regular polyhedra.

5. The famous Euler's theorem states a relation that holds for the number of faces, edges, and vertices of every polyhedron. See if you can discover Euler's theorem. Count the vertices (V), faces (F), and edges (E) in each of your models of the polyhedra.

Figure	*Number of*			
	(V)	(F)	$V + F$	(E)
Tetrahedron				
Hexahedron				
Octahedron				
Dodecahedron				
Icosahedron				

Do you see a pattern in your table? What is it?

6. Perhaps the pattern only holds for the regular figures. Add some more entries to the table. You will find drawings of a triangular prism, square pyramid, pentagonal prism, and others in this chapter. Add these to the table in Exercise 5. Does the pattern persist? What is Euler's theorem?

7. Imagine a solid block of wood in the shape of a cube. Suppose that you sliced off one small corner to form another face. What are the number of vertices, faces, and edges in the larger piece? How has the number of faces changed from the number in the cube? How has the number of vertices changed from those in the cube? The number of edges? If Euler's theorem holds in the cube, what about this new figure? Repeat this investigation by cutting off two corners, then three corners, then four corners.

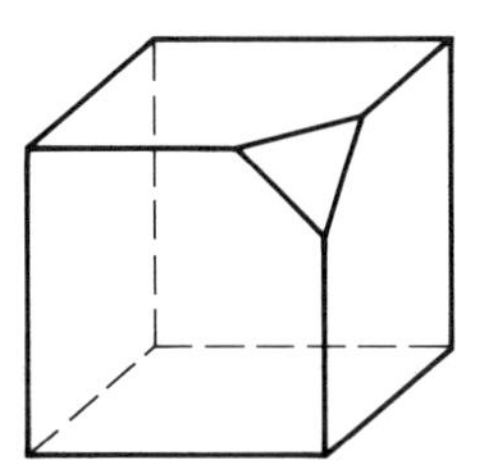

8. The figure in the sketch consists of six equilateral triangles.

a) How many such triangles meet at a vertex?
b) Is it a regular polyhedron? Why?
c) Does Euler's theorem hold for this figure?

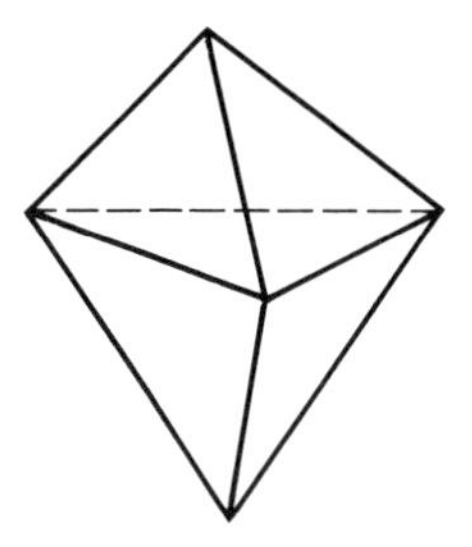

9. Visualize a solid cube of wood. How can it be cut so that the face exposed as a result of the cut is a:

a) square b) rectangle c) triangle
■ d) hexagon ■ e) rhombus

If you have trouble visualizing, make a sketch or actually use a cube of cheese or butter to try out your ideas.

■ **10.** Imagine that the cubic box in the sketch contains a sphere that just fits into the box, that is, the diameter of the sphere is the same length as the edge of the cube. Certain cutting lines are indicated on the drawing. Make four sketches corresponding to each of the cuts to show what you would see if the cuts were actually made. Assume that the sphere is a solid and that the box is paper thin.

Cut 1. Parallel to the base, through a midpoint of the edge.
Cut 2. Parallel to the base, at a quarter height of the edge.
Cut 3. Contains diagonally opposite edges (a diagonal section).
Cut 4. Goes through the midpoint of the edge of the upper face and parallel to cut 3.

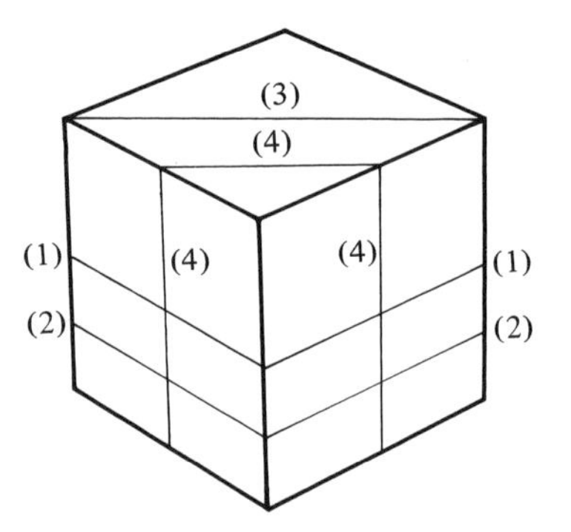

11. A pattern for making a cube is given in the text. There are other patterns consisting of six squares that could be used. The following patterns of six squares are called *hexominoes*. Which of them can be folded to make a hexahedron (a cube)?

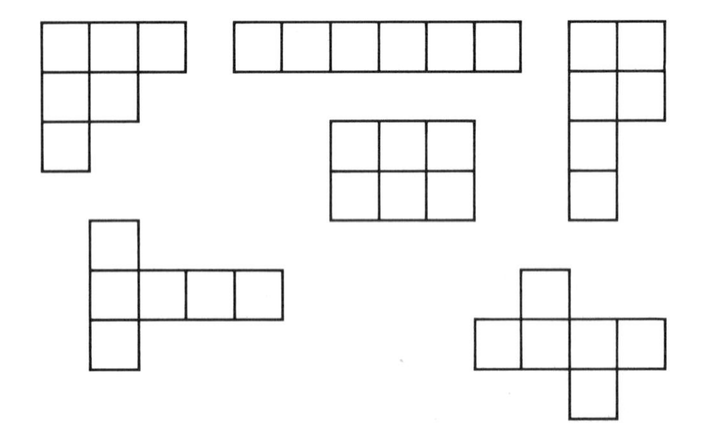

■ **12.** There are 29 other hexominoes possible. Draw some of them and see if any of these could be folded to make a cube.

REVIEW TEST

1. Give a definition of a sphere in terms of congruence.

2. Draw a rhombus that is also a rectangle. What else is this figure called?

3. List the five regular polyhedra. In each case, state which of the regular polygons form a face and how many of them meet at a vertex.

4. Prove that not all parallelograms are similar.

5. There are three angles on the same side of the straight line AOD. These angles are AOB, BOC, and COD; $\measuredangle AOB \cong \measuredangle COD$. Prove that the bisector of $\measuredangle BOC$ is perpendicular to line AOD.

6. Show all the lines of symmetry that exist for a square $ABCD$. The square also has rotational symmetry. How is the center of rotational symmetry related to the lines of symmetry?

7. Draw a polygon that has exactly one line of symmetry.

8. How many different polygons can you find in the given figure?

9. With the Euclidean tools (straightedge and compass) it is possible to construct the image of a point under reflection in a line. Find the image P' of point P in the given line L. Describe the construction.

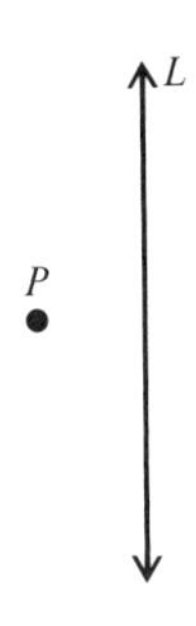

10. If two lines lie in a plane, there are two possibilities for them; they are either parallel or they intersect. Use two pencils as models and find what possibilities there are for two lines in space. Describe them.

11. Construct a triangle congruent to a given triangle ABC by using two sides and the included-angle method (SAS theorem). Explain why this method gives the correct triangle.

12. The sketch represents a block of wood. Make a sketch of the top, side, and front views of this block.

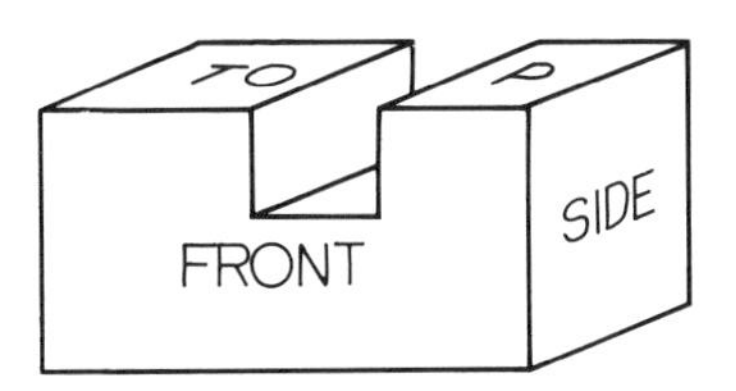

13. A triangle ABC has $\measuredangle CAB \cong \measuredangle ABC$. Show that this triangle also has other congruent parts.

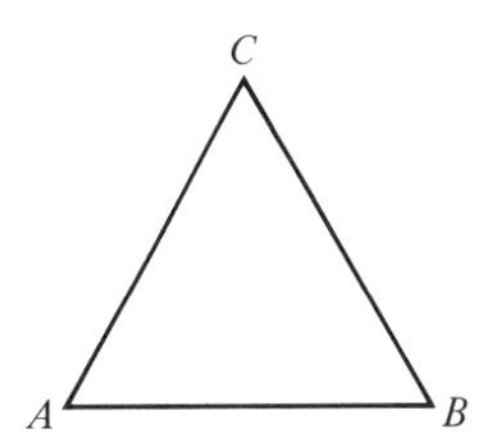

14. Show how to make one cut in the sketch of the parallelogram so it could be reassembled as a rectangle.

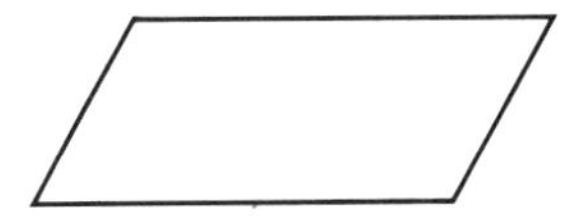

15. Points A and B are two distinct points on a line. Perpendiculars are dropped to points A and B from opposite sides of the line. These perpendiculars are intersected in points C and D, respectively, by a line that passes through the midpoint of $\overline{AB}$. Prove that the perpendicular segments $\overline{AC}$ and $\overline{BD}$ are congruent.

16. In the figure $ABCD$, the following is true: $\overline{AB} \parallel \overline{CD}$, $\overline{AD} \parallel \overline{CB}$.

a) Which sides are congruent?

b) Which angles are congruent?

c) How are the lines $\overleftrightarrow{AC}$ and $\overleftrightarrow{DB}$ related?

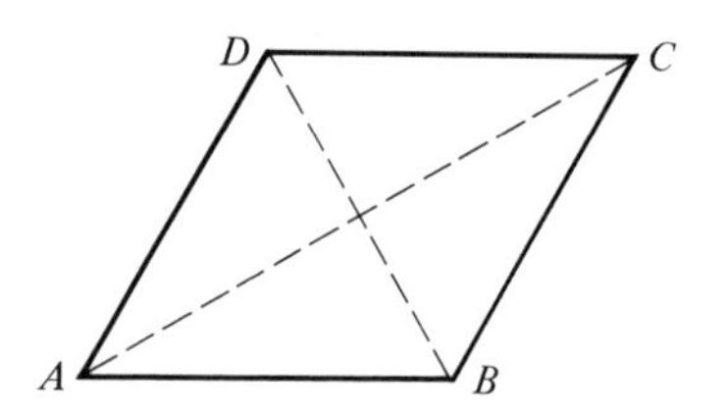

FOR FURTHER READING

1. Pugh, Anthony, *Polyhedra, a Visual Approach*, University of California Press, Berkeley and Los Angeles, 1976, has a first chapter devoted to the Platonic solids. The rest is a storehouse of information about many others. The last chapter is full of practical hints on how to construct the models.

2. "Transformation Geometry and the Artwork of M. C. Escher," in *The Mathematics Teacher*, 69:8, December 1976, pp. 647–652, is by Sheila Haak, who writes ". . . an examination of the mathematical basis for the artwork of M. C. Escher." Most upper-elementary students would find it most interesting to search out the patterns in the designs.

3. Carol Ann Alspaugh has described ". . . an interesting type of mirror geometry that could be utilized to introduce geometrical topics such as regular polygons, coordinates of points in a plane, reflections, and symmetry," in an article, "Kaleidoscopic Geometry," in *The Arithmetic Teacher*, 17:2, February 1970, pp. 116–117. The same issue has several other articles on geometry.

TEXTS

Jacobs, Harold R., *Geometry*. W. H. Freeman and Company, San Francisco (1974).

Kelly, Alice J., and David Logothetti, *Mathematics for Elementary Teachers*. Wadsworth Publishing Company, Belmont, California (1976).

Riedesel, C. Alan, and Leroy G. Callahan, *Elementary School Mathematics for Teachers*. Harper and Row Publishers, New York, New York (1977).

chapter 11 measurement

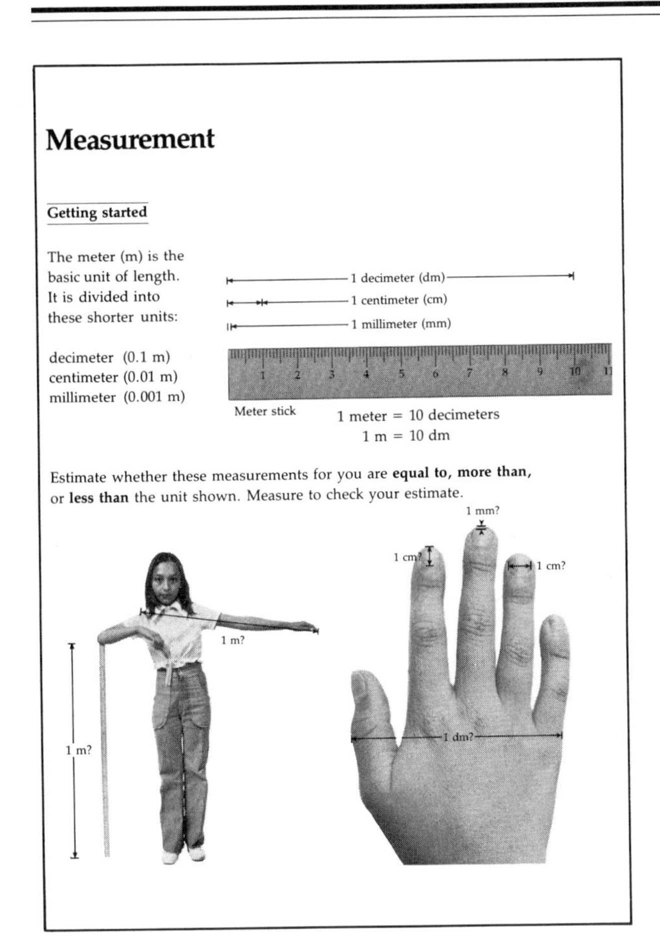

Measurement

Getting started

The meter (m) is the basic unit of length. It is divided into these shorter units:

decimeter (0.1 m)
centimeter (0.01 m)
millimeter (0.001 m)

1 meter = 10 decimeters
1 m = 10 dm

Estimate whether these measurements for you are **equal to, more than,** or **less than** the unit shown. Measure to check your estimate.

Making actual measurements is the best way to get started with the metric system. This sixth-grade lesson works on both estimation and measurement skills.

11.1 WHY MEASURE?

HISTORY. It must not have been long after man had recognized a need for counting that a similar need for measurement was felt in trade and building. Ancient Egyptian papyrus texts show the symbol of a forearm for a cubit, a unit based on the arm length from fingertip to elbow (about 18″). The Great Pyramid of Cheops, for example, was built with this unit of measurement.

Just as the Egyptians used the "standard forearm," people everywhere commonly used a part of the body as a standard. The foot was convenient, but unfortunately, feet do not come in standard sizes.

The British–American system is the hodgepodge result of the units acquired from British kings over a long period of time. For instance, Henry I (1068–1135) decided that a yard was the distance from his nose to his fingertips. In 1305 the English standardized the foot as 36 barleycorns "from the middle of the ear" laid end to end!

Primary youngsters are often given a chance to relive the early measurement history of mankind by being introduced to units of measurement involving their bodies. They count steps as they walk across the room and find, as one might expect, that rarely do two children require the same number of steps. They also might try to measure the width of the desks with "hand" as the unit.

Try Exercise 1

Often the next stage with primary children is to have them work with some familiar object as a unit for the measuring. Any standard item, such as a paper clip, nail, screw, or candy mint, could be used. Our technological society has such excellent techniques for measurement that it produces very uniformly sized objects of this type. The following portion of a lesson for the third grade shows this approach.

From a third-grade textbook

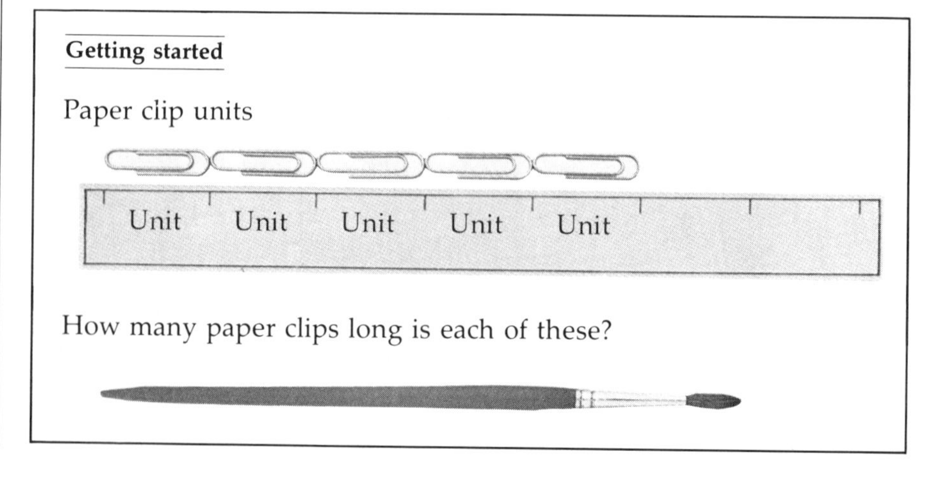

11.1 AIMS

You should be able to:

a) Explain the meaning of *unit* and *subunit* as used in the measure of distance (length).
b) If given a unit segment, use a compass to find the measure of another segment to the nearest unit.
c) Construct a ruler with a specified number of subunits and use it to measure a line segment.
d) State three properties a mathematical measure must have.

1. If you are seated at a desk or a table, use the width of your hand to measure the length or width of the desk in "hands."

"You may not know that seven firkins make a hogshead and four gills make a pint, but don't worry—it's all changing."

Frank Kendig in "*Coming of the Metric System*", Saturday Review, November 25, 1972.

With such units pupils find the length of an object by counting the number of times the chosen unit can be "laid off" along the unknown length.

Try Exercise 2

2. To the nearest "paper clip" unit, what are the dimensions of a standard sheet of typing paper ($8\frac{1}{2}$ by 11")?

In a similar way we could choose, arbitrarily, some *unit* segment as a basis of comparison and assign it a length of 1.

Suppose that we have chosen the length of $\overline{AB}$ as a unit and wish to use it to assign a length to segment $\overline{CD}$. On $\overline{CD}$ we can lay off a segment $\overline{CP_1} \cong \overline{AB}$ with a compass. Then we lay off another one ($\overline{P_1P_2}$), and then still another ($\overline{P_2P_3}$), and so on. Suppose that when we come to P_n, we find that $P_n = D$. Then the number n is assigned to $\overline{CD}$ as its length in terms of the unit length AB.

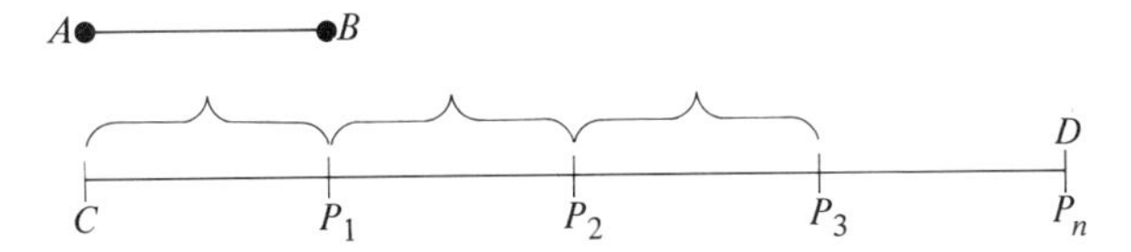

Try Exercise 3

3. What is the length of $\overline{CD}$ to the nearest AB unit?

It would be very unlikely that the unit we chose this way would fit an exact number of times on another line segment. More probably, there would be a bit left over at the end that was not quite one unit. We have several options then:

1. Create a smaller unit and remeasure.
2. Estimate how much is left over and report an estimated part of a unit.
3. Divide the original unit segment into congruent subunits and hope that these subunits will fit exactly.

With the unit segment AB above we might say that the following is "about $3\frac{1}{2}$ units."

4.
a) Make a segment congruent to $\overline{AB}$ and bisect it to get a subunit that is half the length of the unit AB.
b) Lay off the $\frac{1}{2}$-unit and measure $\overline{MN}$ to the nearest half-unit.

If smaller units were decided upon, $\overline{AB}$ would have to be divided into some number of congruent segments. Suppose that the unit segment $\overline{AB}$ was bisected to get a half-unit $\overline{AP}$; then we would try measuring MN with the half-unit, and perhaps it would actually turn out to be $3\frac{1}{2}$ units. It still might not fit exactly and we would be faced with the same problem as before. We could try to solve it in the same way—get still smaller subunits or estimate by "eye" what part of the original subunit the remaining part measured.

Try Exercise 4

No matter what unit is selected, at some point an approximation must still be made for each measurement. An arbitrary unit, such as the one derived

5. Show that the absolute value of the difference of two numbers, 5 and 10, satisfies property 2 of the distance function.

6.
a) Draw a line with three points A, B, and C located so that B is between A and C.
b) Show that the third property of a distance function is still satisfied.

from $\overline{AB}$ above, has the disadvantage that no one else has any idea of what it looks like. The statement $3\frac{1}{2}\ \overline{AB}$ units has no meaning except for readers of these paragraphs.

PROPERTIES OF A MEASURE. Before we introduce standard measures for length, we will discuss what is meant by a *measure.*

Length is a basic system of measure; the ideas and concepts developed in conjunction with length can also be applied to area and volume measure. Mathematically, we can describe the measure called length or distance as a function that has a set of line segments as its domain and the set of nonnegative real numbers as its range.

A *distance function* in an abstract mathematical sense has the following properties. If we let $d(A, B)$ represent "the distance from point A to point B," then:

1. $d(A, B) \geqslant 0$ and $d(A, B) = 0$ if and only if $A = B$;
2. $d(A, B) = d(B, A)$;
3. $d(A, B) \leqslant d(A, C) + d(C, B)$ (the triangle inequality).

Try Exercise 5

For the third property of the distance function, imagine any three points in a plane that are collinear:

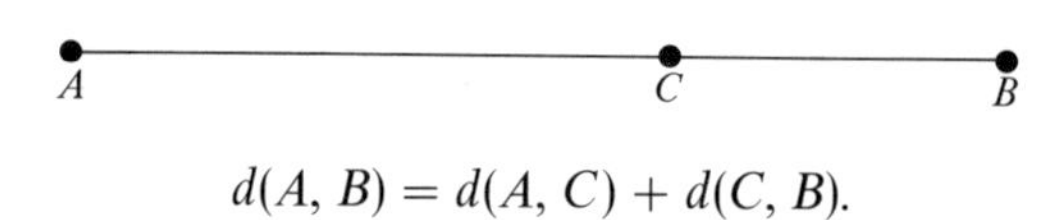

$$d(A, B) = d(A, C) + d(C, B).$$

For three points that are not collinear:

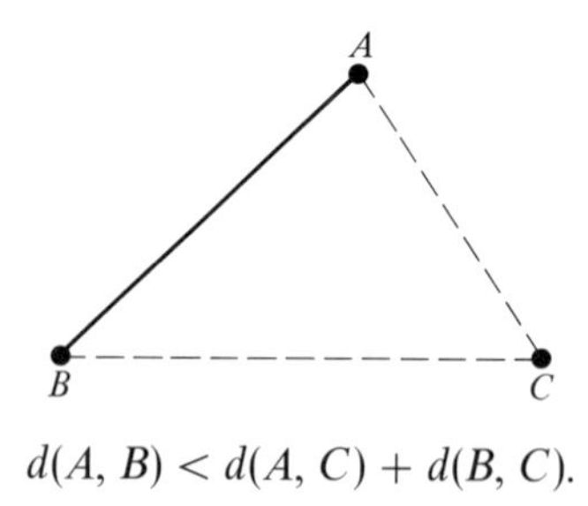

$$d(A, B) < d(A, C) + d(B, C).$$

Try Exercise 6

Pupils in the elementary school are not going to use the notation $d(A, B)$ for distance or length but by junior-high school the length of a line segment AB might be written as mAB (measure of AB) or just as AB, as we have done in Chapter 10.

The following principle underlies all the measuring we do. If a line segment $\overline{AD}$ has a point B between the points A and D, and if $\overline{AB}$ is laid off starting at point

A enough times n on the line segment AD, then there is some n such that one of the laid-off AB lengths will either reach exactly or exceed point D on the line.

Here eight copies of $\overline{AB}$ go just beyond point D.

Standard Units and Rulers

To promote communication, standard measuring instruments such as rulers have been devised. Units and subunits are marked on a straightedge, which is then laid along the segment to be measured, and the length is read off the ruler to the nearest subunit.* The line segment $\overline{MN}$ can be measured in this way with a ruler marked off in standard inches, with subunits of 1/16 of an inch, for example.

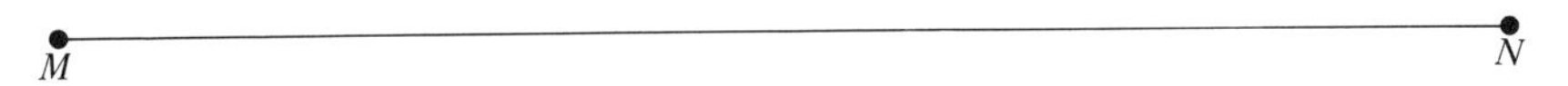

The inch is a standard unit of length in the British–American system of measures, which also contains such measures as feet, yards, rods, and miles for distance or length. Conversion in this system is difficult because the units are not related in a systematic way. In 1790, Thomas Jefferson proposed that the United States adopt a system of measures based on 10; however, Congress rejected the proposal, although they did adopt his suggestion for a decimal monetary system. It was left to the French to create and adopt a system based on 10, the metric system.

Try Exercise 7

7.
a) Measure MN to the nearest inch.
b) Repeat to the nearest 1/2 inch.
c) Repeat to the nearest 1/4 inch.
d) Repeat to the nearest 1/8 inch.
e) Repeat to the nearest 1/16 inch.

* Hold the ruler on edge to help eliminate the error that could be caused by viewing at an angle (error of parallax).

EXERCISE SET 11.1

1. What is meant by a *unit*? a *subunit*? Give an example of a unit in the British–American system of length measurement that has at least two subunits.

2. Use a compass to find the measure of segment $\overline{CD}$ in units of $\overline{AB}$. If AB cannot be laid off evenly:
a) give the answer to the nearest AB-unit;
b) estimate in fractional AB-units.

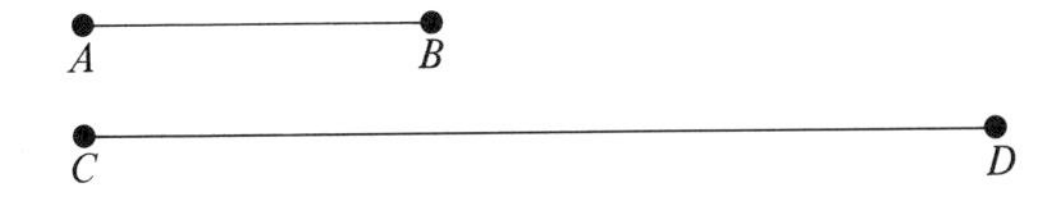

3. Use a compass to find the measure of $\overline{CD}$ in AB-units. *Hint*: Create subunits in AB.

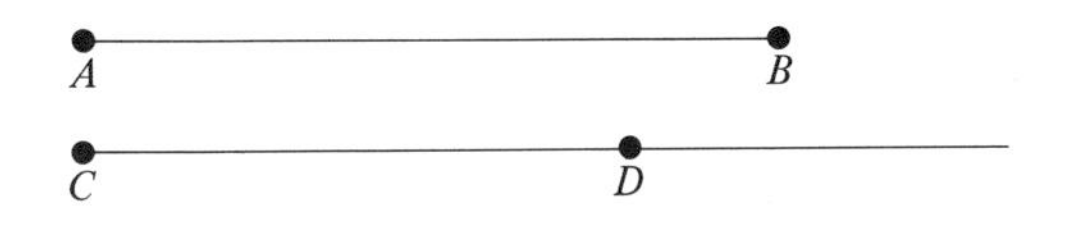

4. Assume that segment $\overline{AB}$ is to be a unit of measure. Use the Euclidean construction methods to subdivide a line segment congruent to $\overline{AB}$ into 10 congruent line segments. *Hint*: Draw a ray $\overline{AD}$ that makes less than a right angle with $\overline{AB}$.

5. Use the subdivided segment found in Problem 4 to make a ruler out of a file card. Use your ruler to measure the following segments:

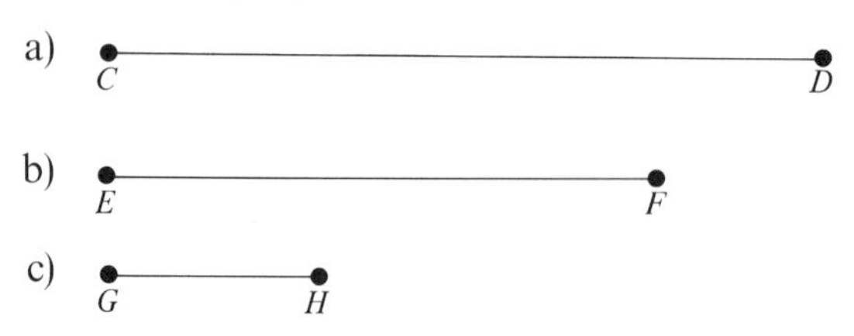

6. What are the advantages of using subunits that are tenth-units over subunits that are half-units?

7. If our numeration system were base 7, what kinds of subunits would be advantageous? Why?

8. A number line embodies the three properties of a measure for distance. The $d(0, 1)$ is the unit.

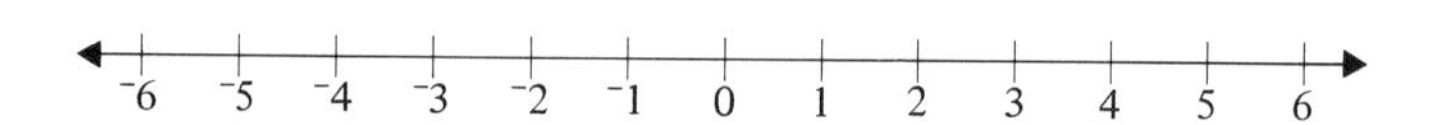

a) What is $d(0, 5)$?
b) What is $d(5, 5)$?
c) What is the sum: $d(0, {}^{-}2) + d(2, 5)$?
d) Let $A = {}^{-}1$, $B = 5$, and $C = 7$ on the number line. Show that property 3 for measure (the triangle inequality) holds for these points.
e) Does property 2 for a distance function hold on a number line?

9. If you think you are very familiar with the British–American system of measurement and would not like to replace it with the metric system, answer the following questions without looking at any reference work to do so:
a) What is the weight of a *keg* of nails?
b) What is the difference in pounds between a *long ton* and a *short ton*?
c) How many ounces are in a *pound troy* weight?
d) Are there more than 120 liquid *ounces* in a *gallon*?
e) In dry measure is 4 *pecks* the same as a *bushel*?
f) Is 6 *quarts* (dry measure) the same as a *peck*?
g) Is a *bushel* more than 36 *quarts*?
h) Are there more or less than 30 *gallons* in a *barrel*?
i) Does it take more, less, or exactly 350 *rods* to make a *mile*?
j) Does the standard U.S. *gallon* have more, less, or exactly 225 cubic inches?

11.2 SYSTEM INTERNATIONAL (SI)—THE METRIC SYSTEM

11.2 AIMS

You should be able to:

a) Explain the four common metric prefixes: kilo, deci, centi, and milli.
b) Make unit conversions within the metric system.
c) Make estimates in the SI units of measure of appropriately selected objects.

CREATION OF THE STANDARD UNITS. Because of the need in 18th-century science for a more reasonable system of measurement, the French Academy of Science in 1795 scrapped the patchwork of old measures and created a completely new, logical system. The unit of length, the *mètre*, approved by the French National Convention in 1795 was originally standardized so that it was one ten-millionth (10^{-7}) of the length of the meridian between the equator and the North Pole. All other units of length were derived from this unit by powers of ten. The Treaty of the Meter was signed in 1875 by the United States, along with the European nations in Paris. In 1893, the United States officially became a metric nation and declared the meter* bar and kilogram weight received from the International Bureau in France to be the fundamental standards for length and mass. The other units were redefined in terms of these units.

However, it was only scientists and engineers who accepted and used the system. The rest of the population continued with the familiar British system, inconvenient as it might be. By 1971 the United States was the only highly industrialized nation in the world still on the British system; not even Great Britain retained it.

On December 23, 1975, President Gerald R. Ford signed the metric conversion act that

1. Called for voluntary conversion to the metric system.
2. Established the mechanism for the creation of a United States Metric Board to coordinate the conversion.

Many manufacturers have begun to convert on their own. For example, the Ford motor car company produced the original Model T with the inch subdivided into subunits of 1/64 inch, the model A was built on the inch subdivided into hundredths, and the Pinto was built using the metric system.†

The original metric system developed by the French has been modified over the years. The present system was established in 1960 by international agreement and is known as SI, which is an abbreviation for the French *Le Système International d'Unités*. In SI there are three categories of units:

1. *The base units*: There are seven of these, only four (length, mass, temperature, and time) are used in everyday affairs.
2. *The derived units*: These are derived from the base units and include those for area, volume, velocity, and so on.
3. *The supplementary units*: These are for angle measurement.

* This is sometimes spelled *metre*. The Educational Materials Sector Committee of the American National Metric Council recommends the spellings *meter* and *liter* (instead of litre) for all publications with copyright dates later than 1982.

† "The worry is greatly overstressed," says P. E. Burke of American Motors. "It turns out to be a myth that it would cost enormous sums." Everett Baugh of General Motors says, "Going metric in the Chevette caused no more than a ripple."

"*How Soon Will We Measure in Metric?*" by Kenneth F. Weaver in National Geographic, August 1977.

Linear Measure

The *meter*, the base unit for length under SI, is defined in a different way than in the older metric systems:

1 meter = 1 m is 1 650 763.73 wavelengths in vacuum
of the orange-red light emitted by krypton-86.

The metric units of length and their relationship to the meter are as follows:

Length Measure

1 *kilo*meter	**= 1 km**	**= 10^3 m**
1 *hecto*meter*	**= 1 hm**	**= 10^2 m**
1 *deka*meter*	**= 1 dam**	**= 10^1 m**
1 meter	**= 1 m**	**= 1 m**
1 *deci*meter*	**= 1 dm**	**= 10^{-1} m**
1 *centi*meter	**= 1 cm**	**= 10^{-2} m**
1 *milli*meter	**= 1 mm**	**= 10^{-3} m**

Try Exercise 8

8.
a) How many meters in a kilometer?
b) How many mm in a meter?
c) How many cm in a meter?
d) How many mm in a cm?

The metric prefixes italicized above are also used with units other than length. The common and most important prefixes and their meanings are:

Prefix	*SI Symbol*	*Multiplication factor*
kilo	k	1000 or 10^3
deci	d	0.1 or 10^{-1}
centi	c	0.01 or 10^{-2}
milli	m	0.001 or 10^{-3}

These few prefixes should certainly be memorized. To convert from one unit to another is just a matter of multiplying by a power of 10.

Youngsters are first given a ruler marked off in centimeters and asked to measure various objects to the nearest centimeter. This is a start in giving them some basis for making estimates of length.

Try Exercise 9

9. Use a centimeter ruler to measure each of the following to the nearest centimeter:
a) The dimensions of the book you are now reading.
b) The dimensions of the paper you are using to do the margin exercises.
c) The length of the pen or pencil you are using to do the margin exercises.

ESTIMATION OF LENGTH. The use of a 32-cm ruler for measuring length is not what bothers people about the conversion to the SI units. It is easy to read such a ruler, since it is similar to a foot-ruler except that the subunits are tenths instead of eighths or sixteenths. However, people have no "feeling" for a measurement reported in centimeters and have no idea if a human hand measures about 10 cm across or 20 cm or perhaps neither. Estimation is then a very

* The American National Metric Council states that the use of the prefixes *hecto* and *deka* should be avoided except as needed to teach the structure of the metric system. *Centi* and *deci* should be used only in centimeter, square centimeter, and cubic centimeter. The decimeter is needed only to establish the cubic decimeter. Facility in these units should not be stressed except for these few exceptions.

important part of learning the SI scheme, and this is just as true for adults as for children. The following lesson is devoted to acquiring experience in estimation in the metric system.

Try Exercise 10

10. Do the exercises of the sixth-grade textbook.

In the exercises above the pupil was asked for the lengths (distance) in both cm and mm. The unit selected should be appropriate to the size of the item being measured. For very small dimensions, such as camera film, nails, nuts, and bolts, millimeters will replace inches. The 35-mm camera has been standard for a long time. A regular paper clip is slightly longer than 30 mm and about 10 mm wide. The diameter of the wire is about 1 mm. Since a paper clip is a familiar item, it makes a good reminder of 3 cm, 1 cm, and 1 mm.

Try Exercise 11

11. Check out the size of a regular paper clip to see how close to the given dimensions it is. Draw a line segment of about 6″. Lay off by "eye" distances of 3 cm. Check your estimates with a centimeter ruler.

For larger dimensions the meter will replace the foot and yard. We will speak of a 3 by 4-meter rug instead of a 9 by 12-foot one. For greater distances the kilometer will replace the mile. We will be able to drive 80 kilometers per hour and not break the speed limit of 55 miles per hour.

Try Exercise 12

12. Find out how tall you are in meters. Although we do not recommend converting from one system to the other, for this exercise use the following:

$$1'' = 2.54 \text{ cm}.$$

The conversions from the metric system to the American system are not convenient, but there is no reason to perform such conversions. Compare in the following examples.

Example 1

a) Convert 80 inches to feet.
b) Convert 76 feet to yards.
c) Convert 5000 yards to miles.

A lesson on possible units of measure

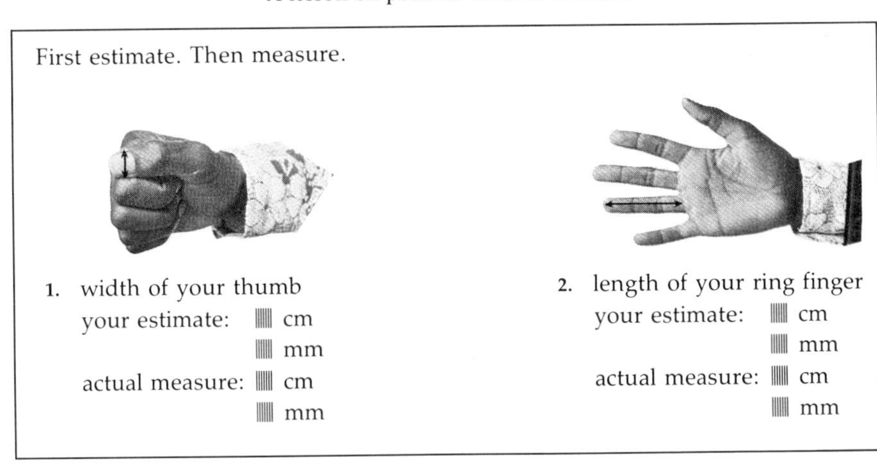
First estimate. Then measure.

1. width of your thumb
your estimate: ▮ cm
▮ mm
actual measure: ▮ cm
▮ mm

2. length of your ring finger
your estimate: ▮ cm
▮ mm
actual measure: ▮ cm
▮ mm

Solution

a) There are 12 inches in 1 foot; hence we form the proportion:

$$\frac{12}{1} = \frac{80}{x}, \qquad 12x = 80, \qquad x = 6\tfrac{2}{3} \text{ feet.}$$

b) There are 3 feet in 1 yard and we form the proportion:

$$\frac{3}{1} = \frac{76}{x}, \qquad 3x = 76, \qquad x = 25\tfrac{1}{3} \text{ yards.}$$

c) There are 3 feet in 1 yard, so 5000 yards is 15 000 feet. We form the proportion:

$$\frac{5280}{1} = \frac{15\,000}{x}, \qquad 5280x = 15\,000, \qquad x \approx 2.84 \text{ miles.}$$

In (a) it was necessary to divide by 12, in (b) to divide by 3, and in (c) to multiply by 3, then divide by 5280.

Example 2

a) Convert 5 m to cm.
b) Convert 600 cm to m.
c) Convert 1260 m to km.

Measuring height and weight in the classroom to introduce metric measure

Solution All conversions are done with multiples of 10.

a) 1 m is 100 cm, so 5 m is 500 cm (multiply by 100).
b) 100 cm is 1 m, so 600 cm is 6 m (divide by 100).
c) 1000 m is 1 km, so 1260 m is 1.260 km (divide by 1000).

The metric conversions are so simple they can be done in one's head.

Try Exercise 13

13.
a) Convert 5650 m to km.
b) Convert 0.5 km to m.
c) Convert 0.75 m to cm.
d) Convert 3.5 cm to mm.
e) Convert 70 mm to cm.

Mass (Weight)*

Another of the base units of SI is the *kilogram*, the unit for mass. The kilogram is the only base unit that has a prefix in its name. The kilogram standard is a cylinder of platinum-iridium alloy kept in Paris at the International Bureau of Weights and Measures and a duplicate of this cylinder is in the Bureau of Standards in Washington, where it serves as the standard for the United States.

Mass Measure (Weight)

1 metric ton = 1 t = 100 kg
1 kilogram = 1 kg = 100 g
1 hectogram† = 1 hg = 100 g
1 dekagram† = 1 dag = 10 g
1 gram = 1 g = 1 g
1 decigram† = 1 dg = 0.1 g
1 centigram† = 1 cg = 0.01 g
1 milligram = 1 mg = 0.001 g

For small quantities such as candy, cereal, and the like, the units used will be *grams*:

Item	*Old*	*New*
Box of candy	1 lb 8 oz	680 g
Box of cereal	12 oz	340 g
Jar of instant coffee	4 oz	112 g
Package of crackers	$3\frac{1}{4}$ oz	92 g

For larger items such as a turkey, roast beef, and your own body mass, the unit will be the base unit, the *kilogram*:

Item	*Old*	*New*
Turkey	14 lb	6.3 kg
Beef roast	8 lb	3.6 kg
A person	115 lb	51.8 kg

* In everyday language the word *weight* just about always means the same as *mass*. The instruction to "weigh" means to "find the mass of." In science the word *weight* is used to describe force; hence a distinction must be made. In some elementary texts the SI terminology of *mass* is used; in others the word *weight* is used for the same thing, and this is acceptable for both elementary and general education.

† Prefixes hecto, deka, deci, and centi are still to be avoided and are only included here for completeness.

14.
a) How many g in 3 kg?
b) How many mg in 2 g?
c) How many t in 1250 kg?

For still larger items such as a car the unit will be the *metric ton*:

Item	*Old*	*New*
A car	3000 lb	1.35 t

Try Exercise 14

ESTIMATION. Again, as with length, the difficulty is in not being able to estimate from experience with the units being introduced. This skill comes from the kind of practice shown in the 6th-grade lesson below, which is the type of experience adults will have to acquire before they feel comfortable with SI units.

Try Exercise 15

15.
a) Study the portion of the 6th-grade textbook.
b) Give the correct mass unit g, kg, or t for each of the following:
   1. a liter carton of milk: 1 ____
   2. A small bag of flour: 2.5 ____
   3. A small pickup truck: 1.6 ____
   4. A box of salted peanuts: 190 ____
   5. A nickel: 5 ____
   6. A small loaf of bread: 340 ____

From a sixth-grade textbook

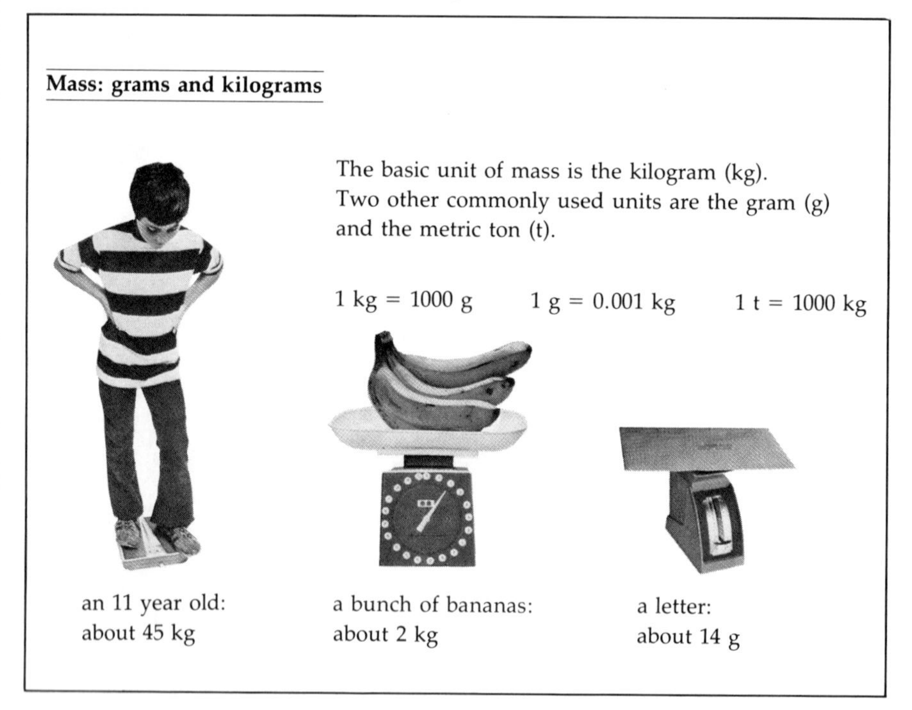

Mass: grams and kilograms

The basic unit of mass is the kilogram (kg).
Two other commonly used units are the gram (g) and the metric ton (t).

1 kg = 1000 g   1 g = 0.001 kg   1 t = 1000 kg

an 11 year old: about 45 kg

a bunch of bananas: about 2 kg

a letter: about 14 g

Temperature

Temperature measure is also one of the base units. In the SI system it is measured in degrees Celsius (°C). On the Celsius scale the following are some important temperatures to remember:

0 °C	**The freezing point of water (32 °F)**
10 °C	**A warm winter day in Chicago (50 °F)**
20 °C	**A mild spring day in New York city (68 °F)**
30 °C	**A quite warm/hot day (86 °F)**
37 °C	**Normal, human body temperature (98.6 °F)**
40 °C	**A summer day in Phoenix, Arizona (104 °F)**
100 °C	**The boiling point of water (212 °F)**

Again, the only difficulty with the Celsius scale is that at 64 °F a person may decide to wear a light sweater but at 22.5 °C would not be sure if a sweater is called for.

The scientist uses the Kelvin temperature scale (K), which is the official SI unit. However, a change of 1 K is the same as a change of 1 °C; the only difference is that the Kelvin scale has its 0 at "absolute zero" (the point at which all atomic vibration ceases). The temperature scale accepted by SI for daily use is the Celsius scale.

Try Exercise 16

Time

Time measure is also one of the SI base units; it uses the same units as the British–American system. No new units will need to be learned. There are 24 hours in the day, 60 minutes in the hour, and 60 seconds in a minute. In the SI measurement scheme, the second is defined in terms of the duration of a specific radiation of a cesium-133 atom.

There have been suggestions for converting to a decimal day instead of the present mixture of conversions of 24 and 60, but for the present this has not been done.

Derived Units

The second classification of units in the SI unit system develop in the course of computation with the base units; they are said to be derived units.

AREA. Area units are the measure of a square region with a specified side length.

Area units	*SI Symbol*	*Side of square*
1 square millimeter	1 mm^2	1 mm
1 square centimeter	1 cm^2	1 cm
1 square decimeter	1 dm^2	1 dm
1 square meter	1 m^2	1 m
1 hectare	1 ha	100 m
1 square kilometer	1 km^2	1 000 m

16. Choose the best estimate for each of the following:

a) A good day for ice skating:

$^-2$ °C 10 °C 25 °C

b) Water temperature for going swimming:

20.5 °C 42 °C 71 °C

c) Call the doctor because your temperature is:

22 °C 39 °C 59 °C

d) A glass of iced tea:

$^-15$ °C 2 °C 15 °C

17.
a) Draw a square that is 1 cm on each side; this is 1 cm^2.
b) Draw a square that is 1 dm on each side; this is 1 dm^2.
c) Find something that is about 1 m long and 1 m wide in the room you are now in. Memorize this image for future comparisons.

Try Exercise 17

VOLUME. Similarly, the units of volume are the measure of a cube of specified edge length.

Volume units	*SI Symbol*	*Edge of cube*
1 cubic meter	1 m^3	1 m
1 cubic decimeter	1 dm^3	1 dm
1 cubic centimeter	1 cm^3	1 cm

When used to measure fluids (either gas or liquid) or dry ingredients in recipes, the cubic decimeter is called a liter (L)*. A liter of water is about the same as a quart of water.

The space inside a car trunk will be measured in liters instead of in cubic feet and milk will be purchased by the liter instead of by the quart. The following relations hold:

$$1 \text{ m}^3 = 1000 \text{ dm}^3 = 1\,000\,000 \text{ cm}^3$$

$$1 \text{ m}^3 = 1000 \text{ L} = 1\,000\,000 \text{ mL}$$

18.
a) How many m^3 in 1520 dm^3?
b) How many liters in 2 m^3?
c) How many cm^3 in one mL?
d) How many m^3 in 3675 L?

Another important property of the SI units is that 1 liter of water at 4 °C has a mass of 1 kilogram.

Try Exercise 18

Supplementary Units

The third and last category of SI units are neither base units nor derived units. These are related to angle measurement and will be discussed in the next section.

A glance toward the classroom

It does not help to learn the metric system if one thinks in terms of converting from the British system and back. To discourage such conversions we are not supplying a conversion chart from British to the SI units. In the same way children should be taught the metric system directly, not as an inconvenient conversion scheme. The conversions are "peculiar" because of the illogic of the British system, and not the other way around.

* The international symbol for liter is the upright lower case l. Since this can be confused with the symbol for one, the United States Department of Commerce has recommended that in the United States the capital L be used as the symbol for liter. The American National Metric Council also recommends the same thing.

EXERCISE SET 11.2

1. Read the opening page of this chapter if you have not already done so. Make the estimates called for and check your estimates with a meterstick.

2. For each of the following make an estimate and then check your estimate with an appropriate ruler or tape.

a) Your own elbow to fingertip measure ____ cm estimate
____ cm actual

b) Your own shoe length ____ cm estimate
____ cm actual

c) The switch plate on the wall for the electric lights
____ mm estimate
____ mm actual

d) The dimensions of a page of your daily newspaper
____ m estimate
____ m actual

3. Find some common objects that have one dimension as given below:

a) three objects that are about 1 mm thick,
b) three that have one measurement about 1 cm,
c) three that have one measurement about 10 cm,
d) three that have one measurement about 50 cm,
e) three that have one measurement about 100 cm.

4. Measure the following line segments to either the nearest mm or the nearest cm (choose the appropriate unit):

a) A ——— B

b) C ——— D

c) E ——— F

d) G ——— H

5. Describe how to build a paper cube of the proper size so that the cube will hold 1 liter of water (assume that the cube is waterproof).

6. Convert the following as indicated:

a) 1.25 km to ____ m b) 15 cm to ____ mm
c) 7536 mm to ____ cm d) 9700 m to ____ km

7. Convert the following as indicated:

a) 1585 kg to ____ t b) 0.65 kg to ____ g
c) 1.5 g to ____ mg d) 250 mg to ____ g
e) 3 t to ____ kg f) 1923 g to ____ kg

8. For each of the following answer TRUE if the statement is always true and FALSE if the statement is not always true.

a) The French Academy introduced the metric system in the late 18th century.
b) The United States officially became a metric nation in 1975.
c) The meter of the SI units is defined in the exact same way as that defined by the French in 1795.
d) The Ford Motor Company used the metric system to build the Pinto.
e) The prefixes *hecto* and *deka* should be learned by every elementary child.
f) There are three categories of units in the SI unit scheme.
g) Conversion within the metric system is simpler than in the British.

h) No base unit of the SI scheme has a prefix in its name.
i) The U.S. Department of Commerce recommends ℓ as the symbol for liter.
j) The temperature scale accepted by SI (metric system) for daily use is the Celsius scale.

9. A car travels 37 km in 40 minutes. At this same rate how far will it travel in 1 hour?

10. In the 1978 qualifying runs for the 5.4 km Watkins Glen Grand Prix, Mario Andretti led for the pole position with an average speed of 199.41 km per hour.
a) What was his speed per second?
b) How long did it take him to cover the 5.4 km track?

11.3 AIMS

You should be able to:
a) Use a protractor, either a standard one or one created from an arbitrary unit angle.
b) Relate the concept of measure to properties of the common polygons.

11.3 ANGLE MEASURE

To measure line segments we chose a unit line segment. Similarly we shall choose some angle as a unit to measure angles. Suppose that ∡ AOB shown is the unit. To measure the ∡ COB we place the unit angle with vertex and ray coinciding in the interior of ∡ COB, as indicated in the sketch. The number of unit angles it takes to cover the interior of ∡ COB will be assigned to the angle as its measure. The measure of the angle COB is denoted m ∡ COB.

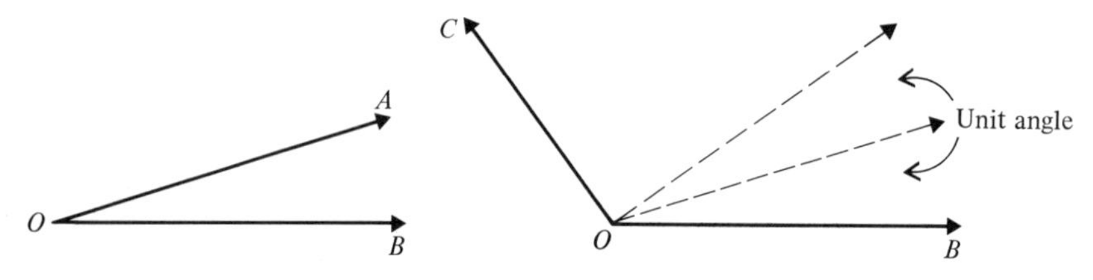

As before, to improve the approximation we can create subunits in the original unit angle. It is understood that if ∡ AOB has unit measure, then any angle congruent to it also measures 1 unit.

Try Exercise 19

STANDARD UNITS. The SI standard unit for angle measure is the radian. To understand what is meant by a *radian*, one must think of an angle having its vertex at the center of a circle. An angle AOB has a measure of 1 radian if the distance along the curve from A to B is the same length as the radius of the circle.

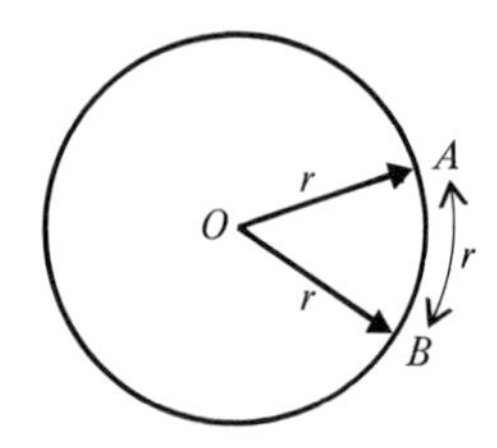

19.
a) Trace AOB and cut it out to use as a unit angle.

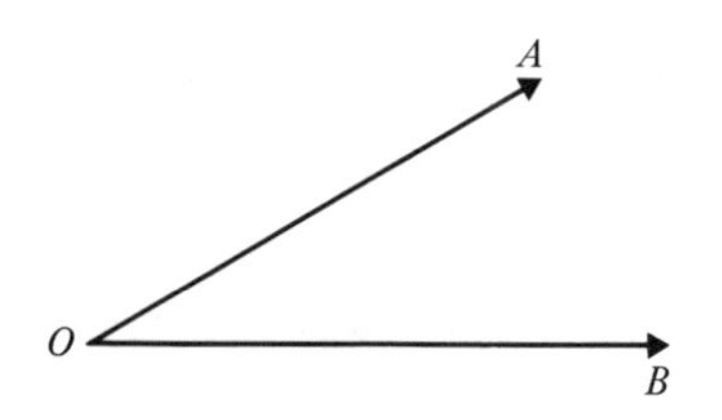

b) Use the unit angle AOB to measure to the nearest unit:

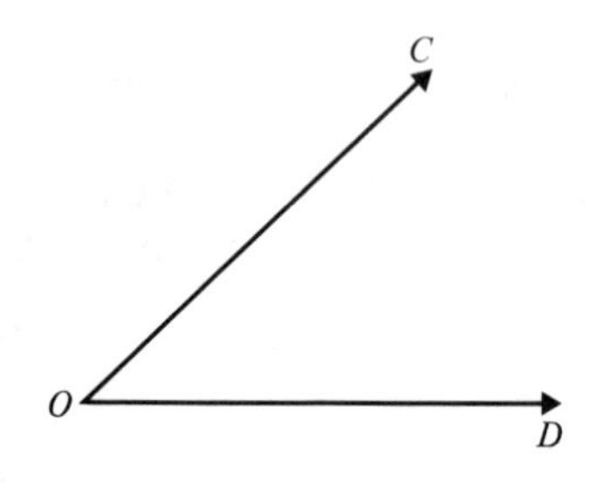

It is also acceptable under SI to use the familiar degree measure. This is the unit that will be used with elementary children. If the radius $\overline{OA}$ is thought of as sweeping around so that the point A travels the complete circumference of the circle, then the angle swept through is one that measures 360°, which means that a right angle must have a measure of 90°, and a straight angle a measure of 180°. The traditional subunits for the degree of angle measure are *minutes* and *seconds*:

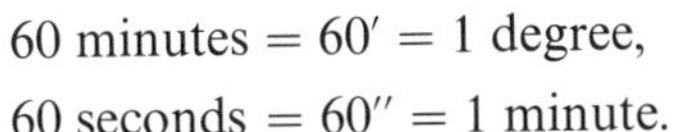

$$60 \text{ minutes} = 60' = 1 \text{ degree},$$
$$60 \text{ seconds} = 60'' = 1 \text{ minute}.$$

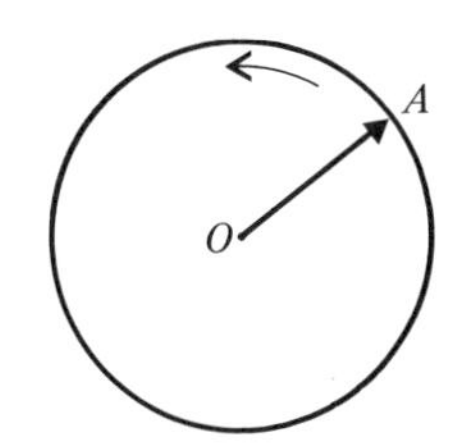

The decimal degree is also sometimes used; the degree is thought of as divided into 10 or 100 subunits, and this may be preferred because of the convenience in calculation. Elementary children will measure to the nearest degree.

PROTRACTORS

From a fifth-grade textbook

Measuring angles

A protractor is an instrument used to measure angles. The unit for measuring angles is the **degree**.

The measure of $\angle CAB$ is 40 degrees. For 40 degrees we write: 40°

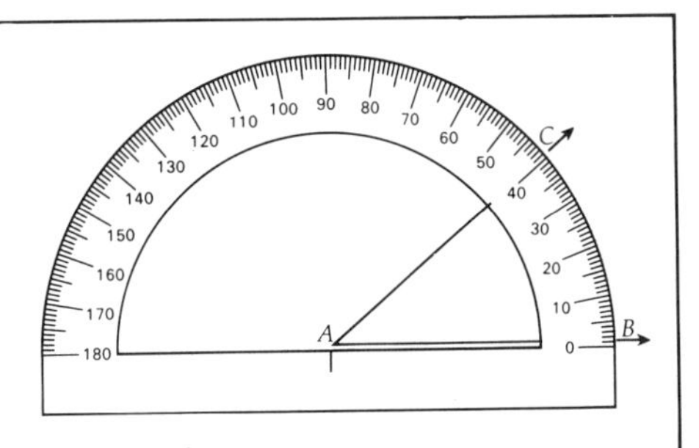

Try Exercise 20

20. Use a protractor to measure the following angles to the nearest degree.

a)

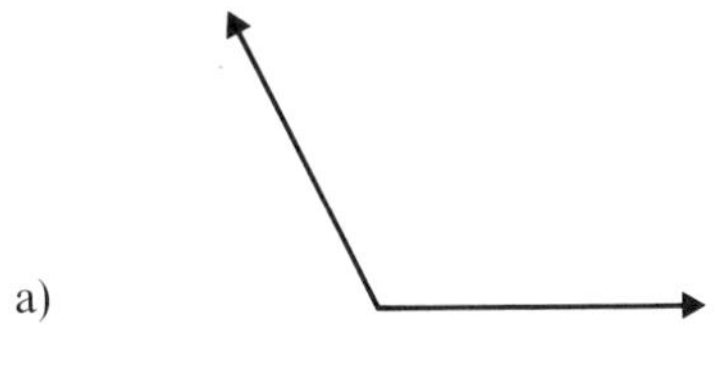

b)

c)

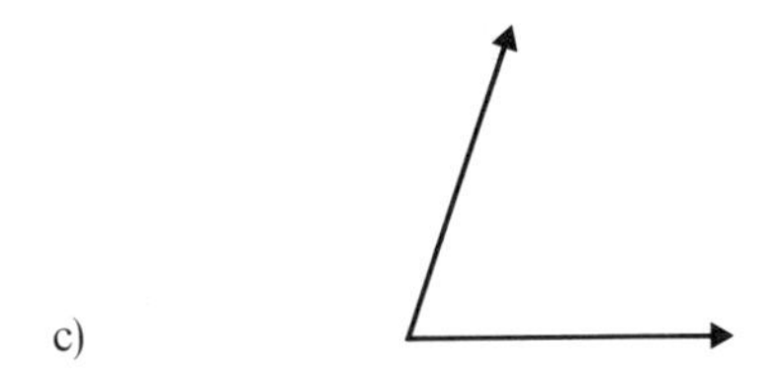

EXERCISE SET 11.3

In Problems 1 and 2, make a unit angle measure that is congruent to $\measuredangle AOB$. It can be folded to make a half-unit. Use it to measure the given angles to the nearest half-unit.

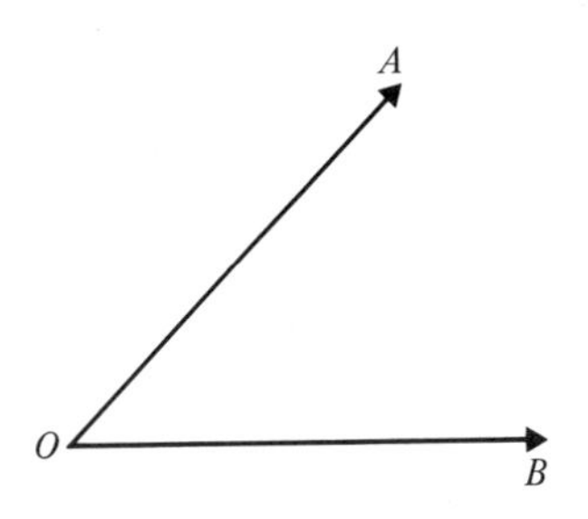

1.

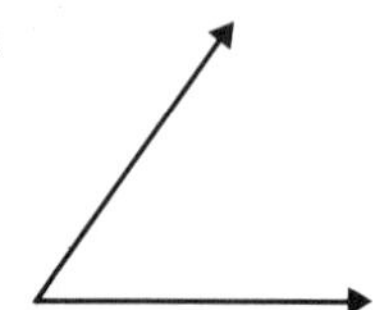

2.

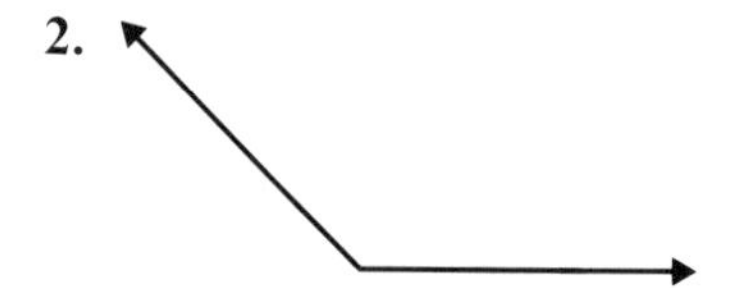

3. What does the measure of congruent angles in the Euclidean constructions have to do with unit angles? with degrees?

4. Use the unit angle AOB above to make a protractor from a file card. Use this protractor for Problem 5.

5. Measure to the nearest half-unit the angles in:
a) Problem 1 b) Problem 2

6. Measure the angles in Problems 1 and 2 to the nearest degree.

7. Assume that you have proved that the three angles of a triangle can be used to form a straight angle, which means the sum of the measures of the angles of a triangle is 180°. Use this to prove that the sum of the measures of the angles of a quadrilateral is 360°. *Hint*: Reduce the problem to one about triangles.

8. What is the measure of the fourth angle in each of the following?

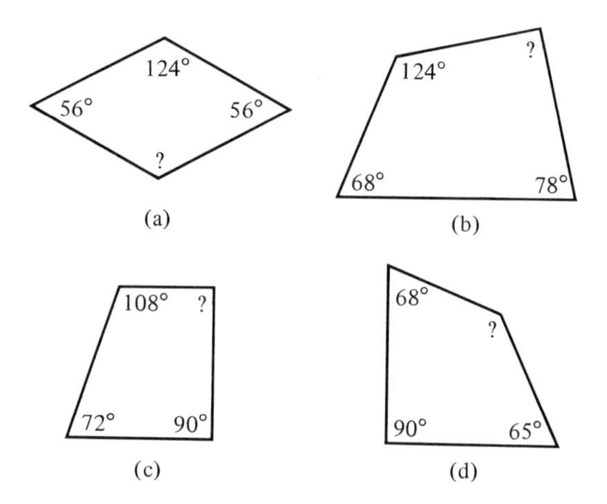

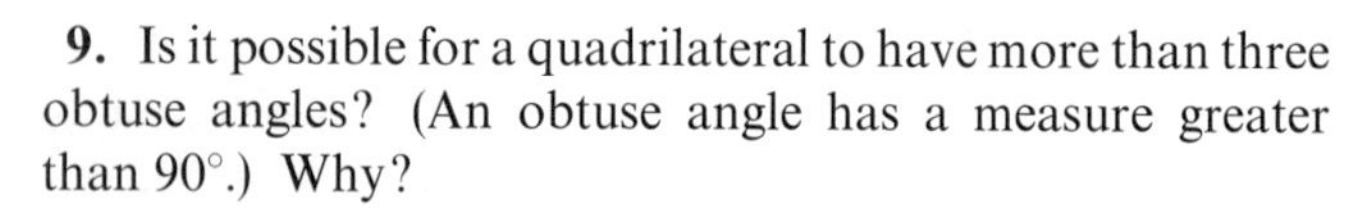

9. Is it possible for a quadrilateral to have more than three obtuse angles? (An obtuse angle has a measure greater than 90°.) Why?

10. What is the sum of the measures of the angles of a pentagon?

11. What is the sum of the measures of the angles of a hexagon?

12. Formulate a general rule for the sum of the measures of the angles of a polygon with n sides. Prove it.

13. Is it possible for a triangle to have more than one angle measuring 90°? Why?

14. What is the measure of each angle of a regular pentagon?

15. What is the measure of each angle of a regular hexagon?

16. Repeat Problem 14 for a regular heptagon.

17. Repeat Problem 15 for a regular octagon.

18. Formulate and prove a general rule for the measure of each interior angle of any regular polygon with n sides.

11.4 CURVE LENGTH AND PERIMETER

Open Curves

A measure of length can be assigned to other figures, such as in the following *polygonal curve* consisting of connected line segments. Its length is the sum of the lengths:

$$AB + BC + CD + DE + EF + FG.$$

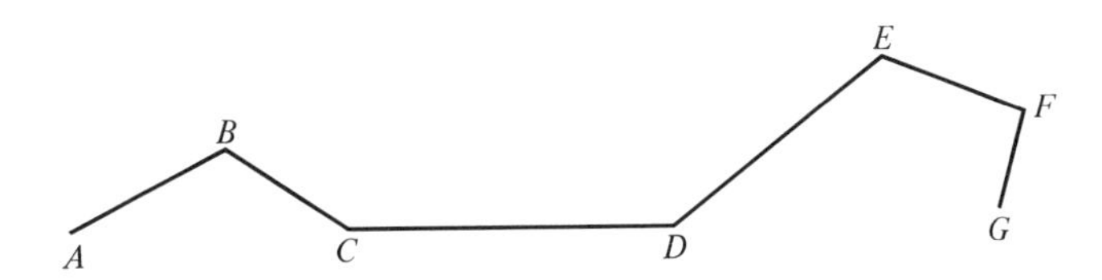

Curves can be approximated with polygonal curves:

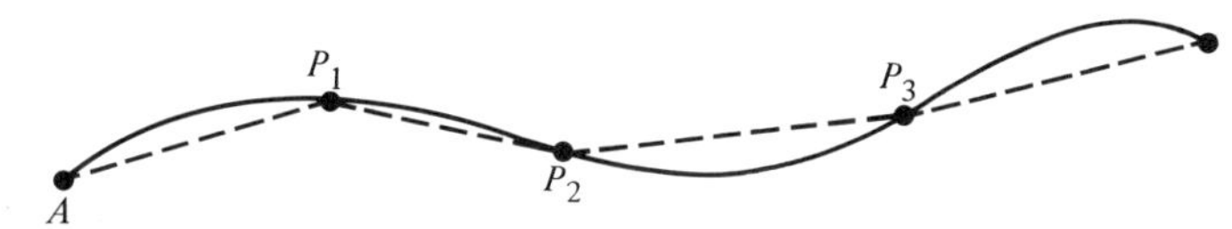

We choose some points on the curve and measure the polygonal curve formed by joining the points with line segments. This length approximates the actual length of the curve. If we next choose additional points on the curve, so that one or more of these points is between each pair of the original points used, then we get a still better approximation. If this process is continued, it is possible to get still better approximations to the length of the curve.

Try Exercise 21

Closed Curves

PERIMETER OF A RECTANGLE. The measure of an open polygonal curve is called its length; however, for a closed curve this measure is usually called the *perimeter*. For example, the perimeter of a rectangle whose sides are the connected segments $\overline{AB}$, $\overline{BC}$, $\overline{CD}$, and $\overline{DA}$ is given by $AB + BC + CD + DA$. Since the lengths AB and CD are usually called l and the side lengths BC and DA are called w, the perimeter of the rectangle can be written as:

$$\text{Perimeter} = p = 2l + 2w.$$

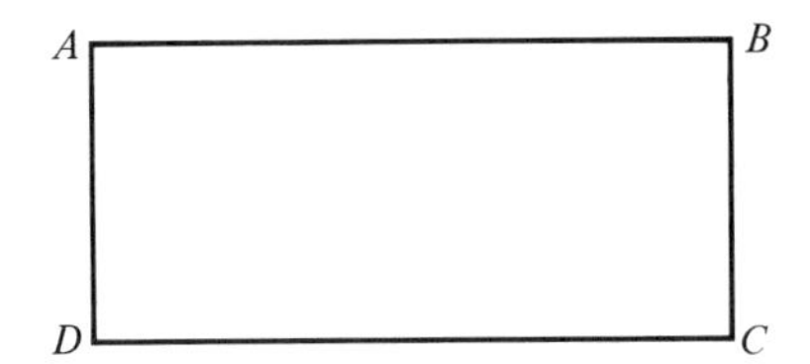

11.4 Aims

You should be able to:

a) Approximate the length of a curve with a polygonal curve.
b) Approximate the circumference of a circle with a regular inscribed polygon perimeter.
c) Given one of the dimensions for the circumference, radius, or diameter of a circle, find the other two.
d) Find the perimeter of a polygon.

21.

a) Draw line segments between the points shown on the curve. Measure the segments and approximate the length of the curve in cm.

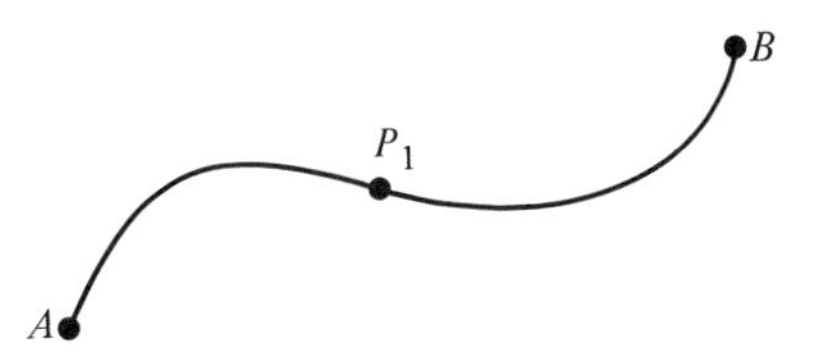

b) Trace the curve in part (a). Add point P_2 between points A and P_1 and point P_3 between points P_1 and B. Find a new approximation to the length of the curve in cm. Compare to part (a).

22.
a) Find the perimeter of a rectangle such that the length $l = 12.2$ m and the width $w = 8.5$ m.
b) Show that if the length of one side of a square is s, then the perimeter is given by $4s$.

A lesson on possible units of measure

Try Exercise 22

CIRCLES—CIRCUMFERENCE. The length of the closed curve known as a circle is called its *circumference*. One way to approximate the circumference of a circle is to calculate the perimeter of a regular polygon inscribed in a circle. For the first approximation we can use a regular hexagon. Since the triangles shown are all equilateral, we see that the sides of the hexagon are all the same length as the radius. This means that the circumference is greater than the length $6r$ (or $3D$).

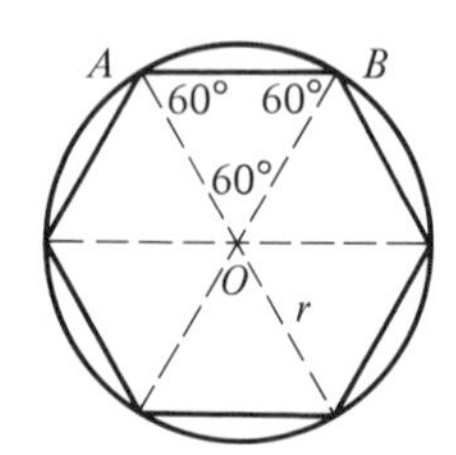

Try Exercise 23

Each time the number of sides of the inscribed regular polygon is doubled, the perimeter of the polygon is a somewhat better approximation of the circumference of the circle. By the time a polygon of 384 equal sides has been inscribed, the length of one side of this regular polygon is only $0.016362r$ and the perimeter of the polygon is $3.141558\ D$. With 768 such equal sides the perimeter is $3.141584\ D$. From these last approximations we can see that the first four digits of π have appeared. This sequence of numbers does not prove but does illustrate an interesting property of circles; namely, the ratio of the circumference to the diameter of any circle is a constant named π*. Therefore the circumference can be represented as:

Circumference $= C = 2\pi r = \pi D$, where r is the radius length and D is the diameter length.

Example 1 A bicycle wheel has a 42 cm radius. Find its circumference.

Solution Since $C = \pi D$ and $D = 84$ cm, we have $C = \pi \cdot 84$ cm.

If 3.14 is used as an approximation for π, we get $C \approx 3.14 \times 84$ or 263.8 cm.

Example 2 The earth travels around the sun in a nearly circular orbit at a distance of 1.495×10^8 km from the sun. It requires a year to make the trip. How far does the earth travel in one year?

$$C = 2\pi r = 2\pi 1.495 \times 10^8 \text{ km} = 2.99 \times 10^8 \pi \text{ km}$$

Use the number 3.14 as an approximation for π:

$$C \approx 9.39 \times 10^8 \text{ km}.$$

Try Exercise 24

23.

a) Why are all three angles of $\triangle AOB$ shown as having angle measure of 60°?

b) If a regular polygon of 12 sides were inscribed in the circle, would you expect the perimeter of the polygon to increase, decrease, or remain the same?

24.

a) The circumference of a circle is 628 cm. Find the diameter in meters. Use 3.14 as an approximation for π.

b) The diameter of an average aspirin tablet is about 10 mm. Use 22/7 as an approximation for π to find the circumference of an average aspirin tablet.

c) The circumference of a watch face is 50 mm; what is the radius of the circular face?

* Ancient mathematicians were aware of this constant and used various approximations for it. The Old Testament (I Kings 7:23) suggests 3 as an estimate for π. In 1873, William Shanks of England computed π to more than 700 decimal places; however modern computers have made it possible to compute many more. The irrational number π is commonly approximated by 22/7 or 3.14159265.

EXERCISE SET 11.4

In Problems 1 and 2, a) approximate the length (in cm) of each curve by connecting the points shown to make a polygonal curve; b) refine the measurement of part (a) by inserting a point on the curve between each two adjacent points. Draw the new polygonal curve and measure the segments in cm.

1.

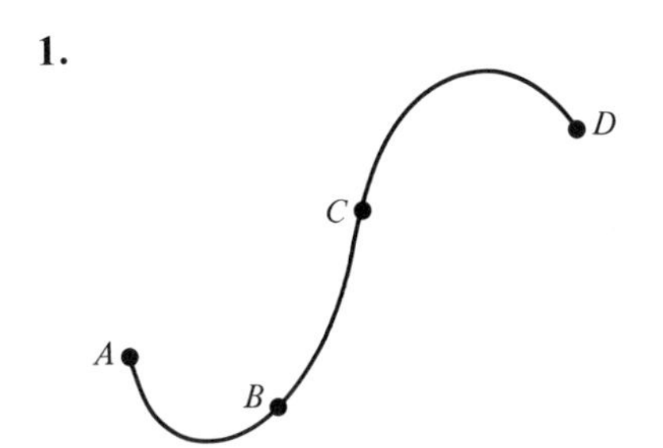

2.

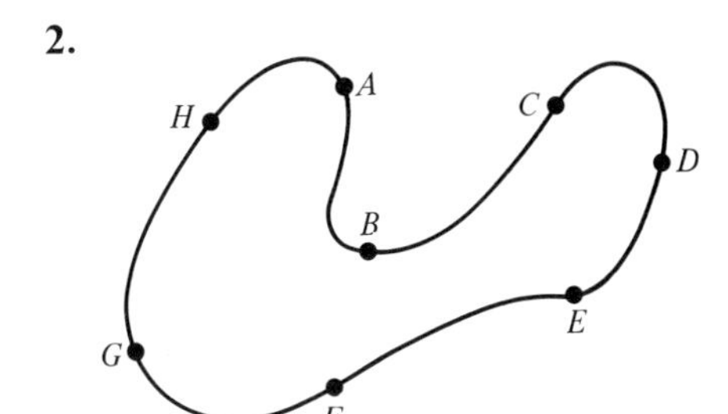

3. Find a formula for the perimeter of a regular polygon of n sides.

4. Find the perimeters of each of the following:

a) An equilateral triangle with one side of length 45 cm. Give the perimeter in m.
b) A parallelogram whose equal pairs of sides are 56 mm and 32 mm. Give the perimeter in cm.
c) A rectangular plot of land 750 m by 350 m. Give the perimeter in km.

5. A measuring wheel has a 15.92 cm radius. How far will it roll in turning 25 revolutions? How could the wheel be used to measure off approximately 10 meters on a line in the playground?

6. An earth satellite is in a circular orbit 8.05×10^2 km high. How far does it travel in making one complete revolution? Assume that the radius of the earth is 6.38×10^3 km.

7. How many turns of the handle are needed to raise the old oaken bucket from the bottom of a 9 m well, if the drum on which the rope is wound has a 30 cm diameter?

8. If the radius of a tractor wheel is 38 cm, how many revolutions does it make as it goes a distance of 1 km?

9. Imagine a band that fits tightly around the equator of the earth. If the length of this band is increased by 10 meters, how will this affect the radius of the band? Suppose the band were increased in length by 100 meters, how would that affect the radius?

10. Use the fact that $C = 2\pi r$ to find an approximation for π. Draw a circle with a 10-cm radius and inscribe a regular hexagon. The perimeter of this hexagon is 60 cm. What is the first approximation to π? Next, in each arc mark the appropriate midpoint as shown and inscribe a dodecagon in the same circle. Measure the sides and compute an approximation to π on the basis of this new perimeter.

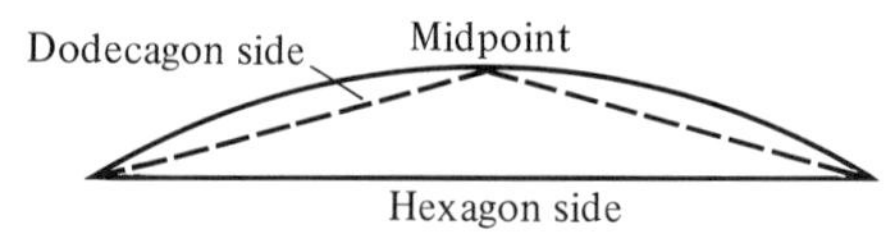

11. Refine the approximation of Problem 10 once more. Again mark the appropriate midpoint of each arc and draw the inscribed polygon of 24 sides. Compute an approximation to π on the basis of the new perimeter. Compare to your previous two estimates of π.

12. The value of π correct to the first eight places is 3.14159265. Find each of the following to five decimal places and compare to π. (Historically, these have all been used as approximations.)

a) 22/7 (Is 3.14 better than this?)
b) $\sqrt{10}$ (Is $\sqrt{10}$ greater or less than π?)
c) 355/113 (How many places are correct?)

11.5 AIMS

You should be able to:

a) Find the area of a rectangle and a triangle.
b) Use the formulas for the area of a rectangle and a triangle to find areas of other geometric figures.
c) Use the method of sweeps to find the area of appropriate figures.
d) Find the area of a circular region.

11.5 AREA

We have already mentioned that the SI units for area are derived from the units for length. The area units are all square regions such as a square centimeter, each side of which measures 1 cm.

To find the area of any region we think of "covering it exactly" with enough of such units. Here is a rectangular region being measured by covering it with square-centimeter units.

1 cm^2

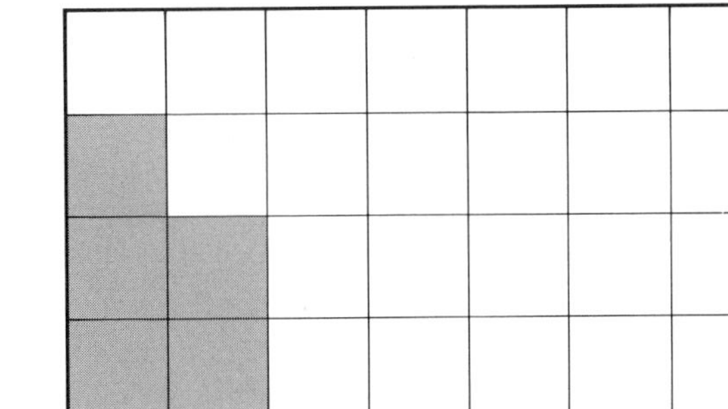

Try Exercise 25

AREA OF A RECTANGULAR REGION. Since exactly 28 units in the above example cover the rectangle completely, the area of this rectangular region is 28 cm^2. The same result can be obtained by multiplying the length in cm by the width in cm.

Example 1 Find the area of a rectangular region whose length is 12 cm and width is 5 cm.

Solution

$$A = l \times w = 12 \times 5 \text{ cm}^2 = 60 \text{ cm}^2.$$

The area of a rectangle* can always be found by multiplying its length by its width. If the length and width are both in cm, then the area is given in cm^2. If both dimensions are in m, then the area is given in m^2 and so on.

The area of any rectangular region is given by $A = l \times w$, where l is the measure of one pair of sides and w is the measure of the other.

Try Exercise 26

SWEEPS. The problem of finding the area of a rectangle can also be approached from another point of view. Imagine a segment sweeping across a rectangular region like this. The length of the sweeping segment is w, the width of the rectangular region. If the sweeping segment sweeps all the way across, it will move a distance l, the length of the rectangular region. The area of the region so "swept" is of course $l \times w$.

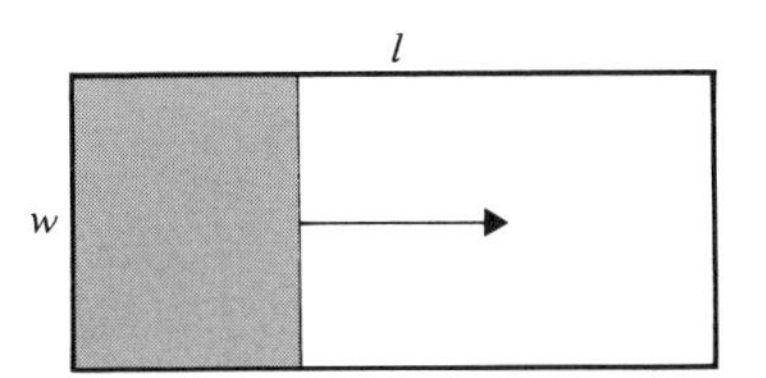

25. How many of the square centimeter units does it take to cover the rectangle exactly?

26.

a) How many square centimeters will fit into a rectangular region that measures 10 cm by 10 cm? Make a sketch.
b) How many square centimeters in a square decimeter?
c) How many square centimeters will fit into a rectangular region 100 cm by 100 cm? Make a sketch.
d) How many square centimeters in a square meter?

* It is common to speak of "the area of a rectangle" when we mean "the area of a rectangular region."

27.

a) Use a sweep to find the area of the following parallelogram.

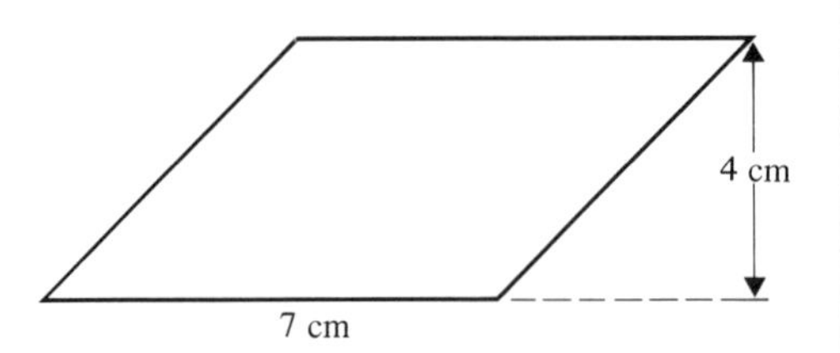

b) How could the figure in (a) be cut and reassembled into a rectangle?

The area is thus the product of the length of the sweeping segment and the distance it goes. The distance it goes must be measured along a line perpendicular to the sweeping segment.

By using sweeps, one can find areas of other kinds of regions. In this example, the region is not rectangular. The sweeping segment still has length w and goes a distance l measured perpendicularly to the segment. Thus the area of the region is $l \times w$.

This works because it is possible to cut the region being swept and to rearrange the pieces to form a rectangular region as shown below.

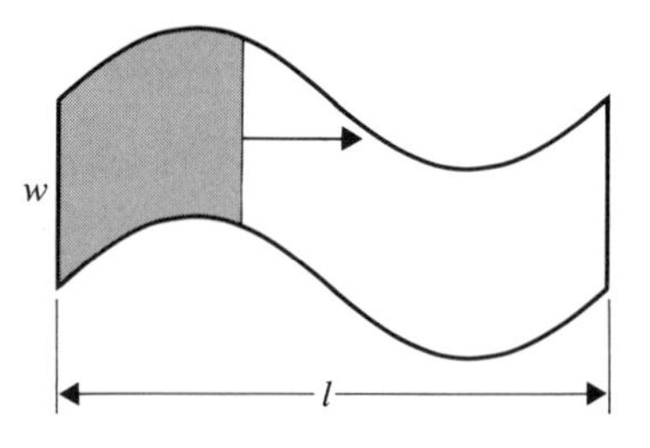

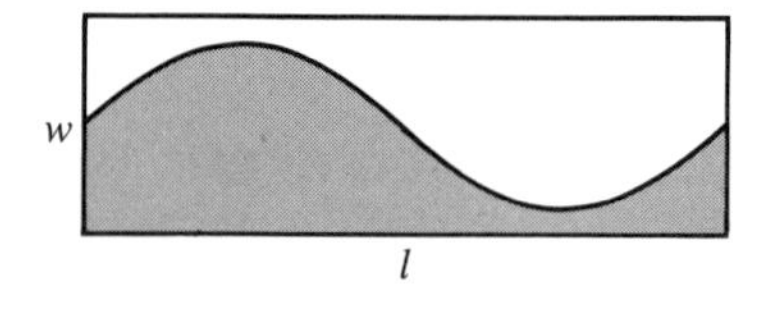

Try Exercise 27

TRIANGLES. Every triangular region can be shown to be associated with a rectangular region, as in the following lesson intended for fifth grade pupils.

From a fifth-grade textbook

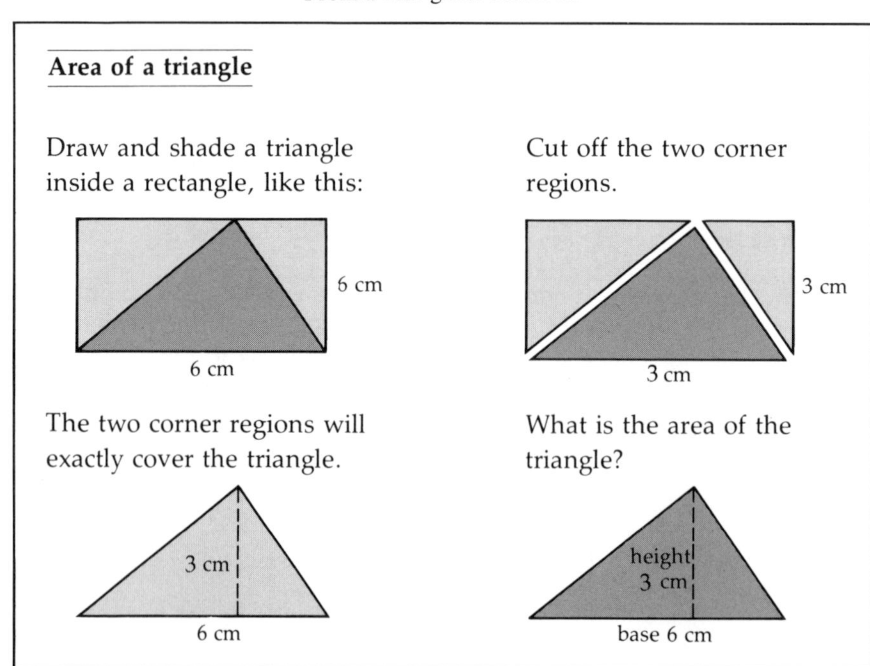

Area of a triangle

Draw and shade a triangle inside a rectangle, like this:

6 cm
6 cm

Cut off the two corner regions.

3 cm
3 cm

The two corner regions will exactly cover the triangle.

3 cm
6 cm

What is the area of the triangle?

height 3 cm
base 6 cm

Formulas for the areas of other polygonal regions can be developed from these basic ones for the rectangle and triangle and are left for the exercises.

Try Exercise 28

CIRCLE AREA. In the same way that we used a regular inscribed polygon to approximate the circumference of a circle, so can such a polygon be used to give an approximation to the area of a circular region.

If we use a regular polygon of 768 sides for the inscribed polygon, the area turns out to be $3.14558r^2$. This is very close to the actual area of a circle πr^2, a formula established by a process similar to our approximating process but carried to the limiting case with the techniques of calculus.

SURFACE AREA. It is frequently necessary to be able to find the surface area of a three-dimensional figure. In order to paint the walls of a room, one must estimate the actual area to be covered with paint. This is a problem in determining the total surface area to be painted. The solid can be thought of as flattened out into the pattern that is needed for its construction.

Example 2 Find the total surface area of a pyramid whose sides are equilateral triangles with an edge length of 4 cm.

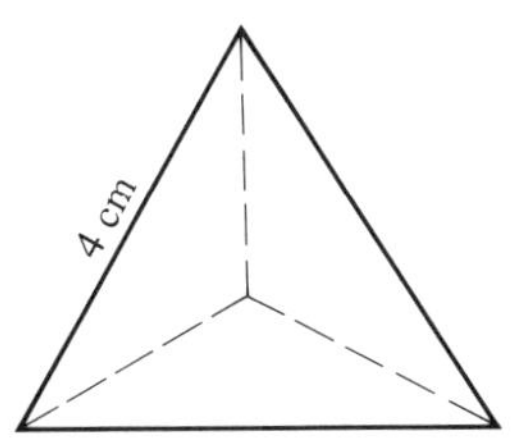

Solution We can think of the pattern for this pyramid as follows:

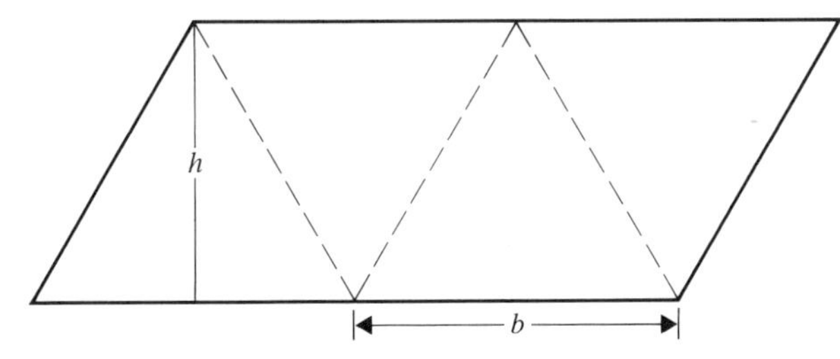

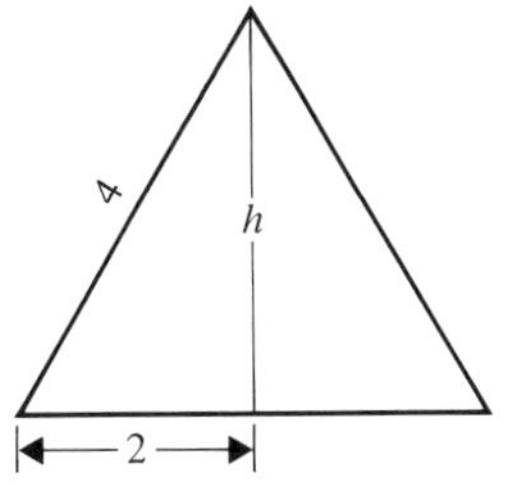

We need to find the total area of this pattern. There are four triangles, and the area of each of them is $\frac{1}{2}b \cdot h$. The base is 4 cm, and we can find the height h using the Pythagorean Theorem. Since $a^2 + b^2 = c^2$, in this case we have

$$2^2 + h^2 = 4^2 = 16, \qquad h^2 = 12 \qquad \text{and} \qquad h = 3.46 \text{ cm}.$$

We can find the area of one triangle easily.

Try Exercise 29

28.

a) Study the diagrams about the area of a triangle. Why do the cut-off corners of the rectangle exactly cover the remaining triangle?

b) What must the area of the triangle be?

c) If in the given triangle ABC, the length AB is b and the height (distance of point C from the line AB) is h, then the area of the triangle in terms of b and h is what?

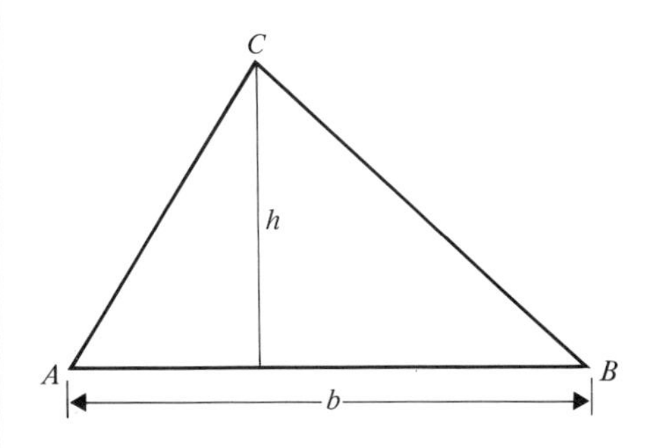

29.

a) What is the area of one of the triangles in cm^2?

b) What is the total surface area of the pyramid in cm^2?

11.5 EXERCISE SET

1. Find the area of each of the following regions. The shaded unit represents 1 square centimeter.

a)

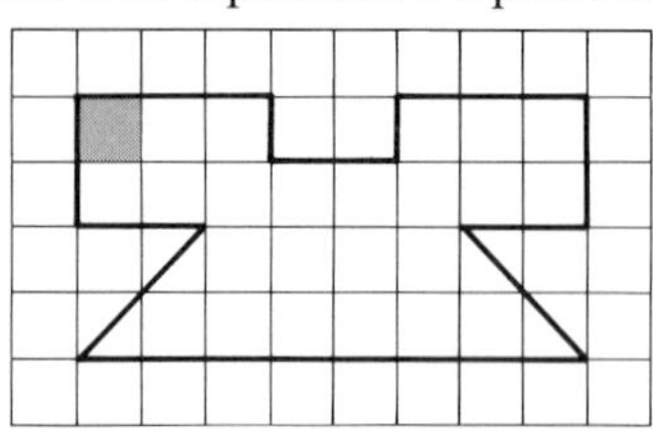

b)

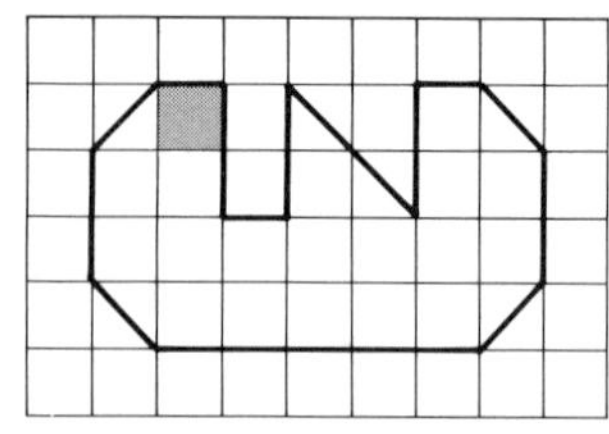

2. A rectangular field is 360 m long and 165 m wide. What is its area in m^2? in hectares? (1 hectare = 10 000 m^2, land or water area only.)

3. A rectangular field is 405 m long and 139 m wide. What is its area in hectares?

4. Show that the area of a trapezoid is half that of a rectangle with the same height as the trapezoid and a base whose length is the sum of the bases of the trapezoid (in the figure, $AB \parallel CD$).

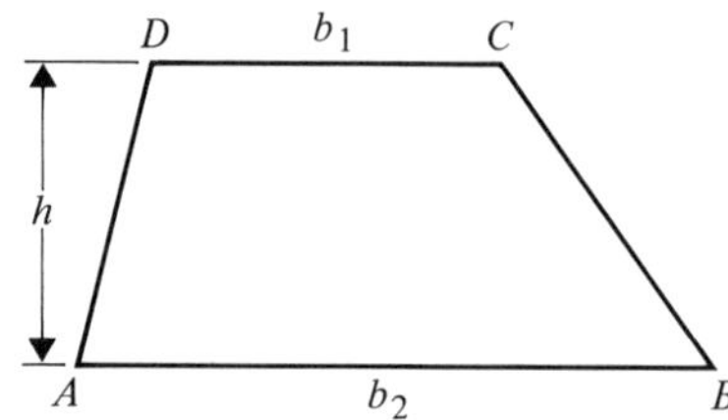

5. Use the formula derived in Problem 4 to find the area of each of the following:

a)

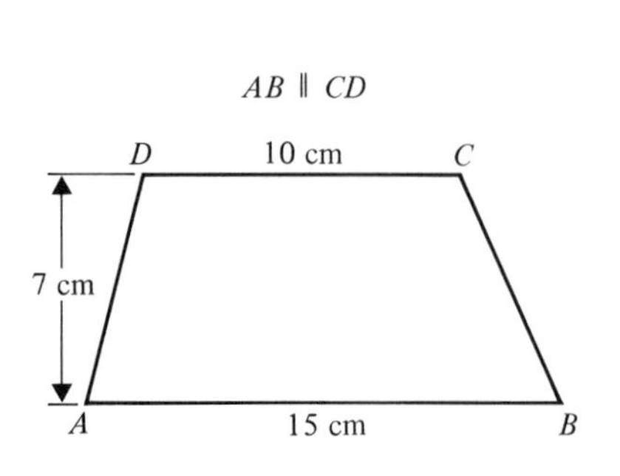

b)

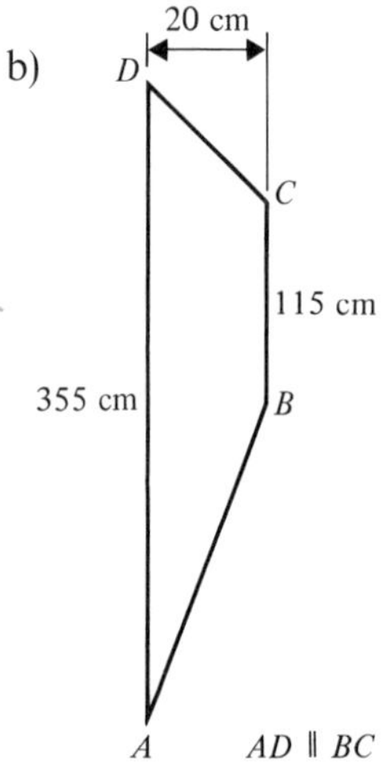

6. In the following make a sketch of the rectangle that is associated with the triangle. Find the area of the rectangle, then of the triangle.

a)

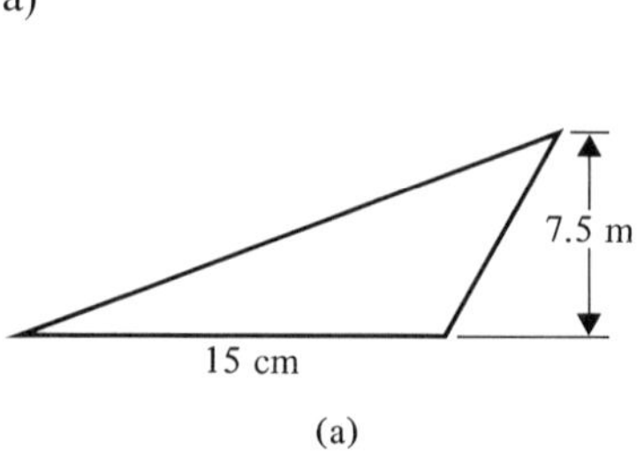

(a)

b)

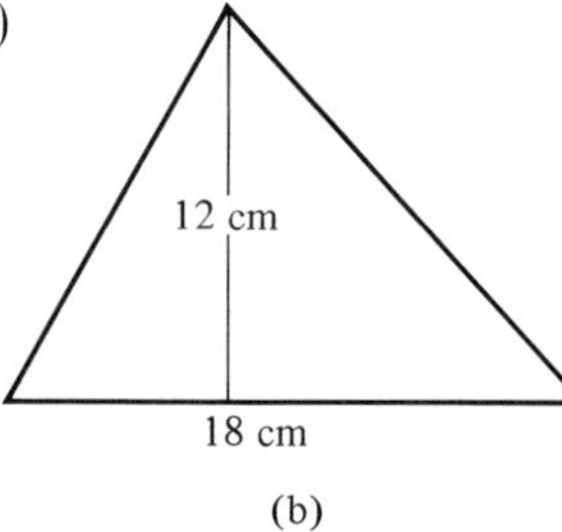

(b)

7. Find the area of the shaded region:

a) Points O and P are the centers of the large and small circles respectively.

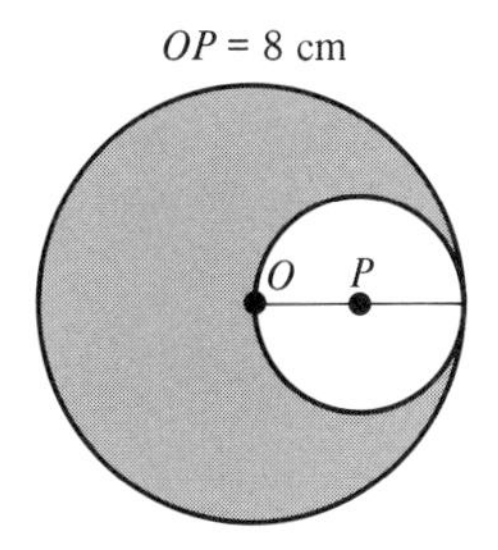

b) Circles A and B have the same radius.

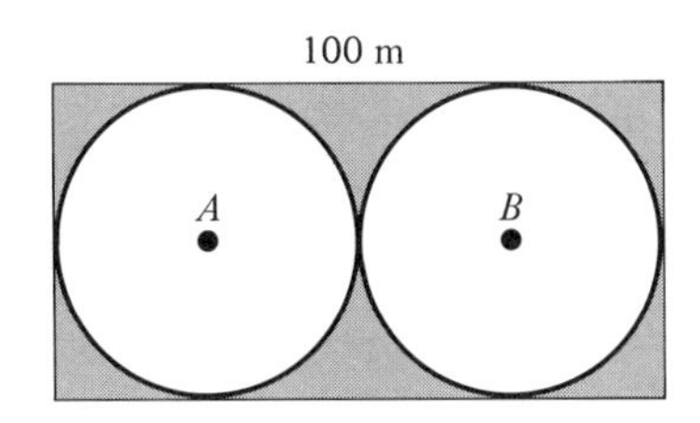

8. Find the area of the shaded region:

a) $ABCD$ is a rectangle inscribed in circle O.

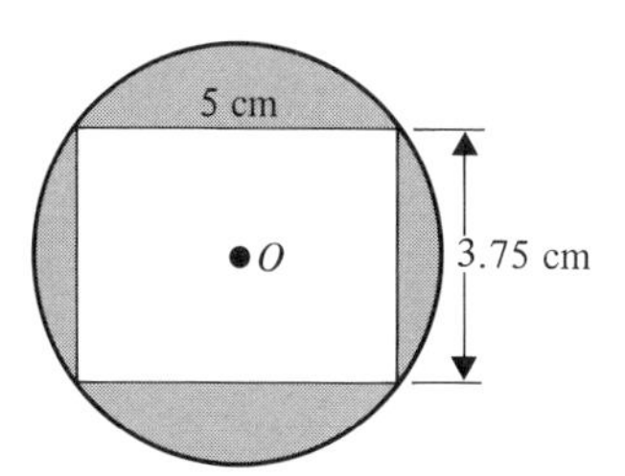

b) Both circles have the same center.

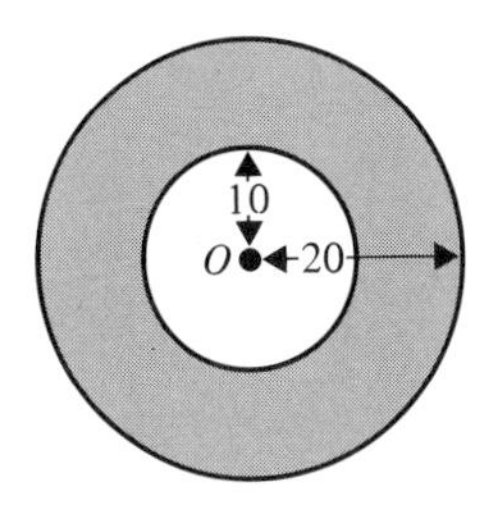

9. Find the area of the shaded region:

a) Point O is the center of the circle.

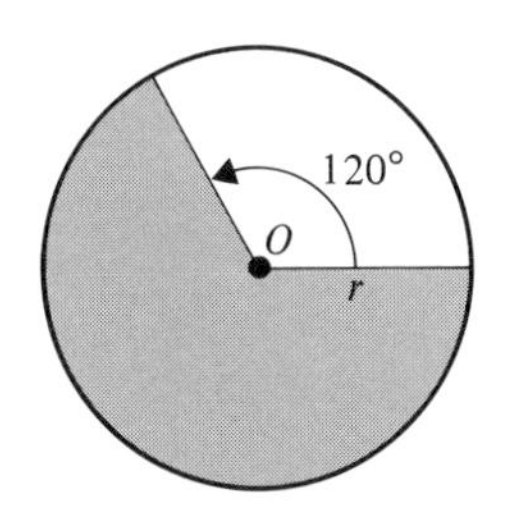

b) The figure is a right circular cylinder.

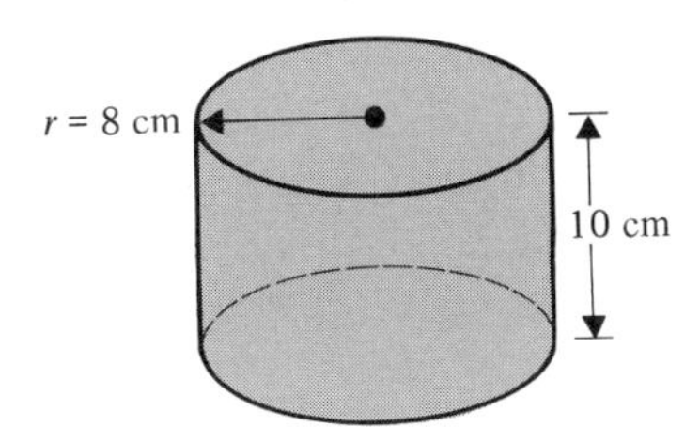

10. Use graph paper ruled in square centimeters if possible. Draw a circle of 5 cm radius.

a) Count the number of square units completely inside this circle.

b) Count the number of square units that are needed to enclose the circle completely.

c) Find the average of the results of (a) and (b) and compare to the calculated value for the area of this circle.

11. Imagine that a regular polygon with six sides has been inscribed inside a circle of 5 cm radius. Calculate the area of this polygon. Compare to Problem 10.

11.6 AIMS

You should be able to:
Find the volumes of the common geometric solids.

11.6 VOLUME

Volume is the measure of a region in three-dimensional space and is found by a method similar to that used for assigning area measure to a plane region. The regions for which we find the volume are bounded by closed surfaces in space. The commonly used units of volume in the SI (metric) system are cubic centimeters (cm^3), cubic meters (m^3), cubic decimeters (dm^3), liters (L), and milliliters (mL).

VOLUME OF A RECTANGULAR REGION. If a right rectangular region is "filled up" with cubic units as shown, then the number of cubic units in the first layer is 4×5. The height gives the number of these layers, in this case 3.

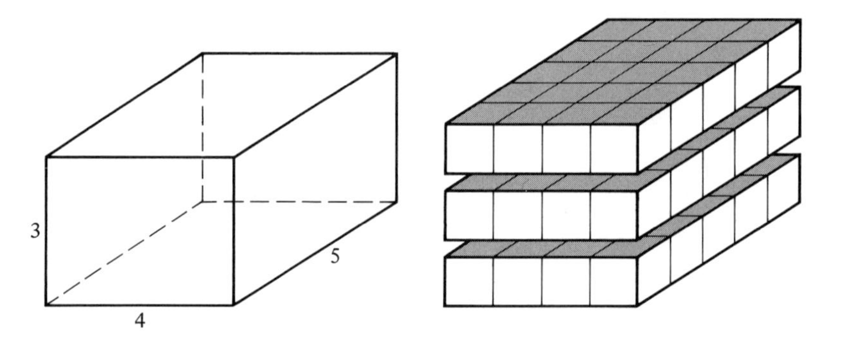

30.

a) If the dimensions of the solid were 3 dm, 4 dm, and 5 dm, what would the volume be in liters?
b) If the dimensions of the solid were 3 cm, 4 cm, and 5 cm, what would the volume be in milliliters?

The total number of cubic units is the product $4 \times 5 \times 3$ or 60 cubic units. The volume of any such solid is defined in terms of the length l, the width w, and the height h.

The volume of any right rectangular prism is

$$V = l \times w \times h.$$

If the length, width, and height are given in cm, the volume is in cm^3. In the figure above if the unit used was meters, the volume would be $60\ m^3$.

Try Exercise 30

CAVALIERI'S PRINCIPLE. The Cavalieri principle depends upon a very plausible and graphic analogy. Imagine a deck of rectangular cards, all congruent to each other. The volume could be calculated with the formula for a rectangular solid. Now think of sliding the deck, so that the top card is slid a slight bit more than the second, which in turn is slid a bit more that the third, and so on. It is reasonable that the volume of the deck does not change under these conditions.

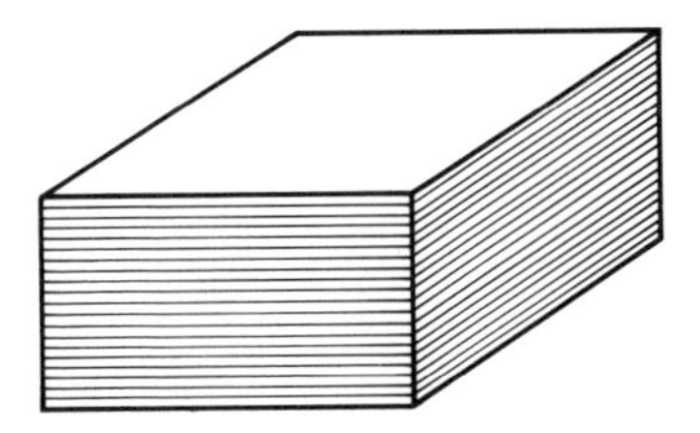

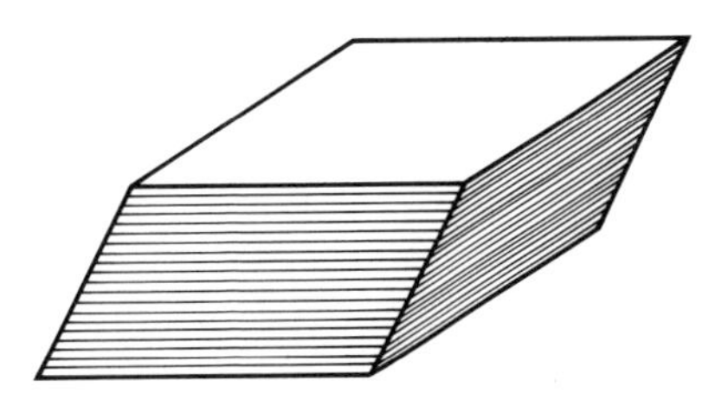

A *cross section* of a geometric region is the figure formed by the intersection of a plane and the solid. In each of the figures of the deck of cards, a cross section made by a plane parallel to the bottom card of the deck is a rectangle, the shape of the individual cards. Similarly the cross section of a right circular cylinder made by a plane parallel to the base of the cylinder is a circle.

Try Exercise 31

Cavalieri's principle is related to this model of the deck of cards. Suppose that two prisms have their bases in the same plane; also, the cross sections formed by every plane intersecting these two prisms and parallel to the base plane have equal areas in every case. Then it is reasonable to assume that the two solids have the same volume.

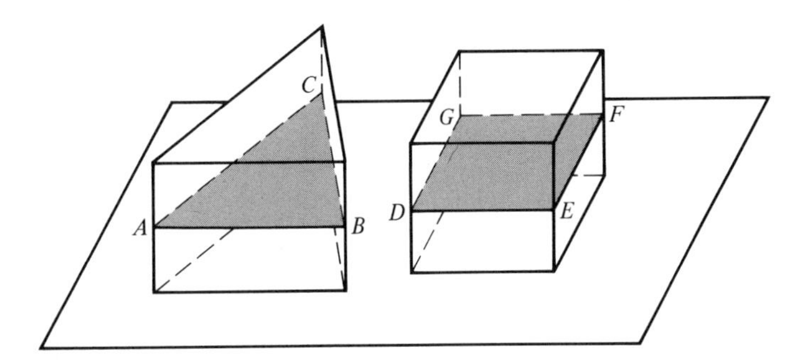

For example, the area of $\triangle ABC$ is equal to the area of rectangle $DEFG$, no matter which plane parallel to the plane P cuts the two solids. With this as a background we state:

CAVALIERI'S PRINCIPLE. If two solid figures lie between two parallel planes and if the two cross sections formed by any plane parallel to these planes are always equal in area, then the volumes of the solids are equal.

We accept this principle as a postulate about volumes. It permits us to find the volume of many common solids.

A TRIANGULAR PRISM. A triangular prism can be thought of as being equivalent in volume to a right rectangular prism with a base area the same

31. Describe the cross section formed by a plane parallel to the base of each of the following:

a) A right circular cone.
b) A cube.
c) A triangular prism with an equilateral triangle for a base.
d) What is the cross section of a sphere always?

32.

a) Find the volume of a prism whose base is an isosceles triangle with sides 10 cm, 10 cm, and 6 cm and whose height is 26 cm.

b) Find the volume of a right circular cylinder with a base radius of 7 cm and a height of 16 cm.

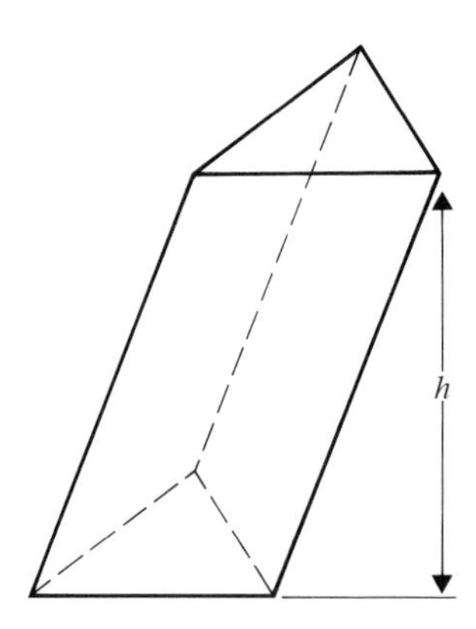

as that of the triangular prism and of the same height. Thus the volume is: $V = B \times h$, where B is the area of the base triangle and h is the height.

A RIGHT CIRCULAR CYLINDER. In the same way, a right circular cylinder can be thought of as equivalent to a right rectangular solid with a cross section equal in area to the base area of the cylinder. The volume is then: $V = B \times h = \pi r^2 \times h$, where r is the radius of the cylinder base and h is the height.

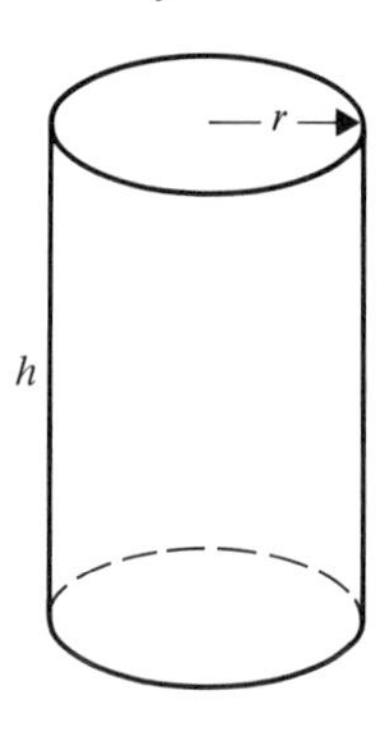

Try Exercise 32

A PYRAMID. The volume of a pyramid can be derived from that of a triangular prism as follows. In the prism $ABCA'B'C'$ below, $AA' = BB' = CC'$ and the plane of ABC is parallel to the plane of $A'B'C'$. In this prism we form three pyramids:

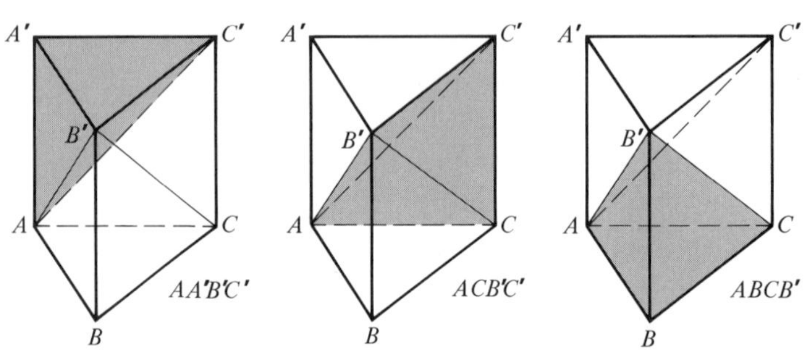

We shall not prove this here, but it is possible to show that these three pyramids have equal volumes. Thus each one must have a volume that is one-third that of the prism. This holds for any pyramid.

$$\text{Volume of a pyramid} = \frac{1}{3}B \times h,$$

where B is the base area and h is the height of the pyramid.

Example 1 Find the volume in cm^3 of a pyramid that is 12 cm high and has a square base measuring 4 cm on one side.

Solution For a pyramid: $V = \frac{1}{3}B \times h$; here the area of the base B is 16 cm^2 and the height h is 12 cm; thus

$$V = \frac{1}{3} \times 16 \times 12 \text{ cm}^3 = 64 \text{ cm}^3.$$

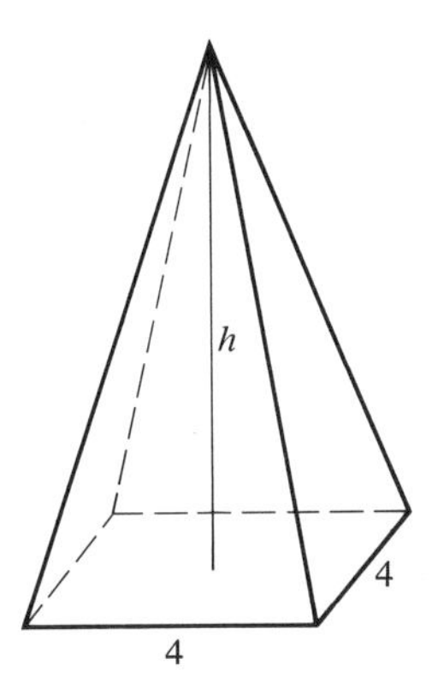

Try Exercise 33

A CIRCULAR CONE. With the help of Cavalieri's Principle one can show that a similar result holds for the volume of a right cone. Here the volume is given by: $V = \frac{1}{3}B \times h$, where B is the area of the circular base and h is the height of the cone.

Example 2 Find the volume of a circular cone having a height of 16 cm and a base radius of 7 cm.

Solution Volume is $\frac{1}{3}B \times h$. The area of the base is B; since the base is a circle, the area B is

$$\pi r^2 = \pi \times 7^2 = 49\pi \text{ cm}^2.$$

33. Find the volume of a pyramid whose base is an isosceles triangle with sides 10 cm, 10 cm, and 6 cm. The height of the pyramid is 26 cm. How is this related to Exercise 32?

34. If milk came in a cone-shaped container, find (to the nearest liter) the amount of milk that would fit in a right circular cone that has a base with a radius of 6 cm and a height of 27 cm.

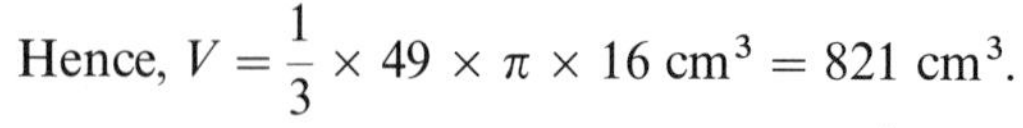

Hence, $V = \frac{1}{3} \times 49 \times \pi \times 16 \text{ cm}^3 = 821 \text{ cm}^3$.

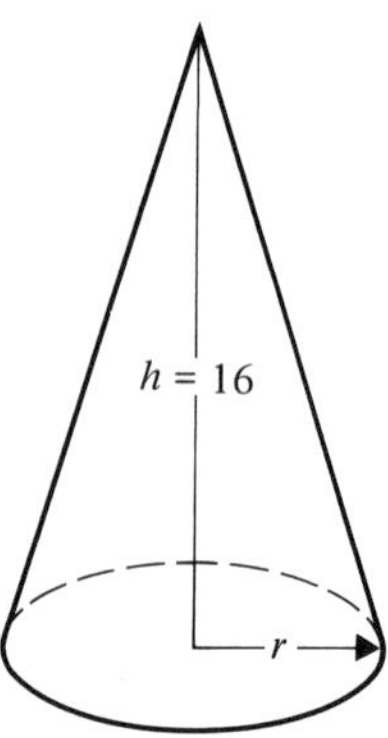

Try Exercise 34

35. Find the volume of the planet Mercury. Assume that it is a sphere and that its diameter is 0.38 times the diameter of the earth.

A SPHERE. Cavalieri's Principle can also be used to find the volume of a sphere. Because the cross section of a sphere is always a circle, it is necessary to find another three dimensional figure with a known volume, that always has the same cross-section area. A right circular cylinder with two cones removed from it proves to be exactly the solid needed (see Problem 13 of Exercise Set 11.7). The volume of the sphere is given by: $V = \frac{4}{3}\pi r^3$, where r is the radius of the sphere.

Example 3 Find the volume of the earth. Assume that it is a sphere with a diameter of 12 756 km.

Solution $D = 12\,756$ km, thus $r = 6378$ km. Therefore

$$V = \frac{4}{3}\pi r^3 = \frac{4}{3} \times \pi \times (6.378 \times 10^3)^3 \text{ km}^3$$
$$= 1.087 \times 10^{12} \text{ km}^3$$

Try Exercise 35

EXERCISE SET 11.6

1. Complete the following chart:

$$1 \text{ cubic meter} = 1 \text{ m}^3$$
$$1 \text{ cubic decimeter} = 1 \text{ dm}^3 = \_\_\_\_ \text{ m}^3$$
$$1 \text{ cubic centimeter} = 1 \text{ cm}^3 = \_\_\_\_ \text{ dm}^3 = \_\_\_\_ \text{ m}^3$$
$$1 \text{ cubic millimeter} = 1 \text{ mm}^3 = \_\_\_\_ \text{ cm}^3 = \_\_\_\_ \text{ m}^3$$

2. A cube that measures 10 cm on one edge will hold a liter of water. Find the edge of a cube that holds 2 liters of water.

3. Find the volume of a rectangular solid that has dimensions 10 cm, 15 cm, and 25 cm. What is the effect on the volume of doubling each dimension?

4. Find the capacity in m^3 of a freezer that is 0.95 m wide, 0.80 m deep, and 1.8 m high. What is the volume in liters?

5. What is the volume in liters of an aquarium that has inside dimensions of 32 cm, 25 cm, and 50 cm?

6. An irregularly shaped rock is placed in a container full of water. This causes 255 mL of water to spill out. What is the volume of the rock in cm^3?

7. The edges of a rectangular box were measured as 8 cm, 10 cm, and 12 cm to the nearest cm. This means a measurement was in error by at most 5 mm. What is the largest possible volume the box could have? the smallest?

8. Suppose the box of Problem 7 had been measured so that the greatest error in a dimension was 1 mm in either direction. What is the largest volume the box could actually have? The smallest it could actually have?

9. A ball just fits snugly into a cubical carton that has inside dimensions on all edges of 30 cm. What percent of the carton is not occupied by the ball?

11.7 PROBLEM SOLVING

11.7 AIMS

You should be able to:
Apply the five-step problem solving process to appropriate problems dealing with measurement and geometry.

For many problems* of daily life, it is necessary to use the properties of well-known geometric figures. The following is a practical problem and it will be treated as such.

Example 1 Find the cost of enough tiles to cover a floor, as sketched. Tiles are 20 cm square and sell for $11.75 per package of 100. Packages will not be split.

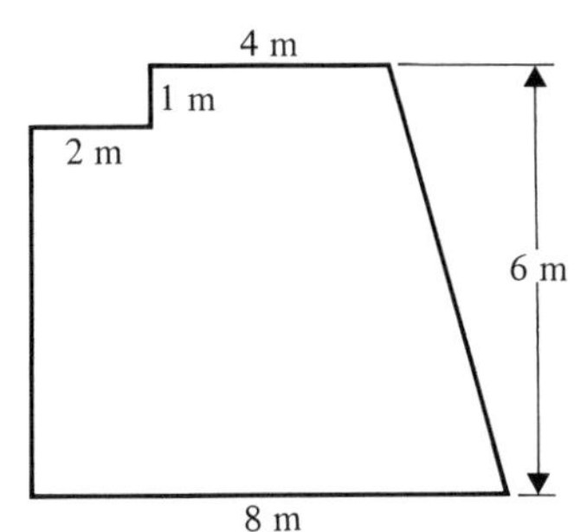

Try Exercise 36

36.
a) List the facts.
b) Decide what the problem is. Are there any subproblems that must be solved?

In your problem statement you may have included the possibility of breakage and the fact that parts of tiles will be used. There is also the subproblem of how many tiles fit on a square meter of floor.

What are some *ideas* for solving this problem?

Idea 1. Approximate the area by a 6 m × 8 m rectangle and find how many tiles are needed.

Idea 2. Cut the region into two rectangles and a triangle.

Idea 3. Think of the area as a trapezoid with a rectangular area removed.

Idea 4. Think of the area as two joined trapezoids.

Idea 5. Think of the area as a trapezoid joined to a rectangle.

* A reminder:

THE FIVE-STEP PROCESS
1. What are the *relevant facts*?
2. What actually is the *problem*?
3. What are my *ideas*?
4. How do I get a *solution*?
5. *Check* and *review* the solution.

37.
a) Make a sketch that corresponds to each of the given ideas.
b) Which would you choose to find the solution?

Try Exercise 37

Idea 2 is as easy as any of the others; we start with it. The sketch for idea 2 is:

We need the area of rectangles $ABCD$ and $BEFG$ and triangle $\triangle EHF$.

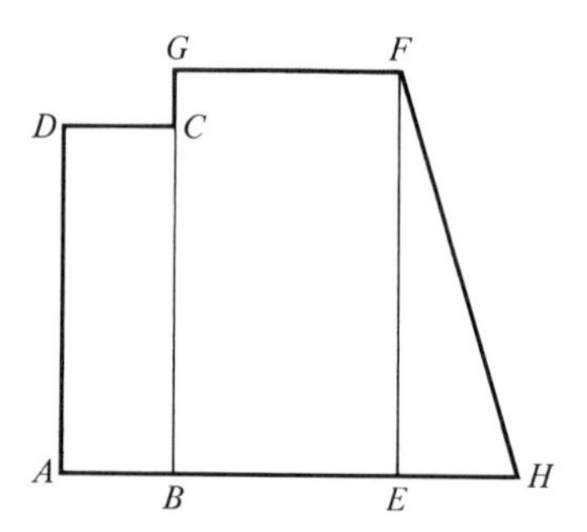

38.
a) Find the dimensions and then the area of $ABCD$.
b) Repeat (a) for $BEFG$.
c) Repeat (a) for EHF.

Try Exercise 38

According to our calculations, the total area is $40\ \text{m}^2$; however, we are not finished, since we have several other problems to solve. How many tiles does it take to cover $1\ \text{m}^2$ of floor?

39. How many tiles does it take to cover $1\ \text{m}^2$ of the floor?

Try Exercise 39

Therefore, we need 40×25 or 1000 tiles. Along the slanting side, the tiles will have to be cut to fit, as in this sketch. There may be some waste because we may not be able to use the cut-off parts in each case. It looks as if one extra row of tiles will be needed. Since 5 tiles fit in a distance of 1 meter, we will need 6×5 or 30 extra tiles. This brings the total number to 1030.

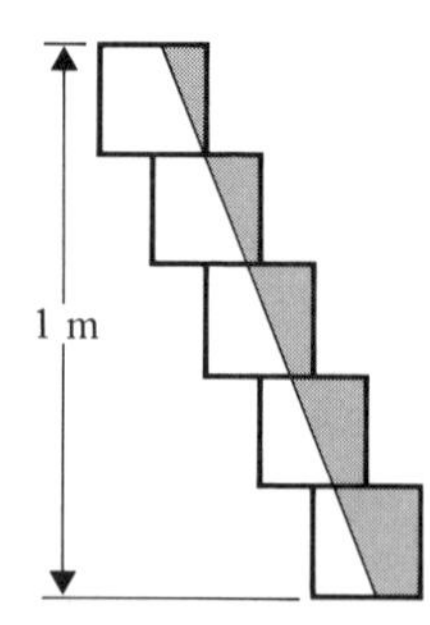

40. How much money did this solution save over the first idea to approximate the area by a rectangle $6\ \text{m} \times 8\ \text{m}$?

To allow for some breakage, we should plan for perhaps another 50 extra tiles (this is close to 5% and we hope enough). This means that 1080 tiles should be ordered; therefore 11 packages will be needed. The total cost is then $\$11.75 \times 11$ or \$129.25.

Try Exercise 40

EXERCISE SET 11.7

1. A circular cake 20 cm in diameter sells for \$2.25, while a cake of the same height but 23 cm diameter sells for \$3.00. Which gives you more cake for your money?

2. The radius of the earth is about 6378 km. In order to go around the world in 80 days, how fast would you have to travel? Assume you travel at constant speed along a great circle.

3. The base of the great pyramid at Gizeh is a square covering about 5 hectares. The pyramid was originally about 155 meters high. Find the approximate volume of the pyramid in m^3.

4. Make a pattern for a triangular prism that has a triangular base with sides 6 cm, 6 cm, and 8 cm. The prism is to be 10 cm high. What is the volume of a prism constructed from the pattern?

5. A rectangular container has the following dimensions: 4.5 cm, 3.75 cm, and 10 cm. Does this container hold more or less than the container in Problem 4?

6. A regular icosahedron has 20 faces, each of which is an equilateral triangle with a side length of 4.75 cm. Find the total surface area of this figure.

Problems 7–10 are suitable for elementary students.

7. Which is worth more, a pile of cents as high as you are tall or a row of dimes laid next to each other along a line that is as long as you are tall? Use metric measurement.

8. How thick is each page of this book (in mm)?

9. Some square floor tiles have a side of 16 cm. Each tile costs \$0.19. How much would it cost to cover the floor of a rectangular room that is 5.6 m long and 4.2 m wide?

■ **10.** Cut three hexagonal shapes of the same size into the parts indicated. Use the 13 pieces to form one complete regular hexagon.

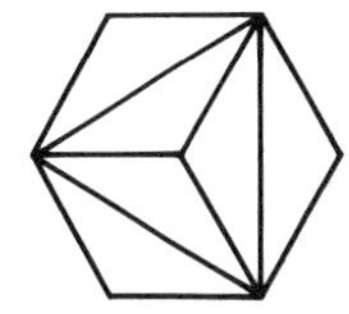

11. A cube can be thought of as consisting of six identical pyramids, each of which has a face of the cube for a base. Make a sketch of a cube with these six pyramids. If the edge of the cube is 12 cm, what is the volume of one pyramid?

■ **12.** The sphere and the cylinder with the two cones removed are resting on the same plane. The radius of the sphere is r, the radius of the cylinder is r and its height is $2r$.

a) Compare the cross section of these two solids at a distance y from the center and show that they are identical.

b) Use Cavalieri's Principle to show that the volume of a sphere is $\frac{4}{3}\pi r^3$. (Assume that you know how to find the volume of the circular cone and the cylinder.)

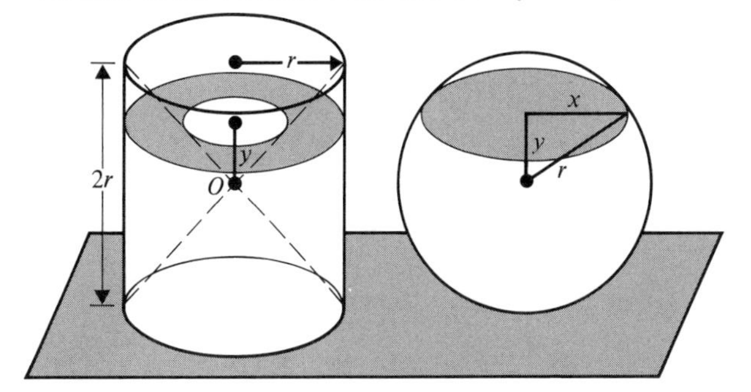

13. Use the given patterns to make models of three pyramids. Use pattern A twice and pattern B once.

a) Fit the three models together to form a triangular prism.

■ b) Prove that the volume of the pyramid made from pattern A is the same as that made from pattern B. What conclusion can be drawn from this result?

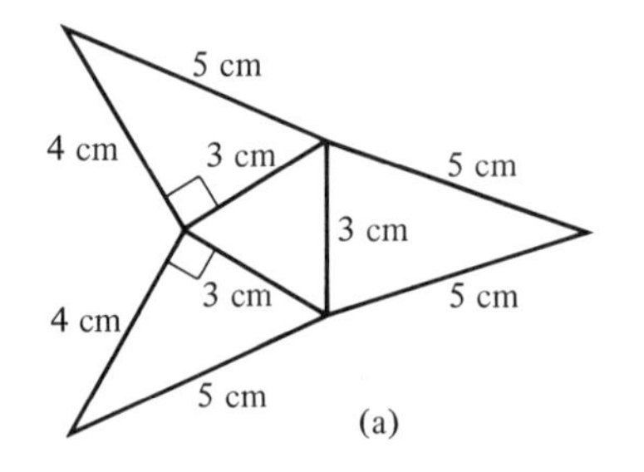

(a)

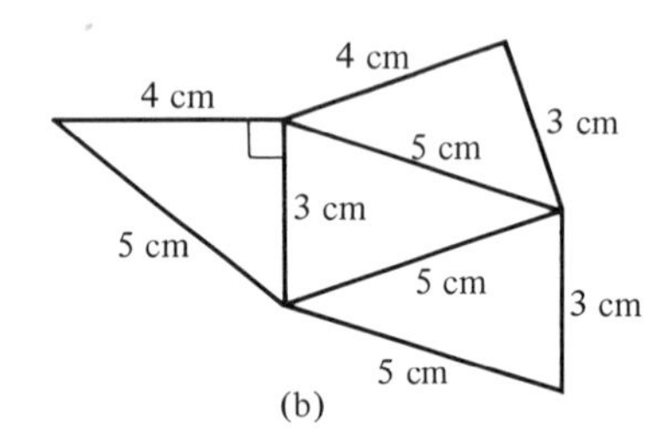

(b)

REVIEW TEST

1. A bug is crawling from point A on the cube to point B by taking the most direct route possible and only along an edge, never a face. How far does the bug have to crawl if an edge of the cube is 9 cm? Suppose that a straight wire has been placed to reach from point A to point B diagonally. If the bug crawls on this wire to get from A to B, how much shorter is the trip?

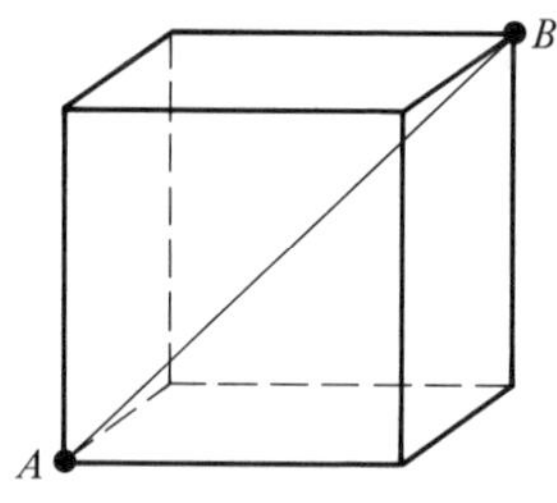

2. A football field, including the end zones, measures 109.728 m by 48.768 m. Find the area of two such fields to the nearest tenth of a hectare.

3. If an average aspirin tablet is 10 mm in diameter, what is the approximate top area of such a tablet in mm^2?

4. A cylindrical glass is 11 cm tall and has a radius of 5.4 cm. Find its volume a) in cm^3; b) in L.

5. A boat travels 1.2 km in 35 minutes. Find its speed in a) km per minute; b) in m per second.

6. Suppose you measured the angles of 50 triangles and you found that the sum for each one was 180°. Could you conclude that this sum for any triangle is 180°? Why? If you found one triangle that measured less than 180°, could you conclude that the sum for any triangle is not 180°? Why?

7. Two elementary students made the following drawings:

a) A triangle with sides of 55 mm, 75 mm, and 90 mm. The student labeled it a "right triangle."

b) A quadrilateral $ABCD$ has $\measuredangle A$ labeled 110°, $\measuredangle B$ labeled 97°, $\measuredangle C$ labeled 80°, and $\measuredangle D$ labeled 73°. The student called this a trapezoid with $AB \parallel CD$. Why are the student's drawings labeled incorrectly?

8. A farmer put a fence around a small square flower garden. The fence post centers were 1 m apart. There were 9 posts on each side when the fence was completed. How many posts were used and what was the area of the garden plot in m^2?

■ **9.** Suppose that the measure of $\measuredangle ABC$ were defined as follows: Locate points A and C so that $AB = BC = 1$ unit. Measure AC; this is the measure assigned to the $\measuredangle ABC$.

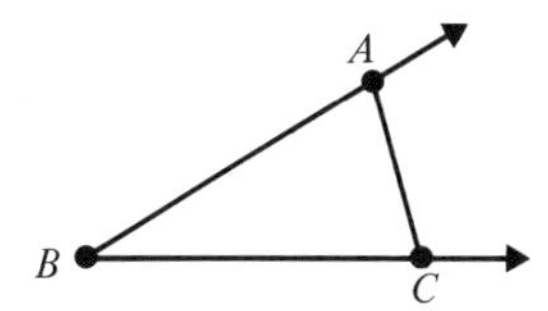

a) With this measure would every angle have a unique measure?
b) Would an obtuse angle have a greater measure than an acute?
c) What would the measure of a 90° angle be? of a 180° angle?
d) What would the sum of the measures of the angles of a quadrilateral be?
e) What are the disadvantages of such a scheme?

10. A right triangular prism has all edges 20 cm in length.
a) Find the area of the base.
b) Find the volume of the prism.
c) Find the total surface area of the prism.

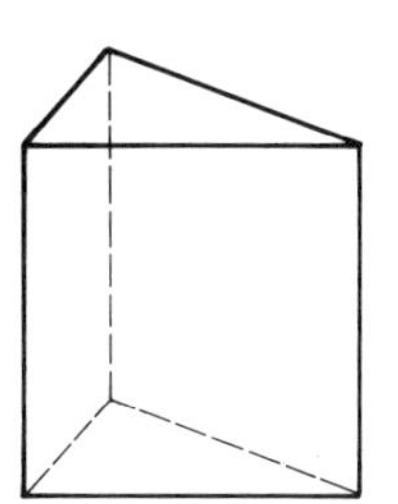

FOR FURTHER READING

1. "Laboratory experiences with perimeter, area, and volume," by Leland Moon, Jr., gives complete instructions for three different activities in a sixth grade. It could be adapted for younger children. This article appears in *The Arithmetic Teacher*, 22:4, April 1975.

2. Considerable motivation for the metric system is given in the article "Introducing the metric system," by Hamilton S. Blum in *The Arithmetic Teacher*, 22:3, March 1975.

TEXTS

Graham, Malcolm, *Modern Elementary Mathematics*, Second Edition. Harcourt Brace Jovanovich, New York, New York (1970, 1975).

Schultz, James E., *Mathematics for Elementary School Teachers.* Bell and Howell Company, Charles E. Merrill Publishing Company, Columbus, Ohio (1977).

chapter 12 probability and statistics

Probability

Experiment	Equally likely outcomes	Chances	Probability
Toss a penny.	Heads Tails	There is 1 chance out of 2 of getting heads.	The probability of getting heads is $\frac{1}{2}$.
Spin the pointer. (1, 2, 3)	1, 2, 3	There are 2 chances out of 3 of getting an odd number.	The probability of getting an odd number is $\frac{2}{3}$.

Give the missing information in each row.

Experiment	Equally likely outcomes	Chances	Probability
1. (1 2 3 4) Draw a card without looking.	1 2 3 4	There is 1 chance out of ■ of getting a 3.	The probability of getting a 3 is ■.
2. Toss a cube that has one of the letters A, B, C, D, E, F on each face.	A B C D E F	There are 2 chances out of ■ of getting a vowel (A or E).	The probability of getting a vowel is ■.
3. (3 4 6) Toss a cube with sides numbered 1-6.	1 2 3 4 5 6	There are ■ chances out of 6 of getting an odd number.	The probability of getting an odd number is ■.

Each of the experiments described in this sixth-grade lesson could easily be carried out in the class. Actually performing the experiments provides an interesting introduction to the concepts of probability and statistics.

12.1 PROBABILITY AND SETS

12.1 AIMS

You should be able to:

a) Describe a probability function.
b) Construct sample spaces and determine events.
c) Find probabilities for appropriate sample spaces.
d) Use the fundamental counting principle for calculating probabilities.

WHAT IS PROBABILITY? The mathematics in this chapter is somewhat different from other kinds of mathematics, since probability deals with uncertainty and chance; in fact, it was games of chance that sparked the initial interest in the subject. Pascal and Fermat are credited with the first investigations that took place in an exchange of letters about some gambling problems supposedly raised in 1654 by the Chevalier de Mere*, who sent word of his difficulties to Pascal.†

Such gambling problems were being solved as early as the middle of the 17th century, but it was not until the 20th century that a complete systematic treatment of the mathematical theory of probability was first published by Kolmogorov.‡

Since the appearance of Kolmogorov's work, the applications of the theory of probability, not related to games of chance, have multiplied rapidly and there is scarcely an area of modern science, industry, or agriculture that is not touched by it. If probability is not used directly, it enters indirectly through the related field of mathematical statistics, so that aside from elementary arithmetic probability is the most universally applied branch of mathematics today.

Probability statements are a fact of our daily life. We hear:

"There is a 60% chance of snow tomorrow in the entire state."
"Jean is a good math student; there is a 90 to 1 chance she will get an A in the test tomorrow."
"I have a 50–50 chance of making the first team."

All such statements involve uncertainty and the future and are interpreted to mean the relative frequency with which the events take place in the long run.

A 60% probability of a snowstorm means that over a long period of time given the current weather conditions it will snow on 60% of such days. The weather predictions are based upon empirical evidence collected by the weather bureau.

When a coin is tossed, we say the chances are 1 out of 2 that it will fall heads, or that the probability that it will fall heads is $\frac{1}{2}$. In the case of the coin we might reason as follows: There are only two ways for it to fall, and there is no reason to suspect that one way would be more likely than the other. Therefore, in a large number of trials, it should fall heads half of the time. This is a *theoretical probability*.

Theoretical determinations of probability are very important and constitute one of the most important branches of mathematics. Does this mean that if one tosses a coin 10 times, it will fall heads exactly 5 times? Of course not. However, for a very large number of such throws, we expect that close to half will be heads.

* One of the Chevalier de Mere's problems was supposedly this: He knew that it was desirable to bet that at least one six would turn up in four tosses of a die. He argued to himself that it should be just as desirable to bet that at least one double six would occur in 24 tosses of a pair of dice. Unfortunately he lost money on the basis of his reasoning and complained to Pascal.

† Blaise Pascal (1623–1662)

‡ *Foundations of The Theory of Probability* by A. N. Kolmogorov (1903–), published in 1933 in German. The axioms in this monograph form the basis of most current research in probability theory.

Since theoretical probabilities refer to events that have not yet occurred, we need to make more rigorous the meaning of the word *event*.

SAMPLE SPACES. The use of sets helps to simplify the study of theoretical probabilities. The act of tossing a coin is referred to as an *experiment* in which the only outcomes are heads (H) and tails (T), so the set of all possible outcomes is {H, T}. This set of all possible outcomes is called the *sample space* for the experiment of tossing a coin. The individual outcomes are referred to as the *points* of the sample space.

Suppose that the experiment consists of tossing two coins, a dime and a nickel. The possible outcomes, or points, in the sample space are ordered pairs (d, n), where d is the result of the dime toss, and n is the result of the nickel toss. These are all of the possible ways for the two coins to fall, and the sample space is the set: {(H, H), (H, T), (T, H), (T, T)}.

Another way to look at this is as the cross product of the sample space for the toss of the dime: {H, T} and that for the toss of the nickel: {H, T}.

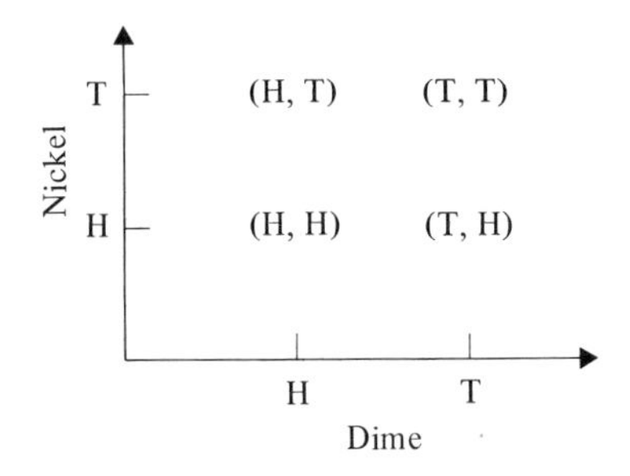

We summarize this discussion in the following definition:

> **The *sample space S* is the set of all possible outcomes of an experiment; each outcome is referred to as a *point* of the sample space.**

A third way of visualizing the points of a sample space is by means of a *tree diagram.* The tree diagram for the dime–nickel experiment is:

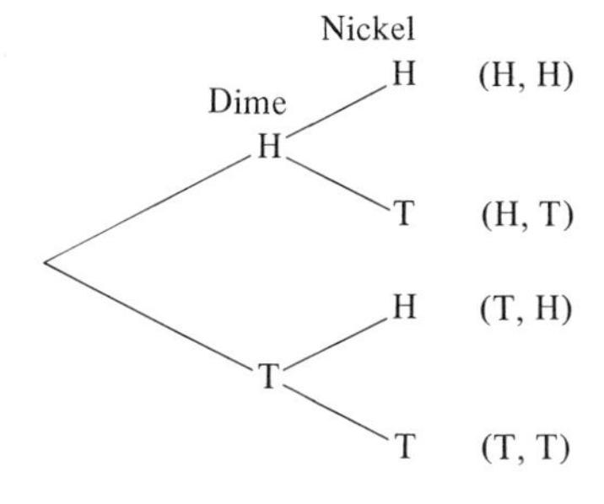

Each branch of the tree corresponds to one outcome of the experiment.

Try Exercise 1

Example 1 An experiment consists of tossing a coin and rolling a die. What is the sample space for this experiment?

The sample space for the coin is $S_1 = \{H, T\}$.
The sample space for the die is $S_2 = \{1, 2, 3, 4, 5, 6\}$.
The sample space for the experiment is the Cartesian product

$$S_1 \times S_2\colon \{(H, 1), (H, 2), (H, 3), (H, 4), (H, 5), (H, 6), (T, 1), (T, 2), (T, 3), (T, 4), (T, 5), (T, 6)\}.$$

Try Exercise 2

EVENTS. In the experiment of Example 1, suppose we wished to find the probability of obtaining a head on the coin and an even number on the die. The ordered pairs that satisfy the requirements are:

$$\{(H, 2), (H, 4), (H, 6)\},$$

so that this set of points constitutes the *event*: "a head on the coin and an even number on the die." We are led to the definition:

An *event* is any subset of a sample space.

Try Exercise 3

THEORETICAL PROBABILITY. When the outcomes of an experiment are equally likely, then each has the same probability. In this case it is easy to assign the probabilities if we interpret probability as the ratio of the number of points in the event to the total number of trials (relative frequency).

If A is an event in a sample space of n equally likely points, then the probability of the event A is given by

$$P(A) = \frac{\text{Number of points in } A}{\text{Number of points in } S}.$$

If a sample space consists of n points, or outcomes, then the probability for each one is $\frac{1}{n}$. If the event consists of k of these points, then $P(A)$ is $\frac{k}{n}$. The probability of the entire sample space is 1 since

$$P(S) = \frac{\text{Number of points in } S}{\text{Number of points in } S}.$$

1.
a) Draw a tree diagram for the following experiment: There are equal numbers of red, white, and blue marbles in a bag. One marble is drawn and its color recorded.
b) Repeat (a) for the experiment in which the first marble is withdrawn and its color recorded; it is then returned to the bag. After mixing, a second marble is drawn and the color recorded.

2. An experiment consists of tossing a nickel with a sample space $\{H, T\}$ and rolling a regular tetrahedron with sample space $\{1, 2, 3, 4\}$. Construct the sample space of the experiment.

3. The experiment consists of tossing a single die. Given the roster notation for each of the following events:
a) E_1: The number on the toss is even.
b) E_2: The number on the toss is odd.
c) E_3: The number on the toss is divisible by 3.
d) E_4: The number on the toss is greater than 6.

Probability can be thought of as a function whose domain consists of sets (events) and whose range is the set of real numbers from 0 to 1 inclusive. The following postulates hold for probability functions whether or not the sample space has the same probability assigned to each point.

POSTULATES FOR PROBABILITY

1. **The probability of any event A is nonnegative and never greater than 1.**
2. **The samples space S has a probability of 1.**
3. **If A and B are two disjoint events, then the event described by "either A occurs or B occurs" has a probability that is the sum of the individual probabilities for event A and event B.**

Try Exercise 4

We return to the event $\{(H, 2), (H, 4), (H, 6)\}$ that led off this discussion. There are 12 points in the sample space; each has the same assigned probability (see Example 1). Thus

$$P(\{(H, 2), (H, 4), (H, 6)\}) = \frac{\text{Number of points in } A}{\text{Number of points in } S} = \frac{3}{12} = \frac{1}{4}.$$

Example 2 What is the probability that when a coin is tossed and a die is rolled, a head will be obtained?

Solution We look at the sample space and find the ordered pairs that constitute the event. There are 12 points in the sample space:

$$S = \{(H, 1), (H, 2), (H, 3), (H, 4), (H, 5), (H, 6), (T, 1), (T, 2), (T, 3), (T, 4), (T, 5), (T, 6)\}.$$

If E is the event "A head will be obtained," then

$$E = \{(H, 1), (H, 2), (H, 3), (H, 4), (H, 5), (H, 6)\}.$$

Six points of the sample space are in this set; thus the probability is:

$$P(E) = \frac{6}{12} = \frac{1}{2}.$$

Try Exercise 5

Example 3 A bag contains equal numbers of red, white, and blue marbles. One marble is drawn, then replaced, and a second one is then drawn. What is the probability that at least one of them is white?

4. Write out the corresponding mathematical statements for each of the probability postulates. Use $P(A)$ to represent the "probability of event A."

5. Use the answer to Exercise 3 of p. 408. Find the following:

a) $P(E_1)$ b) $P(E_2)$
c) $P(E_3)$ d) $P(E_4)$

Use the experiment of Example 2 of the text:

e) Describe the event: "a tail and a 3 occur."
f) Find the probability of the event in (e).
g) What is the probability of the event: "A tail occurs"?

6. A bag contains equal numbers of marbles: red, white, blue, and yellow.
a) Construct the sample space for drawing two marbles, replacing the first before drawing the second.
b) Give roster notation for the event E: the two marbles are the same color.
c) Find the probability of event E.

7.
a) Make a tree diagram for Example 4 of the text.
b) Find the probability that the first two marbles drawn are red.

Solution The sample space for drawing one marble is {R, W, B}. The sample space for drawing two of them is the Cartesian product:

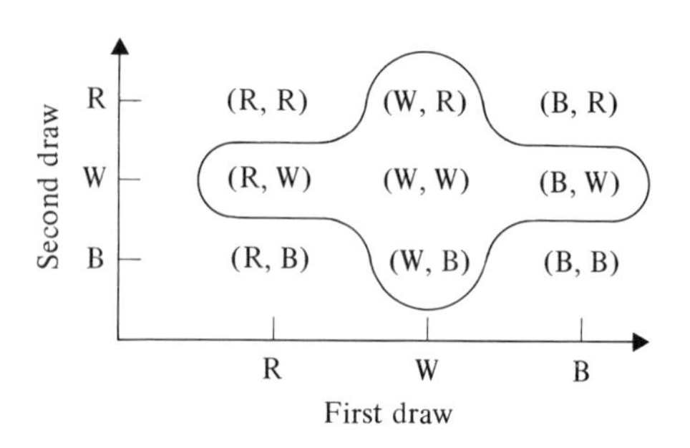

The event of drawing at least one white marble is circled. The probability of this event is 5/9.

Try Exercise 6

Example 4 A bag contains equal numbers of red and blue marbles. One marble is drawn and then replaced and a second one is drawn and replaced. A third one is drawn. What is the probability that they are the same color? The sample space is now a set of ordered triples. In fact, it is the Cartesian product $\{R, B\} \times \{R, B\} \times \{R, B\}$:

$$S = \{(R, R, R), (R, R, B), (R, B, R), (R, B, B), (B, R, R), (B, R, B), (B, B, R), (B, B, B)\}.$$

The probability that all three marbles are the same color is

$$P(\{(R, R, R), (B, B, B)\}) = 2/8 \quad \text{or} \quad 1/4.$$

Try Exercise 7

A COUNTING PRINCIPLE. Constructing sample spaces in this way is easy when the number of points is small; however, for larger sets, a counting principle is useful. This principle can be seen by considering the number of elements in a Cartesian product. If set A has m elements and set B has n elements, then $A \times B$ has $m \cdot n$ elements.

FUNDAMENTAL COUNTING PRINCIPLES. If an element can be chosen from set E_1 in n_1 ways, and an element can be chosen from set E_2 in n_2 ways, and one from E_3 in n_3 ways, and so on, then the total number of ways the choices can be made is

$$n_1 \cdot n_2 \cdot n_3 \cdots.$$

Example 5 How many two-digit numbers are there (decimal notation)?

Solution The first digit of a two-digit number can be selected from the set $\{1, 2, 3, 4, 5, 6, 7, 8, 9\}$, hence in 9 ways. (We do not use 0 for a first digit.) The second can be selected from the set $\{0, 1, 2, 3, 4, 5, 6, 7, 8, 9\}$, hence in ten ways. The total number of ways of selecting from both is, by the counting principle, $9 \cdot 10$ or 90.

Try Exercise 8

Example 6 Two dice are rolled, one red and one green. Use the fundamental counting principle to find the probability that the same number appears on both of them.

Solution The red die can be rolled in 6 ways. The green die can be rolled in 6 ways. The total number of ways they can be rolled is thus $6 \cdot 6$ or 36. This is the number of points in the sample space. Now the red die can be rolled in 6 ways. For each of these, there is just one way the green can be rolled to match it. Thus by the counting principle there are $6 \cdot 1 = 6$ ways in which the same number will appear on both dice. The probability is then 6/36 or 1/6.

The problem of Example 6 could also be solved by constructing a Cartesian product as follows:

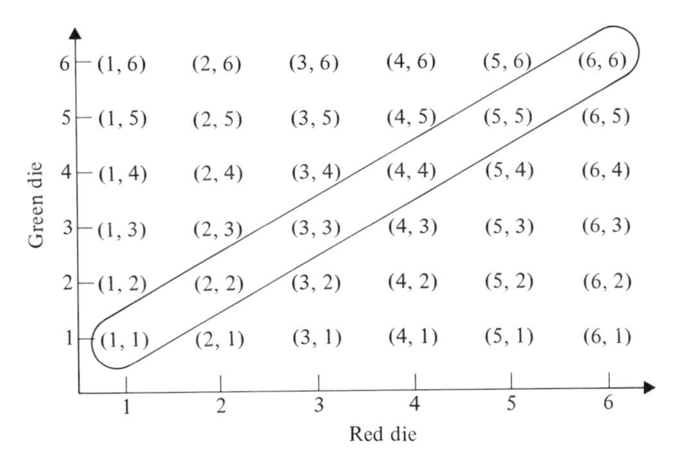

The desired set is indicated on the diagram. Since each point of the sample space can be regarded as a subset consisting of 1 point, the probability for each is 1/36. This fact with the third postulate for probability allows us to find the probability of the event we are interested in by adding the probability for each of the points contained in the event. Thus the probability that the same number appears on both dice is

$$P = 1/36 + 1/36 + 1/36 + 1/36 + 1/36 + 1/36 = 6/36 = 1/6.$$

Try Exercise 9

8. How many three-digit numbers are there (decimal notation)?

9.

a) Make a tree diagram for the experiment of Example 6.

b) On the roll of two dice, one red and one green, what is the probability that you will get an odd number on the red one and an even number on the green one? (Show the set on your diagram.)

c) What is the probability that you will get an odd number on one of them and an even number on the other? (Show the set on your diagram.)

EXERCISE SET 12.1

1. Use a tree diagram to construct a sample space for the experiment of tossing three coins.

2. Use the Cartesian product to construct a sample space for the experiment of rolling two regular tetrahedrons with faces labeled 1, 2, 3, and 4. The number on the bottom face is recorded.

3. In the sample space of Problem 1, exhibit the following events:
a) There are exactly two heads.
b) There is a head on the first toss.
c) There is at least 1 head.
d) There are exactly 3 tails.

4. In the sample space of Problem 2, exhibit the following events:
a) Both tetrahedrons have the same number on the bottom.
b) One has an odd number and the other an even number.
c) The sum of the numbers is 7.

5. Calculate the probabilities for the events of Problem 3.

6. Calculate the probabilities for the events of Problem 4.

Problems 7–10 might be posed to elementary pupils:

7. Sue has two T-shirts, three blouses, one skirt, and one pair of jeans. How many different outfits can she put together from these items? (Assume that all the clothing is color coordinated.)

8. How many ways can Art, Bob, and Carol be assigned to pour juice, pass cookies, and hand out paper cups and napkins if there is just one person per job?

9. In how many ways can Darlene, Lisa, and Frank use a teetertotter if one person per end is allowed?

10. The school cafeteria offers ice cream, pudding, cake, fruit, and jello for dessert each day. If a pupil decides to eat a different dessert each day for a week, how many possible ways could this be done?

11. Jean is on the student council, which has 12 members. A committee of 3 is going to be chosen at random from this council. What is the probability that Jean will be on the committee.

12. A box contains 100 slips of paper on which are written the numbers from 1 to 100. Each slip has the same chance of being drawn from the box. What is the probability of drawing a number divisible by 17? by 5?

13. If four coins are tossed, what is the probability that exactly one will be a head?

14. If four coins are tossed, what is the probability that exactly three will be tails?

15. Assume that when a child is born it is equally likely that it will be a boy or a girl (this is only approximately true). If a family is to have three children, what is the probability that:
a) They will all be girls?
b) There will be two boys and one girl?

16. If a license plate is to have three letters followed by three digits, how many different license plates can be made?

17. In Problem 16, if no letter or digit can be repeated and the first digit cannot be zero, how many different license plates can be made?

12.2 PROBABILITY THEOREMS; UNIONS (DISJUNCTIONS) OF EVENTS

Probabilities of certain compound events can be computed when those of the individual events comprising them are known.

COMPLEMENTARY EVENTS. An event is defined to be any subset of a sample space which, in turn, can be thought of as a universal set. Then every event A has a *complement* $\bar{A}$; it is the event that *A does not* occur:

$$A \cap \bar{A} = \varnothing, \qquad A \cup \bar{A} = \text{Sample space } S.$$

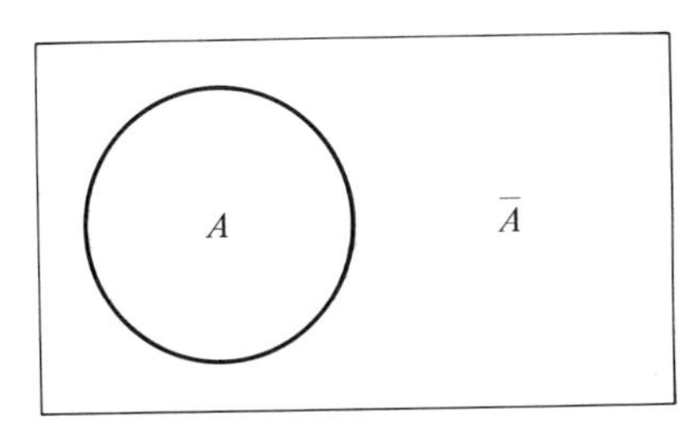

Events that are related as in the diagram are called *complementary events.* We can prove the following theorem:

The probability that an event A will not occur is $1 - P(A)$.

We know that $P(A \cup \bar{A}) = P(S)$ by the definition of $\bar{A}$ and equality for sets. Since $A \cap \bar{A} = \varnothing$, by the third probability postulate we get:

$$P(A \cup \bar{A}) = P(A) + P(\bar{A}) = P(S) = 1.$$

Therefore $P(\bar{A}) = 1 - P(A)$.

The empty set is a subset of any sample space, and it follows from this theorem that the probability of this event is 0. Since $\varnothing = S$, then

$$P(\varnothing) = 1 - P(S) = 1 - 1 = 0$$

and we conclude that $P(\varnothing) = 0$.

Example 1 What is the probability that two tossed coins will *not* both fall heads?

Solution There is just one outcome out of four that both will be heads; call it A. Then $P(A) = 1/4$.

By the theorem for complementary events, the probability for the event "*not* both fall heads" can now be found:

$$P(A) = 1 - \frac{1}{4} = \frac{3}{4}.$$

12.2 AIMS

You should be able to:

a) Calculate the probabilities of complementary events.

b) Calculate the probabilities of unions (disjunctions) of events.

10. Use the rule for complementary events and find the probability of getting a 1, 2, 3, 4, or 5 on a single throw of a die (this is the same as not getting a 6).

Try Exercise 10

UNIONS—DISJOINT EVENTS. Let us consider an event that is the union of two events. If A and B are two events, then $A \cup B$ is the event that A occurs or that B occurs. If A and B are disjoint (have the empty set as their intersection), then the probability of their union is the sum of their individual probabilities, as established by the third postulate of probability.

Example 2 What is the probability that two tossed coins are both heads or both tails?

Solution The sample space for this experiment is $\{(H, H), (H, T), (T, H), (T, T)\}$.

If event A is getting two heads, then $P(A) = 1/4$.
If event B is getting two tails, then $P(B)$ is also $1/4$.
These are disjoint events since $\{(H, H)\} \cap \{(T, T)\} = \varnothing$.

Hence the probability that one *or* the other will occur is

$$\frac{1}{4} + \frac{1}{4} = \frac{1}{2}.$$

Disjoint events are also called *mutually exclusive events*, meaning that they cannot occur at the same time.

UNIONS—EVENTS THAT ARE NOT DISJOINT. To find the probability of the union of events that are not disjoint we prove the theorem:

For any events A and B, the probability of the event A or B is

$$P(A \cup B) = P(A) + P(B) - P(A \cap B).$$

Proof

$$P(A \cup B) = \underset{\substack{\uparrow \\ \text{shaded} \\ \text{horizontally}}}{P(A \cap \bar{B})} + P(A \cap B) + \underset{\substack{\uparrow \\ \text{shaded} \\ \text{vertically}}}{P(B \cap \bar{A})}.$$

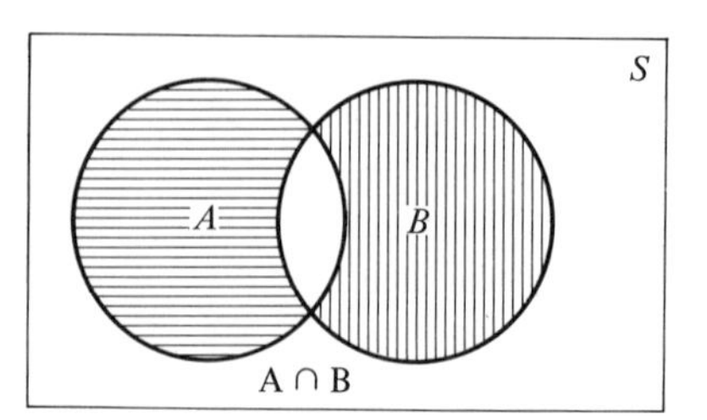

Now, by the third postulate for probability, we have:

$$P(A) = P(A \cap \bar{B}) + P(A \cap B), \qquad P(B) = P(A \cap B) + P(B \cap \bar{A}).$$

Therefore,

$$\begin{aligned} P(A) + P(B) &= P(A \cap \bar{B}) + P(A \cap B) + P(A \cap B) + P(B \cap \bar{A}) \\ &= P(A \cap B) + P(A \cup B) \quad [\text{by renaming}]. \end{aligned}$$

We subtract $P(A \cap B)$ and finally get

$$P(A) + P(B) - P(A \cap B) = P(A \cup B).$$

This theorem takes care of both cases, disjoint and not disjoint events. If A and B are disjoint events, then $P(A \cap B) = 0$.

Example 3 In a bakery, it is found that 75% of the customers buy bread. It is also found that 45% of them buy doughnuts, and 35% of them buy both bread and doughnuts. What is the probability that a customer will buy bread or doughnuts?

Solution Let us call the event of buying bread B, buying doughnuts D. Then we need the probability of the event B or D. By the theorem above we have:

$$\begin{aligned} P(B \cup D) &= P(B) + P(D) - P(B \cap D) \\ &= 0.75 + 0.45 - 0.35 = 0.85. \end{aligned}$$

Example 4 A card is drawn at random from a deck of 52 cards. What is the probability that it is a king or a red card?

Solution Let K be the event "A king is drawn;" then $P(K) = 4/52$.

Let R be the event "A red card is drawn;" then $P(R) = 26/52$.

Since two of the kings are red, $P(K \cap R) = 2/52$ and we can now find the desired probability:

$$P(K \cup R) = P(K) + P(R) - P(K \cap R) = \frac{4}{52} + \frac{26}{52} - \frac{2}{52} = \frac{28}{52}.$$

Try Exercise 11

11. Find the probability that a card drawn at random from a deck of 52 is a seven or a black card.

EXERCISE SET 12.2

1. What is the probability that a rolled die is not a 5?

2. What is the probability that a card drawn from a deck is not an ace?

3. Two marbles are drawn, with replacement, from a bag containing equal numbers of red, white, and blue marbles. What is the probability that they are both blue? that they are not both blue?

4. In Problem 3, what is the probability that they are not the same color?

5. Two blocks are drawn, with replacement, from a bag containing equal numbers of green, blue, and white blocks. What is the probability that:
a) At least one of them is green?
b) None of them is green?
c) At least one of them is white?
d) One green and one white are obtained?
e) At least one of them is blue or white?
f) Neither of them is blue or white?

6. What is the probability that a card drawn at random from a deck of 52 playing cards is
a) A red card?
b) A face card?
c) A red face card?
d) A red or a face card?
e) Neither a red nor a face card?
f) An ace?
g) Both, an ace and a face card?
h) An ace or a face card?
i) Neither an ace nor a face card?

7. If A is the event that a student selected from the sophomore class is taking probability and statistics and B is the event that the student is taking English, write in words what is meant by each of the following probabilities:
a) $1 - P(A)$ b) $P(A \cup B)$ c) $P(\bar{A} \cap B)$

8. Let the experiment be the tossing of three coins.
a) Give the following events, using the roster method:

H_1: a head occurs on the first toss,
H_2: a head occurs on the second toss.

b) What is the set $H_1 \cap H_2$? What is $P(H_1 \cap H_2)$?
c) Find the probability of the event H_1 or H_2.

■ **9.** The addition rule for two events can be generalized to three events. Use a diagram to find a formula for

$$P(A \cup B \cup C).$$

10. Use a Venn diagram or other argument to show that $P(A) \geq P(A \cap B)$.

11. A college student was hired to do some interviewing by a market research company. The student interviewed 100 shoppers and turned in the following report:

65% regularly purchase plain yogurt,
34% regularly purchase flavored yogurt,
23% regularly purchase both kinds,
35% do not purchase yogurt at all.

The supervisor looked at the data and said: "Something is wrong here." Why were the data questioned?

12.3 AIMS

You should be able to:
a) Determine, in simple cases, whether events are dependent or independent.
b) Calculate the probability of the event $A \cap B$.

12.3 CONJUNCTIONS OF EVENTS

How can we calculate the probability of the event "A occurs *and* B occurs" if we know the probabilities of A and B individually? We look at some examples.

INDEPENDENT EVENTS

Example 1 If we roll two dice, one red and one green, what is the probability that we get a 1 on the red die and an even number on the green one?

Solution There are 6 ways for the red die to come up and 6 ways for the green; thus the sample space contains 36 points.

$$S = \{(R, G) \mid R \text{ is the outcome of the red die, } G \text{ is that for the green}\}.$$

If A is the event "1 on the red die," then $P(A) = \frac{6}{36}$ (Why?)

If B is the event "Even number on the green die," then $P(B) = \frac{18}{36}$ (Why?)

We want the probability of the event "1 on the red die and an even number on the green one." There is one way to get a 1 on the red die, and there are 3 ways to get an even number on the green one, so there are 3 ways to get the event 1 on the red *and* even on the green:

$$\{(1, 2), (1, 4), (1, 6)\} = A \cap B.$$

Thus the event A and B has the probability 3/36 or 1/12. In this case the probability of the compound event A and B is the product of the individual probabilities:

$$P(A \cap B) = \frac{1}{12} = P(A) \cdot P(B) = \frac{1}{6} \cdot \frac{1}{2} = \frac{1}{12}.$$

Try Exercise 12

Example 2 What is the probability that when a nickel, a dime, and a quarter are tossed, there will be a head on the nickel and a tail on the quarter?

Solution The sample space for tossing these three coins contains 8 points (N, D, Q), since there are two possibilities for the nickel (N), the dime (D), and the quarter (Q). If the event A is "Head on the nickel" and the event B is "Tail on the quarter" then:

$$A = \{(\mathrm{H, T, T}), (\mathrm{H, T, H}), (\mathrm{H, H, T}), (\mathrm{H, H, H})\} \text{ and } P(A) = \frac{1}{2},$$

$$B = \{(\mathrm{H, T, T}), (\mathrm{H, H, T}), (\mathrm{T, T, T}), (\mathrm{T, H, T})\} \text{ and } P(B) = \frac{1}{2}.$$

$$A \cap B = \{(\mathrm{H, T, T}), (\mathrm{H, H, T})\} \text{ and } P(A \cap B) = \frac{1}{4}.$$

Again in this case the probability of the event $A \cap B$ is the product of the individual probabilities because:

$$P(A \cap B) = \frac{1}{4} = \frac{1}{2} \cdot \frac{1}{2} = P(A) \cdot P(B).$$

12. For the experiment of rolling a red die and a green die:

a) What are the points of the event E_1: a 1 on the red die?

b) What are the points of the event E_2: an odd number on the green die?

c) Find $P(E_1)$, $P(E_2)$.

d) What points are in $E_1 \cap E_2$?

e) What is $P(E_1 \cap E_2)$?

f) Compare $P(E_1) \cdot P(E_2)$ to $P(E_1 \cap E_2)$.

13. The experiment consists of tossing the three coins: a nickel, a dime, and a quarter. Let C be the event that a tail shows on the dime; let D be the event that a tail shows on the quarter.

a) What are the points of event C? event D?
b) Find $P(C)$, $P(D)$.
c) What are the points of the set $C \cap D$? What is $P(C \cap D)$?
d) Compare $P(C \cap D)$ with $P(C) \cdot P(D)$.
e) Are the events C and D independent?

14. Use the same reasoning as in the text to find:

a) The probability of red on the first draw and red on the second.
b) The probability of blue on the first draw and red on the second.
c) The probability of blue on the first draw and blue on the second.

Try Exercise 13

In both Examples 1 and 2, the probability of the conjunction of the two events can be found by multiplying the probabilities of the individual events. Events like these are called *independent* and form the basis of the definition of such events.

Event A and event B are said to be *independent events* if and only if the following is true:

$$P(A \text{ and } B) = P(A \cap B) = P(A) \cdot P(B)$$

DEPENDENT EVENTS, CONDITIONAL PROBABILITY

Example 3 There are 10 red marbles in a bag and 6 blue ones. A marble is drawn and not replaced. A second one is drawn. What is the probability that the first one is red and the second is blue?

Solution A tree diagram will help in visualizing the sample space.

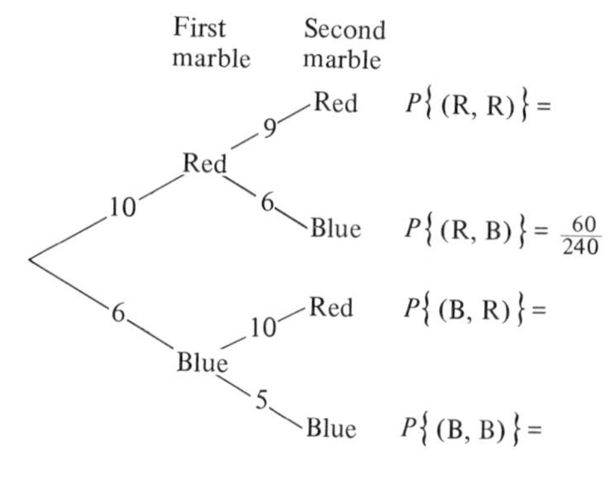

The branches are labeled with the number of marbles of each color that are in the bag at the time of the draw. Each branch represents a single point of the sample space. The first marble can be chosen in a total of 16 ways, since there are 16 marbles in the bag. The marble is not replaced, and the second can be chosen in 15 ways. Thus the total number of ways is $16 \cdot 15 = 240$. If the first marble is red and the second blue, we have a success. In how many ways can we do this? There are 10 ways to get a red marble the first time, and 6 ways to get a blue one the second time, or 60 ways. The probability is $\frac{60}{240}$ or $\frac{1}{4}$.

Try Exercise 14

From the tree diagram we see that the event A: red on the first draw consists of the union of the sets $\{(\text{R, R})\}$ and $\{(\text{R, B})\}$ which are disjoint. Thus the probability of the event A is:

$$P\{(\text{R, R})\} + P\{(\text{R, B})\} = \frac{90}{240} + \frac{60}{240},$$

$$\frac{3}{8} \quad + \quad \frac{1}{4} \quad = \frac{5}{8}.$$

In a similar fashion we find that the event B: blue on the second draw consists of the union of the events $\{(\text{R, B})\}$ and $\{(\text{B, B})\}$ which are also disjoint events. The probability of event B is given by:

$$P\{(\text{R, B})\} + P\{(\text{B, B})\} = \frac{60}{240} + \frac{30}{240} = \frac{2}{8} + \frac{1}{8} = \frac{3}{8}.$$

In this case we see that the events A and B do not satisfy the definition for independent events, since the probability of the event "Red on the first draw and blue on the second" is

$$P(A \text{ and } B) = P(A \cap B) = \frac{1}{4} \quad \text{but} \quad P(A) \cdot P(B) = \frac{5}{8} \cdot \frac{3}{8} = \frac{15}{64} \neq \frac{1}{4}.$$

Events like those in Example 3 are called *dependent*; the occurrence of one of them affects the probability of the other. We say that the probability of one is *conditional* upon the other. In Example 3, the probability of getting a blue marble on the first draw was 6/16. However, once a blue marble has been withdrawn and not replaced, the probability for getting a blue marble on the second draw has changed to 5/15.

If we take the probability for blue on the first draw and multiply it by the probability for blue on the second draw, given that a blue one has already been removed, it will be the probability for blue on both draws:

$$P(\text{B on first draw}) = \frac{6}{16} = \frac{3}{8},$$

$$P(\text{B on second draw, given B on first}) = \frac{5}{15} = \frac{1}{3},$$

$$P(\text{B on both draws}) = \frac{3}{8} \cdot \frac{1}{3} = \frac{1}{8}.$$

Try Exercise 15

The discussion for dependent events can be summarized as:

For *dependent events* A and B, the probability of the event A and B is

$$P(A \cap B) = P(A) \cdot P(B, \text{given } A \text{ has happened})$$

or, for short, $P(A) \cdot P(B|A)$.

We use this in the next example.

15. Use the second method of Example 3 to find the additional probabilities:

a) The probability of red on the first draw and red on the second.

b) The probability of blue on the first draw and red on the second.

c) The probability of red on the first draw and blue on the second.

16.
a) What is the sample space for the experiment of Example 4?
b) Find the probability of getting a red block on the first draw and a blue on the second.

Example 4 There are 5 red blocks and 4 blue ones in a bag. A child draws one out and does not replace it. Another child then draws a second one. What is the probability that both blocks are red?

Solution We shall do this two ways.

1. There are 9 ways to select the first block and 8 ways to select the second, so the number of ways is $9 \cdot 8 = 72$. How can we get 2 red blocks? We draw one on the first try, and there are 5 ways to do this. We must also draw one on the second try, and there are 4 ways to do this. Thus the number of ways to select two red blocks is $5 \cdot 4 = 20$. The probability of drawing two red blocks is $\frac{20}{72} = \frac{5}{18}$.

We use another approach.

2. The probability of drawing a red block on the first try is 5/9. The block is not replaced, and the probability of drawing a red on the second try changes. This is clear from the tree diagram.

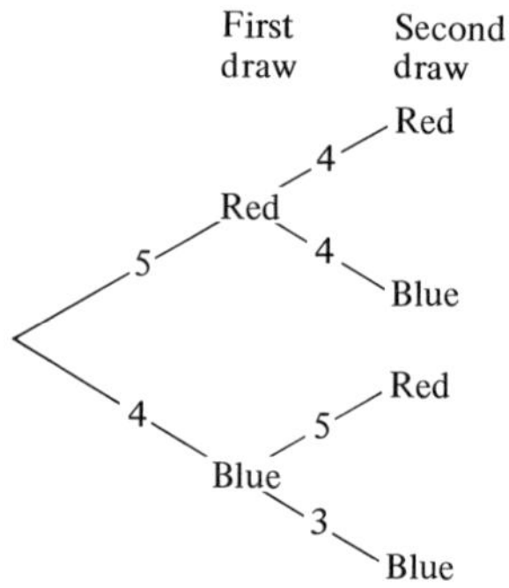

The probability for (R, R) is the product of P(red on first draw) and P(red on second, given the first was red):

$$P\{(\text{R, R})\} = P(\text{1st is red}) \cdot P(\text{2nd is red, given 1st is red}) = \frac{5}{9} \cdot \frac{1}{2} = \frac{5}{18}.$$

This agrees with the answer found by the first method.

Try Exercise 16

EXERCISE SET 12.3

1. Which of the following are independent events?
a) A coin is tossed, then tossed again.
b) A card is drawn from a deck and replaced; then another is drawn.
c) A card is drawn from a deck and not replaced; another is drawn.

2. Which of the following are dependent events?
a) A marble is drawn from a bag and replaced; another is drawn.
b) A block is drawn from a bag and not replaced; another is drawn.
c) A coin is tossed; then a die is rolled.

3. Events B and C are independent; $P(B) = 0.14$ and $P(C) = 0.56$. Find $P(B \text{ and } C)$, that is, find $P(B \cap C)$.

4. Events F and G are dependent; $P(F)) = 0.41$. If F happens, then $P(G) = 0.54$. Find the probability of F and G.

5. A coin is tossed four times. What is the probability of:
a) Getting all heads?
b) Getting one tail and three heads?

6. Suppose a coin has been bent, and the probability it falls heads is actually 0.55. If this coin is tossed three times, what is the probability of:
a) Getting all tails?
b) Getting at least two heads?

In Experiments 7 and 8, a bag contains 100 beads, identical except for the color. Sixty of the beads are red, 30 are blue, and 10 are white. Beads are selected at random and replaced before the next is drawn.

7.
a) What is the probability of two red beads in succession?
b) What is the probability of a red followed by a white?

8.
a) What is the probability of three red beads in succession?
b) What is the probability of red, white, and blue in that order?

Experiments 9 and 10 are the same as Experiments 7 and 8, except that after a bead is removed it is not replaced before the next is removed.

9.
a) What is the probability of two red beads in succession?
b) What is the probability of a red bead followed by a white?

10.
a) What is the probability of three red beads in succession?
b) What is the probability of a red, white, and blue in that order?

In Experiments 11–14, a box contains four balls labeled 1, 2, 3, 4. A ball is removed and not replaced; a second ball is then drawn.

11. Find the probability that ball 2 is drawn at least once.

12. Find the probability that the same ball is drawn twice.

13. Find the probability that both balls are labeled with odd numbers.

14. Find the probability that number 2 is not drawn either time.

15. Your friend has only two different kinds of socks. They are all alike except for the color: five pairs are black and five pairs are maroon. They are jumbled in a drawer.
a) If two socks are drawn at random, what is the probability that they will match?
b) How many socks must be drawn to guarantee that at least two of them match?

■ **16.** Your friend from Problem 15 buys three more pairs of socks; this time they are blue. If two are drawn out at random as before,
a) What is the probability of getting a matched pair?
b) How many socks must be taken from the drawer to guarantee that at least one matched pair will show up?

17. There are three spoiled eggs among the dozen eggs in the refrigerator. What is the probability that you choose those very eggs to make a 3-egg omelet for lunch?

18. In Problem 17, what is the probability that you do not get any of the spoiled ones?

19. The probability that a certain thumbtack falls point up is 0.3.

a) What is the probability that it falls point up twice out of two trials?

b) What is the probability that it falls point up on the first trial and on its side on the second?

20. A bag contains 4 red marbles, 4 white marbles, and 4 blue marbles. Three marbles are drawn at random, without replacement. What is the probability that:

a) They are all the same color?

b) They are all of different colors?

21. You have a penny, a nickel, a dime, and a quarter in a pocket. You pull out two at random. What is the probability that the combined value is greater than 15 cents?

12.4 AIMS

You should be able to:

a) Organize a suitable set of data into a table and draw a histogram to illustrate it.

b) Find the mean, median, range, and standard deviation for a suitable set of ungrouped data.

12.4 DESCRIPTIVE STATISTICS

Descriptive statistics refers to the methods used to condense in some way the main characteristics of a set of data. For example, suppose that the 25 first-graders of a classroom are weighed and the average weight is found to be 21 kg. This average describes the weight of the entire class in some way.

ORGANIZATION OF DATA. It is very difficult to make sense out of a mass of data, even a relatively short list of 56 items. In order to make the information more accessible it can be organized into tables or displayed graphically.

What features can you discover in the following list of numbers?

Average Number of Students in Courses Taken by Freshmen in a Liberal Arts College

30	108	37	31	24	83	74	109	44	23	35	28	45	54	48
41	78	52	48	35	49	33	55	95	21	37	37	49	65	39
48	72	80	34	30	95	30	49	24	27	44	38	57	71	66
57	76	93	57	49	55	67	28	30	49	56				

It is very difficult to absorb the information present in this table. If the list is merely put in order it starts to make a little more sense:

21	28	30	35	39	48	49	55	57	72	83	109
23	28	31	37	41	48	49	55	65	74	93	
24	30	33	37	44	48	49	56	66	76	95	
24	30	34	37	44	49	52	57	67	78	95	
27	30	35	38	45	49	54	57	71	80	108	

For convenience, a table can be condensed to a more useful size if the data are grouped into categories (or classes); ordinarily we use no fewer than five nor more than twelve of these. We do this for the table just above:

Average number of students	*Frequency*
20— 29	7
30— 39	14
40— 49	12
50— 59	8
60— 69	3
70— 79	5
80— 89	2
90— 99	3
100—109	2
Total	56

The histogram that illustrates this table is given below:

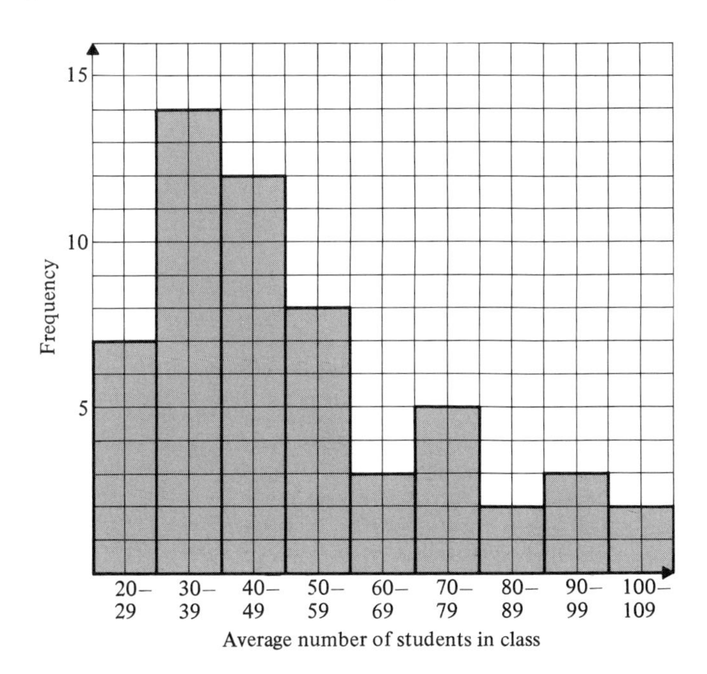

Try Exercise 17

17. Condense the data for the number of students still more. Make a table using the indicated class limits.

Average number of students	*Frequency*
20—34	
35—49	
50—64	
65—79	
80—94	
95—109	

18. The following are grades for a series of math tests for two pupils. Find the mean test score for each pupil.

Pupil *A*	Pupil *B*
69	75
71	77
70	74
71	78
99	76

Measures of Central Tendency

For many purposes the organization and graphical display of the data are all that is necessary. Further analysis of the data makes it possible to present it even more compactly than in a table or with a graph.

THE ARITHMETIC MEAN. Measures of central tendency give some information about where the data are centered or "bunched." There are several ways to measure such an "average" value. The one usually considered first is the arithmetic mean.

The *arithmetic mean* M of n numbers $x_1, x_2, x_3, \ldots, x_n$ is the sum of the numbers divided by n:

$$M = \frac{x_1 + x_2 + x_3 + \cdots + x_n}{n}.$$

The mean can be computed without putting the data in order. However, it has the disadvantage that it can be influenced by a few very large or very small numbers that are not really typical.

Example 1 Find the mean grade for the following set of test grades:

74, 61, 92, 81, 74, 71, 80, 75, 79, 76.

To find the arithmetic mean, the scores are added and the sum is divided by 10, the number of grades:

$$M = \frac{74 + 61 + 92 + 81 + 74 + 71 + 80 + 75 + 79 + 76}{10} = 76.3.$$

Try Exercise 18

THE MEDIAN. In Exercise 18 we see the effect that one grade of 99 has. On the basis of the average alone one might conclude that the students are alike, yet four of pupil *A*'s grades are below those of pupil *B*. In this case it might have been better to use the middle score, or median.

The *median* of a set of n numbers arranged according to size is the middle number. If n is even, the median is taken halfway between the two middle numbers.

The median has the same number of items above it as below it.

Example 2 Find the median for the following set of scores:

99, 69, 71, 70, 71.

We put them in order: 69, 70, 71, 71, 99. The median is 71, the middle number.

Example 3 Find the median of the following set of scores:

74, 61, 92, 81, 74, 71, 80, 75, 79, 76.

We arrange them in order:

$$61, 71, 74, 74, 75, 76, 79, 80, 81, 92.$$

Since there are ten items, there is no middle score. The median is the average of the two nearest middle; it is $\frac{75 + 76}{2} = 75.5$.

Try Exercise 19

19. Find the median of each of the following sets of scores:
a) 75, 77, 74, 78, 76
b) 53, 72, 69, 88, 75, 86, 89, 72, 69, 74

Measures of Variation (Dispersion)

The mean and median, while useful, may not always be enough to distinguish between two sets of data. Examine the following test scores for two different junior-high math classes:

Class *A*	67 69 70 73 74 76 76 78 78 79
Class *B*	47 55 68 74 74 76 77 77 92 100

Try Exercise 20

20. Find the median test scores for class *A* and class *B* of the text.

Close scrutiny of the grades shows that the classes are not really alike even though the mean and the median are identical. The scores for class *B* are more spread out.

RANGE. The *range* is a simple way to describe this variability. To get the range, subtract the lowest number from the highest number.

Example 4 Find the range for the scores of class *A* and class *B* given above.

Class *A*	Low score 67 High score 79	Range $79 - 67 = 12$
Class *B*	Low score 47 High score 100	Range $100 - 47 = 53$

The range is only a rough measure of the spread, or dispersion, since it is determined by only two items of the data.

A glance toward the classroom

Even in the primary grades youngster can get a taste of the collection and organization of data and thus be introduced to statistical methods. Such information as height, weight, hair color, and eye color can be collected and summarized for a class. Colorful pictorial graphs can be made to display and compare the results.

STANDARD DEVIATION. There is another measure of how widely spread a set of data is; it uses all the items in the set. It is called the *standard deviation* and is based on the following reasoning. If the numbers lie far from the mean, there is more variation than if they are all close to the mean. If all the deviations from the mean are added, some will be positive and some negative. So we square the deviations and then average. Then we take the square root of this average.

The *standard deviation* σ for a set of n numbers $x_1, x_2, x_3, \ldots, x_n$ is defined as follows:

$$\sigma = \sqrt{\frac{(x_1 - M)^2 + (x_2 - M)^2 + \cdots + (x_n - M)^2}{n}}.$$

Collecting the birthday month statistics

Example 5 Find the standard deviation for the set of test scores for class A and class B above.

Solution A table is helpful to organize the computation; x_i is the test score, M is the mean.

Class A

Score x_i	$x_i - M$	$(x_i - M)^2$
67	−7	$(-7)^2 = 49$
69	−5	$(-5)^2 = 25$
70	−4	$(-4)^2 = 16$
73	−1	$(-1)^2 = 1$
74	0	$(0)^2 = 0$
76	2	$(2)^2 = 4$
76	2	$(2)^2 = 4$
78	4	$(4)^2 = 16$
78	4	$(4)^2 = 16$
79	5	$(5)^2 = 25$
740	0	156

$$M = 740 \div 10 = 74 \qquad \sigma = \sqrt{\frac{156}{10}} = 3.95$$

Class B

Score x_i	$x_i - M$	$(x_i - M)^2$
47	−27	729
55	−19	361
68	−6	36
74	0	0
74	0	0
76	2	4
77	3	9
77	3	9
92	18	324
100	26	676
740	0	2148

$$M = 740 \div 10 = 74 \qquad \sigma = \sqrt{\frac{2148}{10}} = 14.66$$

These standard deviations (3.95 for class A and 14.66 for class B) show that the test scores for class B vary more than do those for the other class. It is not necessary to have the original data to see this.

21. Find the standard deviation of the following sets of data:

a) 59, 71, 70, 71, 99
b) 73, 75, 72, 76, 74
c) 74, 74, 74, 74, 74

Try Exercise 21

The following graph from a sixth-grade text book shows how data might be presented.

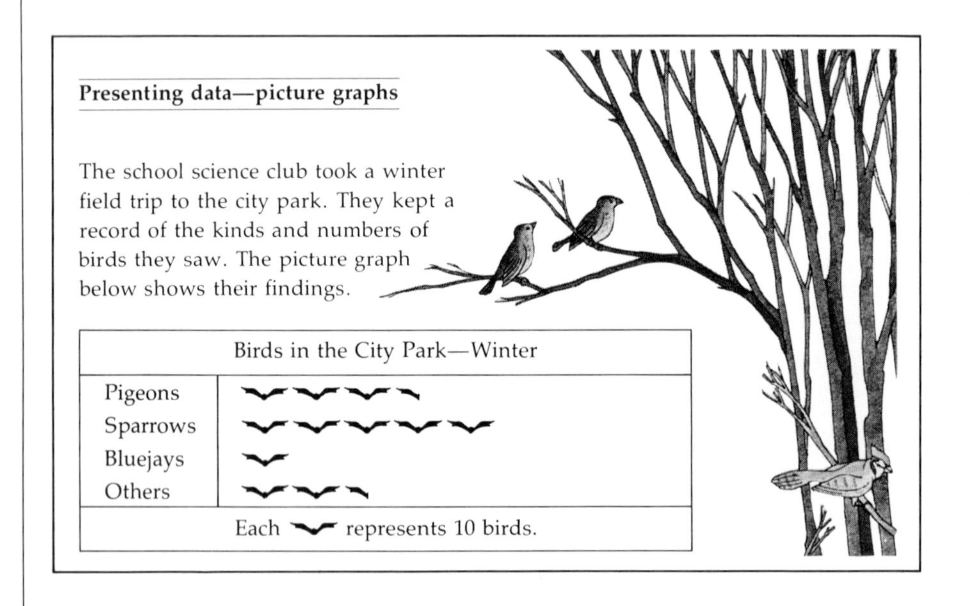

Presenting data—picture graphs

The school science club took a winter field trip to the city park. They kept a record of the kinds and numbers of birds they saw. The picture graph below shows their findings.

Birds in the City Park—Winter	
Pigeons	
Sparrows	
Bluejays	
Others	
Each represents 10 birds.	

EXERCISE SET 12.4

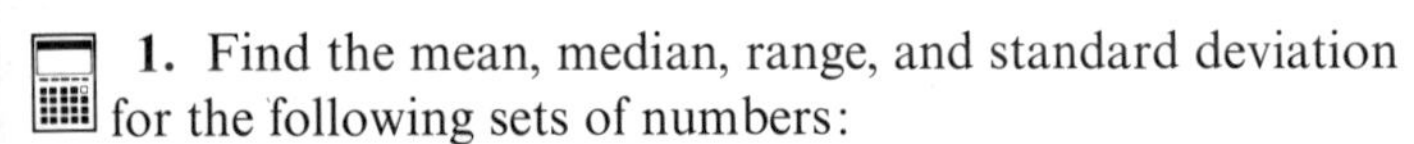

1. Find the mean, median, range, and standard deviation for the following sets of numbers:
a) 13, 3, 7, 4, 8 b) 3, 7, 15, 8, 2

2. Find the mean, median, range, and standard deviation for the following sets of numbers:
a) 8, 5, 11, 2, 59, 1 b) 3, 1, 2, 49, 21, 31, 27, 22

3. The standard deviation, as calculated by the definition, can be laborious to find, particularly if the set of data is large. An equivalent formula for the standard deviation is as follows:

$$\sigma = \sqrt{\frac{x_1^2 + x_2^2 + x_3^2 + \cdots + x_n^2}{n} - M^2}.$$

Use this formula with Problem 1 to show that the result is the same as that computed by the definition.

4. Repeat Problem 3 but this time use the data of Problem 2.

5. The following grades are a random selection from the first-semester grades of freshmen at a small college:

2.2	3.8	2.5	2.5	2.8	2.8	1.2	3.3
2.3	3.7	3.7	3.4	3.2	2.8	3.4	2.3
2.6	2.9	2.8	2.2	3.1	2.8	4.0	3.5
3.1	3.4	3.7	3.3	3.2	2.8	3.5	1.9
1.5	0.75	2.7	2.1	3.2	2.9	2.8	3.1

a) Organize these grades by means of a table. Use the classes: 0.05—0.55, 0.55—1.05, 1.05—1.55, 1.55—2.05, 2.05—2.55, 2.55—3.05, 3.05—3.55, 3.55—4.05.
b) Draw a histogram for this frequency distribution.

6. The data on eye color were collected in an undergraduate class in statistics:

Eye Color

Dark brown	2	Hazel	2
Brown	12	Green	4
Blue-green	1	Blue	6

a) Illustrate this information with a bar graph.
b) This type of information is qualitative rather than numerical; it does not make sense to talk about the mean or median. However, if there is a category (or more than one) for which the frequency is the greatest in the set, it is called the *mode.* What is the mode for these data?

7. The data on hair color were collected in the same class. Repeat Problem 6 for these data:

Hair Color

Dark brown	1	Blond	4
Brunette	1	Light brown	3
Black	3	Brown	15

8. The following table summarizes data given in a recent *World Almanac.*

The Heights (in Stories) of 90 Buildings in New York City

102	77	67	71	70	60	59	60	54	57
54	52	52	50	50	60	50	56	53	44
47	50	44	44	47	48	50	40	46	37
41	47	34	42	45	43	35	48	40	39
39	45	39	44	41	45	42	41	39	42
41	45	45	44	38	42	31	43	41	41
32	43	42	41	41	44	43	39	51	33
40	41	38	40	38	36	46	40	47	40
34	38	42	40	33	44	41	40	34	31

a) Find the mean and the median heights of these 90 buildings.
b) Condense the data by grouping into the following classes: 30—39, 40—49, 50—59, 60—69, 70—79, and so on.
c) Draw a histogram to illustrate these grouped data.

9. Write a flowchart for calculating the arithmetic mean of a set of data.

10. Write a flowchart for the standard deviation of a set of n numbers.

11. You are told that the average depth of a river is $2\frac{1}{2}$ feet. Is it safe to wade across? Why?

12. You live in a cold climate and are moving to Balmy City. The Chamber of Commerce there says the average daily temperature is 71.3 °F. Is it safe to dispose of all your heavy clothing? Why?

Problems 13–16 could be posed to upper elementary students.

13. Find the arithmetic mean of 1/4, 2/5, 5/8.

14. A person has an average weekly salary of $217. What is the yearly salary of this person?

15. The mean of a list of 12 numbers is 76.4. Find the sum of the numbers.

16. Find the median of this set of numbers if the mean is 47:

22, 33, 47, ____, 50, 56, 72.

In Problems 17 and 18 use the following data on sex and weight collected in a statistics class:

Female	*Weight, kg*
x_1	59.0
x_2	57.1
x_3	53.5
x_4	47.6
x_5	44.4
x_6	60.8
x_7	70.3
x_8	56.7
x_9	61.3
x_{10}	50.8
x_{11}	62.1
x_{12}	52.6
x_{13}	61.2
x_{14}	95.2
x_{15}	48.5

Male	*Weight, kg*
y_1	77.1
y_2	74.8
y_3	77.1
y_4	79.4
y_5	69.8
y_6	77.1
y_7	88.4
y_8	81.6
y_9	83.9
y_{10}	72.6
y_{11}	74.8
y_{12}	79.4

17.
a) Find the mean weight and then the median weight for the females.
b) Group the data into classes: 43.5—53.4 kg, 53.5—63.4 kg, 63.5—73.4 kg, and so on.
c) Make a histogram of the grouped data.

18. Repeat Problem 17 for the males. In part (b) use classes: 69.5—73.4 kg, 73.5—77.4 kg, 77.5—81.4 kg, and so on.

■ **19.** Are the mean and the median always close to each other? Give examples to support your argument.

12.5 AIMS

You should be able to:
a) Distinguish between theoretically determined probability and experimentally determined probability.
b) Carry out simple experiments to determine probabilities experimentally.

12.5 INFERENTIAL STATISTICS

While it is important to be able to discuss theoretical probabilities and methods for their computation, we are often interested in experimentally determined probabilities. For the theoretical toss of a balanced coin it is assumed that there are just two outcomes and that both have the same probability of occurring. However, we might also ask: "Do, in fact, real coins, when tossed, behave in the way the theoretical one does?" This is a statistical question rather than a probabilistic one.

Much of the everyday problem solving and decision making of modern society depends upon the collection of data which is organized and analyzed so that inferences can be made from it. For example, what is the probability that a given family will be watching program X at 8:30 on this coming Saturday evening? Such questions are answered by interviewing a sample of the population to find what proportion have viewed this program in the past. This information is used to estimate the probability that a given family will view the program in the future.

EXPERIMENTALLY DETERMINED PROBABILITIES. If we actually toss a coin a large number of times we can determine the proportion of the tosses that show heads. This is the *experimentally determined* probability for a head to turn up on the toss of a coin.

The *experimental probability* $P'(A)$ of an event A (for a large number of trials) is the *ratio*:

$$\frac{\textbf{The number of times } A \textbf{ occurs}}{\textbf{The number of trials}}.$$

This definition is consistent with the way theoretical probability was defined earlier.

Example 1 A light-bulb manufacturer has found that 2 out of every 100 light bulbs are defective and burn out in only a few hours. The school department orders 1000 of these bulbs. How many should the maintenance staff expect will probably burn out in this short time?

Solution The probability that a bulb will have a short life is given by the ratio:

$$\frac{2}{100} = \frac{\text{Number of short-life bulbs}}{\text{Number of trials}}.$$

This means we can expect that on the average 2% of the bulbs purchased will have a short life. That means for 1000 bulbs we could expect

$$(0.02)(1000) = 20 \text{ bulbs}$$

to have a short life.

Example 2 There are 110 marbles in a bag, some red and some blue. Children are to estimate, without looking in the bag, the number of each color. They take turns choosing a marble without looking in the bag. They note the color and return the marble to the bag. They select a total of 43 red ones and 27 blues ones. What is the probable number of each color marble in the bag?

22. In an experiment similar to the one of Example 2, children choose 48 green cubes and 144 white ones from a box. About how many of the cubes are green if there are 500 of them in the box altogether?

Solution The total number of trials was $43 + 27 = 70$. The probability of getting a blue one is

$$\frac{\text{Number of blue ones chosen}}{\text{Total number chosen}} = \frac{27}{70}, \text{ or about } 0.39.$$

Thus they estimate that about 39% of the marbles, or 43 of them are blue and about 67 are red.

Try Exercise 22

RANDOM SAMPLES. When making such inferences on the basis of a sample from the population, it is desirable that the sample be *random.*

A *random sample* (subset) of a population is one where each element of the population has the same chance of being included in the sample.

It is important to use such a sample so as not to influence or bias the results. In Example 2 the blue marbles might be a bit heavier, for example, and tend to sink to the bottom of the bag. Therefore, if the children reach to the bottom every time and do not stir the marbles thoroughly, this will influence the result.

The greater the number of trials, the more times we repeat an experiment, the greater will be our confidence in our results.

In Example 2 the randomness of the sample can be assured by a thorough mixing of the marbles in the bag each time. It is also possible to select a random sample with the assistance of a table of random numbers*. In such a table you find page after page of the digits 0 through 9. You can imagine that such a table has been generated by recording the results obtained with a "spinner" of the type found in children's games. The spinner has ten equal sections each labeled with one of the digits from 0 to 9.

A random-number table may also be used in the Monte Carlo method for simulating statistical problems that are very difficult to analyze any other way. Monte Carlo methods and experimental determination of probability are ideal for introducing youngsters to this branch of mathematics.

* *A Million Random Digits with 100 000 Normal Deviates*. RAND Corporation, Free Press, Glencoe, Illinois (1966, 1955).

EXERCISE SET 12.5

1. Children draw at random with replacement 63 red checkers and 27 black checkers from a bag containing 253 checkers. Estimate the number of each color in the bag.

2. A sample is taken from a 200 kg bin of freshly harvested cherries. The sample weighs 2.1 kg and contains 375 cherries of which 59 are found to be culls. Estimate the weight of culls in the bin (in kg).

3. Obtain ten pennies.

a) Place them in a cup or box, shake the container, roll the coins onto a flat surface. Count the number of heads.
b) Repeat part (a) 50 times; record the results of each trial.
c) Make a frequency table using the results of part (b):

Number of heads	0	1	2	3	4	5	6	7	8	9	10
Number of occurrences											

d) Make a histogram of the data in the table. The horizontal axis is the number of heads that showed on a single toss and the vertical is the number of times such a trial happened.
e) From this experiment, what would you say is the most probable number of heads to show up in a toss of 10 pennies?
f) In this experiment you have tossed a coin 500 times. What was the total number of heads that showed up? What is your estimate for the probability that a head shows on the toss?

4. Make a triangular pyramid from cardboard, using this pattern. Color the three isosceles faces. Determine experimentally the probability that it will land on its base when tossed. Carry out 150 trials.

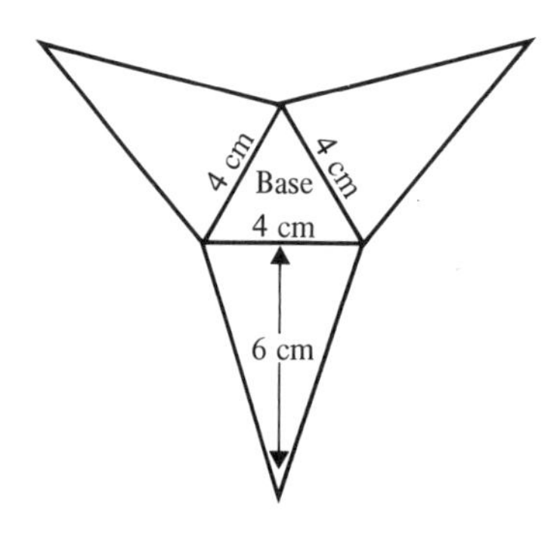

5. Obtain 10 thumbtacks, all the same kind. Repeat Problem 4 by tossing the thumbtacks to determine the probability that a tack will fall point up.

6. Roll a pair of dice 180 times.

a) Make a frequency table and a histogram for this experiment.
b) Estimate the probability that a 7 will be obtained; that a 2 will be obtained.
c) Experimentally, what number seems to have the greatest probability of being rolled?
■ d) Compare your experimental results with the theoretical probabilities for this experiment.

7. Spinners from children's games can also be used to generate strings of random numbers. Explain how the spinner illustrated could be used to simulate the experiment of tossing a single coin as often as desired. Conduct such an experiment with a spinner to estimate the probability that a coin falls heads.

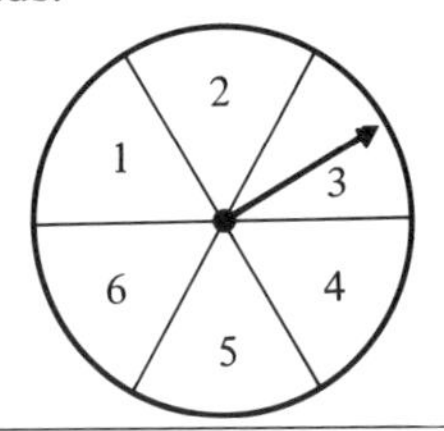

8. Explain how the spinner could be used to simulate the tossing of two dice. How could it be used to simulate the tossing of a coin? Carry out the experiment of Problem 6 with a spinner of this type.

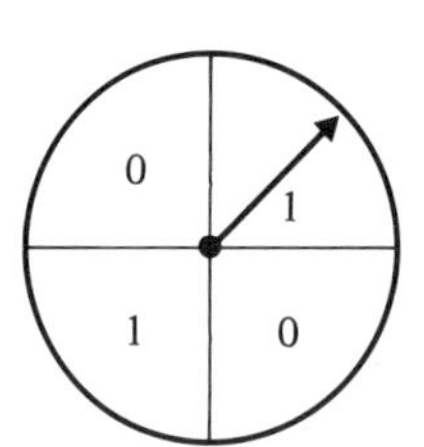

12.6 AIMS

You should be able to:
Use the Monte Carlo method to simulate some appropriate problems.

12.6 PROBLEM SOLVING*

In this section we will solve a classic in statistics known as the "Cookie Problem."

Example 1 You are going to bake a batch of 100 cookies. Because of the high cost of chocolate bits, you are planning to put exactly 150 bits into the batter, which will be well mixed. How many cookies will end up without a chocolate bit?

Try Exercise 23

23.
a) What are the facts?
b) What is the problem?
c) What are my ideas?

There are advanced mathematical techniques by which this problem can be solved. One idea might be: Bake a batch of 25 cookies with 38 chocolate bits, and examine the number of bits that show up in each cookie. Another simpler idea is to merely simulate this procedure and we decide to do this. Such a simulation is an application of the Monte Carlo method.

Solution The first step is to draw a grid representing the 100 cookies.

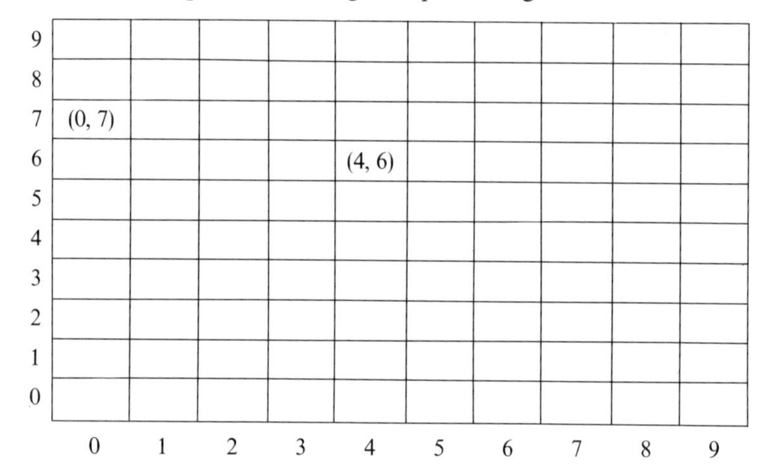

Each cell has a pair of coordinates. The cells labeled (0, 7) and (4, 6) are indicated on the diagram.

The 150 chocolate bits can then be placed in the cookies (cells) with a sequence of two-digit numbers chosen at random by using a spinner (such as that pictured) or from a table of random numbers, where the work of spinning the pointer has already been done.

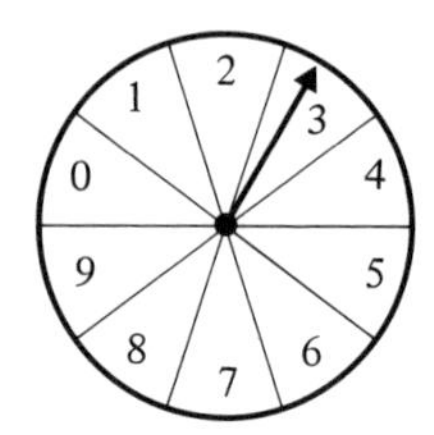

* A reminder:

THE FIVE-STEP PROCESS

1. What are the *relevant facts*?
2. What actually is the *problem*?
3. What are my *ideas*?
4. How do I get a *solution*?
5. *Check* and *review* the solution.

The following two-digit numbers came from such a table. Only 15 of the 150 numbers used to solve the problem are shown here:

46	11	59
15	07	62
92	23	13
30	03	52
79	27	07

Try Exercise 24

When all 150 numbers are plotted, the diagram will show the distribution of the chocolate chips in the 100 cookies. Notice that some cookies end up with no chips at all.

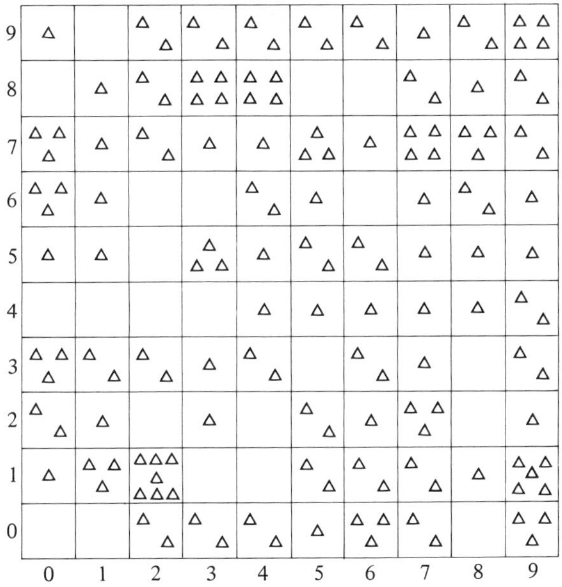

Try Exercise 25

As experimentally determined, the probability that a cookie has two chips, is the number of cookies with two chips divided by the total number of cookies. It is interesting to compare the experimental probabilities with theoretical probabilities determined by advanced mathematics. Even in this simple simulation there is remarkable agreement.

24. Locate the 15 chips that correspond to the numbers in the text on the grid; the points (0, 7) and (4, 6) have already been located.

25.

a) Tally the number of cookies that have 0 chips, 1 chip, 2 chips, 3 chips, 4 chips, and more than 4 chips.

b) Summarize your information in a table.

26. The data from the simulation show that if only 150 chips are used, there will be a large number of cookies without any bits at all. What is a commonsense remedy to this difficulty?

No. of chips per cookie	*Simulation probabilities*	*Theoretical probabilities*
0	0.22	0.223130
1	0.32	0.334695
2	0.30	0.251021
3	0.10	0.125511
4	0.04	0.047066
>4	0.02	0.018579
	1.00	1.00

The Monte Carlo method is a very effective way to determine theoretical probabilities where other methods are either not available or are impractical. It is ideal for use with elementary-school pupils.

Try Exercise 26

EXERCISE SET 12.6

1. Estimate the value of π using probabilistic methods. You will need an area covered with squares, such as a tile floor or a piece of graph paper. You will also need a circular disc, such as a coin or a piece of cardboard. The diameter of the disc should be the same as the length of the side of the squares.

a) Toss the coin or paper disc 100 or more times. Keep a record of the number of hits and misses. A hit means that the disc covers a vertex of a square. A miss means that the disc does not cover a vertex of a square.

b) The center of the disc always falls in some square. If the center of the disc falls inside any of the shaded regions, the toss will still be a hit. This is because the coin center will be less than the distance r from a vertex. The ratio of the number of hits to the number of tosses is the ratio of the area of the shaded regions to the total area of the square. ■ Show that the ratio of hits to total number of tosses is $\pi/4$.

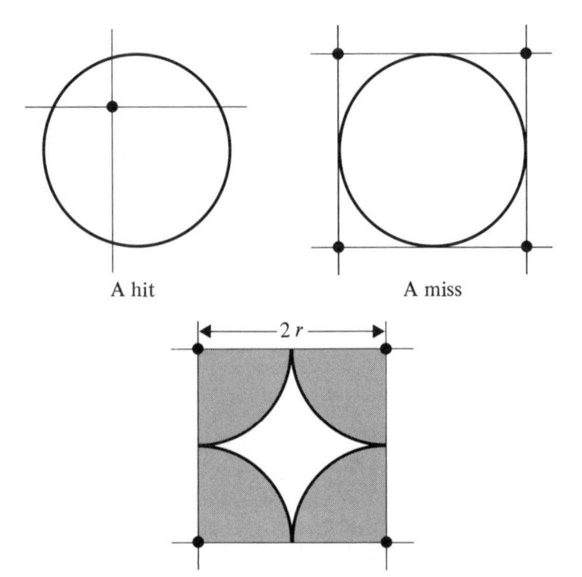

2. Estimate π using the idea from a famous problem of statistics, Buffon's Needle Problem. Draw a number of equally spaced parallel lines on a large cardboard or use lines on the floor. Then find a "needle." This can be a piece of wire, a soda straw, and so on. Make its length less than the distance between the parallel lines.

a) Toss the needle 100 or more times. Keep track of the number of tosses and the number of times it falls across one of the lines.
b) If the distance between the lines is d and the length of the needle is L then:

$$\frac{\text{Number of hits}}{\text{Number of tosses}} = P(\text{Hit}) = \frac{2L}{\pi d}.$$

(The proof of the formula involves mathematics beyond the scope of this book.)

3. A simulation of a game of chance. It is common for three people to "match coins" to determine who pays for the beverages they have just ordered. Each person tosses a coin, which will land *head* or *tail*. If two coins are alike, the person whose coin is different pays the bill. If all the coins are the same everyone retosses. A simulation can help you decide whether this is a fair thing to do. You need a spinner with an even number of equal areas marked off. Half of the areas will represent a head and the others a tail. For each game it is necessary to generate three different results. Then it is possible to see who pays. For example:

HHT	third person pays,
TTT	retoss,
THT	second person pays.

a) Simulate 100 such coin matchings of three people.
b) What percentage of the time did the 1st person pay? the 2nd person? the 3rd person? What percentage were retosses?
c) On the basis of your evidence would you say that this is a fair game (does each of the three people have the same chance of winning)?

4. Describe how the experiment of Problem 3 could be done by using a table of random numbers.

REVIEW TEST

1. The experiment is to roll one green die and one red die. List the points of the following events:

a) A total of 6 is rolled and both numbers are odd.
b) The green die shows 4 and the total rolled is 6.
c) The numbers are both odd or the green die shows 4.

2. Find the probability for the events in Problem 1.

3. Make an appropriate tree diagram for 4 successive tosses of a coin. List the sample space for the experiment. What is the probability of getting at least 1 head in a toss of 4 coins?

4. In how many ways can 7 pupils be seated in one row?

5. Is it possible to have a sample space in which the event A has a probability of 0.74 and the event $\bar{A}$ has a probability of 0.36? Explain your answer.

6. What is the probability of drawing two black cards, without replacement, from a well-shuffled deck of cards?

7. A box contains 100 slips of paper each labeled with one of the numbers from 1 to 100. All slips have the same chance of being drawn from the box. What is the probability of drawing a number that is divisible by 13? by 3?

8. What is the probability of drawing 13 spades from a deck of 52 well-shuffled cards?

9. Find the probability that a card drawn at random from a deck of 52 cards is a 7 or a black card. Event A is "A 7 is drawn" and event B is "A black card is drawn." Are A and B independent events?

10. Find the mean, median, and mode of the set of heights in meters collected for the female members of a statistics class:

1.70, 1.72, 1.60, 1.65, 1.74, 1.65, 1.55, 1.63,
1.50, 1.68, 1.63, 1.65, 1.65, 1.62, 1.68.

11. Find the range and the standard deviation for the data of Problem 10.

12. Describe how to select 6 children at random from a class of 36.

■ **13.** A student decides to answer a multiple-choice test using the spinner below. If there are eight questions with 4 choices each, what is the probability that the student can get 7 or more answers correct in this way?

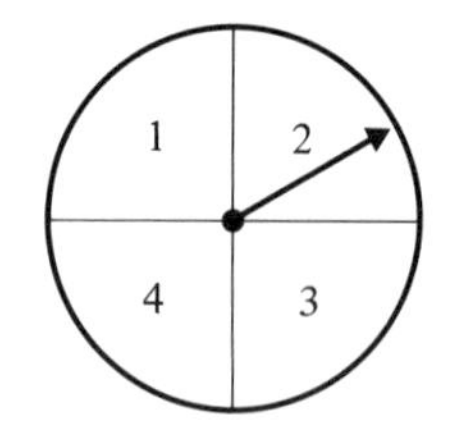

■ **14.** Simulate 60 turns of play for the following game, which is suitable for elementary students. Use a table of random numbers or a spinner to do this. There are three teams; A, B, and C. There are 30 green cubes and 30 white cubes in a bag and a spinner divided into two equal sections: white and green. One player from team A chooses a cube from the bag and another spins the spinner. If the colors are the same, the team scores. Now team B has a turn, and so on.

a) From your simulation, is the game a fair one (that is, are the scores for the three teams nearly equal)?

b) Is it more likely that the two colors agree and the team scores or that the team does not score?

c) Find the theoretical probabilities for this game and compare to your simulation.

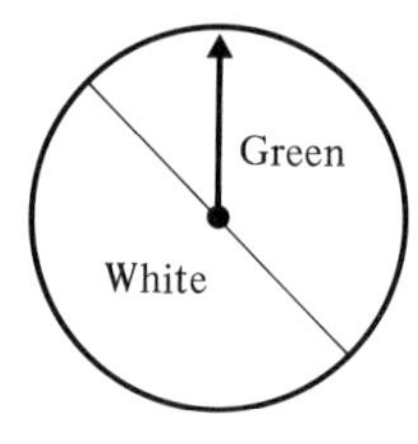

FOR FURTHER READING

1. Some interesting activities in simulation and sampling are given in the article "Quantifying chance" by Randall J. Souviney in *Arithmetic Teacher*, 25:3, December 1977, pp. 24–26.

2. Some activities and games that are not the usual toss-a-coin type are described in "A fun way to introduce probability" by Rick Billstein in *The Arithmetic Teacher*, 24:1, January 1977, pp. 39–42. In the same issue there is a short article called "M & M's candy: a statistical approach."

TEXTS

Dubisch, Roy, *Basic Concepts of Mathematics for Elementary Teachers.* Addison-Wesley Publishing Company, Reading, Massachusetts (1977).

University of Maryland Mathematics Project, *Unifying Concepts and Processes in Elementary Mathematics.* Allyn and Bacon, Inc., Boston, Massachusetts (1978).

answers

CHAPTER 1

Margin Exercises

1. 1, 3, 5, 7, 9, 11, 13, . . . ; 1, 3, 5, 7, 9, 1, 3, 5, . . . ; 1, 3, 5, 7, 9, 7, 5, 3, 1, . . . , and many others.

2. 1, 3, 5, 3, 7, 3, 9, 3, . . . ; 1, 3, 5, 3, 1, 3, 5, 3, . . . ; 1, 3, 5, 3, 1, 1, 3, 5, 3, 1, 1, 3, 5, 3, 1, . . . ; 1, 3, 5, 3, 5, 7, 5, 7, 9, 7, . . . , and many others.

3.

n	6	7	8	9
k	3	4	5	6
d	9	14	20	27

4. All check.

5. By looking for patterns that appear in the observations of particular cases, i.e., by inductive reasoning.

6. Counterexample.

7. Be systematic. Try 0, 1, 2, 3, etc.; when $n = 4$, we get $4^2 - 4 + 13 = 25 = 5^2$, which is a counterexample.

8.

N	Cost, $		
	to open	*to close*	*total*
8	0.80	1.20	2.00
7	0.70	1.05	1.75
6	0.60	0.90	1.50
5	0.50	0.75	1.25
4	0.40	0.60	1.00
3	0.30	0.45	0.75
2	0.20	0.30	0.50
1	0.10	0.15	0.25

9. Cut three links and use these to join the rest.

10. Who was ahead at the end of the first lap? The second lap? Is it possible the race was a tie? etc.

11. Make a sketch of the cars on the route at the end of the first lap, at the end of the second; make the racetrack a straight line; find out how long it takes to go around each lap for each car, compare the times, etc.

12. Check your calculator instruction book.

13. Flowchart

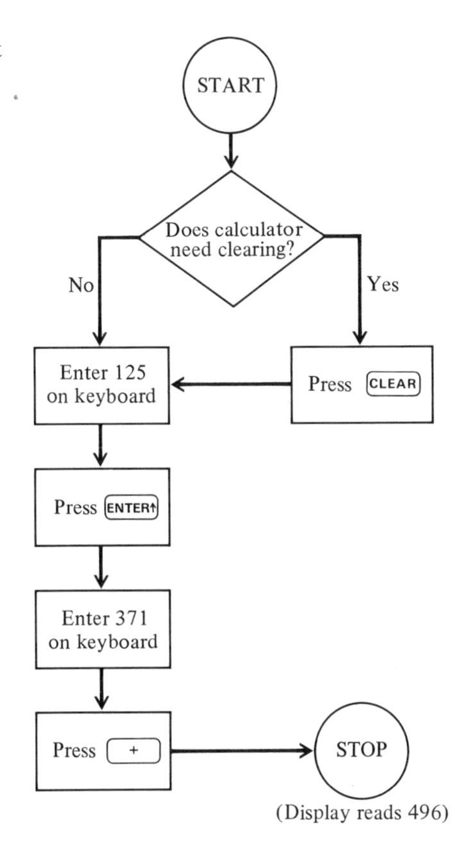

14. *a*, *b*

15. a) $\exists x: x - 3 = 5$. True since at least one number x makes the statement true; for example $x = 8$.
b) $\forall x: x - 3 = 5$. False since at least one x does not make statement true; for example $x = 10$.

16. a) $\underset{\text{F}}{1 + 6 = 2}$ *and* $\underset{\text{T}}{5 > 3}$; conjunction is false.
b) July has 15 days (F) *and* Thanksgiving day is in November (T); conjunction is false.

17. a) $(2 + 5 = 4 + 3) \vee (9 - 5 = 2 \quad 3)$; true b) False.

18. a) $\overline{q}$: Two added to five is not seven. b) $\overline{q}$: $3 + 6 \neq 9$.

19. a) It is not true that all college students study a lot. Not all college students study a lot. Some college students do not study a lot.
b) There is an x for which x^2 is not greater than 0. For some x, x^2 is not greater than 0.

20. a) It is not true that some TV programs are very informative. No TV programs are very informative. Every TV program is such that it is not informative.
b) It is not the case that at least one of my students is home with the flu. None of my students are home with the flu.

21. a) If it rains, then this hat will be ruined. Antecedent: it rains; consequent: this hat will be ruined.
b) If a prime number is odd, then it is greater than 2. Antecedent: a prime number is odd; consequent: a prime number is greater than 2.

22. a) 1, 3, and 4 b) 2 **23.** a) T b) T c) F d) T **24.** T, F, T, T

25. If triangle ABC is equiangular, then it is equilateral. If triangle ABC is equilateral, then it is equiangular.

26. (a) (b) same, T, T, T, F. **27.** a) If the shape is triangular, then the object is green. b) Original true, converse false.

28. Converse: If a number is greater than 10, then it is larger than 15.
Inverse: If a number is not larger than 15, then it is not greater than 10.
Contrapositive: If a number is not greater than 10, then it is not larger than 15.

29. Yes (T, F, T, T) **30.** No **31.** Not valid.

Exercise Set 1.1, p. 7

1. 15, 18, 21 **3.** 15, 17, 19 **5.** 25, 32, 43 **7.** 3, 6, 9, 12 (and others) **9.** 2, 3, 5, 7, 11, 13 **11.** 2, 4, 11, 23

13. a) Short rod, long rod b) Square rod, long rod

15. $17 + 18 + 19 + 20 + 21 + 22 + 23 + 24 + 25 = 64 + 125 = 189$. If n is the row number, then first term in each line is $(n - 1)^2 + 1$, last term in row is n^2, number of terms in the row is $2n - 1$, and the sum is $(n - 1)^3 + n^3$.

17. $n = 5$: $1 + 3 + 5 + 7 + 9 = 25 = 5^2$,
$n = 6$: $1 + 3 + 5 + 7 + 9 + 11 = 36 = 6^2$, etc.

19. 0 | 0 | 0 0 0 0 0 0 can all be regrouped in two's and hence are even.
0 | 0 | 0 0 0 0 0 0
Proof: $2m + 1$ is an odd number, $2n + 1$ is an odd number, but $2m + 1 + 2n + 1 = 2m + 2n + 2 = 2(m + n + 1)$ has a factor of 2 and is even.

21. a) Let $n = 3$, then $n^2 + 1 = 9 + 1 = 10$(even). b) No.

23. Label each student $S_1, S_2, S_3, \ldots, S_{25}$. Label each section $P_1, P_2, P_3, \ldots, P_{15}$. Assume that each section has one of the students S in it. Only 15 of the students can be accommodated before a section will have two or more students. If each section P does not have at least one student S to start, then the two students will appear in a section even earlier.

Exercise Set 1.2, p. 11

1. Closed in a circle. Three cuts are needed.

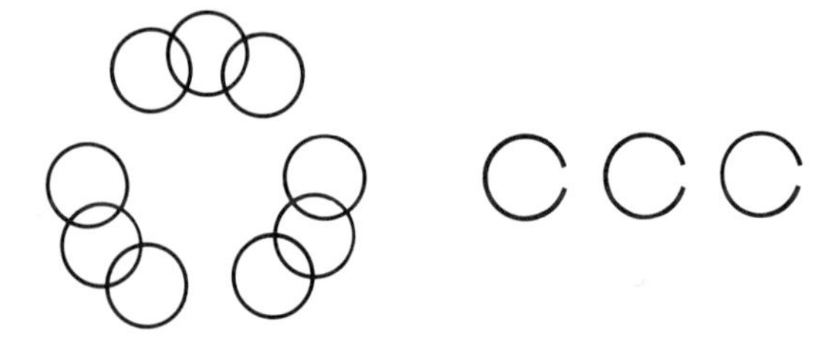

3. 6 **5.** Save $0.17 by getting only the needed ones. **7.** They cannot give $1.00 in dimes. Yes.

9. Weigh two coins on the balance, one on each side. If both are the same, the heavy coin is off the scale. If they are not the same, the heavy one can be identified.

11. a) 20 b) $(n)(n + 2)(n + 1)/6$, where n is the number of rows in the figure.

12. *Hint*: Some are: CAT = 3 + 1 + 20, SAD = 19 + 1 + 4, ARE = 1 + 18 + 5.

Exercise Set 1.3, p. 15

1. The flowcharts will vary depending upon the calculator. a) 617 b) 10 215 c) 18
Flowchart for algebraic calculator (assume the machine is cleared in each case).

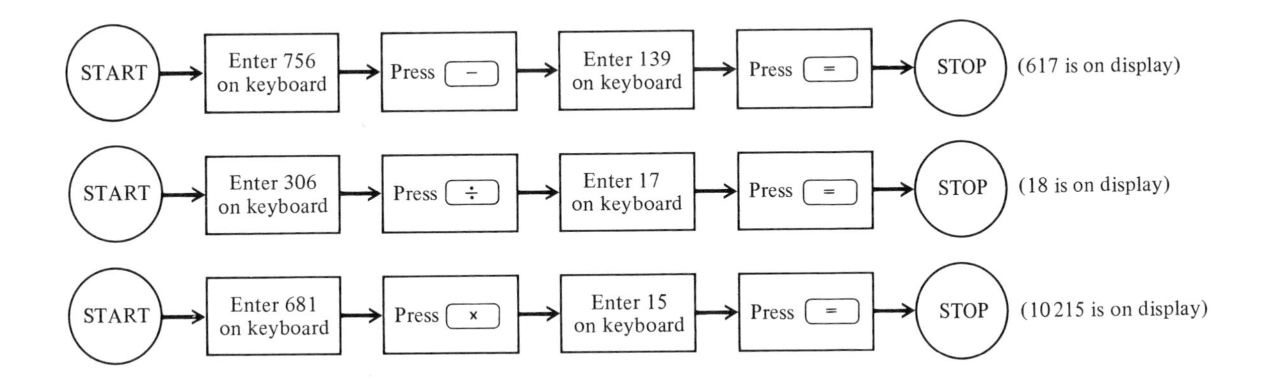

3. $^{-}1 + (987654321 \times 9) = 8888888888$ *Hint*: Look at $9 \times 9, 9 \times 8, 9 \times 7, 9 \times 6$, etc., and see what is needed to make a total of 8 at the right-hand digit. Then look at $9 \times 6, 9 \times 70, 9 \times 800, 9 \times 9000$ and see the patterns developed.

5. Smallest 1.9; largest 10.

Exercise Set 1.4, p. 25

1. a) Statement. c) Statement. d) Not a statement; can't use a quantifier.
g) If replacements are college students, this is a statement already quantified. i) For some y, $17 - y = 9$, where y is a whole number.

2. a) If you exercise regularly, then you don't eat junk food. c) If you don't keep healthy, then you don't exercise regularly.
e) You don't exercise regularly or you don't keep healthy. g) You eat junk food and you don't exercise regularly.
i) If you eat junk food, then you won't keep healthy.

3. p: Rick said he would mail a letter; q: Rick said he would call me next week; $p \vee q$.

5. p: Sue gets the report finished; q: Sue will bring it to you tomorrow; $p \rightarrow q$.

7. p: x is a professor; q: x expects you to do some work outside of class; $\forall x$: $p \rightarrow q$.

9. p: x is a mathematics problem; q: x is interesting; $\exists x$: $p \wedge q$.

11. a) There is no prime number that is divisible by 2; every prime number is not divisible by 2.
c) It is not the case that all numbers are integers; some numbers are not integers; there is at least one number that is not an integer.

12. a) Yes c) Yes **13.** a) T **14.** a) T **15.** a) T **16.** a) T

17. Told the truth in (a), (c), and (d). **19.** Both truth values are possible.

21. a) Always false. b) Always true. **23.** a) Statement must be true. c) Can be true or false.

Exercise Set 1.5, p. 32

1. Converse: If a number is a multiple of 7, then it is a multiple of 14.
Inverse: If a number is not a multiple of 14, then it is not a multiple of 7.
Contrapositive: If a number is not a multiple of 7, then it is not a multiple of 14.

3. Converse: If I am able to get my term paper finished, then I'll stay home from the party.
Inverse: If I don't stay home from the party, then I won't get my term paper finished.
Contrapositive: If I am not able to get my term paper finished, then I will not have stayed home from the party.

5. (b). **6.** a) F, F, F, T for both. **7.** a) Yes c) Yes e) Yes

8. a)

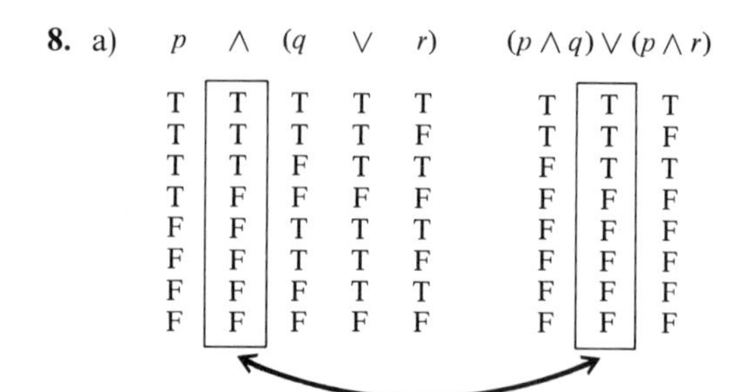

p	$\wedge$	$(q$	$\vee$	$r)$	$(p \wedge q)$	$\vee$	$(p \wedge r)$
T	T	T	T	T	T	T	T
T	T	T	T	F	T	T	F
T	T	F	T	T	F	T	T
T	F	F	F	F	F	F	F
F	F	T	T	T	F	F	F
F	F	T	T	F	F	F	F
F	F	F	T	T	F	F	F
F	F	F	F	F	F	F	F

9. a) p: It snows.
q: I shall go skiing.
$((p \to q) \wedge p) \to q$.

p	q	$((p \to q) \wedge p)$	$\to q$
T	T	T	T
T	F	F	T
F	T	F	T
F	F	F	T

Valid

10. a) p: I go to Toronto.
q: I shall travel by plane.
$((p \to q) \wedge \bar{q}) \to \bar{p}$.

$((p$	$\to q)$	$\wedge \bar{q})$	$\to \bar{p}$
T	T	F	T
T	F	F	T
F	T	F	T
F	F	T	T

Justified

c) p: I have a bad cold.
q: I go to work.
$((p \to \bar{q}) \wedge \bar{p}) \to q$.

p	q	$((p \to \bar{q})$	$\wedge \bar{p})$	$\to q$
T	T	F	F	T
T	F	T	F	T
F	T	T	T	T
F	F	T	T	F

Not valid

11. Valid **13.** No **15.** No

17. The statements assumed true in the argument could actually be false. Correct reasoning from a false premise can give either a true or false conclusion.

CHAPTER 2

Margin Exercises

1. $\{11, 12, 13, 14, 15, 16\}$ **2.** $\{x \mid x$ is a whole number and $10 < x < 17\}$.

3. a) p does not belong to (is not a member of) the empty set. b) q is a member of set U.

4. a) $\not\subset$, $\not\subseteq$ $\notin$ b) $\in$, $\not\subseteq$ c) $\subset$, $\subseteq$ d) $\notin$ $\not\subseteq$ e) $\subseteq$, $=$ f) $\subseteq$ g) $\subseteq$

5. $\{1, 3, 5, 9, 2, 4, 10\}$ **6.** a) $\{5, 9\}$ b) $\varnothing$ or $\{\ \}$ **7.** $\{2, 4, 6, 8, 7\}$

8. a) $\left\{\begin{matrix}(4, 3) & (5, 3)\\(4, 2) & (5, 2)\\(4, 1) & (5, 1)\end{matrix}\right\}$ b) $\left\{\begin{matrix}(1, 5) & (2, 5) & (3, 5)\\(1, 4) & (2, 4) & (3, 4)\end{matrix}\right\}$ c) No.

9. a) $\{b, e\}$ b) $\{b, e\}$ c) $A \cap B = B \cap A$

10. a) $A \cup B = \{c, p, n, q, r, s, t\}$; $(A \cup B) \cup C = \{c, p, q, n, r, s, t, f\}$
b) $B \cup C = \{c, p, q, r, s, t, f\}$; $A \cup (B \cup C) = \{c, p, n, q, r, s, t, f\}$

11. $\{(2, 1), (3, 1), (4, 1), (3, 2), (4, 2), (4, 3)\}$

12. a) $\{(0, 0), (1, 6), (2, 12), (3, 18), (4, 24), (5, 30)\}$ b) $\{0, 6, 12, 18, 24, 30\}$ c) Yes

13. (b) and (d) are functions **14.** Yes **15.** No **16.** No

17. a) $\{a, b\}$, $\varnothing$, $\{a, b\}$ b) $\{a, b\}$, $\{a, b\}$, $\{a\}$, $\{a, b\}$

18. a)

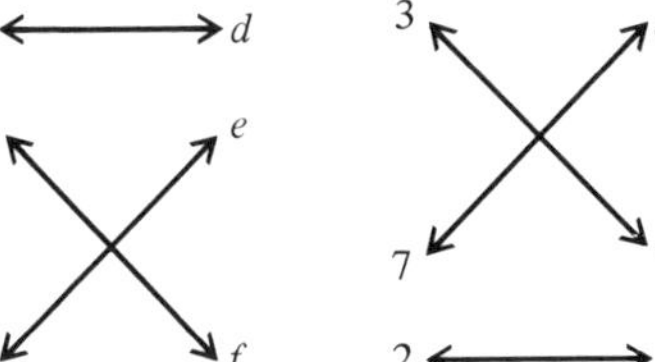

b)

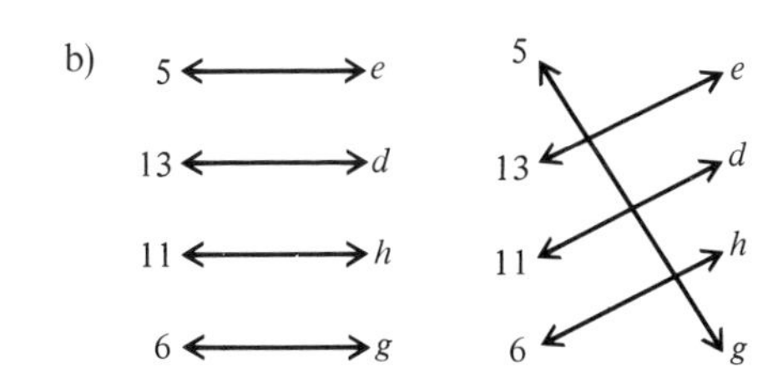

19. Yes.

20.

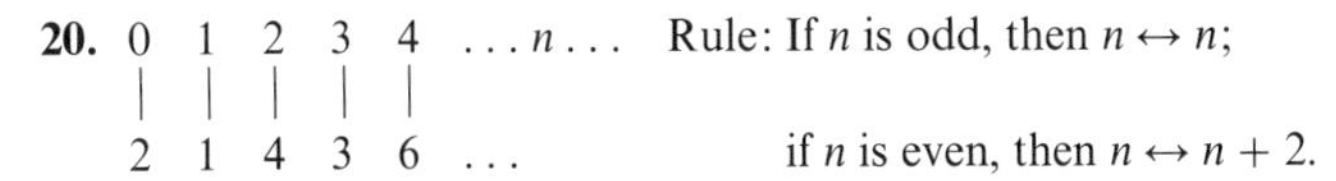

0 1 2 3 4 . . . n . . . Rule: If n is odd, then $n \leftrightarrow n$;

2 1 4 3 6 . . . if n is even, then $n \leftrightarrow n + 2$.

21. Use the numbers 1, 2, 3, 4, 5, 6 three at a time to find three different ways to get the sum 9. Half the numbers must be used once, the other half must be used twice. For example, $3 + 4 + 2 = 9$ is one such sum.

22.

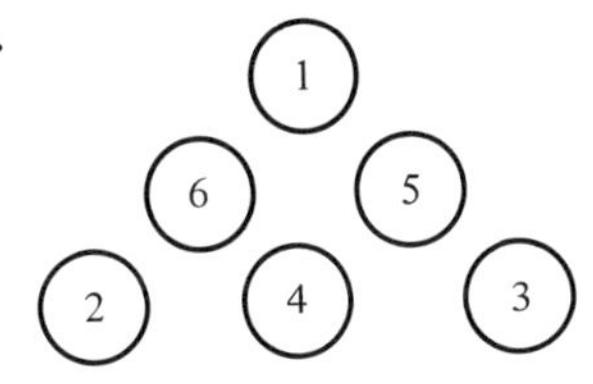

23. a) The definition of $\cup$ and $\cap$. The elements of each set were carefully described. b) $p \wedge (q \vee r) \equiv (p \wedge q) \vee (p \wedge r)$

24. a) 103 purchased ice cream, 66 purchased ice milk, 35 purchased yogurt, 9 made no purchase of the three items, 20 purchased all three, 36 purchased ice cream and ice milk, 23 purchased ice milk and frozen yogurt, 24 purchased ice cream and frozen yogurt.
b) How many households were interviewed altogether? How many used only ice cream? Only ice milk? Only frozen yogurt?

25. c) Total households 150, only ice cream 63, only ice milk 27, only frozen yogurt 8.

Exercise Set 2.1, p. 42

1. $\{0, 1, 2, 3, 4, 5, 6, 7, 8, 9, 10, 11\}$ **3.** {Cent, nickel, dime, quarter, half-dollar, dollar} **5.** $\{x \mid x$ is an integer and x is even$\}$

7. $\{x \mid x$ is rational and x is greater than $2\}$ **9.** $\subseteq, =$ **11.** $\notin, \not\subset, \nsubseteq$ **13.** $\nsubseteq, \not\subset$ **15.** a) F c) T e) F

17. $\{\ \}, \{p\}, \{n\}, \{d\}, \{p, n\}, \{p, d\}, \{n, d\}, \{p, n, d\}$ **19.** 32; adding one more element doubles the number of subsets.

21. 8; the set has three elements and we want all possible subsets.

Exercise Set 2.2, p. 49

1. a) both are $\{1, 5\}$ b) both are $\{0, 1, 2\}$

3. a) Both are $\{0, 1, 2, 5, 10\}$ b) Both are $\{0, 1, 2, 5, 7\}$ c) Both are $\{1, 2, 5\}$ d) Both are $\{0, 1, 2\}$

5. Both give the same region:

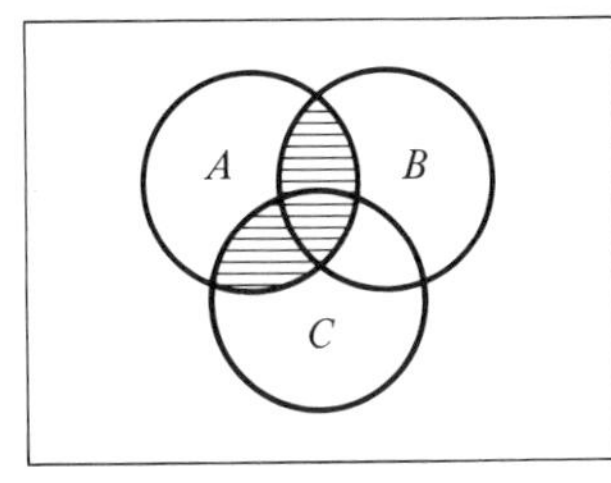

7. a) Both are $\{10\}$
b) Both are $\{2, 7, 10\}$

9. a)

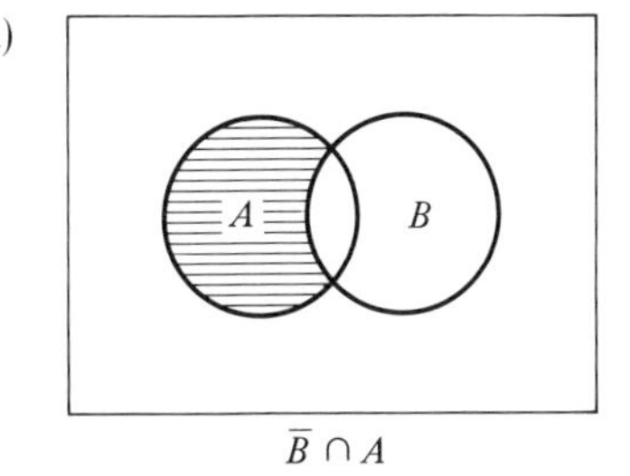

$\overline{B} \cap A$

12. a) T c) T e) F, $B \cup \bar{B} = U$ always g) T **13.** a) A = B c) A = B e) B ⊆ A g) B = ∅

14. a) {(0, 5), (1, 5), (2, 5), (5, 5), (10, 5), (0, 1), (1, 1), (2, 1), (5, 1), (10, 1), (0, 0), (1, 0), (2, 0), (5, 0), (10, 0)} c) { }

e) {(0, 10), (1, 10), (0, 5), (1, 5), (0, 2), (1, 2), (0, 1), (1, 1), (0, 0), (1, 0)}

16. Females: $\{N, R, B\}$ Males: $\{J, G, E\}$
$F \times M = \{(N, E), (R, E), (B, E), (N, G), (R, G), (B, G), (N, J), (R, J), (B, J)\}$

18. *Outline*: $A \cap B$ consists of all sets such that $x \in A$ is true and also $x \in B$ is true. Thus the form of the statement for the set members is $p \wedge q$. Also $B \cap A$ consists of all elements such that $x \in B$ is true and $x \in A$ is true, which means the statement is of the form $q \wedge p$. We know that $p \wedge q \equiv q \wedge p$, so the descriptions for the sets are equivalent and hence mean the same thing and will pick out exactly the same elements.

Exercise Set 2.3, p. 53

1. a) {(0, 1), (1, 2), (2, 3), (3, 4), (4, 5), (5, 6), (6, 7)} b) {0, 1, 2, 3, 4, 5, 6}; {1, 2, 3, 4, 5, 6, 7} is the range c) Yes

3.

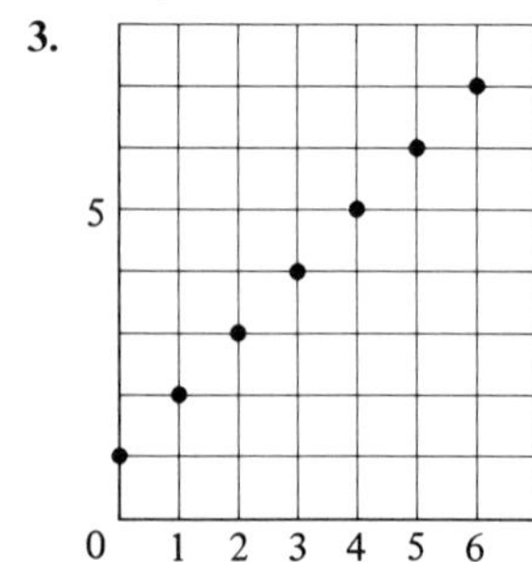

5. a) No c) Yes **6.** a) {1, 2} c) {8, 9, 10, 11, 12, 13} **7.** a) Range: {1, 2} c) Range: {7, 8, 9}

8. a) Yes c) No

9. b) No **11.** a) {(1, 1), (3, 3), (5, 5), (7, 7)} b) Yes

13. 1 is related to all elements of B, 3 is related to 4 and 6 by "less than" and 9 is not related to any element of B.

Exercise Set 2.4, p. 60

1. If R stands for reflexive, S stands for symmetric, T for transitive, then the answers are in corresponding order:
a) R, S, T → F, F, F, c) R, S, T → F, F, F, e) R, S, T → F, T, F

2. a) F, F, T c) F, T, F e) T, T, T

3. a) On a set of sets: "not disjoint from" c) On the set of subsets of {0, 1, 2, 3, 4, 5, 6, 7, . . . , 19, 20}: "has at least the element 5 in common"

5.
$$\begin{array}{cccccccc} 1 & 2 & 3 & 4 & 5 & 6 & 7 & 8 \\ \downarrow & \downarrow & \downarrow & \downarrow & \downarrow & \downarrow & \downarrow & \downarrow \\ a & b & c & d & e & f & g & h \\ b & c & d & e & f & g & h & a \end{array}$$

7. *Outline*: Logical equivalence means the two statements have identical truth values under all assignments of truth values, hence $p \equiv p$ is obvious. If $p \equiv q$, then, since they have same truth values, q is another name for p and vice versa, so $p \equiv q$ means it is also true that $q \equiv p$. If $p \equiv q$, we can replace q by p; if $q \equiv r$, we can replace q by p and get $p \equiv r$.

9.
$$\begin{array}{cccccc} 0 & 2 & 4 & 6 & 8 \ldots & 2n \\ \downarrow & \downarrow & \downarrow & \downarrow & \downarrow & \downarrow \\ 1 & 3 & 5 & 7 & 9 \ldots & 2n+1 \end{array}$$

11. If $A = B$, then the sets are really the same set, so each element can be put in a one-to-one correspondence with itself and hence the sets are equivalent. A is equivalent to itself even when it is called B. **13.** No

15. *Hint*: The ordered pairs still consists of all possible pairs of type: (An element from A, an element from B), even though the elements from B are now sets. One such pair is $(a, \{a, b\})$. There are a total of 24 pairs.

17. *Hint*: Let $A = \{a_1, a_2, a_3, \ldots, a_j, \ldots\}$, $B = \{b_1, b_2, b_3, \ldots, b_i, \ldots\}$, $C = \{c_1, c_2, c_3, \ldots, c_m, \ldots\}$. $A \times (B \times C)$ has points $(a_j, (b_i, c_m))$; $(A \times B) \times C$ has points $((a_j, b_i), c_m)$. It can be shown that these points can be the basis for a one-to-one correspondence.

Exercise Set 2.5, p. 67

1. *Hint*: Restate the problem; start with the solution for the total of 9 and see how it might be modified.
2. *Hint*: The smallest total occurs when the numbers used twice are the smallest.
3. *Hint*: Change the numbers used, make a larger triangle, use another figure, ask what the largest total on a side is for a given sequence of numbers, and so on.
4. *Hint*: Start with two points, then three points, and look for a pattern; make a table to summarize the results.
5. a) 16 b) 26 c) 4 **6.** *Hint*: Make an appropriate Venn diagram and label each section on the basis of the provided facts.
7. *Outline* for (17): The elements of set A make true a statement p; the elements of $\bar{\bar{A}}$ make true a statement $\bar{\bar{p}}$. But $p \equiv \bar{\bar{p}}$, therefore the properties for the elements that get into each of the sets are identical.
8. *Hint*: In how many ways can 6 liters be formed? Make a drawing of liters to scale; how could 1 liter be measured? How could 2 and 5 liters be measured?
9. *Ideas*: Use part of a truth table, T means person is going, F means person is not going; A, B, C, D, E are the people. Use the clues to determine what the possibilities are.

CHAPTER 3

Margin Exercises

1. 234 567 **2.**

3. 32 **4.** 1723 **5.** 49 **6.** 1984 **7.** DLXXIX **8.** MMMMMMMCDXLIX

9. a) 576

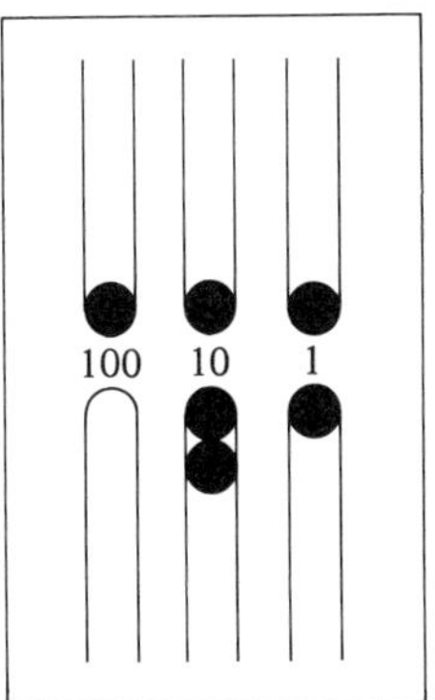

b) 2372

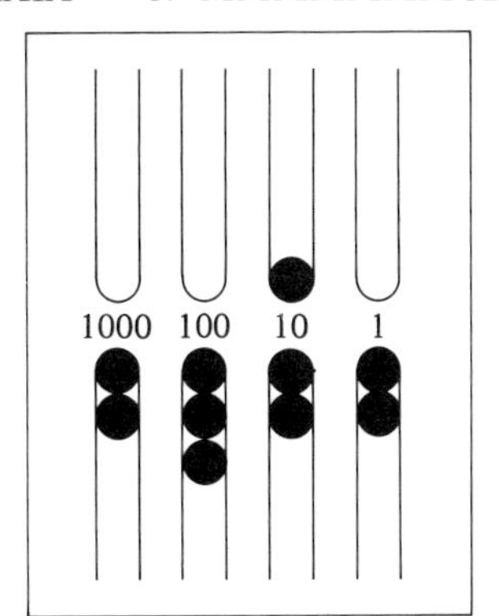

10. 576 + 20

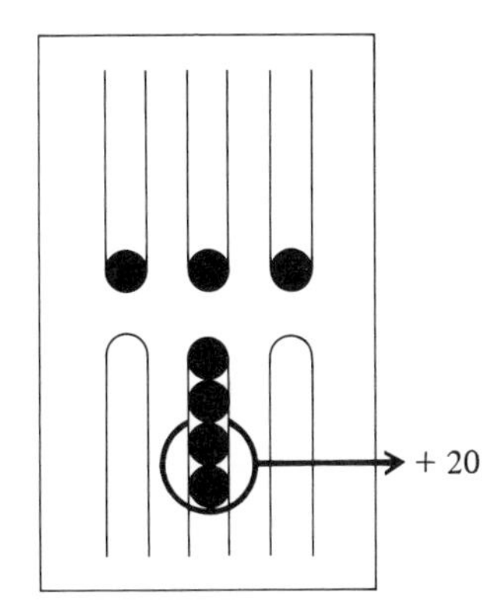

11. 8 ten thousands + 7 thousands + 9 hundreds + 2 tens + 4 ones

12.

Millions	*Hundred thousands*
1 000 000 10^6	100 000 10^5

13. 176

14. a) 500 000 + 70 000 + 3000 + 400 + 20 + 8 b) $5 \times 100\,000 + 7 \times 10\,000 + 3 \times 1000 + 4 \times 100 + 2 \times 10 + 8 \times 1$
c) $5 \cdot 10^5 + 7 \cdot 10^4 + 3 \cdot 10^3 + 4 \cdot 10^2 + 2 \cdot 10^1 + 8 \cdot 10^0$

16.

Trillions			*Billions*		
10^{14}	10^{13}	10^{12}	10^{11}	10^{10}	10^9

17. $3 \cdot 5^3 + 2 \cdot 5^2 + 1 \cdot 5^1 + 4$ **18.** 432_{five} **19.**

5^6	5^5
15 625	3125

20. Base ten: $3 \cdot 5^4 + 2 \cdot 5^3 + 1 \cdot 5^2 + 3 \cdot 5^1 + 2$.
Base five: $3 \cdot 10^4 + 2 \cdot 10^3 + 1 \cdot 10^2 + 3 \cdot 10^1 + 2$.

21. 542 **22.** 1667 **23.** 1102_{three} **24.** 10032_{four} **25.** 1234_{five}

26.

2^6	2^5	2^4
64	32	16

27. 109 **28.**

12^4	12^3
20 736	1728

29. 17 848

30. 13 white cubes represent the logs left, 7 red cubes represent the logs burned; replace the burned logs.

31. How many children in the math corner to start? How many before the teacher reorganized? etc. **32.** One **33.** a) 2 b) 3

34. When B = 4, then R = 8 and M = 4, therefore there were nine in the math area to start.

Exercise Set 3.1, p. 77

1. 27 **3.** 46 **5.** 1 224 **7.** 1 444 **9.** DCXLIX **11.** MMMMMCCCXLII

13. MMMMMMMMMMMMMMCCXLIX **15.** 10 132 **17.** 140 134

19.

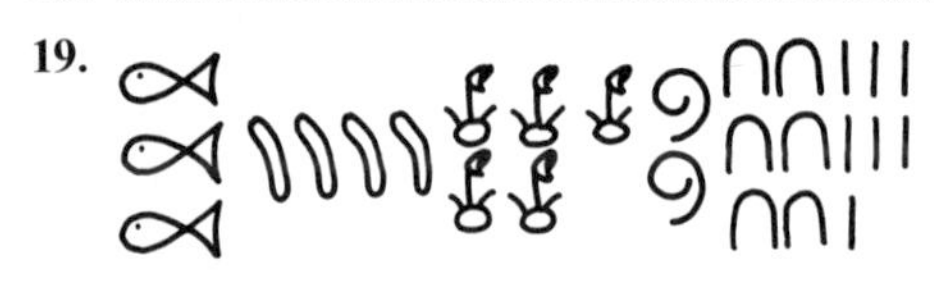

21.

24. a) XXIII
+XXVII
XXXXVIIIII = XXXXX = L

29. In an additive system there is no need to have a zero to "keep a place" since there are no places. There is no need to indicate that a particular symbol might be lacking since its omission indicates that it is not to be added in any case. Similarly on a counting board, the fact that there are no counters on a particular place on the board is all the information that is needed.

Exercise Set 3.2, p. 80

1. 7 hundred thousands + 5 ten thousands + 6 thousands + 4 hundreds + 2 tens + 5

3. 119 **5.** Thirty-eight **7.** Eight hundred seventy-nine

9. One thousand four hundred ninety-two **11.** Ninety-four thousand six

13. Seventy-five million nine hundred twenty thousand four hundred eighty-three

15. $10\,000 + 2000 + 400 + 10 + 8$, $1 \cdot 10\,000 + 2 \cdot 1000 + 4 \cdot 100 + 1 \cdot 10 + 8 \cdot 1$, $1 \cdot 10^4 + 2 \cdot 10^3 + 4 \cdot 10^2 + 1 \cdot 10^1 + 8 \cdot 10^0$

17. $400\,000 + 90\,000 + 2000 + 100 + 40 + 8$,
$4 \cdot 100\,000 + 9 \cdot 10\,000 + 2 \cdot 1000 + 1 \cdot 100 + 4 \cdot 10 + 8 \cdot 1$,
$4 \cdot 10^5 + 9 \cdot 10^4 + 2 \cdot 10^3 + 1 \cdot 10^2 + 4 \cdot 10^1 + 8 \cdot 1^0$

19. Fish are grouped in 10, 10, and 4, so as to write 2 tens and 4. Bundles of tens are counted along with the units to develop understanding of numerals such as 24, 43, and so on.

Exercise Set 3.3, p. 88

1.

6^5	6^4	6^3	6^2	6^1	6^0
7776	1296	216	36	6	1

3. Base ten: $1 \cdot 6^4 + 3 \cdot 6^3 + 2 \cdot 6^2 + 4 \cdot 6^1 + 5$.
Base six: $1 \cdot 10^4 + 3 \cdot 10^3 + 2 \cdot 10^2 + 4 \cdot 10^1 + 5$.

5. 2045 **7.** 239 **9.** 312_{four} **11.** a) 105_{six} b) 147_{eight}

13. a) 123_{ten} **14.** a) 988142_{ten} **15.** a) 100111_{two} **16.** a) 126_{twelve}

17.

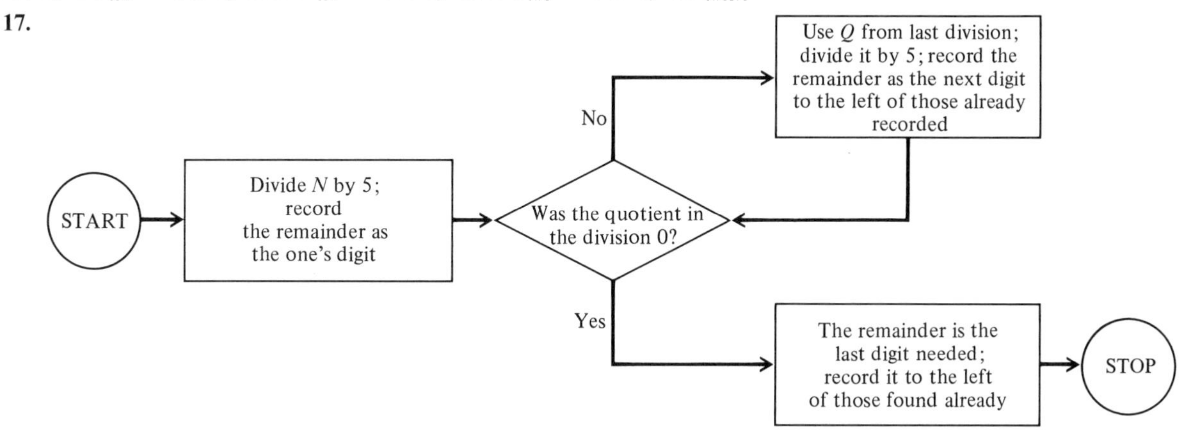

21. Partial answer:

Base 10	10	11	12	13	14	15	16	17	18	19	20
Base 5	20	21	22	23	24	30	31	32	33	34	40

Exercise Set 3.4, p. 92

1.

Monday	Tuesday	Wednesday	Thursday	Friday	Saturday
00	000	000	000	000	000
0	00	000	000	000	000
3	5	0	000	000	000
		7	9	00	000
				11	0
					13

3. *Hint*: How old are the mother and the daughter now? How old will they be in 5 years? Make a chart to compare the ages. Can you use trial and error?

5. *Hint*: The first card has all the numbers with a 1 in the units place.

7. *Hint*: How many digits are used for pages 1 to 9? How many are used for pages 10–99? How many are left for pages that require three digits?

9. *Hint*: The conditions mean that $4 \cdot b + 6$ must be an odd number; is this possible?

11. *Hint*: The conditions of the problem mean that $2b + 1 = 5$, or 10, or 15, and so on.

13. a) 21_{ten} c) 63_{ten}

15. *Hint*: What is the effect of replacing the powers of five $5^4, 5^3, 5^2, 5^1, 5^0$ by an arbitrary sequence of powers of some base b?

CHAPTER 4

Margin Exercises

1. a) 4 b) 7 c) 6 **2.** $n(A) = n(B)$

3. Let $A = \{c, d, e, f, g\}$, $B = \{h, j\}$; then $n(A \cup B) = a + b$, $n(A \cup B) = n\{c, d, e, f, g, h, j\} = 7 = 5 + 2$.

4. a) Associative (+) b) Commutative (+) **5.** a) $\{(a, 4), (b, 4), (c, 4),$ $(a, 3), (b, 3), (c, 3),$ $(a, 2), (b, 2), (c, 2),$ $(a, 1), (b, 1), (c, 1)\}$ b) 12 c) 3×4

6. Let $A = \{a, b\}$ and $C = \{5, 6, 7\}$; $n(A \times C) = 6 = 2 \cdot 3$

7. a) $A \times B = \{(1, c), (2, c), (1, b), (2, b), (1, a), (2, a)\}$; $B \times A = \{(a, 2), (b, 2), (c, 2), (a, 1), (b, 1), (c, 1)\}$
   c) Let $(1, c) \leftrightarrow (c, 1)$, etc., and establish a one-to-one correspondence to show that $A \times B \sim B \times A$.

8. $(A \times B) \times C = \{((1, 4), e), ((1, 3), e), ((2, 3), e), ((2, 4), e),$ $((1, 4), d), ((1, 3), d), ((2, 3), d), ((2, 4), d),$ $((1, 4), c), ((1, 3), c), ((2, 3), c), ((2, 4), c)\}$;
   $n(A \times B) = 4$, $n(C) = 3$, $n(A \times B) \times C = 12$

9. a) Commutative (+) b) Commutative (·) c) Commutative (+) and (·)

10. a) Commutative (+) b) Associative (·) and commutative (·) twice c) Associative (+), commutative (+)

11. $(6 + 4) + (27 + 3) + (15 + 5) = 60$ **12.** $(25 \cdot 4) \cdot (5 \cdot 2) \cdot (6 \cdot 7) = 42\,000$ **13.**

```
  00|0000         00   0000
  00|0000         00 + 0000
  00|0000         00   0000
3·6 = 3·(2 + 4)   3·2 + 3·4
```

14. a) $2 \cdot (9 + 7) = 2 \cdot 16$ b) $3 \cdot (4 + 11) = 3 \cdot 15$ **15.** Commutative (+)

16. Let $A = \{a, b, c\}$ and $B = \varnothing$; then $n(A \times B) = 0$, $n(A) \cdot n(B) = 3 \cdot 0$ **17.** Commutative (·)

18. No. The sum $1 + 6$ is not in the set. **19.** a) 5 b) 11 c) 18 **20.** Double the second number and add the result to the first.

21. $a \star b = a + 2b$ **22.** 3 **23.** $24 \div 6 = c \leftrightarrow 24 = c \cdot 6$; thus $c = 4$ **24.** b) $0 \div 9 = 0$

25. a) $>, \not<$ b) $\not<, \not>$ c) $<, \not>$ d) $>, \not<$ **26.** a) $x < y + 2$ b) $2 < 8$ c) $4 < 6 \cdot 4$

27. a) $30 < 45$ b) $x < 3$ c) $x < 4$ **28.** a) Associative (+) b) Distributive (·) over (+)

29.

$$\begin{array}{r} 87 \\ 92 \\ \underline{76} \\ 15 \\ \underline{240} \\ 255 \end{array}$$

30. For step 3: associative (·), renaming; for step 4: distributive law, all for whole numbers

31. a)

$$\begin{array}{rl} 300 + 20 + 5 & \\ \times\ 4 & \\ \hline 20 & [4 \times 5] \\ 80 & [4 \times 20] \\ 1200 & [4 \times 300] \\ \hline 1300 & [\text{Addition}] \end{array}$$

b)

$$\begin{array}{rl} 300 + 20 + 4 & \\ \times\ (30 + 5) & \\ \hline 20 & [5 \times 4] \\ 100 & [5 \times 20] \\ 1500 & [5 \times 300] \\ 120 & [30 \times 4] \\ 600 & [30 \times 20] \\ 9000 & [30 \times 300] \\ \hline 11340 & [\text{Addition}] \end{array}$$

32. For step 2, commutative (+); for step 3, associative (+) for whole numbers

33.

$$\begin{array}{r} 7000 + 200 + 40 + 6 \\ -(6000 + 500 + 10 + 9) \\ \hline \end{array} \rightarrow \begin{array}{r} 6000 + 1200 + 30 + 16 \\ -(6000 + 500 + 10 + 9) \\ \hline 700 + 20 + 7 = 727 \end{array}$$

34.

(XXXX) —1st→ $20 - 4 = 16$

(XXXX) —2nd→ $16 - 4 = 12$

(XXXX) —3rd→ $12 - 4 = 8$

(XXXX) —4th→ $8 - 4 = 4$

(XXXX) —5th→ $4 - 4 = 0$

Since 4 is removed 5 times, 5 is the quotient.

35. a) $Q = 44$, $R = 16$ b) $1738 = (42) \cdot (41) + 16$

36.

$$\begin{array}{r} 2 \\ 50 \\ 47\overline{)2453} \\ \underline{2350} \\ 103 \\ \underline{94} \\ 9 \end{array}$$

$Q = 52$,
$R = 9$
$2453 = (47) \cdot (52) + 9$

37.

3rd Q	4th Q
5	2
$32 \cdot 5 = 160$	$32 \cdot 2 = 64$
$160 < 226$	$64 < 66$
Yes	Yes
226	66
-160	-64
66	2
$66 < 32$	$2 < 32$
No	Yes

$Q = 20 + 20 + 5 + 2 = 47$,
$R = 2$,
$1506 = 47 \cdot 32 + 2$

38. $Q = 345$, $R = 4$

39. $3 + 2 = 10,\ 4 + 3 = 12,\ 4 + 4 = 13$
$3 \cdot 2 = 11,\ \ 3 \cdot 4 = 22,\ \ 4 \cdot 2 = 13$
40. 1241_{five} **41.** 1324_{five} **42.** 42_{five} **43.** 23001_{five} **44.** 22322_{five}

45. $Q = 1012_{\text{five}},\ R = 41$ **46.** 10001101_{two} **47.** $101010100101_{\text{two}}$

48. a) $56 + 50 = 106$ b) 721 c) $100 + 100 = 200$ d) 750 e) 77 f) $60 + 75 = 135$

49. a) 53 b) 35 c) $(400 - 100) + (76 - 60) = 316$ d) 178

50. a) 156 b) 188 c) 465 d) $\dfrac{4600 \cdot 2}{10} = \dfrac{9200}{10} = 920$ e) $\dfrac{52 \cdot 100}{4} = \dfrac{5200}{4} = 1300$ **51.** a) 180 000 b) 1300 c) 30

52. 6174 shows up **53.** a) No matter what N is, four times N cannot be 1, 11, 21, or 31. b) $92 = 23 + 23 + 23 + 23$

Exercise Set 4.1, p. 98

1. a) 3 c) 5 e) 18 **2.** a) F; $n\{a, b, c\} = n\{1, 2, 3\}$ but $\{a, b, c\} \neq \{1, 2, 3\}$. b) T **3.** Yes
5. *Hint*: Look at a smaller case; imagine the class has three members.

Exercise Set 4.2, p. 107

1. $A \cap B = \varnothing,\ n(A \cup B) = n\{a, b, c, 1, 2\} = 5,\ 3 + 2 = 5$

3. All sets are disjoint; $n(B) = 2$, $n(C) = 5$, $n\{1, 2, p, q, r, s, t\} = 7$, $n(A \cup (B \cup C)) = n(A) + n(B \cup C) = 3 + 7 = 10$; similar argument shows $5 + 5 = 10$.

5. $A \times C = \{(a, t), (b, t), (c, t),$
$(a, s), (b, s), (c, s),$
$(a, r), (b, r), (c, r),$
$(a, q), (b, q), (c, q),$
$(a, p), (b, p), (c, p)\};$

$n(A \times C) = n(A) \cdot n(C);$
$15 = 3 \cdot 5$

7. $2 \cdot 5 = 5 + 5 = 10$

9.

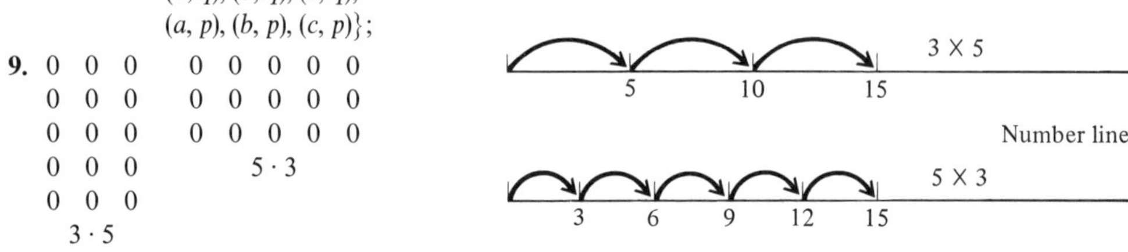

11. *Hint*: It must be shown that $a \cdot (b \cdot c) = (a \cdot b) \cdot c$. Compare sets $A \times (B \times C)$ and $(A \times B) \times C$ and show that they are equivalent, hence have the same cardinality n.

12. a)

```
00000|000000        00000     000000
00000|000000        00000  +  000000
00000|000000        00000     000000
00000|000000        00000     000000
```

$4 \cdot (5 + 6) = 4 \cdot 11 = 44$ $\quad 4 \cdot 5 + 4 \cdot 6 = 44$

13. a) $3 \cdot 6 + 3 \cdot 12 = 3 \cdot (6 + 12) = 3 \cdot (18)$ **15.** a) Let $A = \{1, 2, 3, 4, 5\}$, $A \times \varnothing = \varnothing$, hence $n(A \times \varnothing) = 0 = n(A) \cdot n(\varnothing) = 5 \cdot 0$.

17. *Hint*: The only set that has cardinality 0 is the empty set. Hence $A \times B$ must be the empty set. If both sets A and B contain elements, what must the cross product be?

19. Outline of proof: $a \cdot c = a \cdot c$; $a = b$ means that b is another name for a, so by renaming a in the first equation, $a \cdot c = b \cdot c$.

Exercise Set 4.3, p. 111

1. a) $12 \neq 13$; no b) $40 \neq 44$; no
c) If $a \in W$, then $3 \cdot a$ is also in W; if $b \in W$, then $2b$ is also in W. The sum of two elements of W is also an element of W.

3. b) $8 \neq 9$; not commutative **5.** No; Problem 4 is a counterexample. **6.** a)

```
00      00|000      3·3 = 9
00      00|000     15 − 9 = 6
00      00|000
 6     3·5 = 15
```

8. a) $1 = 1$, same **9.** a) $19 = d + 7$, hence $d = 12$ c) $11 = d + 11$, hence $d = 0$

11. Outline of argument: $a - c = a - c$ is true. If $a = b$, it means a can be renamed b, or b can be substituted for a. This gives the conclusion.

13. *Hint*: $a - b = k \leftrightarrow a = k + b$; add c to both of these equal numbers: $a + c = k + b + c$ and use the definition of subtraction to arrive at the conclusion.

15. a) $14 = c \cdot 7 \rightarrow c = 2$ c) $21 = c \cdot 3 \rightarrow c = 7$ **17.** a) $2 + 6 = 8$

Exercise Set 4.4, p. 115

1. a) $<$ c) $<$ e) $<$ **2.** a) $\{0, 1, 2\}$ b) $\{9, 10, 11, 12, \ldots\}$

3. a) Not reflexive, not symmetric, transitive c) Reflexive, not symmetric, transitive **5.** a) $13 < 14$

7. *Hint*: This follows immediately from Problem 6. If $a < b$, then $b > a$, and similarly for the conclusion.

9. Yes. *Hint*: $a < b \leftrightarrow a + k = b$, $c < d \leftrightarrow c + p = d$, where k and p are nonzero whole numbers. These equalities can be added and the definition for "less than" applied again to give the desired result.

Exercise Set 4.5, p. 126

1. a)

```
 200 +  80 +  6     Regroup and rearrange: (900 + 100) + 10 + 5
 700 +  20 +  9
 --------------
 900 + 100 + 15     Rename, then use place-value notation: 1000 + 10 + 5 = 1015
```

2. a)

```
        125
        276
        132
        ---
         13   [add ones]
        120   [add tens]
        400   [add hundreds]
        ---
          3   [add ones]
         30   [add tens]
        500   [add hundreds]
500 + 30 + 3 = 533 [place-value notation]
```

3.

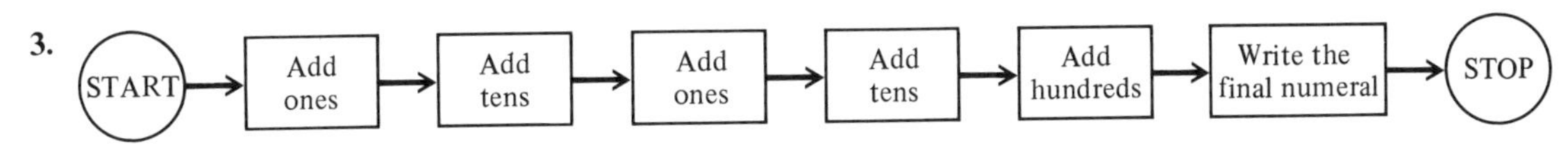

6. a)

```
 7621
 1973
 2059
 8722
-----
18265
2037
20 375
```

7. a)

```
700 + 80 + 6
          ×4
------------
  24 [4 × 6]
 320 [4 × 30]
2800 [4 × 700]
3144 [by addition]
```

8. a)

```
 539
  ×6
  54 [6 × 9]
 18  [6 × 30 = 180, 0 omitted]
30   [6 × 500 = 3000]
----
3234
```

11. a)

```
 359
  ×6
1804 [6 × 300 = 1800]
 35  [6 × 50 = 300]
----
2154 [6 × 9 = 54]
```

13. a)

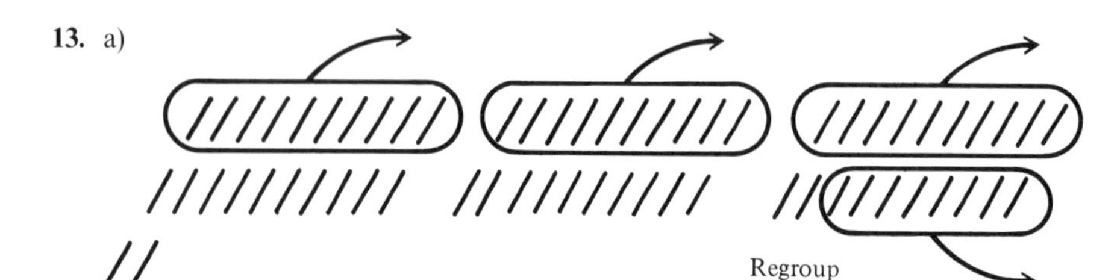

$62 - 38 = 24$

15. The chart depends upon the calculator. Follow your chart exactly and see if you get the correct answer.

17. a)

$$\begin{array}{r} \overrightarrow{6\ 0\ 3\ 2\ 2} \\ -5\ 9\ 9\ 9\ 9 \\ \hline \not{1}\ \not{1}\ \not{4}\ \not{3}\ 3 \\ 0\ 0\ 3\ 2\ \end{array}$$

Difference is 323

19. $Q = 43$, $R = 22$

21. a)

$$\begin{array}{r} 3 \\ 20 \\ 20 \\ \hline 35\overline{)1527} \\ 700 \\ \hline 827 \\ 700 \\ \hline 127 \\ 105 \\ \hline 22 \end{array}$$

$Q = 43$, $R = 22$

$1527 = 43 \cdot 35 + 22$

23. Essentially follow the flowchart of the text, except that the calculator is to be used at each stage possible. All operations on the calculator are confined to [+] and [−].

$$\begin{array}{r} 3 \\ 10 \\ \hline 279\overline{)3754} \\ 2790 \\ \hline 964 \\ 279 \\ \hline 685 \\ 279 \\ \hline 406 \\ 279 \\ \hline 127 \end{array}$$

If possible, use multiples of 10 since this multiplication can be done mentally.

$Q = 13$, $R = 127$

25. *Hint*: Look at the difference $9 - 3$ and compare it to the sum $9 + (9 - 3)$. What has been done to the difference? What effect does this have in the tens place, the hundreds, the total effect?

Exercise Set 4.6, p. 133

1.

+	0	1	2	3
0	0	1	2	3
1	1	2	3	10
2	2	3	10	11
3	3	10	11	12

·	0	1	2	3
0	0	0	0	0
1	0	1	2	3
2	0	2	10	12
3	0	3	12	21

2. a) 1020_{four} **3.** a) 2202_{four} **4.** a) $Q = 123_{\text{four}}$

5.

+	0	1	2	3	4	5
0	0	1	2	3	4	5
1	1	2	3	4	5	10
2	2	3	4	5	10	11
3	3	4	5	10	11	12
4	4	5	10	11	12	13
5	5	10	11	12	13	14

·	0	1	2	3	4	5
0	0	0	0	0	0	0
1	0	1	2	3	4	5
2	0	2	4	10	12	14
3	0	3	10	13	20	23
4	0	4	12	20	24	32
5	0	5	14	23	32	41

6. a) 441_{six} **7.** a) 15_{six} **8.** a) 3300_{six}

10. a) The base cannot be determined since 11011 can be a numeral of any base greater than 1 and the products of 0 and 1 are always the same in any base. c) Any base greater than 6.

11. a) Base seven c) Base five **13.** Base six

$1 \cdot 12$	12
$2 \cdot 12$	24
$3 \cdot 12$	40
$4 \cdot 12$	52
$5 \cdot 12$	104
$10 \cdot 12$	120
$100 \cdot 12$	1200
$20 \cdot 12$	240
$200 \cdot 12$	2400

$Q = 340$, $R = 5$ in base six.

15. *Hint*: The process of dividing by 2 has the effect of converting to base-two notation. Meanwhile the process of doubling in the other number has the effect of introducing powers of two and these mesh in the product.

Exercise Set 4.7, p. 138

1. a) 84 c) 200 **2.** a) 61 c) 62 **3.** a) 116 c) 24 **5.** Multiply by 10 and subtract the number from the product. **7.** Multiply by 100 and add the number to this product. **9.** Double the number and divide by 10. **11.** a) 90 000 c) 1500 **12.** a) 1600 c) 1500 **13.** a) 2100 c) 300 **14.** a) 40 c) 50

15. *Hints*: Calculate the number on the basis of an 8-hour day, a 5-day week, and a 52-week year.

17. $23 \cdot 23 = 529$, $22 \cdot 22 = 484$, $22.5 \cdot 22.5 = 506.25$, $22.3 \cdot 22.3 = 497.29$ **19.** Add 119 or 229

21. Estimate and think before using the calculator: $143 + 80 = 223$. **23.** Subtract 70, 71, 72, 73, or 75.

Exercise Set 4.8, p. 142

1. *Hints*: The product 3A must end in the same digit A; how many are there that have this property?

3. *Hints*: N + N + Y is a number whose final digit is Y, hence N is 0 or 5; E + E + T is a number whose final digit is T, hence E is also 0 or 5. Thus, no other letters can be either 0 or 5.

5. *Hint*: The letter A represents 1 since $7 \times 200\,000 = 1\,400\,000$ and the result here has only six digits. This means that C × 7 must have the final digit of 1. How does this fix the value of C?

7. *Hint*: $30 < 3x < 39$ since x can vary from 0 to 9. How many of these possibilities will give a product X1X? If there is only one such product, this gives the divisor.

9. *Hint*: Let x be the starting number; perform the manipulations. Alternatively let a blue chip be the unknown number and add white and blue chips as necessary to represent the stages of the problem.

11. *Hint*: The sought number must be of the form $8 \cdot k$, where k is a whole number. How can $8k$ be written in terms of a quotient and a remainder when 7 is the divisor?

15. *Ideas*: Make a sketch for each circumstance and label, find the time one way for each and compare, and so on.

17. Many solutions are possible; don't confine yourself to just one.

CHAPTER 5

Margin Exercises

1. a) 13 860 b) All products are 13 860 **2.** a) True, $p \cdot 1 = p$ b) False, there is no c such that $7 \cdot c = 37$

3. a) $7|35$ b) $4|3200$ c) $3|108$ **4.** a) $2|24 \cdot 3$ b) $7\nmid(70 + 23)$ c) $7|203$

5. a) Mark out: 4, 6, 8, 10, 12, . . . , 98, 100. b) Mark out: 6, 9, 12, 15, . . . , 96, 99.

6. a) Mark out: 10, 15, 20, 25, . . . , 95, 100. b) Mark out: 14, 21, 28, . . . , 91, 98.

7. b) Primes less than 100 are: 2, 3, 5, 7, 11, 13, 17, 19, 23, 29, 31, 37, 41, 43, 47, 53, 59, 61, 67, 71, 73, 79, 83, 89, 97.

8. $3 \cdot 3 \cdot 2 \cdot 2 \cdot 2$ **9.** a)

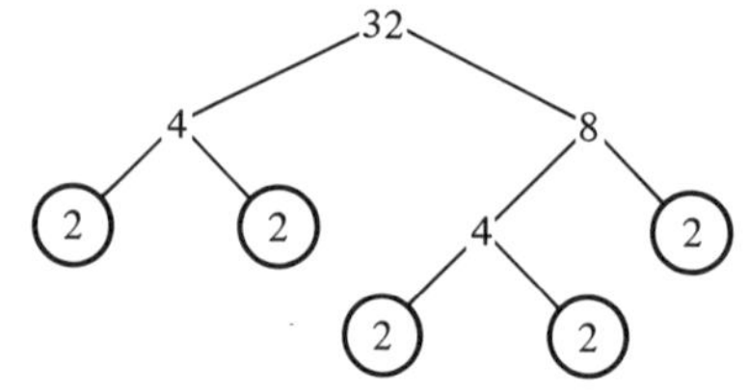

b)

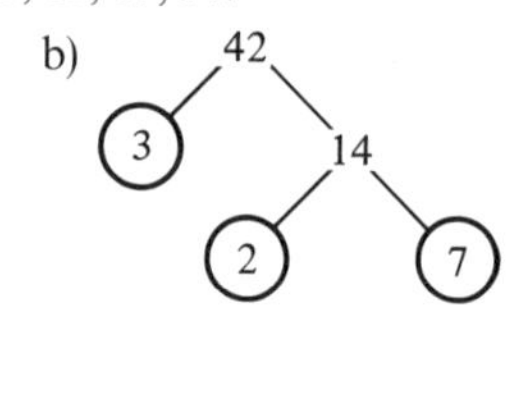

10. $5^2 \cdot 6^3 \cdot 7 \cdot 11 \neq 5^3 \cdot 6^3 \cdot 7 \cdot 11$ **11.** $175 = 1 \cdot 10^2 + 7 \cdot 10 + 5$
$2|10 \rightarrow 2|(1 \cdot 10^2 + 7 \cdot 10)$
$2\nmid 5 \rightarrow 2\nmid 175$

12. $4 \cdot 10^4 + 5 \cdot 10^3 + 6 \cdot 10^2 + 24$
$4|10^2 \rightarrow 4|(4 \cdot 10^4 + 5 \cdot 10^3 + 6 \cdot 10^2)$; $4|24 \rightarrow 4|45624$

13. a) $467 = (4 \cdot 99 + 6 \cdot 9) + 4 + 6 + 7$
$9|99$ and $9|9 \rightarrow 9|(4 \cdot 99 + 6 \cdot 9)$; $9\nmid(4 + 6 + 7) \rightarrow 9\nmid 467$

b) $669 = 6 \cdot (99 + 1) + 6 \cdot (9 + 1) + 9 = 6 \cdot 99 + 6 \cdot 9 + (6 + 6 + 9)$
$3|(6 \cdot 99 + 6 \cdot 9)$; $3|(6 + 6 + 9) \rightarrow 3|669$

14. a) Divisors of 25: $\{1, 5, 25\}$; divisors of 35: $\{1, 5, 7, 35\}$; GCD(25, 35) = 5 b) 1
c) Divisors of 17: $\{1, 17\}$; divisors of 0: $\{1, 2, 3, \ldots\}$; GCD(17, 0) = 17

15. 5 **16.** 15 **17.** a) Yes b) Yes c) No d) No **18.** 6 **19.** a) Yes b) Yes c) Yes d) No

20. a) $M_3 = \{0, 3, 6, 9, 12, 15, 18, \ldots\}$
b) $M_9 = \{0, 9, 18, 27, \ldots\}$
c) $M_9 \subset M_3$

21. a) 264, 288, 312, . . .
b) 288, 324, 360, . . .
c) 288

22. a) 72 b) 51 **23.** a) $2 \cdot 24$ b) $2 \cdot 37 + 1$ **24.** 15, 21, 28 **25.** The first 13 natural numbers.

26. Next three: 25, 36, 49. Add: 9, 11, 13 **27.** It is the sixth square number, hence 36.

28. $R_5 = 5 \cdot 6 = 30$, $R_6 = 6 \cdot 7 = 42$, $R_7 = 7 \cdot 8 = 56$ **29.** 171 **30.** a) $T_4^2 = 1^3 + 2^3 + 3^3 + 4^3 = 100$ b) 20

31. $2k$, $2k + 1$, $2k + 2$; total $6k + 3 = 3(2k + 1)$

32. a) Rewrite the sum as $2 \cdot (1 + 2 + 3 + \cdots + 100) = 2 \cdot [(1 + 2 + 3 + \cdots + 50) + (51 + 52 + \cdots + 100)]$
$= 2 \cdot [2(1 + 2 + \cdots + 50) + 50 \cdot 50]$

b) 10 100

Exercise Set 5.2, p. 152

1. a) F c) F e) F g) F **2.** a) 3, 4 c) 24, 36 e) 3, 9 g) 6, 12 **3.** a) 2, 7, 14 c) 6, 11, 3 e) 3, 57, 171

7. Outline of the argument: If $a|b$, then $b = p \cdot a$ by the definition of "divides." Similarly, if $b|c$, then $c = q \cdot b$. Therefore $c = q \cdot b$; replacing b by $p \cdot a$ we get $c = q \cdot (p \cdot a) = (q \cdot p) \cdot a$ and we see that a also divides c.

Exercise Set 5.3, p. 158

1. a) $3 \cdot 2 \cdot 2 \cdot 2$ c) $5 \cdot 2 \cdot 2 \cdot 2 \cdot 2 \cdot 3 \cdot 3$

2. a)

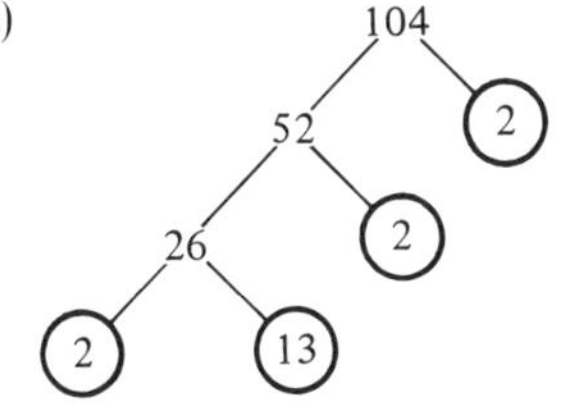

3. a) Three ways because there are three distinct ways to factor into two factors.

5. Primes found: 101, 103, 107, 109, 113, 127, 131, 137, 139, 149, 151, 157, 163, 167, 173, 179, 181, 191, 193, 197, 199

7.

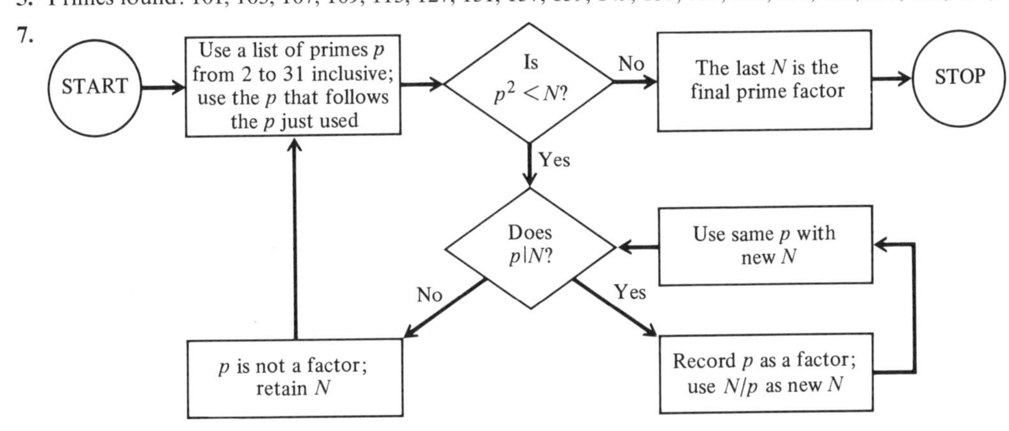

9. a) Prime b) $3 \cdot 67$ c) $11 \cdot 29$ d) Prime e) Prime f) $11 \cdot 37$ **12.** a) True c) False

15.

·	1	2	3	11
2	2	4	6	22
3	3	6	9	33
5	5	10	15	55
7	7	14	21	77

17.

·	1	4	5	13
1	1	4	5	13
3	3	12	15	39
2	2	8	10	26
7	7	28	35	91

19. *Hint*: Only x values from 1 to 22 need be considered. Use your table of primes to test those up to 200. Use other techniques for the rest.

Exercise Set 5.4, p. 162

1. Outline of the argument: Since $8 \mid 1000$ (because $1000 = 8 \cdot 125$), 8 will divide any power of 10 higher than 3. Therefore we need only worry about the number specified by the first three digits at the right of the numeral.

3. By 5 **5.** By 2 **7.** By 2, 3, 4, 5, and 10 **9.** By 2, 3, 4, 5, 9, 10 **11.** Yes

13. *Hint*: If $6 \mid p$, then $p = 6 \cdot q$, which is also $2 \cdot 3 \cdot q$ or $3 \cdot 2 \cdot q$, so both 2 and 3 divide this number. Now suppose a number is divisible by both 2 and 3; this means it can be written as $2 \cdot 3 \cdot k$, which makes it divisible by 6.

15. *Hints*: $10^2 = 100 = 99 + 1$, $10^4 = 1000 = 999 + 1$, and so on. $10^1 = 11 - 1$, $10^3 = 1000 = 1001 - 1$, and so on. Rewrite the place-value notation using these ideas.

16. For 2: a) Yes c) Yes e) No g) Yes

17. For 3: a) Yes c) No e) Yes g) No For 6: a) Yes c) No e) No g) No For 11: a) Yes c) No e) Yes g) Yes For 22: a) Yes g) Yes For 33: a) Yes e) Yes

19. Same test as for base ten, since 2 divides all powers of 4, and we need only look at the final-unit's place.

21. a) Write each power of 10 as $9 \cdot k + 1$, $10 = 9 + 1$, $100 = 99 + 1$, and so on. By rearranging and using the fact that 9 divides part of this sum, it is always possible to find the remainder.

c)
$4326 \to 6 \to 6 + 4 = 10 \to 1$
$8536 \to 13 \to 4$
$12862 \to 19 \to 10 \to 1$ (same)

d)
$8109 \to 18 \to 9$
$-7824 \to 21 \to 3$ ($9 - 3 = 6$)
$285 \to 15 \to 6$ (same)

e) *Hint*: Look at $(a \cdot 10 + b) \cdot (c \cdot 10 + d)$ as a start.

Exercise Set 5.5, p. 167

1. a) 3 c) 75 **2.** a) $\{1, 2, 4, 8, 16, 32\}$ c) $\{1\}$ e) $\{5\}$ **3.** a) 2 e) 3 e) 1 **5.** a) 420

7. *Hint*: In the first line of the table, $a \cdot b = 7560$, GCD$(a, b) = 6$, LCM $= 1260$; $7560 = 6 \cdot 1260$

9. The GCD of two primes is 1, the LCM of two primes is their product. **11.** The two numbers must be the same.

13. $M_7 \cap M_3 = M_{21} = \{21, 42, 63, \ldots\}$ **15.** a

Exercise Set 5.6, p. 173

1. c) Any even number is one that has a factor of 2, hence $2k$ and $2m$ are such numbers, their sum $2k + 2m = 2 \cdot (k + m)$ is also even. The odd numbers are $2k + 1$ and $2m + 1$; their sum $2k + 1 + 2m + 1 = 2(k + m + 1)$ is even.

3. No **3.** 650 **7.** 12656 **9.** 153 **11.** 36

13. $S_1 = T_0 + T_1$, $S_2 = T_1 + T_2$, $S_3 = T_2 + T_3$, etc.; in general, $S_k = T_{k-1} + T_k$

15. b) 1, 6, 15, 28, 45, 66, 91 c) $H_1 = P_1 + T_0$, $H_2 = P_2 + T_1$, etc.; in general, $H_k = P_k + T_{k-1}$

Exercise Set 5.7, p. 175

1. *Hint*: Either k or $k + 1$ is an even number. **3.** $R_n = 2 + 4 + 6 + \ldots + 2n$; $R_n = T_n + T_n = 2T_n$; $R_{100} = 100 \cdot 101 = 10\,100$

5. *Hints*: Consider five numbers in succession k, $k + 1$, $k + 2$, $k + 3$, $k + 4$. How many of these are divisible by 2? The fewest possible; the most possible? How many are divisible by 3? How many by 4, and so on.

7. a) $5 + 3 = 8$ c) $5 + 7 = 12$ e) $5 + 83 = 71 + 17 = 29 + 59 = 41 + 47 = 88$ **9.** What ideas do you get from Problem 5?

11. a) 2016 has 36 factors and their sum is 5536 (excluding 2016) **13.** a) Yes b) Triangular numbers

15. 267 250; divide by 5: 53 450; subtract 50×20: 52 450; add 6×52: 52 762 (done).

CHAPTER 6

Margin Exercises

1. ● ● ¦ ● ● ¦ 0 0 2/3
● ● ¦ ● ● ¦ 0 0

2. 3/8

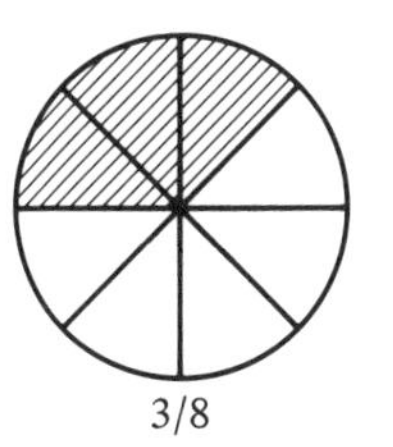

3/8

3. 8:12 or 2:3

4. 0 $\frac{4}{5}$ 1

5. $\frac{1}{3}$ $\frac{2}{6}$ $\frac{4}{12}$

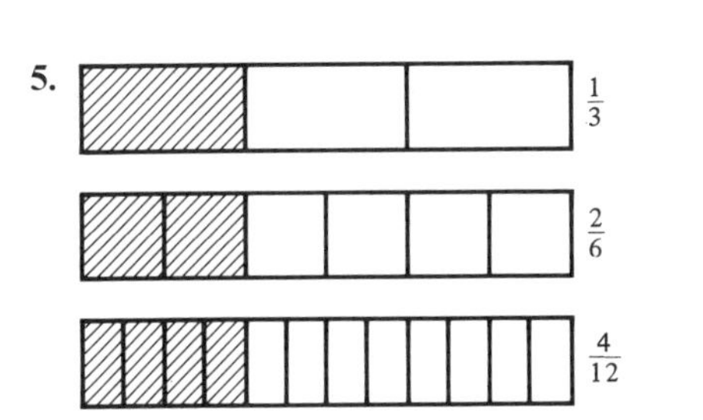

6. 0 $\frac{3}{4}$ 1

0 $\frac{6}{8}$ 1

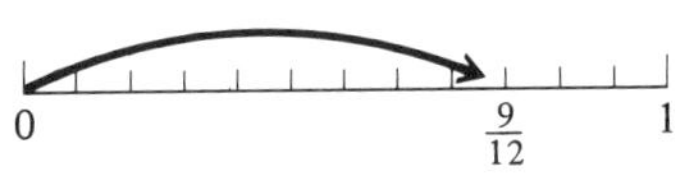

7. a) $5 \cdot 24 = 15 \cdot 8 = 120$ b) $7 \cdot 72 = 36 \cdot 14 = 504$ **8.** $\frac{12}{14}, \frac{18}{21}, \frac{24}{28}, \frac{30}{35}$ **9.** $\frac{2}{9}$ **10.** a) $\frac{30+40}{100} = \frac{70}{100}$ or $\frac{7}{10}$ b) $\frac{7}{15}$

11. a) $\frac{5}{7} = \frac{10}{14}$ b) $\frac{21}{18} = \frac{7}{6}$ **12.** $\frac{46}{35}$ **13.** $\frac{134}{210}$ or $\frac{67}{105}$ **14.** a) $\frac{7}{9} + \frac{1}{9} = \frac{8}{9}; \frac{2}{9} + \frac{6}{9} = \frac{8}{9}$ b) $\frac{6}{12} + \frac{1}{12} = \frac{7}{12}; \frac{4}{12} + \frac{3}{12} = \frac{7}{12}$ **15.** $\frac{2}{9}$

16. a) $\frac{9}{17} = \frac{7}{17} + \frac{2}{17}$ b) $\frac{2}{3} = \frac{1}{6} + \frac{3}{6}$ **17.** $\frac{1}{24}$ **18.** a) $\frac{43}{8}$ b) $2\frac{3}{4}, \frac{11}{4}; 3\frac{1}{2}, \frac{7}{2}$ **19.** $16\frac{2}{3}$ or $\frac{50}{3}$

20.

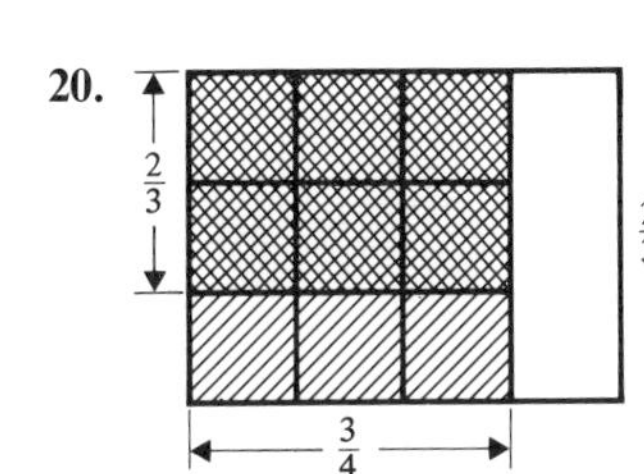

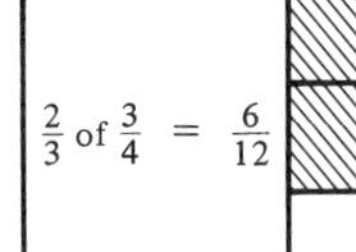

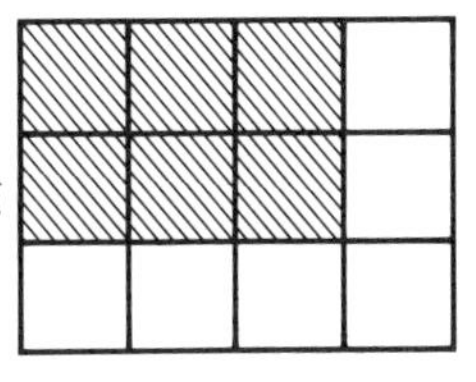

21. a) $\frac{24}{77} = \frac{8}{21} \cdot \frac{9}{11}$ b) $\frac{2}{3} \cdot \frac{36}{77} = \frac{72}{231} = \frac{24}{77}$ c) The same **22.** Both are $\frac{63}{55}$ **23.** $\frac{3}{35}$ **24.** a) $\frac{13}{8}$ b) $\frac{1}{5}$ c) $\frac{1}{a \cdot b}$ **25.** $x = 9$

26. $\frac{1}{5} \cdot \frac{5}{3} = \frac{1}{3} = \frac{1}{15} + \frac{4}{15}$ **27.** 36/15 **28.** $\dfrac{\frac{3}{4} \cdot \frac{8}{3}}{\frac{3}{8} \cdot \frac{8}{3}} = \dfrac{\frac{24}{12}}{\frac{24}{24}} = 2$ **29.** $\frac{63}{99} > \frac{55}{99}$, hence $\frac{7}{11} > \frac{5}{9}$ **30.** a) 5/12 b) 8 c) None

31. a) $\{1, 1/2, 1/4, 1/8, \ldots, 1/2^{n-1}, \ldots\}$ b) 1/512 c) No

32. Suzanne and Charles are going to sell candy bars.
First customer bought one-half of their stock and $\frac{1}{2}$ bar more;
second customer bought one-half of the remaining stock and $\frac{1}{2}$ bar more;
third customer bought one-half of the remaining stock and $\frac{1}{2}$ bar more.
Suzanne and Charles ate 1 bar.
After all this, they had 8 bars left to sell.

33. Did you think of a diagram? Representing the bars with physical models? Trying a smaller problem? Trying to find a number that satisfies any *one* of the transactions?

34. Start with 79.
First got $\frac{1}{2}$ of 79 + $\frac{1}{2}$, or 40; there are 39 left.
Second got $\frac{1}{2}$ of 39 + $\frac{1}{2}$, or 20; there are 19 left.
Third got $\frac{1}{2}$ of 19 + $\frac{1}{2}$, or 10; there are 9 left.
They ate 1, so 8 are left.

35. Facts: one jar holds 7 liters (exactly); one jar holds 3 liters (exactly); jars are not marked; there is an infinite supply of water.

36. Problem: How can one measure out exactly 5 liters of water? How can one use the 7 and 3 measures to get 5 liters? etc.

37. Solution: Fill the 7-liter jar; pour out 3 liters into the 3-liter jar and dump this; pour out three more into the 3-liter jar and dump this also. Now there is one liter left in the 7-liter jar; put it in the now empty 3-liter jar. Fill the 7-liter jar, put 2 liters of it into the 3-liter jar that still has 1 liter in it. There are now 5 liters in the 7-liter jar.

Exercise Set 6.1, p. 185

1. Cannot partion 12 in 11, 8, or 9 parts

● ● | 0 0 | 0 0
● ● | 0 0 | 0 0

1/3

● ● ● | 0 0 0
0 0 0 | 0 0 0

1/4

3.

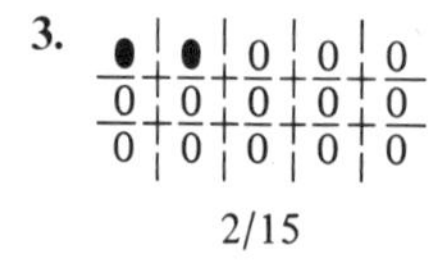

2/15

5. a)

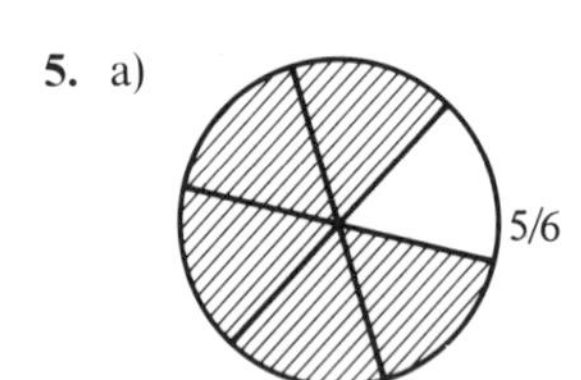

6. a)

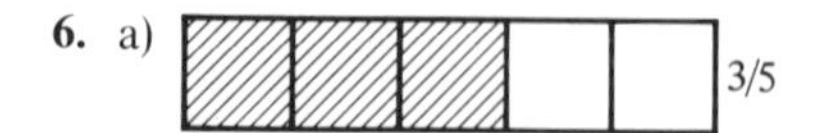

7.

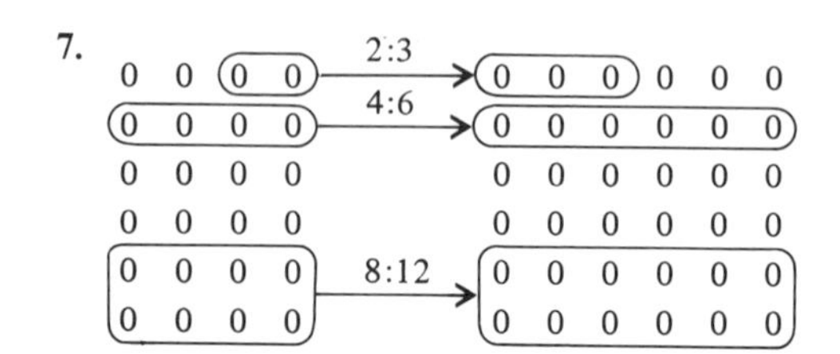

11. a)

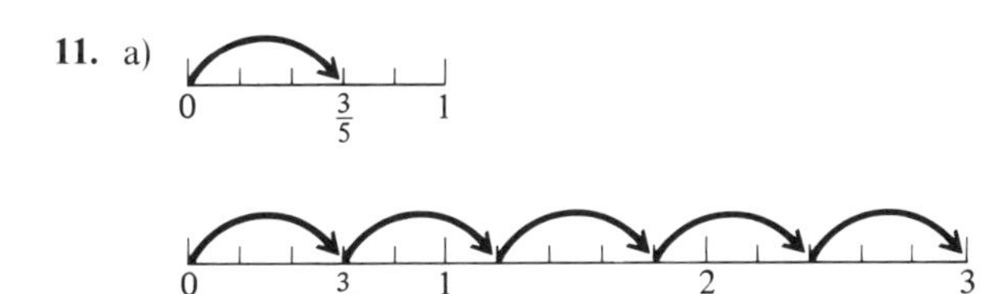

13. $\dfrac{\text{Number of steps taken by } A}{\text{Number of steps taken by } B} = \dfrac{20}{24}$

Exercise Set 6.2, p. 191

1.

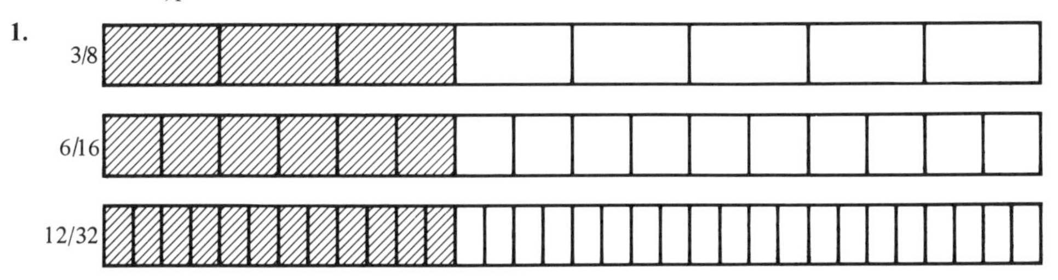

3. a) $\frac{6}{22}, \frac{9}{33}, \frac{12}{44}, \frac{15}{55}, \frac{18}{66}$ c) $\frac{4}{26}, \frac{6}{39}, \frac{8}{52}, \frac{10}{65}, \frac{12}{78}$ e) $\frac{0}{2}, \frac{0}{3}, \frac{0}{4}, \frac{0}{6}, \frac{0}{7}$ g) $\frac{4}{1}, \frac{8}{2}, \frac{12}{3}, \frac{20}{5}, \frac{24}{6}$ **5.** 9/45 **7.** 50/99

8. a)

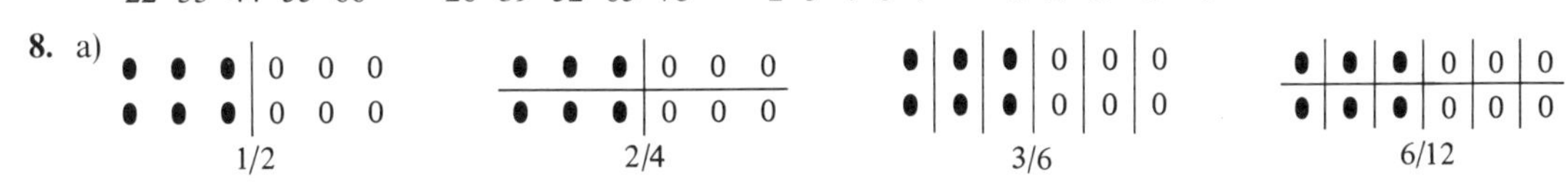

9. 1:2, 2:4, 6:12

11. a) No **13.** No, 15/16 is a lowest-terms fraction, yet neither is a prime.

Exercise Set 6.3, p. 198

1. a) $\frac{10}{10}$ c) $\frac{126}{64}$ **2.** a) $\frac{7}{8}$ c) $\frac{11}{16}$ **3.** a) $\frac{139}{210}$ **4.** a) $\frac{357}{630}$ or $\frac{17}{30}$ **5.** a) Both are $\frac{59}{40}$

6. a) $\frac{3}{15} + \frac{9}{15} = \frac{5}{15} + \frac{7}{15} = \frac{12}{15}$ **7.** a) $\frac{11}{3}$ c) $\frac{209}{12}$ **9.** $38\frac{2}{11}$

11. Outline of proof: By definition of addition, $\frac{a}{b} + \frac{c}{d} = \frac{a \cdot d + b \cdot c}{b \cdot d}$. Similarly, $\frac{c}{d} + \frac{a}{b} = \frac{c \cdot b + d \cdot a}{d \cdot b}$. By the properties of whole numbers, $c \cdot b = b \cdot c$, $a \cdot d = d \cdot a$, and $b \cdot d = d \cdot b$; hence these are actually the same rational number.

13. Outline of argument: Any number $a > b$ can be expressed as $a = b \cdot Q + R$, where $R < b$. Thus, $\frac{a}{b} = \frac{b \cdot Q + R}{b}$. By the definition of addition of rational numbers, we can write $\frac{b \cdot Q + R}{b}$ as $\frac{b \cdot Q}{b} + \frac{R}{b}$ or else we can write $\frac{b \cdot Q}{b}$ as Q, so the result is $Q + \frac{R}{b}$.

15. 9 **17.** $\frac{17}{24}$

Exercise Set 6.4, p. 207

1. a) 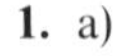

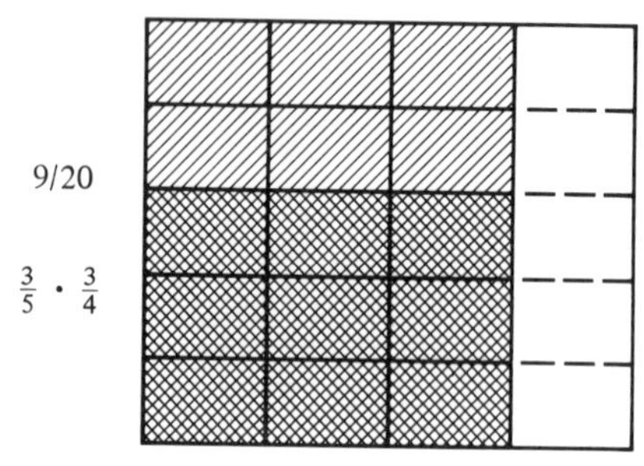

b)

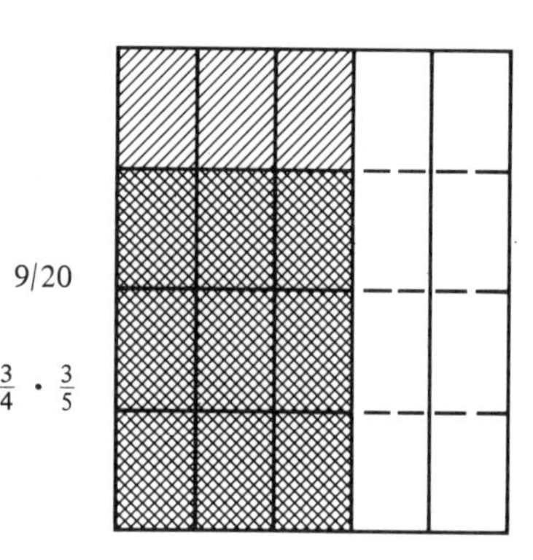

3. Outline of argument: a is a whole number, c is a whole number, and $a \cdot c$ is a whole number; similarly for b, d, and $b \cdot d$. Thus the result $\frac{a \cdot c}{b \cdot d}$ satisfies the requirements for a rational number since we also know that $b \neq 0$ and $d \neq 0$.

5. Outline: We must show that $\frac{a}{b} \cdot \frac{c}{d} = \frac{a \cdot c}{b \cdot d}$ is the same as $\frac{c}{d} \cdot \frac{a}{b} = \frac{c \cdot a}{d \cdot b}$. Whole-number properties tell us that $a \cdot c = c \cdot a$ and $b \cdot d = d \cdot b$; thus these are in fact identical.

7. a) $\frac{7}{6}$ **9.** a) $\frac{5}{8} \cdot \frac{2}{7} = \frac{5}{28} = \frac{15}{56} - \frac{5}{56}$ **11.** a) 10 **13.** a) Both are $\frac{4}{11}$ **15.** $3\frac{2}{3}$ **17.** 6

Exercise Set 6.5, p. 211

1. a) 5/8 **2.** a) 31/34

3. Outline of argument: $\frac{a}{b} < \frac{c}{d} \leftrightarrow \frac{a}{b} + k = \frac{c}{d}$, where k is not zero but some nonnegative rational number. Now we multiply both of these equal numbers by $\frac{1}{2}$ and get $\frac{1 \cdot a}{2 \cdot b} + \frac{1 \cdot k}{2} = \frac{1 \cdot c}{2 \cdot d}$ and then add $\frac{1 \cdot a}{2 \cdot b}$ to both of these numbers in the equality. This gives: $\frac{1 \cdot a}{2 \cdot b} + \frac{1 \cdot a}{2 \cdot b} + \frac{1 \cdot k}{2} = \frac{1 \cdot a}{2 \cdot b} + \frac{1 \cdot c}{2 \cdot d}$. When this is simplified, we have $\frac{a}{b} + \frac{1 \cdot k}{2} = \frac{1}{2}\left(\frac{a}{b} + \frac{c}{d}\right)$. Since the last term is the average of $\frac{a}{b}$ and $\frac{c}{d}$, we can see that it is actually larger than $\frac{a}{b}$. A similar argument holds for showing that $\frac{c}{d}$ is greater than the average.

5. $\frac{97}{144}, \frac{98}{144}, \frac{99}{144}, \frac{100}{144}$; many others are possible. **7.** $\frac{127}{90}, \frac{128}{90}, \frac{129}{90}, \ldots, \frac{146}{90}$; many others are possible.

9. a) 1/2 b) 3/4 c) nth term: $\frac{2^{n-1} - 1}{2^{n-1}}$ **11.** 3/14

Exercise Set 6.6, p. 214

1. Ann had 64. **3.** 2 gallons **5.** *Hint*:

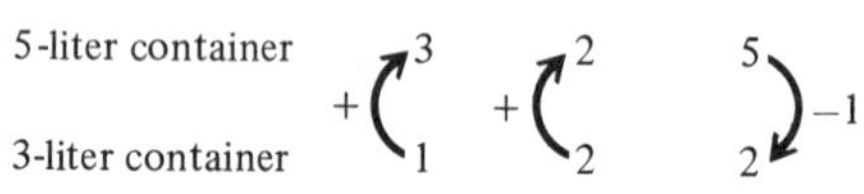

7. 4

9. *Hint*: Finding a digital root is very similar to casting out nines. Any number can be written as $9 \cdot n + k$, where n is a whole number and k is any whole number from 0 to 9. What does this mean when the numbers are squared?

CHAPTER 7

Margin Exercises

1.

Ten-thousandths	Hundred-thousandths
$\frac{1}{10^4}$ or $\frac{1}{10\,000}$	$\frac{1}{10^5}$ or $\frac{1}{100\,000}$

2. 0.627 **3.** 0.225 **4.** a) 6/100 b) 56/1000 c) 24375/100000 **5.** a) 0.06 b) 0.056 c) 0.24375

6. (a) and (d) are terminating decimals **7.** a) 375/10 b) 23/100 c) 12/10000 d) 65315/1000

8. a) $27\frac{6}{10}$ b) $\frac{37}{100}$ c) $\frac{12}{1000}$ d) $\frac{26}{10000}$ **9.** 333.69 **10.** $\frac{841 + 1737 + 1294}{100} = \frac{3872}{100} = 38.72$ **11.** 268.27 **12.** 582.342

13. a) 3.375 b) 63.9845 c) 23.694715 **14.** a) 5.145 b) 65.625 c) 208.6875 **15.** 0.1375 **16.** $\dfrac{\frac{43054}{1000}}{\frac{412}{100}} = \frac{43054}{4120}$

17. a) 0.5125 b) 0.365 c) 12.13 **18.** $2.6\overline{363}$ **19.** 12 **20.** a) 2341/9999 b) 69/11 **21.** 2681/495 **22.** a) 2 b) 4

23. a) $\left.\begin{array}{l} 0.6666666667|0000 \\ 0.6666666666|6666 \end{array}\right\}$ differ by less than 0.0000000001 b) 2.001, 2.0001, 2.00001, and so on **25.** a) 0.0735 b) 1.27 c) 0.005

26. a) 243% b) 71% c) 3.2% d) 75% e) 625% **27.** a) 0.0025 b) $\frac{1}{3}$ c) 6.25% d) 335% **28.** a) 48 b) 6

29. Facts: The team played 19 games; the team won 14 games

30. Write a fraction that describes the ratio of those that had never purchased to the total number of people. 27 out of 37 is how many out of 100?

31. A nurse needs to give a 2500 unit dose; 1 cc has 5000 units of the drug

32. a) The number of cc multiplied by 5000 gives the total amount of drug: $x \cdot 5000 = 2500$

b) $2500 = x\% \cdot 5000$; $\frac{2500}{5000} = x(0.01)$

In either case the amount of the drug is 0.5 cc.

Exercise Set 7.1, p. 224

1. a) 12.9 c) 0.26 e) 0.0379 **2.** a) 0.45 c) 0.065 e) 0.088 g) 0.1575

3. Outline of argument: Assume that 5/6 can be expressed as a terminating decimal $c/10^k$. This means that $5 \cdot 10^k = c \cdot 6$. Now, 3 must divide $5 \cdot 10^k$, since it divides $c \cdot 6$. But we know that 3 does not divide 5, it does not divide 10, and does not divide 10^k; hence we have a contradiction and the original assumption (that there is a terminating decimal) must be false.

6. There are 15. **9.** a) 0.517 b) 500.17 **10.** a) 368/1000 c) 4127/1000 e) 90165/1000

11.

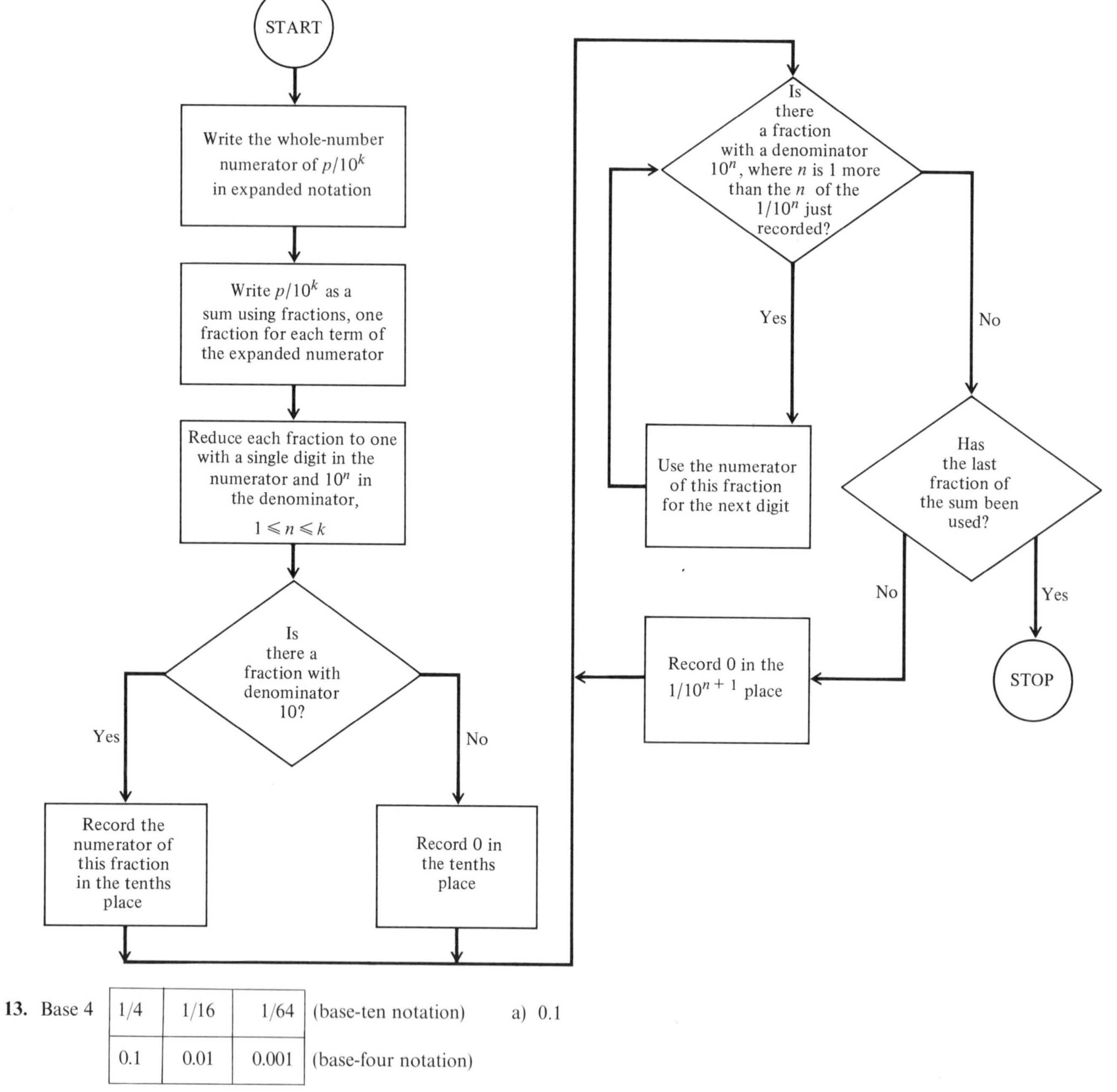

13. Base 4

1/4	1/16	1/64	(base-ten notation)
0.1	0.01	0.001	(base-four notation)

a) 0.1

15. a) 11/64 c) 14/125

Exercise Set 7.2, p. 234

1. a) $\frac{1612}{100} + \frac{753}{100} = 23.65$

2. a)
$$\begin{array}{r} 37.45 \\ +\ 6.37 \\ \hline 12 \\ 70 \\ 1300 \\ 3000 \\ \hline 43.82 \end{array}$$

3.

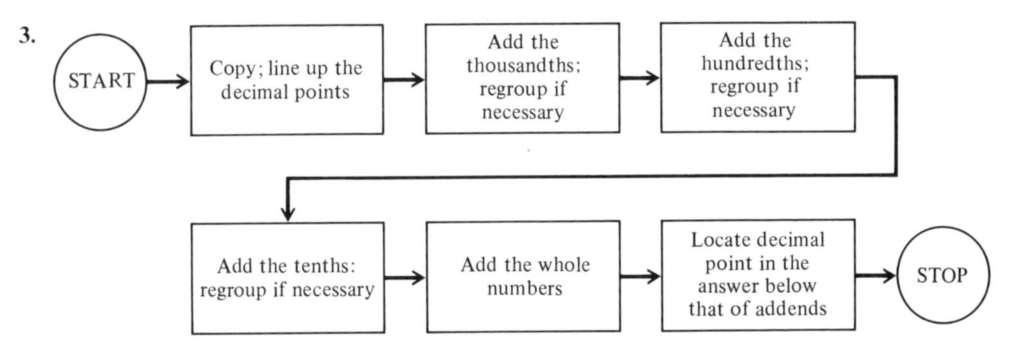

(This can be extended to include ten thousandths, etc.)

5. a) 534/100 = 5.34 c) 1798/100 = 17.98 **9.** a) 227.513 b) 50.0004 **11.** a) 0.45 c) 0.065 **12.** a) 5.13 c) 6
13. a) 513 c) 4.21 **15.** \$0.71 per person

Exercise Set 7.3, p. 239

1. a) 5/9 b) 7/9 **2.** a) 6451/990 b) 22919/1665 **3.** 0.5130000 . . .

5.

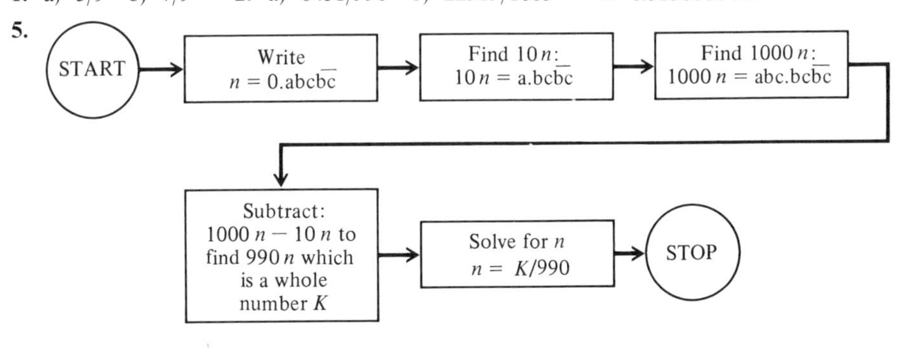

7. $1/3 \neq 33/100$ since $100 \neq 99$

9. a) Not closed under addition, 1/3 and 2/3 are not terminating decimals, yet 1/3 + 2/3 = 1 is represented by a terminating decimal.

c) No, $\frac{7}{3} \cdot \frac{3}{7} = 1$; the product of two nonterminating decimals is a terminating decimal.

13. $1/13 = 0.\overline{076923}$, $2/13 = 0.\overline{153846}$, $3/13 = 0.\overline{230769}$, $4/13 = 0.\overline{307692}$, $5/13 = 0.\overline{384615}$, $6/13 = 0.\overline{461538}$, $7/13 = 0.\overline{538461}$, $8/13 = 0.\overline{615384}$, $9/13 = 0.\overline{692307}$, $10/13 = 0.\overline{769230}$, $11/13 = 0.\overline{846153}$, $12/13 = 0.\overline{923076}$

15. a) $3333/10000 \neq 1/3$ since $9999 \neq 10000$ b) $33333/100000 \neq 1/3$ since $99999 \neq 100\,000$

Exercise Set 7.4, p. 243

1. Use the intermediate algorithm, add the hundredths and tenths to get 2.12. Add the units and tens columns to get 217. Now add the hundredths and tenths, then the ones and tens, then the hundreds. Answer is 219.12.

3. Treat as a whole-number problem and find partial products with the calculator: 5×75, 4×75 and annex a 0, 2×75 and annex appropriate zeros, 1×75 and annex appropriate zeros. Now add by considering two columns at a time. Finally locate the decimal point in 9337.5.

5. Treat the problem as the usual long-division algorithm, except that the calculator can be used to find many products and differences.

```
     0.0340909
88)3.0000
     2 64      [3 × 88 on calculator]
       360     [difference on calculator]
       352     [4 × 88 on calculator, and so on]
```

7. 0.0434782608695652173913 is the repeating cycle **9.** 0.2452830188679 is the repeating cycle **11.** a) 0.02 b) 0.15

13. a) 0.2222 b) 0.6825 **15.** $1/11 = 0.0909\overline{09}$, $2(1/11) = 2(0.09)$, and so on. **17.** a) $0.0\overline{66}$ b) $0.013\overline{3}$ c) $0.002\overline{6}$

19. a) $0.\overline{09}$ b) $0.81\overline{81}$ c) $0.\overline{153}$

Exercise Set 7.5, p. 247

1. a) 0.75 is 75/100, so we already have a number compared to 100, which is what is meant by percent, hence this is accomplished automatically by moving the decimal point.

2. a) 0.5, 50% c) 2/3, $0.6\overline{6}$ e) 0.75, 75% g) 2/5, 40% i) 4/5, 80% k) 3/8, 0.375 m) 0.875, 87.5%

5. Many such patterns can be found. Here is one: multiply 0.367 in turn by 3, 6, 9, 12, and so on.

7. Better to have 30% all at once. **9.** Won 62.8%, lost 30.2%

Exercise Set 7.6, p. 251

1. 11 **3.** 31, $83\frac{1}{3}\%$ **5.** 1700 **7.** Protein: 16.67%, 72.22%

CHAPTER 8

Margin Exercises

2. a) $^{-}5$ b) 7 c) $^{-}18$ **3.** a) $^{-}6$ b) 10 c) $x + 1$

4. a)

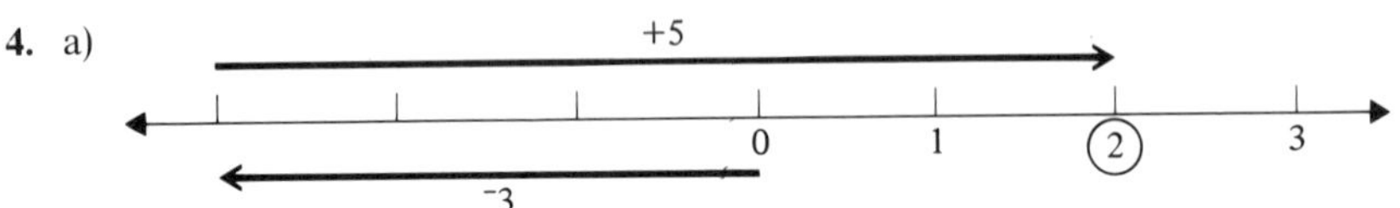

b)

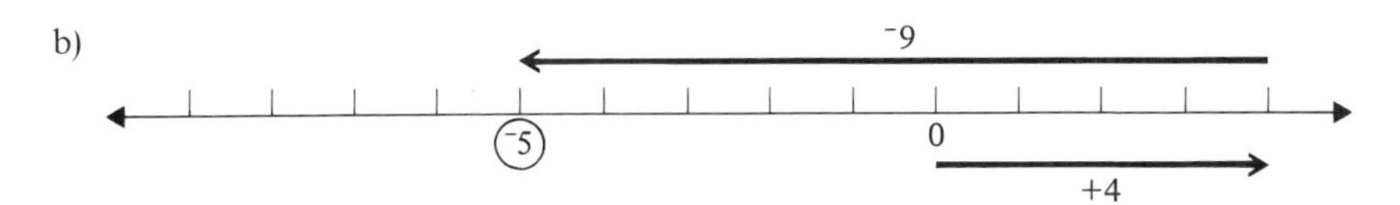

c)

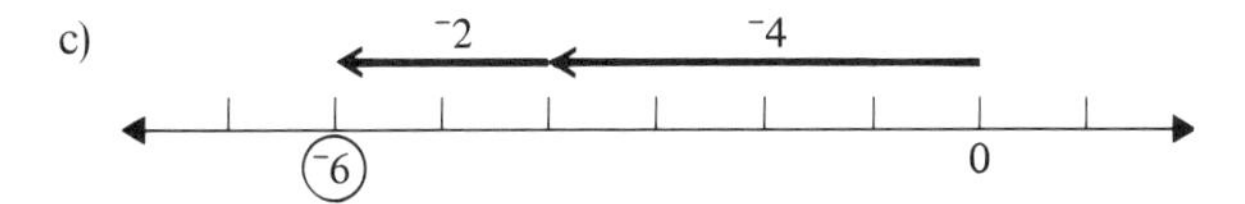

5. a) $^-4 + (4 + 3) = (^-4 + 4) + 3 = 3$ b) $(^-3 + {}^-5) + 5 = {}^-3 + (^-5 + 5) = {}^-3$ **6.** a) 13 b) 3 c) $^-5$ d) $^-9$

7. a) $^-6 + (^-4 + {}^-8) = {}^-18$ b) $7 + (^-3 + 10) = 7 + 7 = 14$ **8.** a) 8, 17, 0 b) 6, 1, 3 **9.** a) $^-8$ b) 11

10. a) $x = 6$ b) $x = {}^-2$ c) $x = {}^-9$ **11.** a) $15 + 7 = 22$ b) $^-12 + {}^-4 = {}^-16$ c) $^-7 + 8 = 1$ **12.** 6, 12, 18

13. a) $5(^-3) + 15 = 0 \rightarrow 5(^-3) = {}^-15$ b) $^-3 \cdot (^-2 + 2) = 0 \rightarrow {}^-3 \cdot {}^-2 = 6$ **14.** a) $^-36$ b) 132 c) $^-13$

15. a) Both $^-72$ b) Both 66 c) Both $^-40$ **16.** a) 7 b) $^-6$ c) $^-2$ d) 0

17. a) $^-16 + k = {}^-5, k = 11$ b) $^-30, {}^-17, {}^-2, 0, 1, 2, 17, 45$ **18.** Same as 5, and ^-k has an additive inverse k.

19. a) is less than b) equals c) is greater than

20. 1. Each integer has an additive inverse.
2. Associative (+) for integers.
3. Definition of additive inverse, 0 property for (+), renaming $^-30$.
4. Unique factorization.

21. $^-3, {}^-2, {}^-1, 0, 1, 2, 3$ **22.** a) $^-6 \cdot 12 = 8 \cdot {}^-9 = {}^-72$ b) $6 \cdot {}^-16 = {}^-8 \cdot 12 = {}^-96$ **23.** a) $^-1/2$ b) 1/4 c) $^-1/2$

24. a) $^-5/11$ b) $^-5/12$ c) 1/40 **25.** $^-2/3$, 2/55, 2/9, 2/9 **26.** a) 38/35 b) $^-5/6$ **27.** a) $^-9/8$ b) $^-9/8$ c) $^-3/5$ d) $^-7/8$

28. a) 1 b) 1 c) 81 **29.** a) $^-7/12$ is greater b) $^-5/45$ is their average, many others are possible. **30.** 0.23223222322223 . . .

31. 1^2 is 1 and is too small, 2^2 is 4 and hence too large. **32.** a) 1.4140, 1.4142 b) 1.99996164

33. a) 3.1397, 3.1427 b) 3.141592654 (nine decimal places.)

34. a) 15 794 959 684 (Remember that your calculator can be used to do calculations that do not fit on the screen.)
b) If you have a scientific calculator it will convert to this notation. c) 1.579 495 968 10

35. Some calculators cannot accept such large numbers, they may or may not show error signals.

36. $1 = \frac{10^5}{10^5}$, 8.415×10^5 **37.** $1 = \frac{10^3}{10^3}$, 8.952×10^{-3} **38.** 2.356×10^6 **39.** 3.296×10^{11}

40.

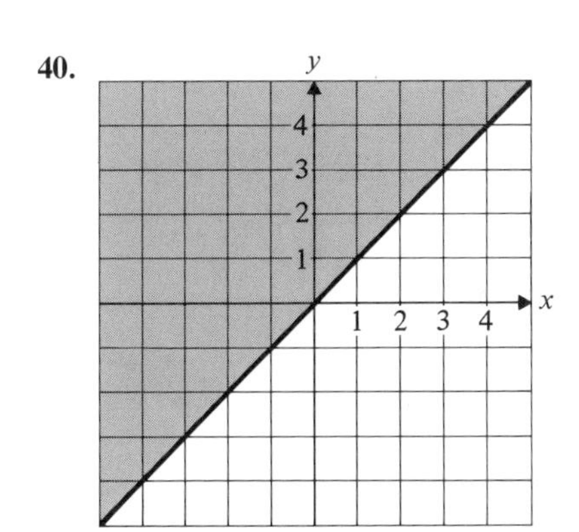

41. $\sqrt{4}$, $\sqrt{9}$ are exact; the others are irrational, hence approximate values. **42.** a) $^{-}1$ b) 8 c) $^{-}7$

43.

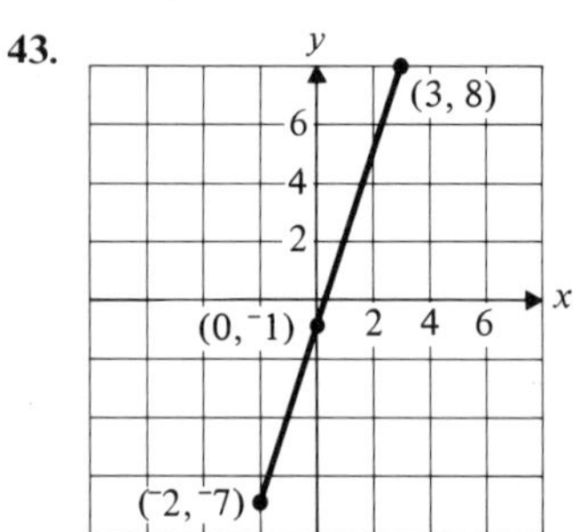

44. a) $y = \pm\sqrt{5}$ b) $y = \pm\sqrt{8}$ c) $^{-}3 \leqslant x \leqslant 3$, $^{-}3 \leqslant y \leqslant 3$
d) Domain: the real numbers from $^{-}3$ to 3 inclusive. Range: the real numbers from $^{-}3$ to 3 inclusive.

45. a) Yes b) No c) Yes d) No

46. a) There are four people, there are pieces of fruit to share, one person gets 1/3 of the fruit, one gets 1/5 of the fruit, third person gets 6 pieces, last person gets 1 piece.
b) How many pieces of fruit are needed? Assume that fruit is not cut.

47. $\frac{1}{3}x$ and $\frac{1}{5}x$ are rational numbers and hence each has an additive inverse; $x = 15$.

Exercise Set 8.1, p. 262

1. 0, since it satisfies the definition $0 + 0 = 0$.

3. Outline of argument: Suppose that a is a whole number, then part (1) of the definition of addition gives the result; suppose that a is a negative integer, then $a = {}^{-}b$, where b is now a whole number, and the result is now given by part (3) of the definition of addition.

5. If a is $^{-}6$, then $^{-}(^{-}6) = 6$. **6.** Both are $^{-}9$, both are $^{-}8$, both are 1, both are $^{-}24$, both are $^{-}5$. **8.** a) $5 - (3 - 2) \neq (5 - 3) - 2$

10. a) $^{-}8 + {}^{-}3 = {}^{-}11$ c) $^{-}12 + 2 = {}^{-}10$ e) $15 + 3 = 18$ **12.** a) $^{-}30$

14. a)

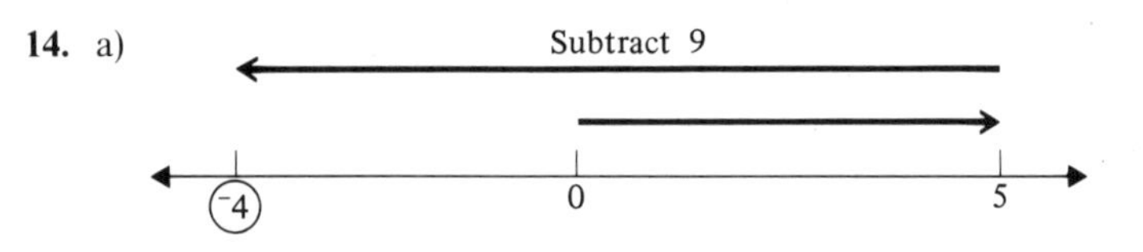

c)

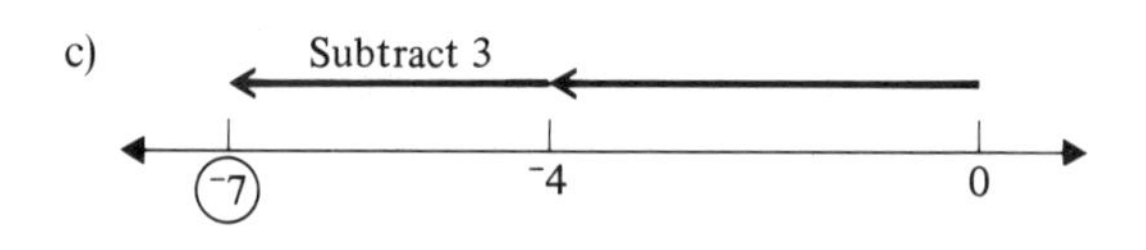

e)

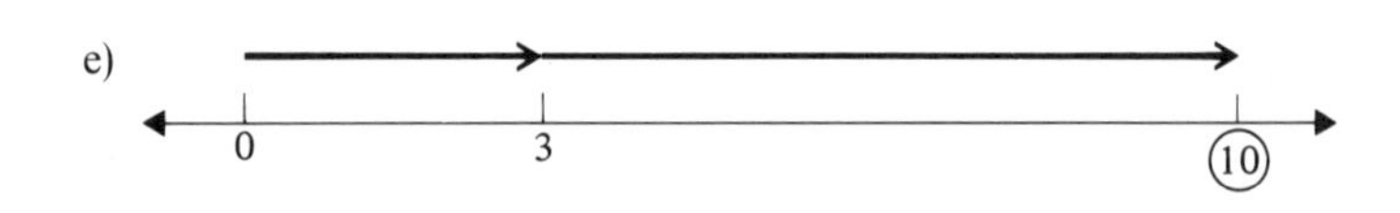

g)

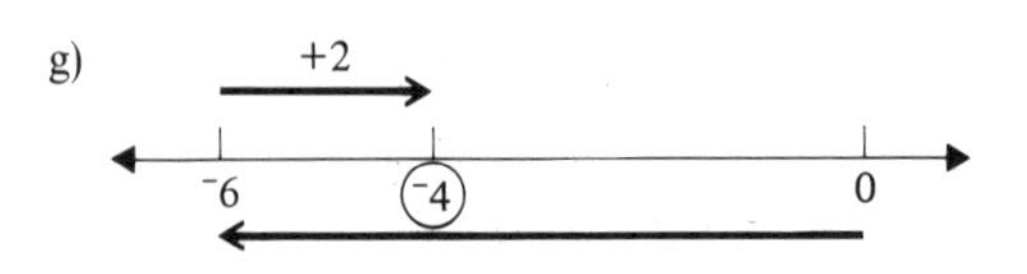

15. a) Commutative (+) for integers c) Associative (+) for integers e) Definition of addition for integers g F

18. a) Both are 2 c) Both are 6

21. a) (W)(W)(W) + (W)(W) ⟶ 5

c) (W)(W)(W) + (R)(R)(R) (R)(R)(R) ⟶ (R)(R)(R) = $^{-}3$

22. c)

```
 R   R   R   R   R
 R   R   R   R   R    Hence ⁻5
 R   R   R   R   R
 R   R   R   R   R
```

e)

```
W  W  W  W  W       R   W       R   W
W  W  W  W  W   +   R   W   +   R   W    Hence 20
W  W  W  W          R   W       R   W
```

Exercise Set 8.2, p. 265

1. Outline of argument: If a is a whole number, then $a \cdot 0 = 0$ by whole-number multiplication. If a is a negative integer, then there is a whole number b such that $a = {}^{-}b$. So the product we want is $^{-}(b) \cdot 0$. By the definition for multiplication this is $^{-}(b \cdot 0)$ or $^{-}(0) = 0$.

3. a) Both $^{-}25$ c) Both 18 e) Both 40 **5.** $-, \div$ **7.** $\times, -$ **9.** $\times, -, +, \times, +$ **11.** $+, \div, \div, +$

14. a) Both columns are the same in each row; reading down: 6, $^{-}14$, 28, 8, 22.

16. Both columns read the same in each row; reading down 2, 10, $^{-}12$. **22.** a) Yes b) No c) No d) No

23. a) 21 c) $^{-}2$ e) $^{-}4$ g) 18

29.

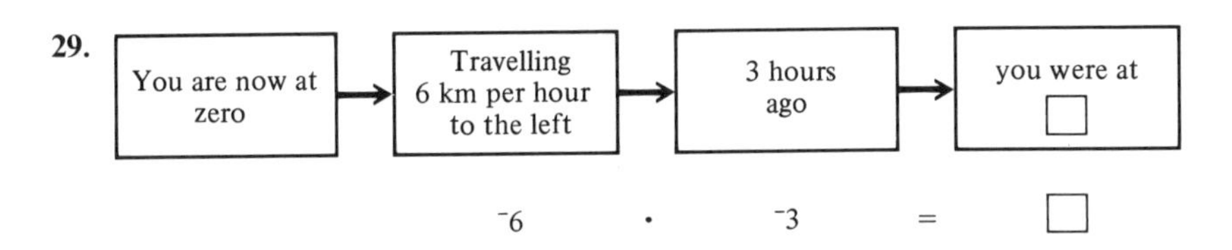

Exercise Set 8.3, p. 272

1. a) $5 + 0 < 13 + 0$ d) $0 < 17$

3. Yes. The argument is identical to that for whole numbers. Use the definition of "less than" to write an equality for each inequality $a < b$ and $c < d$. Then add the equations and use the definition for "less than" again.

4. The first three lines: $5 < 8$ and $^{-}15 > {}^{-}24$; $^{-}3 < 4$ and $6 > {}^{-}8$; $^{-}7 < {}^{-}1$ and $28 > 4$.

7. a) $\{^{-}6, {}^{-}5, {}^{-}4, \ldots 3, 4, 5, 6\}$ b) $\{^{-}2, {}^{-}1, 0, 1, 2, 3, 4, 5, 6\}$ **8.** a) Any x that is greater than 5 or less than 3 will work.

9. a) Any x that is greater than 7 or less than $^{-}8$ will work. **10.** a) Any x such that $^{-}3 < x < {}^{-}6$ **11.** a) Any x such that $1 < x < 2$.

Exercise Set 8.4, p. 280

1. a) $\frac{3}{16}$ (lowest terms), $\frac{6}{32}, \frac{9}{48}, \frac{12}{64}$ c) $\frac{^{-}2}{3}$ (lowest terms), $\frac{^{-}4}{6}, \frac{^{-}6}{9}, \frac{8}{^{-}12}$ **4.** a) $^{-}3/16$, $16/3$ c) $3/8$, $^{-}8/3$ e) $^{-}17$, $1/17$

5. a) F b) T c) T d) F e) T f) F g) F **12.** $\frac{^{-}1}{4}, 0, \frac{1}{4}$; many others **13.** $\frac{3}{4} - \left(\frac{1}{4} - \frac{5}{4}\right) \neq \left(\frac{3}{4} - \frac{1}{4}\right) - \frac{5}{4}$ **16.** a) 18

18. a) $\left\{x \mid x \text{ is rational and } x < \frac{11}{8}\right\}$ c) $\left\{x \mid x \text{ is rational and } x < \frac{^{-}9}{8}\right\}$

Exercise Set 8.5, p. 288

1. This follows the argument in the text. Assume that $\sqrt{3}$ is rational and represented by a/b in lowest terms; then follow the same steps.

3. Assume the $\sqrt{4} = a/b$, as before; this time $4 \cdot b^2 = a^2$ and both sides have an even number of factors so it is impossible to arrive at a contradiction.

5. a) $\sqrt{3} \approx 1.73205$ c) $\sqrt{23} \approx 4.79583$

6. [1, 2] [1.5, 2.0] [1.714, 1.750] [1.7320, 1.7321]

13. No for (+), since $^{-}\sqrt{3} + \sqrt{3} = 0$; no for (×), since $\sqrt{3} \cdot \sqrt{3} = 3$. **15.** a) 6 300 000 c) 0.0000725

16. a) 6.14×10^{7} c) 8.95×10^{-3} **17.** 1.323×10^{3} **19.** 1.530516×10^{3}

Exercise Set 8.6, p. 293

1. a) A rectangle b) A right triangle c) A parallelogram d) A hexagon

3. $\{(x, y) \mid x$ and y are real numbers and $x < y + 1\}$ or $\{(x, y) \mid x$ and y are real numbers and $y > x - 1\}$

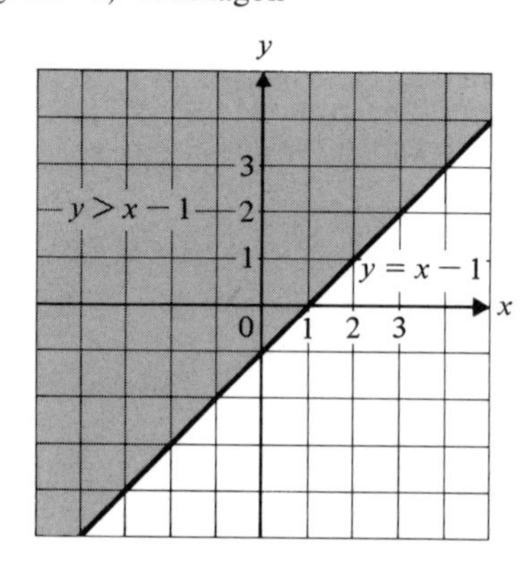

5. a) See Problem 3. b) Yes c) Domain: all real numbers, range: all real numbers.

7. a) 3 b) 7 c) 1/2 d) $^{-}7/2$ e) 2.7 f) 3.32

9. $f = \{(x, y) \mid y = 2$ and x and y are real$\}$

11. It is a function: the range is the nonnegative real numbers.

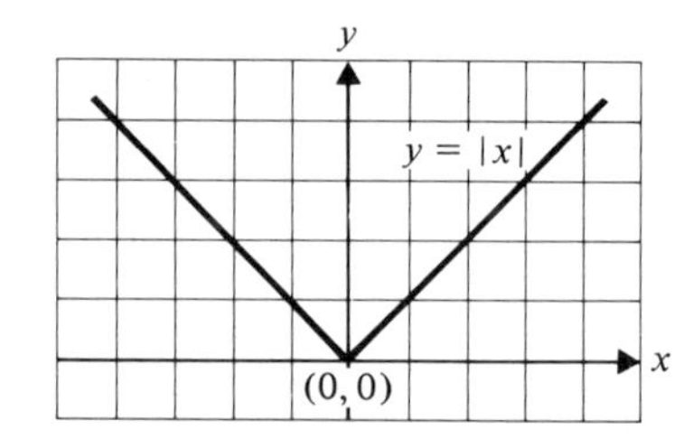

16.

x	$^{-}8$	$^{-}6.24$	$^{-}4$	$^{-}3$	$^{-}2$	$^{-}1$	0	2.24	4
$f(x)$	9	$^{-}0.011$	$^{-}7$	$^{-}8.5$	$^{-}9$	$^{-}8.5$	$^{-}7$	$^{-}0.011$	9

Exercise Set 8.7, p. 297

1. a) $2 \cdot \square$ c) $\frac{1}{2} \cdot \square$ **3.** *Hint*: For each row, column, and diagonal the sum is 261.75.

5. *Hint*: Make a drawing. How many cuts are needed to get the pieces talked about?

11. a) 3.4922475 approximately b) 3.5 exactly

CHAPTER 9

Margin Exercises

1. a) 0 b) 1 **2.** a) 0 b) 3 c) 1 **3.** Yes **4.** Yes; for all pairs $a \otimes b$ the answer is the same as $b \otimes a$.

5. If k is the identity, then for any element a, $k \otimes a = a \otimes k = a$. The element 1 acts as the identity. **6.** a) Both are 2 b) Both are 1 **7.** Both are 2 **8.** a) Both are 1 b) Both are 0 **9.** a) 1 b) 2 c) 3 d) 0 **10.** a) $3 = c + 2$, thus $c = 1$ b) $1 = c + 3$, $c = 2$ **11.** Pairs are the same; in order: 3, 2, 3 **12.** a) 4 b) 2 **13.** a) 3 b) 4 c) 1 **14.** Pairs are same; 4, 3 **15.** a) 3 b) 1(mod 7), hence Sunday **16.** On 3; 3 **17.** a) $14 = 3 \cdot 4 + 2$; $42 = 10 \cdot 4 + 2$ b) 40; 46 **18.** a) $20 \equiv 32 \bmod 4$ **19.** $4 \equiv 14 \bmod 5$ **20.** $266 \equiv 230 \bmod 3$

21. 40 → 1 mod 3
16 → 1 mod 3 — Sum is 2 mod 3
56
2 mod 3 ←

22. a) $y = {}^-10, {}^-210$, and neither are possible b) 6, 10, 14, 18, 22

23. a) Facts: There are two positive integers, both have a remainder of 2 when divided by 3, both have a remainder of 3 when divided by 5, both have a remainder of 2 when divided by 7. b) Find the smallest integers that satisfy the conditions.

24.
$$\begin{aligned} 3n + 2 &\equiv 3 \bmod 5 \\ 3 &\equiv 3 \bmod 5 \\ \hline 3n + 5 &\equiv 6 \bmod 5 \\ 3n &\equiv 1 \bmod 5 \end{aligned}$$

25. $2 + 0 \cdot y \equiv 2 \bmod 3$; $3 + 0 \cdot y \equiv 3 \bmod 5$ **26.** {23, 128, 233, 338, . . .}

Exercise Set 9.1, p. 305

1.

⊕	0	1	2	3	4
0	0	1	2	3	4
1	1	2	3	4	0
2	2	3	4	0	1
3	3	4	0	1	2
4	4	0	1	2	3

3. a) Closure for ⊕, for ⊗.
b) Commutative for ⊕, ⊗.
c) For ⊕ it is 0, for ⊗ it is 1.

5. *Hint*: Follow the argument used for the proof of associativity of addition in J_4. To find $a \oplus (b \oplus c)$ we need to count off $a + (b + c)$ spaces on the clock. For $(a \oplus b) \oplus c$ we need to count off $(a + b) + c$ spaces on the clock. However these are counting numbers which are associative.

6. a) $2 \otimes 3 = 1 \otimes 1 = 1$

Exercise Set 9.2, p. 309

1. a) Both are 2 b) Both are 2 **3.** Yes

5.

⊕	0	1	2	3	4	5	6
0	0	1	2	3	4	5	6
1	1	2	3	4	5	6	0
2	2	3	4	5	6	0	1
3	3	4	5	6	0	1	2
4	4	5	6	0	1	2	3
5	5	6	0	1	2	3	4
6	6	0	1	2	3	4	5

7. All have additive inverses, all except 0 have a multiplicative inverse. **8.** a) Both are 4 b) Both are 2 **11.** a) 4 b) 4
13. a) 4 b) 4 **17.** Just by 3, by 1 **19.** a) 0 c) 0 **21.** For Problem 18: a) 5 b) 2 c) 4
23. The zero is called 8, symbols such as ⊕ and ⊖ are not used. 1. 6 3. 8 5. 2 7. 4 9. 6

Exercise Set 9.3, p. 316

1.

⊗	0	1	2	3	4	5
0	0	0	0	0	0	0
1	0	1	2	3	4	5
2	0	2	4	0	2	4
3	0	3	0	3	0	3
4	0	4	2	0	4	2
5	0	5	4	3	2	1

3. Arithmetic J_{11}; the row for the element 3.
5. $\{0, 1, 2, 3, 4, 5, 6, 7, 8\}$

7. $R = 0: \{\ldots, {}^{-}21, {}^{-}14, {}^{-}7, 0, 7, 14, 21, \ldots\}$
$R = 1: \{\ldots, {}^{-}20, {}^{-}13, {}^{-}6, 1, 8, 15, 22, \ldots\}$
$R = 2: \{\ldots, {}^{-}12, {}^{-}5, 2, 9, 16, 23, \ldots\}$
$R = 3: \{\ldots, {}^{-}11, {}^{-}4, 3, 10, 17, 24, \ldots\}$
$R = 4: \{\ldots, {}^{-}17, {}^{-}10, {}^{-}3, 4, 11, 18, \ldots\}$
$R = 5: \{\ldots, {}^{-}16, {}^{-}9, {}^{-}2, 5, 12, 19, 26, \ldots\}$
$R = 6: \{\ldots, {}^{-}15, {}^{-}8, {}^{-}1, 6, 13, 20, 27, 34, \ldots\}$

11. a) 4 mod 8 b) 7 mod 8 **13.** $w^3 \equiv 1 \bmod 9$ **15.** a) $372 \equiv 1 \bmod 7$; $513 \equiv 2 \bmod 7$; $1 + 2 \equiv 3 \bmod 7$; $886 \equiv 4 \bmod 7$, No check.
15. c) $316 \equiv 1 \bmod 7$; $21 \equiv 0 \bmod 7$; Product $\equiv 0 \bmod 7$, $6636 \equiv 0 \bmod 7$; checks.
17. a) $x = 6, y = 3$ b) No solution c) $x = 2, y = 4$ d) No solution
19. *Hint*: $10^1 \equiv 3 \bmod 7$; $10^2 \equiv 9 \equiv 2 \bmod 7$. What is $10^6 \bmod 7$?
23. Outline of proof: We must show that the product is unique and this uniqueness forces some product to be 1. Assume that $a \cdot m \equiv a \cdot n \bmod p$, which means $p \mid (a \cdot m - a \cdot n)$ or $p \mid a(m - n)$. Since p is prime and a is less than p, $p \nmid a$. Thus $p \mid (m - n)$ and $m \equiv n \bmod p$. If both of them are less than p, then m must be n. (Note: This part of the proof fails if p is not a prime number.) This means that if an element a is multiplied by an element b_1, the only time another product of a with b_2 is the same is if b_1 is identical to b_2. Now, a can be one of the set $\{0, 1, 2, 3, 4, \ldots, p - 1\}$ and b can be any one of the set $\{0, 1, 2, 3, 4, \ldots, p - 1\}$. As a is multiplied by each b in turn, the results cannot be equal if the b's used are different. Thus each answer must be a distinct member of the set $\{0, 1, 2, 3, \ldots, p - 1\}$ and hence at some point the product $a \cdot b = 1$.

Exercise Set 9.4, p. 320

3. *Hint*: By one condition, the number must belong to the set $\{7, 16, 25, \ldots,\}$; by the other it must belong to $\{4, 11, 18, \ldots\}$. This is an alternative to looking at the congruences.
6. *Hint*: Try to get an equation that involves only knives and fishhooks in terms of the currency of coconuts. **7.** 16 mod 23, 18 mod 23
9. *Hint*: Try to get an equation in dimes and nickels only and then use congruence. **12.** *Hint*: Make a drawing, apply congruence.

CHAPTER 10

Margin Exercises

1. All are in plane ABC. **2.** The line segment CD does not intersect line L. **3.** $\overrightarrow{BA} \cup \overrightarrow{BC}$ **4.** a) 1, 4, 5, 6 b) 1, 3, 6 c) 1, 6

5. (b) and (d) **6.** a)

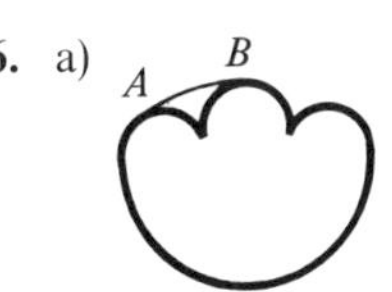

b) A B

7. a) False b) True c) True d) True

8. In 1: $\overline{AB} \cong \overline{FD}$, $\overline{AC} \cong \overline{FE}$, $\overline{BC} \cong \overline{DE}$, $\measuredangle A \cong \measuredangle F$, $\measuredangle B \cong \measuredangle D$, $\measuredangle C \cong \measuredangle E$
In 2: $\overline{IG} \cong \overline{JK}$, $\overline{GH} \cong \overline{KL}$, $\overline{IH} \cong \overline{JL}$, $\measuredangle I \cong \measuredangle J$, $\measuredangle G \cong \measuredangle K$, $\measuredangle H \cong \measuredangle L$

9. $\measuredangle CAD \cong \measuredangle DAB$

10. Counterexample: The angles marked are congruent; the lines are clearly not parallel.

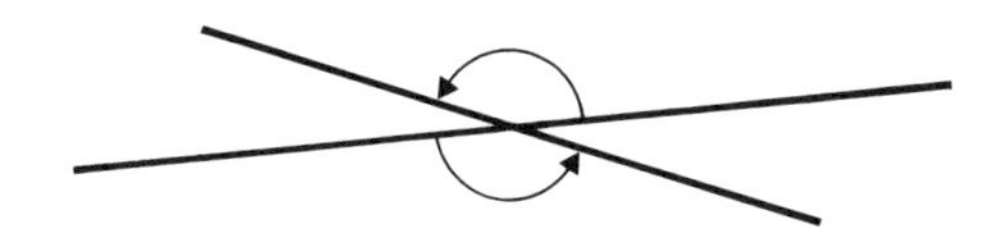

11. Unless the rectangle is also a square, point B will not fall on D.

12. 1. 2 2. 4 3. 0 4. 2 5. 5 6. 6

13. a) All sides are not the same length b) All angles are not congruent c) All angles are not congruent

15. C' is (2, 10), B' is (14, 4), A' is (6, 4); the triangles are similar.

16. $\dfrac{D'C'}{DC} = \dfrac{9}{6}$ but $\dfrac{C'B'}{CB} = \dfrac{6}{3}$, and these are not the same ratios

17. a) $\overline{AA'}$ is parallel to $\overline{BB'}$ b) Each point that is translated moves the same distance as every other point in the translation.

18. a) Yes b) $\overline{AC} \cong \overline{A'C'}$, $\overline{BC} \cong \overline{B'C'}$, $\overline{AB} \cong \overline{A'B'}$ c) P must be on segment $A'B'$

19. a) A' is (4, 2), B' is (8, 2), C' is (8, 5), D' is (4, 5) b)

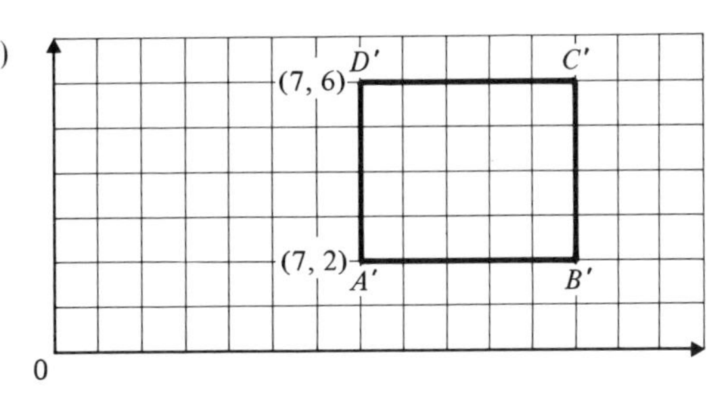

c) Yes

20. Your $\triangle ABC$ and its reflection should coincide.

21. a)

5
2
0
7

b) X' is (5, 4), Y' is (6, 3), Z' is (5, 1), C' is (7, 5), B' is (5, 4), D' is (7, 3), A' is (5, 2).
The y coordinates remain the same; the sum of the x coordinates for the pairs of old and new is 8.

22.

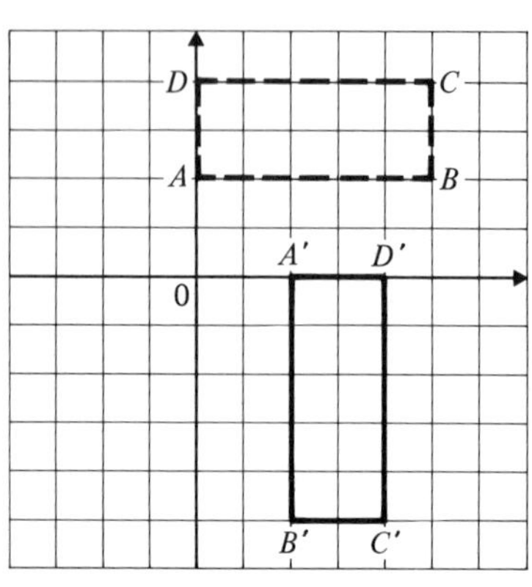

23. a) Facts: A geometric figure of 6 by 5 units, except that two units are cur out at the corner of one long side.
b) Must cut the figure so the pieces are congruent, shade part of the figure so it has the same shape and size as the unshaded part, and so on.

24. a) Cut the figure into more than two pieces and try to reconstruct the two figures. Cut the original into 28 parts and reconstruct, only one side has 6 units so both pieces cannot end up with 6; the most squares that can appear on one side is 5?
b)

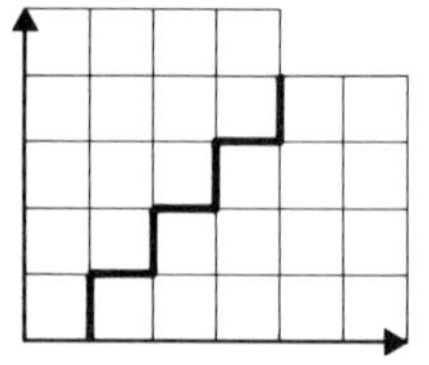

c) Make a model, cut it apart and superimpose the parts.

25. a) Add an arbitrary point A to the line, lay off segment PB with compass so $\overline{PB} \cong \overline{AP}$, we need to find point C so that $\overline{AC} \cong \overline{CB}$, etc.
b) If you cannot solve the problem, continue reading.

26. b) We have assumed that if corresponding sides of two triangles are congruent, then the triangles are congruent; SSS = SSS
c) $\measuredangle APB \cong \measuredangle BPC$ since parts of $\cong$ triangles are congruent. **27.** a) On B b) Yes; they are radii of the same circle c) Yes

28. a) $\overline{AC} \cong \overline{A'Q}$, $\overline{AB} \cong \overline{A'P}$, $\overline{BC} \cong PQ$ (SSS) b) Corresponding parts of congruent triangles are congruent.

29. a) Yes; the same radius used b) SSS for congruent triangles c) Corresponding parts of congruent triangles

30. P_5 in conjunction with the construction. $\measuredangle RPQ \cup \measuredangle PQL$ is a straight angle so the lines cannot intersect on that side, similarly they cannot intersect on the other side.

33. The edges that were hidden are now closest to the viewer. Sketch:

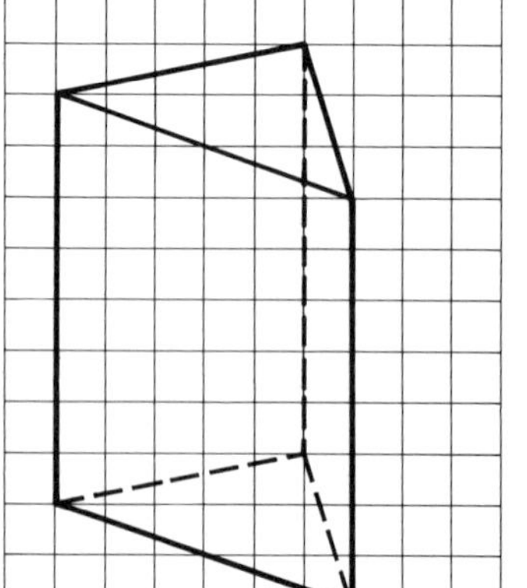

Many others possible

Exercise Set 10.1, p. 325

1. Transitive **3.** If $a = b$, $c = d$, then $a - c = b - d$ **5.** A_5 **7.** P_2 **9.** L will interesect M

Exercise Set 10.2, p. 332

1. $\overleftrightarrow{AB}$, $\overleftrightarrow{BC}$, $\overleftrightarrow{CD}$, $\overleftrightarrow{AC}$, $\overleftrightarrow{AD}$, $\overleftrightarrow{BD}$ **3.** a) No, yes b) Yes, yes **5.** a) $\overline{MN}$ b) $\overleftrightarrow{MN}$ **7.** a) $\overline{CD}$ b) $\overline{BC}$ c) $\overline{BC}$

9. a) No b) No c) Yes d) No e) Yes **11.** 1 point, 0 points, infinite number of points

15. $\triangle ABC$ has three vertices that determine a plane. Points A and B determine a line that also lies in the plane, similarly this holds for the points A and C and the points B and C.

17. Some of them are: ADG, ADC, $DFHO$, FHC, AGE, DEB, DFB, $EBFD$, etc.

Exercise Set 10.3, p. 340

1. Yes, provided you rotate the sheet at the same time.

3. Label the intersection of $\overline{AC}$ and $\overline{DB}$ as O. $\triangle ADB \cong \triangle DCB$, $\triangle ACD \cong \triangle ACB$, $\triangle DOA \cong \triangle AOB$, $\triangle DOC \cong \triangle COB$. In $\triangle ADB$ and $\triangle DCB$, $\overline{AD} \cong \overline{DC}$, $\overline{AB} \cong \overline{BC}$, $\overline{DB} \cong \overline{DB}$, $\measuredangle ADB \cong \measuredangle CDB$, $\measuredangle DAB \cong \measuredangle DCB$, $\measuredangle ABD \cong \measuredangle DBC$

5. a) b)

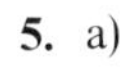

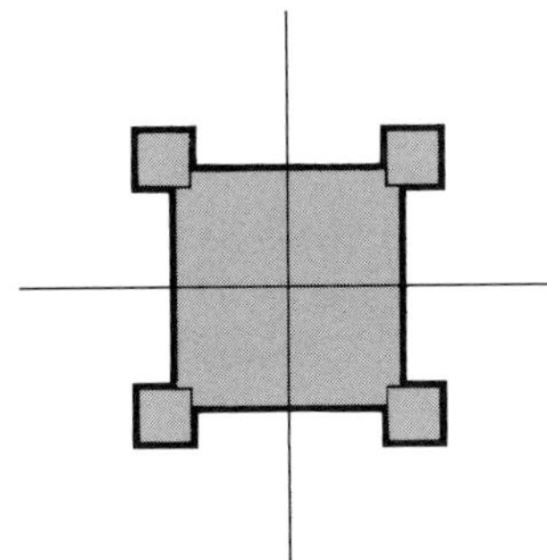

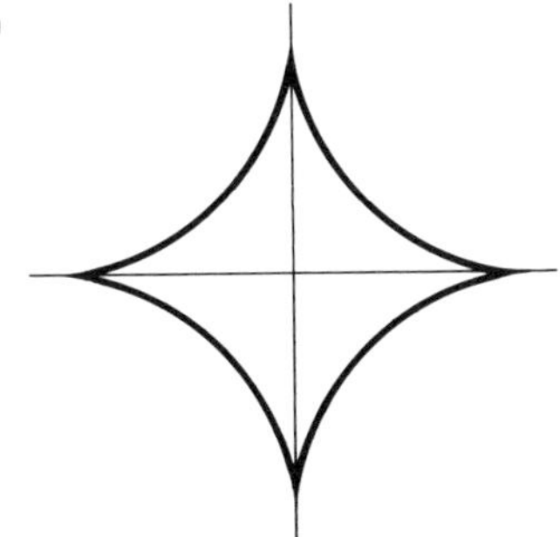

7. $\overrightarrow{OA}$ and $\overrightarrow{OB}$ are perpendicular to each other because adjacent angles that form a straight angle are congruent.

9.

Regular figure	*No. of lines of symmetry*
Triangle	3
Square	4
Pentagon	5
Hexagon	6
Heptagon	7

Conjecture: the number of sides for a regular figure is the same as the number of lines of symmetry.

11. In the new figure, $\measuredangle CBA$, $\measuredangle BB_1C_1$, $\measuredangle CBB_1$ are all right angles, therefore the interior angles on both sides of line AB_1 form a straight angle. (Imagine lines CB and C_1B_1 extended through line AB_1.) By Euclid's postulate P_5, the lines must be parallel, since the condition for intersection is not met on either side of $\overleftrightarrow{AB_1}$.

13. Yes **15.** A counterexample: corresponding angles are not congruent. **17.** a) 3 b) 3 c) 5 d) 2

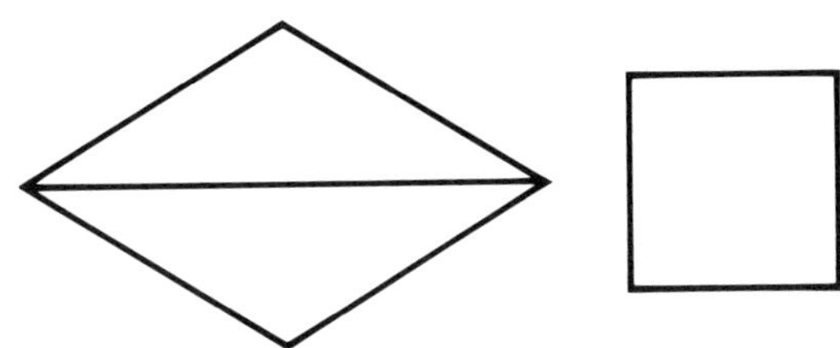

Exercise Set 10.4, p. 348

1. Check your reflection by folding on the line L; the image should be superimposed on the original triangle.

3. Fold the paper, so P coincides with P'. The fold is the line of reflection.

5. a) Under reflection a geometric figure is transformed to a congruent image, thus the triangles are congruent.
b) Corresponding parts of congruent triangles are congruent. c) same as (b). **7.** c) Yes, a single translation would do it.

9. a) One possibility is a reflection followed by a translation. c) A reflection followed by rotation or translation followed by reflection.

Exercise Set 10.5, p. 353

1. Use P as the center of a circle and draw a circle with a radius long enough, so the circle intersects the line L in the points A and B. Keep the radius the same and use point A and point B in turn as the center of two circles. Draw arcs that intersect at P'. The line determined by P and P' is the required perpendicular.

3. *Hint*: Draw a triangle ABC and imagine that side AC determines a translation along the line AC extended. Point C has its image on the original point A. The parallelogram formed will lead to the desired result.

5. Triangle ABC reflects to ADC in line AC. Therefore $\overline{BO}$ reflects to $\overline{OD}$ and hence $\overline{BO} \cong \overline{OD}$. Similarly $CO \cong \overline{OC}$. Now $\triangle ABO$ reflects to $\triangle BOC$ in line BO, but diagonal AC is a straight line and the angles formed must be right angles.

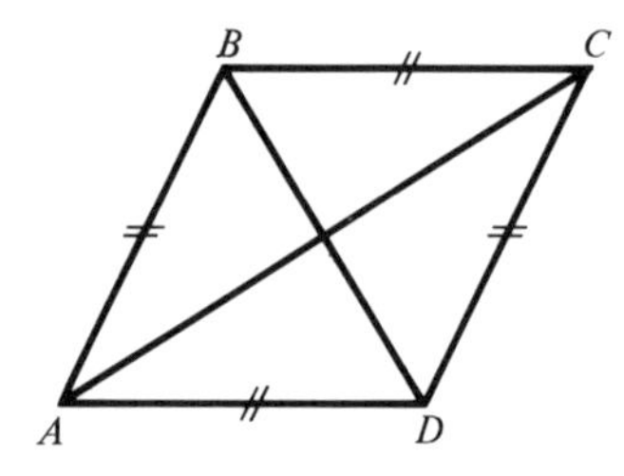

9. No, such triangles are similar.

11. *Hint*: How many triangles have a pentagon side as a side? How many sides? How many triangles are interior, i.e., have no side that is a side of the large pentagon?

Exercise Set 10.6, p. 357

1. Draw a line where $A'C'$ is to be located. Lay off the length of $\overline{AC}$. Use point A' as the center and then point A as center to transfer the angle at A to A'. Repeat for the angle at C. Extend the sides of the angles to meet at the point B'.

7. b) 6 c) Yes. **12.** c) The interior quadrilateral seems to be a parallelogram.

Exercise Set 10.7, p. 362

5.

Figure	V	F	V + F	E
Tetrahedron	4	4	8	6
Hexahedron	8	6	14	12
Octahedron	6	8	14	12
Dodecahedron	20	12	32	30
Icosahedron	12	20	32	30

8. a) Three meet at some and four meet at others.
b) No. Since different numbers of triangles meet, the polyhedral angles are not the same.
c) Yes

9. a) Cut with a plane parallel to any face. b) Cut through with a plane containing two parallel edges that are not in the same face.
c) Cut off a corner as in Problem 7.

11.

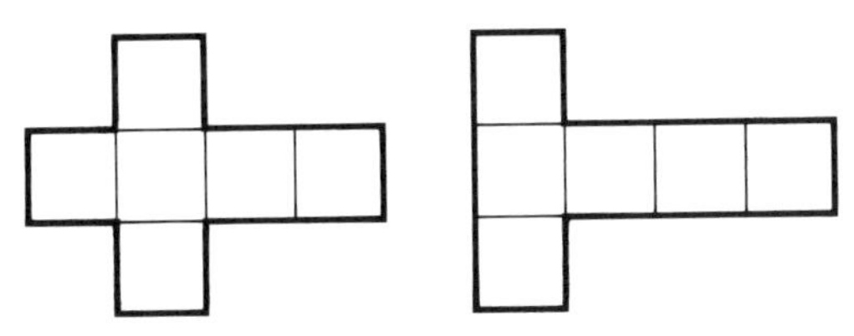

CHAPTER 11

Margin Exercises

1. Answers vary. **2.** $6\frac{7}{8}$ by $9\frac{1}{8}$ **3.** 5 **4.** b) Between 6 and 7 **5.** $|5 - 10| = |10 - 5| = 5$

6.

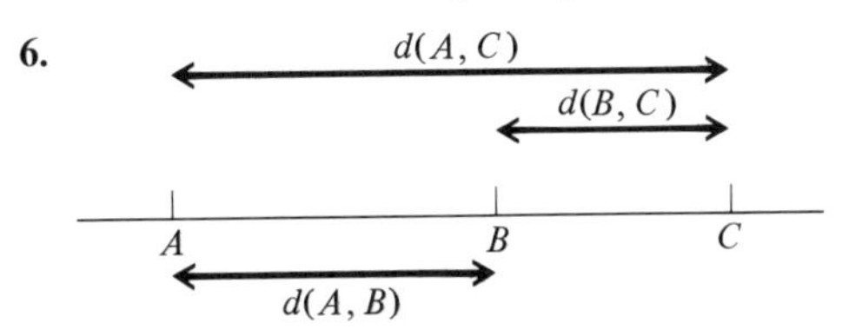

7. a) 4 b) $4\frac{1}{2}$ c) $4\frac{1}{4}$ e) $4\frac{2}{8}$ e) $4\frac{6}{16}$ **8.** a) 1000 b) 1000 c) 100 d) 10

13. a) 5.65 m b) 500 m c) 75 cm d) 35 mm e) 7 cm **14.** a) 3000 g b) 2000 mg c) 1.25 t

15. 1. kg 2. kg 3. t 4. g 5. g 6. g **16.** a) $^{-}2°$C b) 20.5°C c) 39°C d) 2°C

17. a) 1 cm^2 **18.** a) 1.52 b) 2000 c) 1 d) 3.675 **19.** b) 2 **20.** a) 117° b) 162° c) 71°

21. a) 7 cm b) 7.6 cm **22.** a) 41.4 m b) $2s + 2s = 4s$

23. a) There are six equal angles about the center, so each has 60°. Since each triangle has equal radii for sides, the base angles measure the same and hence are also 60°. b) Increase

24. a) 2 m b) 31.43 mm c) 7.96 mm **25.** 28 **26.** a) 100 b) 100 c) 10 000 d) 10 000

27. a) 28 cm^2 b)

28. a) $\triangle AED \cong \triangle AEF$, SSS, $\overline{DA} \cong \overline{EF}$, $\overline{DE} \cong \overline{AF}$. Similarly, $\triangle FEB \cong \triangle ECB$, SSS. b) Half the area of the rectangle. c) $\frac{1}{2}h \cdot b$

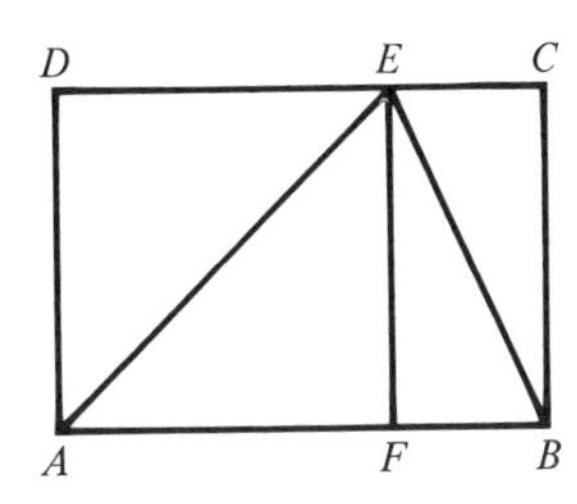

29. a) 6.93 cm^2 b) 27.71 cm **30.** a) 60 L b) 60 mL **31.** a) Circle b) Square c) Equilateral triangle d) Circle
32. a) 744 cm^3 b) 28.618 cm^3 **33.** 248 cm^3. It is 1/3 of the volume of the prism. **34.** 1 L **35.** 5.9634×10^{10} km^3
36. a) Facts: A diagram of the floor is provided with the dimensions; tiles are 20 cm square; tiles cost $11.75 per 100; must buy in lots of 100.
b) Find the cost in advance, find the area of the floor, estimate the losses due to irregular shape.

37. a)

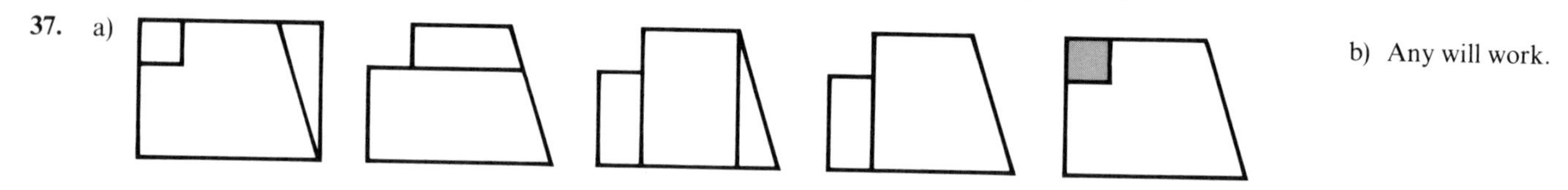

b) Any will work.

38. a) 10 m^2 b) 24 m^2 c) 6 m^2 **39.** 25 tiles **40.** Saved $11.75

Exercise Set 11.1, p. 371

1. A unit of length is an arbitrarily chosen measure of length. A subunit is some fractional part of the selected unit. 3 feet = 1 yard, 36 inches = 1 yard.

3. 3 of the $\frac{1}{4}$ AB units **5.** Units are $\frac{1}{10}\overline{AB}$ a) 20 b) $15\frac{3}{4}$ c) $3\frac{1}{2}$

7. If numeration were base 7, subunits equal to 1/7 of the unit would help since 1 unit and 1 subunit would be 1.1.

9. a) 100 lbs b) 240 lb c) 12 ounces d) Yes e) Yes f) No g) No h) More i) Less j) More

Exercise Set 11.2, p. 381

5. Each edge must be 10 cm; then the cube will hold 1000 cm^3 = 1 dm^3 = 1 L. **6.** a) 1250 m c) 753.6 cm **7.** a) 1.585 t c) 1500 mg e) 3000 kg **8.** a) T c) F e) F g) T **9.** 55.5 km

Exercise Set 11.3, p. 383

5. a) 3 b) 8 **7.** Any quadrilateral can be thought of as consisting of two triangles so the total measure is 360°. **8.** a) 124° c) 90°
11. 720° **13.** No. If two angles measure greater than 90°, the total for the figure would exceed 180°. **15.** 120° **17.** 135°

Exercise Set 11.4, p. 387

1. a) 72 mm b) Should be greater **3.** If s is one side, then perimeter is $n \cdot s$.

5. 25 m; 10 revolutions will be very close to 10 m.

9. 1.59 m, 15.9 m **11.** 3.1326

Exercise Set 11.5, p. 392

1. a) 26 square units b) 22 square units **3.** 5.63 ha **5.** a) 87.5 cm^2

6. a) 56.25 m^2

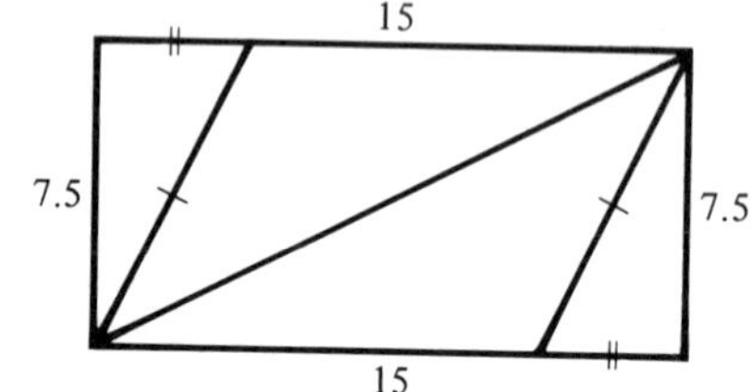

7. a) 603.2 cm^2 **8.** a) 11.93 cm^2

9. a) $\frac{2}{3}\pi r^2$ for any radius r **11.** 64.95 cm^2

Exercise Set 11.6, p. 398

1.

mm^3	cm^3	dm^3	m^3
10^9	10^6	10^3	1
		1	10^{-3}
	1	10^{-3}	10^{-6}
1	10^{-3}		10^{-9}

3. 3750 cm^3; multiplied by 8, i.e., 30 000 cm^3 **5.** 40 L

7. Largest: 1115.6 cm^3; smallest: 819.4 cm^3 **9.** 47.64%

Exercise Set 11.7, p. 401

1. The 20 cm cake is slightly less expensive; if the 23 cm cake were sold at the cheaper price it would cost \$2.976. **3.** 2 583 333 m^3

5. Holds less **8.** *Hint*: Measure 100 sheets at a time by holding them firmly.

9. If whole tiles are used for the 27th row that is needed, then \$179.55. If fractional parts are used carefully, the cost could be slightly cut.

11. Volume is 1/6 that of the cube, hence 288 cm^3 (which is consistent with the formula for a pyramid).

CHAPTER 12

Margin Exercises

1. a) b)

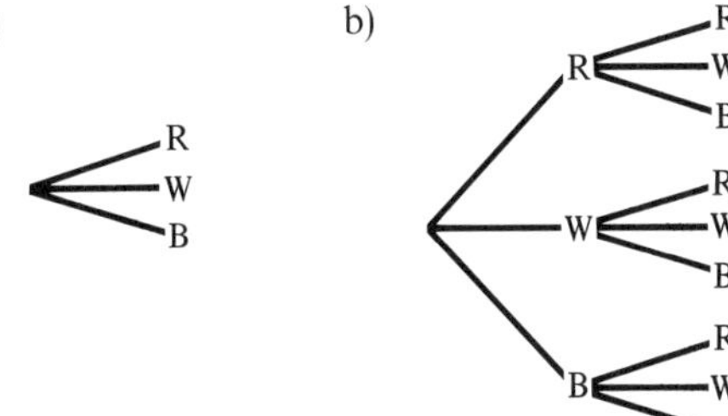

2. {(H, 1), (H, 2), (H, 3), (H, 4), (T, 1), (T, 2), (T, 3), (T, 4)}

3. a) {2, 4, 6} b) {1, 3, 5} c) {3, 6} d) { }

4. (1) $0 \leqslant P(A) \leqslant 1$ (2) $P(S) = 1$
(3) If $A \cap B = \varnothing$, then $P(A \text{ or } B) = P(A) + P(B)$

5. a) 1/2 b) 1/2 c) 1/3 d) 0 e) {(7, 3)} f) 1/12 g) 1/2

6. a) {(R, Y), (W, Y), (B, Y), (Y, Y), (R, B), (W, B), (B, B), (Y, B), (R, W), (W, W), (B, W), (Y, W), (R, R), (W, R), (B, R), (Y, R)}
b) {(R, R), (W, W), (B, B), (Y, Y)}

7. a)

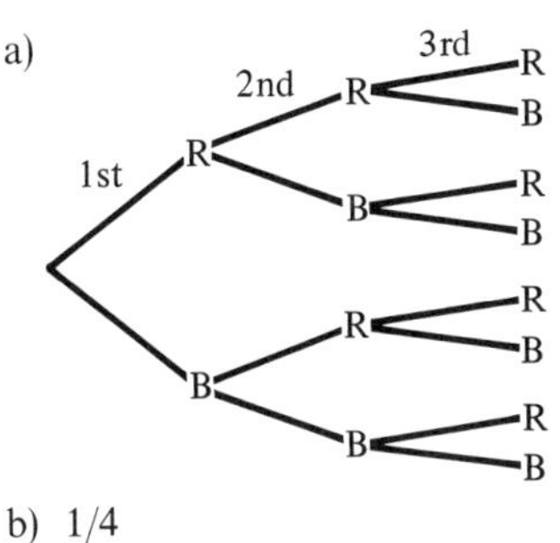

b) 1/4

8. 900

9. a)

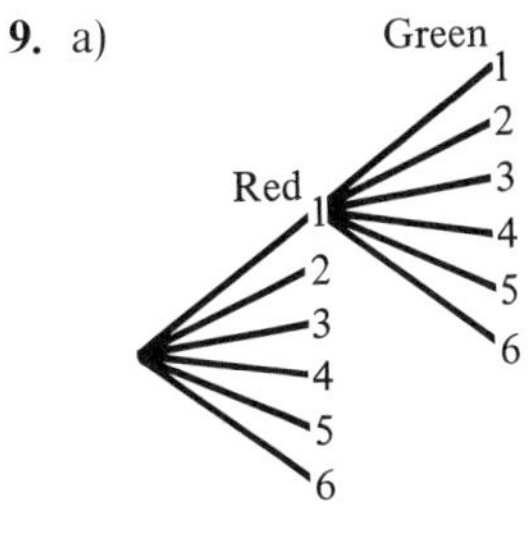

The six branches that come out of the first branch appear on each of the other five branches; a total of 36 branches.

b) 1/4 c) 1/2

10. $1 - \frac{1}{6} = \frac{5}{6}$ **11.** $\frac{7}{13}$

12. a) {(1, 1), (1, 2), (1, 3), (1, 4), (1, 5), (1, 6)} b)

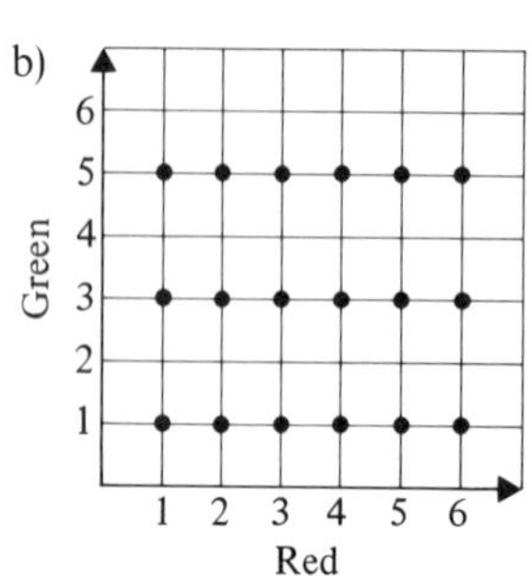

c) 1/6, 1/2 d) {(1, 1), (1, 3), (1, 5)} e) 1/12 f) Same, 1/12

13. a) {(H, T, T), (H, T, H), (T, T, H), (T, T, T)} = C b) 1/2, 1/2 c) 1/4 d) Same e) Yes
{(H, H, T), (H, T, T), (T, H, T), (T, T, T)} = D

14. a) 3/8 b) 2/8 c) 1/8 **15.** a) 3/8 b) 1/4 c) 1/4

16. a) {(R, R), (R, B), (B, R), (B, B)} b) 5/18 **17.** Frequencies in order: 14, 19, 8, 8, 3, 4

18. Both 76 **19.** a) 78 b) 73 **20.** Both 75 **21.** a) 13.297 b) 1.414 c) 0 **22.** 1/4

23. a) Facts: 150 bits for the batter, 100 cookies will be made, b) How many cookies will end up without a chocolate chip?
c) We have not studied a formula that will work here. Bake a trial batch of cookies.

24. With only 15 chips many cookies will be without a chip.

9						√		√		
8										
7	√√		√							
6					√					
5		√								
4										
3	√	√	√							
2						√	√			√
1		√								
0				√						
	0	1	2	3	4	5	6	7	8	9

25. b)

No. of chips	*f*
0	22
1	32
2	30
3	10
4	4
>4	2
Total	100

26. Use more chips. If 400 are used, the probability that a cookie has no chips drops to 0.0183.

Exercise Set 12.1, p. 412.

3. a) {(H, H, T), (H, T, H), (T, H, H)} b) {(H, H, H), (H, H, T), (H, T, H), (H, T, T)}
c) {(H, H, H), (H, H, T), (H, T, H), (H, T, T), (T, H, H), (T, H, T), (T, T, H)} d) {(T, T, T)}

5. a) 3/8 b) 1/2 c) 7/8 d) 1/8 **7.** 10 **9.** 3 **11.** 1/4 **13.** 1/4 **15.** a) 1/8 b) 3/8

Exercise Set 12.2, p. 415

1. 5/6 **3.** 1/9, 8/9 **5.** b) 4/9 d) 2/9 f) 1/9 **7.** a) The probability that a student is not taking probability and statistics.
8. b) {(H, H, H), (H, H, T)}; 2/8 **11.** Total percentage of interviews adds up to 111%.

Exercise Set 12.3, p. 420

1. a) Independent c) Not independent **3.** 0.0784 **5.** a) 1/16 b) 1/4 **7.** a) 0.36 b) 0.06
9. a) 59/165 b) 2/33 **11.** 1/2 **15.** a) 0.474 b) 3 **17.** 0.00455 **19.** a) 0.09 b) 0.21

Exercise Set 12.4, p. 428

1. a) Mean 7, median 7, range 10, standard deviation 3.52. **2.** a) Mean 14.333, median 6.5, range 58, standard deviation 20.26
5. Frequencies in order: 0, 1, 2, 1, 7, 11, 13, 5 **6.** a)

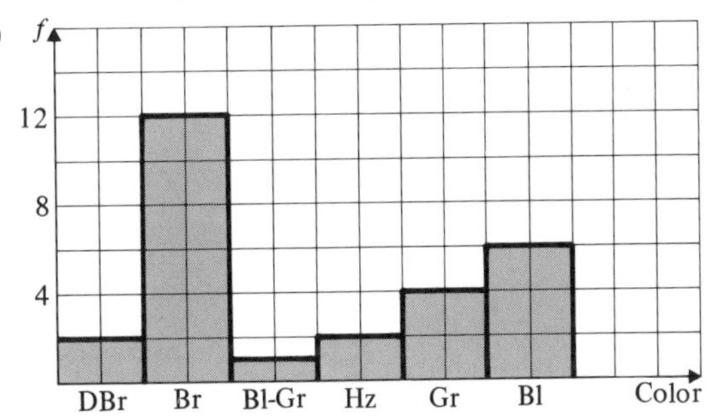

8. a) Mean 45.4, median 43 b) Frequencies in order (last class is 100–109): 20, 48, 14, 4, 3, 0, 0, 1
9.

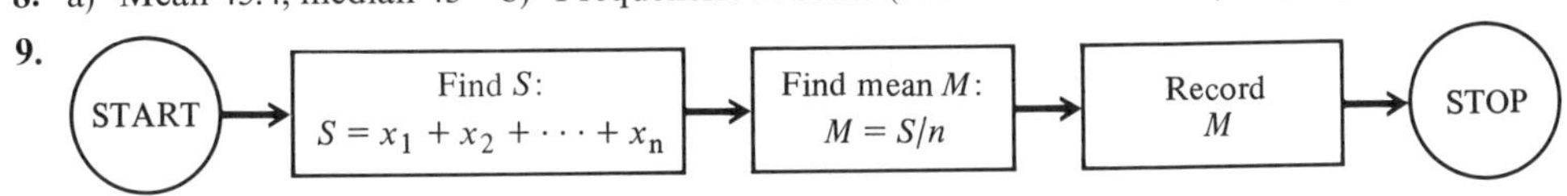

13. 0.425 **15.** 916.8 **17.** a) Mean 58.74 kg, median 57.1 kg

Exercise Set 12.5, p. 432

1. 177 red, 76 black **6.** Theoretical probabilities:

Sum	2	3	4	5	6	7	8	9	10	11	12
Probability	$\frac{1}{36}$	$\frac{2}{36}$	$\frac{3}{36}$	$\frac{4}{36}$	$\frac{5}{36}$	$\frac{6}{36}$	$\frac{5}{36}$	$\frac{4}{36}$	$\frac{3}{36}$	$\frac{2}{36}$	$\frac{1}{36}$
Frequency for $n = 180$	5	10	15	20	25	30	25	20	15	10	5

7. If the spinner lands on 1 it is counted as H, if on 0 it is counted as T; 200 spins can simulate 200 tosses of a coin.

Exercise Set 12.6, p. 436

3. c) Sample space
{(H, H, H), (T, H, H), (H, T, H), (H, H, T), (T, T, H), (T, H, T), (H, T, T), (T, T, T)}
↑ Retoss, ↖ 1, ↑ 2, ↗ 3, ↗ 3, ↑ 2, ↖ 1, ↑ Retoss

Event	*Probability*
Retoss	1/4
1st pays	1/4
2nd pays	1/4
3rd pays	1/4

A Summary of the Formulas for Perimeter, Area, and Volume

PERIMETER AND AREA

Rectangle: $P = 2l + 2w$

$A = l \times w$

Square: $P = 4s$

$A = s^2$

Parallelogram: $A = b \cdot h$

Triangle: $A = \frac{1}{2} b \cdot h$

Trapezoid: $A = \frac{1}{2}(b_1 + b_2) \cdot h$

Circle: $C = 2\pi r = \pi D$

$A = \pi r^2$

VOLUME

Regular prism:

$V = \text{Area of base} \cdot h$

Right circular cylinder:

$V = \pi r^2 \cdot h$

Pyramid:

$$V = \frac{1}{3} \cdot \text{Area of base} \cdot h$$

Circular cone:

$$V = \frac{1}{3} \pi \cdot r^2 \cdot h$$

Sphere:

$$V = \frac{4}{3} \pi r^3$$

Common Metric Units

LENGTH

1 km = 1000 m
1 m = 10 dm
1 m = 100 cm
1 cm = 10 mm

MASS

1 t = 1000 kg
1 kg = 1000 g
1 g = 1000 mg

AREA

1 km^2 = 1 000 000 m^2
1 ha = 10 000 m^2
1 m^2 = 10 000 cm^2
1 cm^2 = 100 mm^2

VOLUME

1 cm^3 = 1 mL
1000 cm^3 = 1 L
1 dm^3 = 1 L
1 m^3 = 1000 L

index

index